Silkworm and Silk: Traditional and Innovative Applications

Silkworm and Silk: Traditional and Innovative Applications

Editors

Silvia Cappellozza
Morena Casartelli
Federica Sandrelli
Alessio Saviane
Gianluca Tettamanti

MDPI • Basel • Beijing • Wuhan • Barcelona • Belgrade • Manchester • Tokyo • Cluj • Tianjin

Editors

Silvia Cappellozza
CREA - Research Centre for Agriculture and Environment
Italy

Morena Casartelli
University of Milan
Italy

Federica Sandrelli
University of Padova
Italy

Alessio Saviane
CREA - Research Centre for Agriculture and Environment
Italy

Gianluca Tettamanti
University of Insubria
Italy

Editorial Office
MDPI
St. Alban-Anlage 66
4052 Basel, Switzerland

This is a reprint of articles from the Special Issue published online in the open access journal *Insects* (ISSN 2075-4450) (available at: https://www.mdpi.com/journal/insects/special_issues/silkworm_silk).

For citation purposes, cite each article independently as indicated on the article page online and as indicated below:

LastName, A.A.; LastName, B.B.; LastName, C.C. Article Title. *Journal Name* **Year**, *Volume Number*, Page Range.

ISBN 978-3-0365-6858-4 (Hbk)
ISBN 978-3-0365-6859-1 (PDF)

Cover image courtesy of Council for Agricultural Research and Economics, Research Centre Agriculture and Environment

Contents

Preface to "Silkworm and Silk: Traditional and Innovative Applications" . . . . **ix**

Silvia Cappellozza, Morena Casartelli, Federica Sandrelli, Alessio Saviane and Gianluca Tettamanti
Silkworm and Silk: Traditional and Innovative Applications
Reprinted from: *Insects* **2022**, *13*, 1016, doi:10.3390/insects13111016 . . . . **1**

Lincah Andadari, Dhany Yuniati, Bambang Supriyanto, Murniati, Sri Suharti, Asmanah Widarti, et al.
Lens on Tropical Sericulture Development in Indonesia: Recent Status and Future Directions for Industry and Social Forestry
Reprinted from: *Insects* **2022**, *13*, 913, doi:10.3390/insects13100913 . . . . **5**

Domenico Giora, Giuditta Marchetti, Silvia Cappellozza, Alberto Assirelli, Alessio Saviane, Luigi Sartori and Francesco Marinello
Bibliometric Analysis of Trends in Mulberry and Silkworm Research on the Production of Silk and Its By-Products
Reprinted from: *Insects* **2022**, *13*, 568, doi:10.3390/insects13070568 . . . . **31**

Cristian Lujerdean, Gabriela-Maria Baci, Alexandra-Antonia Cucu and Daniel Severus Dezmirean
The Contribution of Silk Fibroin in Biomedical Engineering
Reprinted from: *Insects* **2022**, *13*, 286, doi:10.3390/insects13030286 . . . . **51**

Panomir Tzenov, Silvia Cappellozza and Alessio Saviane
Black, Caspian Seas and Central Asia Silk Association (BACSA) for the Future of Sericulture in Europe and Central Asia
Reprinted from: *Insects* **2022**, *13*, 44, doi:10.3390/insects13010044 . . . . **77**

Gabriela-Maria Baci, Alexandra-Antonia Cucu, Alexandru-Ioan Giurgiu, Adriana-Sebastiana Muscă, Lilla Bagameri, Adela Ramona Moise, et al.
Advances in Editing Silkworms (*Bombyx mori*) Genome by Using the CRISPR-Cas System
Reprinted from: *Insects* **2022**, *13*, 28, doi:10.3390/insects13010028 . . . . **93**

Luca Tassoni, Silvia Cappellozza, Antonella Dalle Zotte, Simone Belluco, Pietro Antonelli, Filippo Marzoli and Alessio Saviane
Nutritional Composition of *Bombyx mori* Pupae: A Systematic Review
Reprinted from: *Insects* **2022**, *13*, 644, doi:10.3390/insects13070644 . . . . **113**

Aurora Montali, Francesca Berini, Alessio Saviane, Silvia Cappellozza, Flavia Marinelli and Gianluca Tettamanti
A *Bombyx mori* Infection Model for Screening Antibiotics against *Staphylococcus epidermidis*
Reprinted from: *Insects* **2022**, *13*, 748, doi:10.3390/insects13080748 . . . . **135**

Andreja Urbanek Krajnc, Tamas Bakonyi, Istvan Ando, Eva Kurucz, Norbert Solymosi, Paula Pongrac and Rebeka Lucijana Berčič
The Effect of Feeding with Central European Local Mulberry Genotypes on the Development and Health Status of Silkworms and Quality Parameters of Raw Silk
Reprinted from: *Insects* **2022**, *13*, 836, doi:10.3390/insects13090836 . . . . **149**

Pauline Nicole O. de la Peña, Adria Gabrielle D. Lao and Ma. Anita M. Bautista
Global Profiling of Genes Expressed in the Silk Glands of Philippine-Reared Mulberry Silkworms *(Bombyx mori)*
Reprinted from: *Insects* **2022**, *13*, 669, doi:10.3390/insects13080669 . . . . **179**

Ivan Y. Dee Tan and Ma. Anita M. Bautista
Bacterial Survey in the Guts of Domestic Silkworms, *Bombyx mori* L.
Reprinted from: *Insects* **2022**, *13*, 100, doi:10.3390/insects13010100 . . . . **197**

Kridsada Unban, Augchararat Klongklaew, Pratthana Kodchasee, Punnita Pamueangmun, Kalidas Shetty and Chartchai Khanongnuch
Enterococci as Dominant Xylose Utilizing Lactic Acid Bacteria in Eri Silkworm Midgut and the Potential Use of *Enterococcus hirae* as Probiotic for Eri Culture
Reprinted from: *Insects* **2022**, *13*, 136, doi:10.3390/insects13020136 . . . . **209**

Giulia Alessandra Bassani, Valentina Vincoli, Marco Biagiotti, Elisa Valsecchi, Marta Virginia Zucca, Claudia Clavelli, et al.
A Route to Translate a Silk-Based Medical Device from Lab to Clinic: The Silk Biomaterials Srl Experience
Reprinted from: *Insects* **2022**, *13*, 212, doi:10.3390/insects13020212 . . . . **223**

Rui Zhang, Yu-Yao Cao, Juan Du, Kiran Thakur, Shun-Ming Tang, Fei Hu and Zhao-Jun Wei
Transcriptome Analysis Reveals the Gene Expression Changes in the Silkworm (*Bombyx mori*) in Response to Hydrogen Sulfide Exposure
Reprinted from: *Insects* **2021**, *12*, 1110, doi:10.3390/insects12121110 . . . . **239**

Huimin Guo, Benzheng Zhang, Xin Zheng, Juan Sun, Huiduo Guo, Gang Li, et al.
Pathogenicity Detection and Genome Analysis of Two Different Geographic Strains of BmNPV
Reprinted from: *Insects* **2021**, *12*, 890, doi:10.3390/insects12100890 . . . . **253**

Manish Singh, Estera S. Dey, Sunil Bhand and Cedric Dicko
Supercritical Carbon Dioxide Impregnation of Gold Nanoparticles Demonstrates a New Route for the Fabrication of Hybrid Silk Materials
Reprinted from: *Insects* **2022**, *13*, 18, doi:10.3390/insects13010018 . . . . **269**

Meirong Zhang, Pingzhen Xu and Tao Chen
Hemolymph Ecdysteroid Titer Affects Maternal mRNAs during *Bombyx mori* Oogenesis
Reprinted from: *Insects* **2021**, *12*, 969, doi:10.3390/insects12110969 . . . . **287**

Li Chang, Zhiqing Li, Hao Guo, Wenchang Zhang, Weiqun Lan, Jue Wang, et al.
Function of Polyamines in Regulating Cell Cycle Progression of Cultured Silkworm Cells
Reprinted from: *Insects* **2021**, *12*, 624, doi:10.3390/insects12070624 . . . . **299**

Kakeru Yokoi, Takuya Tsubota, Akiya Jouraku, Hideki Sezutsu and Hidemasa Bono
Reference Transcriptome Data in Silkworm *Bombyx mori*
Reprinted from: *Insects* **2021**, *12*, 519, doi:10.3390/insects12060519 . . . . **311**

Yasuhiko Matsumoto, Yuki Tateyama and Takashi Sugita
Evaluation of Antibacterial Drugs Using Silkworms Infected by *Cutibacterium acnes*
Reprinted from: *Insects* **2021**, *12*, 619, doi:10.3390/insects12070619 . . . . **331**

Yudai Masuoka, Wei Cao, Akiya Jouraku, Hiroki Sakai, Hideki Sezutsu and Kakeru Yokoi
Co-Expression Network and Time-Course Expression Analyses to Identify Silk Protein Regulatory Factors in *Bombyx mori*
Reprinted from: *Insects* **2022**, *13*, 131, doi:10.3390/insects13020131 . . . . **343**

Salvador D. Aznar-Cervantes, Beatriz Monteagudo Santesteban and José L. Cenis
Products of Sericulture and Their Hypoglycemic Action Evaluated by Using the Silkworm, *Bombyx mori* (Lepidoptera: Bombycidae), as a Model
Reprinted from: *Insects* **2021**, *12*, 1059, doi:10.3390/insects12121059 . . . . **357**

Preface to "Silkworm and Silk: Traditional and Innovative Applications"

Humans have studied the silkworm, *Bombyx mori*, since time immemorial because of its economic relevance to silk production. Due to the complete domestication of this insect, the duration and any parameters of its development can be fully controlled by regulating environmental conditions. For this reason, this insect has become an increasingly useful and suitable laboratory tool for research in genetics, physiology, animal nutrition, medicine, science of materials, and chemistry, among other application fields. Moreover, advances related to this insect have been accelerated by the sequencing of its genome. This impressive growth of knowledge related to the silkworm has encouraged the proliferation of sericultural advances which have been described in scientific journals dealing with different topics. Therefore, there is a need to focus on the unity of the sericultural sciences as a group of disciplines, with the silkworm as a binding element. The scope of this Special Issue provides a comprehensive overview of the fields and applications for which the silkworm can be exploited, and even to reinforce the link between traditional sericulture and new technological horizons. In fact, silkworm rearing techniques and preservation, strain selection, and maintenance, which are the traditional activities regarding silkworms, are the starting basis for any innovative application. In light of these considerations, the final purpose of this Special Issue is to establish a virtual dialogue between traditional sericulture and new trends in silkworm and silk valorisation.

Silvia Cappellozza, Morena Casartelli, Federica Sandrelli, Alessio Saviane, and Gianluca Tettamanti

Editors

insects

MDPI

Editorial

Silkworm and Silk: Traditional and Innovative Applications

Silvia Cappellozza [1,*], Morena Casartelli [2,3], Federica Sandrelli [4], Alessio Saviane [1] and Gianluca Tettamanti [3,5]

1 Council for Agricultural Research and Economics, Research Centre Agriculture and Environment, Sericulture Laboratory, 35143 Padova, Italy
2 Department of Biosciences, University of Milan, 20133 Milano, Italy
3 Interuniversity Center for Studies on Bioinspired Agro-Environmental Technology (BAT Center), University of Napoli Federico II, 80055 Portici, Italy
4 Department of Biology, University of Padova, 35131 Padova, Italy
5 Department of Biotechnology and Life Sciences, University of Insubria, 21100 Varese, Italy
* Correspondence: silvia.cappellozza@crea.gov.it

Citation: Cappellozza, S.; Casartelli, M.; Sandrelli, F.; Saviane, A.; Tettamanti, G. Silkworm and Silk: Traditional and Innovative Applications. *Insects* **2022**, *13*, 1016. https://doi.org/10.3390/insects13111016

Received: 27 October 2022
Accepted: 30 October 2022
Published: 3 November 2022

The various subjects covered in the present Special Issue "Silkworm and Silk: Traditional and Innovative Applications" demonstrate how sericulture, a practice deeply rooted in human history, can act as a bridge to bring together an exceptionally wide range of scientific and technical expertise in both conventional topics and cutting-edge technologies.

As with many other species, *Bombyx mori* was obtained through human-driven selection from a wild ancestor but, in contrast to other animals, the genetic stocks of silkworm strains have been preserved in public research facilities for centuries. Therefore, in nations with a history of sericulture, there are wide collections of various genotypes that must be maintained through a customary activity. Additionally, the majority of sericultural regions in the world continue to practice agricultural production of cocoons through conventional silkworm rearing. Since the mulberry leaf is the only food source for the larvae of this insect (or an artificial diet based on it), there has always been a need for moriculture, and the same organizations dedicated to maintaining *B. mori* genetic resources typically preserve mulberry germplasm collections as well; a significant amount of research is devoted to understanding the traits of these varieties and how they relate to the silkworm. A multilevel study on the effects of regional mulberry genotypes on the health of the silkworm is reported in this issue; the authors demonstrated that the selection of mulberry varieties characterized by specific qualitative and quantitative features can strongly impact larval development, cocoon production, and raw silk parameters [1]. Scientists also learnt how to use mulberry varieties to feed other animals or as a source of raw materials and active chemicals; further, they adopted new technologies to obtain advanced materials from silk. This issue presents a bibliometric paper by Giora et al. [2], which analyzed recent research trends in the fields of sericulture and moriculture and provides an overview of the past 20 years of research in this sector of rising interest. Two other reviews [3,4] aimed to connect advances in or criticisms of sericulture to societal and economic changes in various regions of the world, highlighting how this activity may be heavily dependent not only on technological improvements, but also on many external factors.

One of the newest trends in sericulture is to promote the development of circular economy supply chains to employ silk by-products for feed and food, a topic that is comprehensively discussed and documented in the article by Tassoni et al. [5]. Nowadays, silk by-products are easily exploitable via melting silk with the right solvents to obtain solutions of silk proteins, opening the door to entirely new applications in industry including in the textile sector, which is still of the utmost importance for an economy dependent on traditional sericultural agroindustry. Furthermore, this issue contains numerous articles that address these novel applications and cover both the general role of silk fibroin in biomedicine [6] and methods for creating hybrid materials, for instance, by adding nanoparticles [7]. The paper by Bassani et al. [8] examined particular devices, such as silk nerve conduits, in a study case illustrating how research findings can be translated into medical products for use in clinical settings.

Advances in methodology and knowledge are not limited to the new biotechnological uses of silk proteins. A significant part of the Special Issue is devoted to papers using the most recent techniques to study silkworm physiology to enhance silk production. This can improve the general knowledge of silkworm biology and hasten the potential use of this organism in various applied fields. Specifically, Yokoi et al. [9] undertook an RNA-seq investigation of several silkworm larvae tissues, including the midgut, fat body, testis, ovary, and the different subregions of the silk gland, generating reference transcriptome data for a *B. mori* strain. Masuoka et al. [10] identified novel silk protein regulation factors, using a proprietary software, to conduct co-expression networks and time-course expression analyses on transcriptomic data from various tissues, including different subregions of the silk glands, in reference silkworm strains. Zhang et al. [11] performed a transcriptomic analysis of fat body derived from silkworms treated with hydrogen sulfide, a molecule that, at specific doses, exerts beneficial effects on the developmental and economical traits of *B. mori*. All these results provide data and methods with which to develop strategies for enhancing silk production in *B. mori*. Moreover, de la Peña et al. [12] compared RNA-seq transcriptional profiles of silk glands belonging to various silkworm strains commonly reared in the Philippines and maintained under different temperature conditions, producing a set of information useful for the genetic improvement of *B. mori* strains and to increase their silk productivity in this geographical area. Finally, the article by Baci et al. [13] reports a comprehensive overview of the most advanced genome-editing techniques performed in *B. mori*, using the CRISPR/Cas9 methodology. In their paper, the authors illustrated strategies used to obtain both knock-out and knock-in *B. mori* strains. They also focused on the possible applications of CRISPR/Cas9 to generate silkworm strains characterized by a higher resistance to viral infections, highlighting the importance of this technology for both basic research and applied purposes in *B. mori*. Since *B. mori* is considered a model species among Lepidoptera, it has been extensively employed in studies of physiology, genetics, molecular biology, and pathology. Zhang et al. [14] made use of this lepidopteron to investigate how hemolymph ecdysteroid titer affects maternal genes during oogenesis, demonstrating that maternal mRNAs begin accumulating in the ovary during the larval period, at the wandering stage. In the article by Chang et al. [15], the role of polyamines in regulating cell cycle progression and DNA replication was explored as well as the expression of genes coding for enzymes involved in the polyamine pathway in different tissues. Since it is known that the administration of polyamines to *B. mori* larvae increases the expression of the gene coding for fibroin heavy chain and improves the quantity and quality of silk, this study opens the door to manipulating the expression of key genes involved in the polyamine pathway in silk glands to improve silk production. The use of silkworms and insects in general as model organisms began in response to growing concerns about animal health and welfare, with the goal of preventing, or at least reducing, the usage of higher animals in all sectors of the life sciences. In fact, the principles of the 3Rs (Replacement, Reduction, and Refinement), developed in the 1950s, offer a foundation for conducting more compassionate animal research. Since then, they have become an integral part of national and international laws and regulations governing the use of animals in research and are progressively leading to the development of suitable alternative models, among invertebrates, for experimental aims. Holometabolous insects represent a promising option due to their lower costs, convenience in handling, lack of ethical restrictions, and features that can successfully reproduce different biological mechanisms occurring in mammals. In this Special Issue, three articles pave the way for the development of the silkworm as an alternative model for screening antimicrobial compounds against nosocomial pathogens, such as *Staphylococcus epidermidis* [16], and bacteria that cause skin diseases and systemic infection, such as *Cutibacterium acnes* [17], as well as for studying the effects of natural compounds with hypoglycemic activity [18]. Although it is still up for debate whether these insect-based models are applicable in clinical settings, the findings of these three studies are undoubtedly encouraging. Another article focused on the health status of the silkworm and its pathologies. In particular, the study dealt with nuclear polyhedrosis, a major viral disease in sericulture that causes high larval lethality and great economic losses. By using a genomic approach that considered different BmNPV strains, the authors demonstrated that mutation and rearrangement of the genome can lead to differential pathogenicity of this virus [19]. The holobiont concept, which has recently emerged as a theoretical and experimental context in which to study the interactions between hosts and their associated microbial communities, is worth mentioning while discussing the silkworm as a laboratory model. It is becoming increasingly clear that the development, growth, and health of macro-organisms are influenced by the complex microbial communities they host. The article by Dee Tan and Bautista [20] deals with this topic; the authors studied the gut microbial composition of four *B. mori* strains reared in the Philippines to construct a database of useful information fostering an improved understanding of whether and how the gut microbiota affects silk production and to develop strategies for future improvement of the strains. Another topic that has attracted increasing attention in recent years is the administration of probiotics to provide health benefits, generally by improving or restoring the gut microbial community. Unban et al. [21] isolated from the midgut of *Samia ricini*, another lepidopteran species used for silk production, a xylose-utilizing lactic acid bacterium, *Enterococcus hirae* SX2, which acts as a probiotic in this insect due to its tannin tolerance and antimicrobial activity against insect pathogens. The results presented in this article demonstrate that the oral administration of *E. hirae* SX2 to Eri silkworm not only improves its growth and reduces mortality, therefore positively influencing its economic traits, but also represents a starting point for developing strategies to select novel probiotics that can positively affect *B. mori* larvae and improve sericulture research.

Taken together, the articles in this Special Issue cover the transversality of all the topics encompassed by sericulture, which goes far beyond silkworms and silk. This collection of many articles from different countries in a relatively limited period of time testifies to the great interest the scientific community takes in sericulture, encouraging the guest editors to repropose the subject in the near future.;

Author Contributions: Conceptualization, S.C., M.C., F.S., A.S. and G.T.; writing—original draft preparation, S.C., M.C., F.S., A.S. and G.T.; writing—review and editing, S.C., M.C., F.S., A.S. and G.T. All authors have read and agreed to the published version of the manuscript.

Funding: This research received no external funding.

Conflicts of Interest: The authors declare no conflict of interest.

References

1. Urbanek Krajnc, A.; Bakonyi, T.; Ando, I.; Kurucz, E.; Solymosi, N.; Pongrac, P.; Berčič, R.L. The Effect of Feeding with Central European Local Mulberry Genotypes on the Development and Health Status of Silkworms and Quality Parameters of Raw Silk. *Insects* **2022**, *13*, 836. [CrossRef] [PubMed]
2. Giora, D.; Marchetti, G.; Cappellozza, S.; Assirelli, A.; Saviane, A.; Sartori, L.; Marinello, F. Bibliometric Analysis of Trends in Mulberry and Silkworm Research on the Production of Silk and Its By-Products. *Insects* **2022**, *13*, 568. [CrossRef]
3. Tzenov, P.; Cappellozza, S.; Saviane, A. Black, Caspian Seas and Central Asia Silk Association (BACSA) for the Future of Sericulture in Europe and Central Asia. *Insects* **2022**, *13*, 44. [CrossRef]
4. Andadari, L.; Yuniati, D.; Supriyanto, B.; Murniati, M.; Suharti, S.; Widarti, A.; Steven, E.; Sadapotto, A.; Winarno, B.; Minarningsih, M.; et al. Lens on Tropical Sericulture Development in Indonesia: Recent Status and Future Directions for Industry and Social Forestry. *Insects* **2022**, *13*, 913. [CrossRef]
5. Tassoni, L.; Cappellozza, S.; Dalle Zotte, A.; Belluco, S.; Antonelli, P.; Marzoli, F.; Saviane, A. Nutritional Composition of *Bombyx mori* Pupae: A Systematic Review. *Insects* **2022**, *13*, 644. [CrossRef]
6. Lujerdean, C.; Baci, G.M.; Cucu, A.A.; Dezmirean, D.S. The Contribution of Silk Fibroin in Biomedical Engineering. *Insects* **2022**, *13*, 286. [CrossRef]
7. Singh, M.; Dey, E.S.; Bhand, S.; Dicko, C. Supercritical Carbon Dioxide Impregnation of Gold Nanoparticles Demonstrates a New Route for the Fabrication of Hybrid Silk Materials. *Insects* **2022**, *13*, 18. [CrossRef]
8. Bassani, G.A.; Vincoli, V.; Biagiotti, M.; Valsecchi, E.; Zucca, M.V.; Clavelli, C.; Alessandrino, A.; Freddi, G. A Route to Translate a Silk-Based Medical Device from Lab to Clinic: The Silk Biomaterials Srl Experience. *Insects* **2022**, *13*, 212. [CrossRef]

9. Yokoi, K.; Tsubota, T.; Jouraku, A.; Sezutsu, H.; Bono, H. Reference Transcriptome Data in Silkworm *Bombyx mori*. *Insects* **2021**, *12*, 519. [CrossRef]
10. Masuoka, Y.; Cao, W.; Jouraku, A.; Sakai, H.; Sezutsu, H.; Yokoi, K. Co-Expression Network and Time-Course Expression Analyses to Identify Silk Protein Regulatory Factors in *Bombyx mori*. *Insects* **2022**, *13*, 131. [CrossRef]
11. Zhang, R.; Cao, Y.Y.; Du, J.; Thakur, K.; Tang, S.M.; Hu, F.; Wei, Z.J. Transcriptome Analysis Reveals the Gene Expression Changes in the Silkworm (*Bombyx mori*) in Response to Hydrogen Sulfide Exposure. *Insects* **2021**, *12*, 1110. [CrossRef] [PubMed]
12. de la Peña, P.N.O.; Lao, A.G.D.; Bautista, M.A.M. Global Profiling of Genes Expressed in the Silk Glands of Philippine-Reared Mulberry Silkworms (*Bombyx mori*). *Insects* **2022**, *13*, 669. [CrossRef] [PubMed]
13. Baci, G.M.; Cucu, A.A.; Giurgiu, A.I.; Muscă, A.S.; Bagameri, L.; Moise, A.R.; Bobiș, O.; Rațiu, A.C.; Dezmirean, D.S. Advances in Editing Silkworms (*Bombyx mori*) Genome by Using the Crispr-Cas System. *Insects* **2022**, *13*, 28. [CrossRef] [PubMed]
14. Zhang, M.; Xu, P.; Chen, T. Hemolymph Ecdysteroid Titer Affects Maternal Mrnas during *Bombyx mori* Oogenesis. *Insects* **2021**, *12*, 969. [CrossRef]
15. Chang, L.; Li, Z.; Guo, H.; Zhang, W.; Lan, W.; Wang, J.; Shen, G.; Xia, Q.; Zhao, P. Function of Polyamines in Regulating Cell Cycle Progression of Cultured Silkworm Cells. *Insects* **2021**, *12*, 624. [CrossRef]
16. Montali, A.; Berini, F.; Saviane, A.; Cappellozza, S.; Marinelli, F.; Tettamanti, G. A *Bombyx mori* Infection Model for Screening Antibiotics against *Staphylococcus epidermidis*. *Insects* **2022**, *13*, 748. [CrossRef]
17. Matsumoto, Y.; Tateyama, Y.; Sugita, T. Evaluation of Antibacterial Drugs Using Silkworms Infected by *Cutibacterium acnes*. *Insects* **2021**, *12*, 619. [CrossRef]
18. Aznar-Cervantes, S.D.; Santesteban, B.M.; Cenis, J.L. Products of Sericulture and Their Hypoglycemic Action Evaluated by Using the Silkworm, *Bombyx mori* (Lepidoptera: Bombycidae), as a Model. *Insects* **2021**, *12*, 1059. [CrossRef]
19. Guo, H.; Zhang, B.; Zheng, X.; Sun, J.; Guo, H.; Li, G.; Zhao, G.; Xu, A.; Qian, H. Pathogenicity Detection and Genome Analysis of Two Different Geographic Strains of Bmnpv. *Insects* **2021**, *12*, 890. [CrossRef]
20. Dee Tan, I.Y.; Bautista, M.A.M. Bacterial Survey in the Guts of Domestic Silkworms, *Bombyx mori* L. *Insects* **2022**, *13*, 100. [CrossRef]
21. Unban, K.; Klongklaew, A.; Kodchasee, P.; Pamueangmun, P.; Shetty, K.; Khanongnuch, C. Enterococci as Dominant Xylose Utilizing Lactic Acid Bacteria in Eri Silkworm Midgut and the Potential Use of *Enterococcus hirae* as Probiotic for Eri Culture. *Insects* **2022**, *13*, 136. [CrossRef] [PubMed]

insects MDPI

Review

Lens on Tropical Sericulture Development in Indonesia: Recent Status and Future Directions for Industry and Social Forestry

Lincah Andadari [1], Dhany Yuniati [2], Bambang Supriyanto [3], Murniati [4], Sri Suharti [4], Asmanah Widarti [5], Eden Steven [6], Andi Sadapotto [7], Bondan Winarno [4], Minarningsih [8], Retno Agustarini [1,*], Nurhaedah Muin [4], Wahyudi Isnan [4], Yetti Heryati [1], Yelin Adalina [9], Irma Yeny [4], Rosita Dewi [4], Ari Nurlia [5], Septiantina Dyah Riendriasari [1], Kun Estri Maharani [8], Luthfan Meilana Nugraha [4] and Budi Hadi Narendra [4]

1 Research Center for Applied Zoology, Research Organization for Life Sciences and Environment, National Research and Innovation Agency (BRIN), Bogor 16911, Indonesia
2 Research Center for Behavioral and Circular Economics, Research Organization for Governance, Economy and Community Welfare, National Research and Innovation Agency (BRIN), Jakarta 10340, Indonesia
3 Directorate General of Social Forestry and Environmetal Partnership, The Ministry of Environment and Forestry (KLHK), Jakarta 10270, Indonesia
4 Research Center for Ecology and Ethnobiology, Research Organization for Life Sciences and Environment, National Research and Innovation Agency (BRIN), Bogor 16911, Indonesia
5 Research Center for Society and Culture, Research Organization for Social, Sciences and Humaniora, National Research and Innovation Agency (BRIN), Jakarta 10340, Indonesia
6 Emmerich Research Center, Jakarta 14450, Indonesia
7 Faculty of Forestry, Hasanuddin University, Makasar 90245, Indonesia
8 Center for Standardization of Sustainable Forest Management Instruments, The Ministry of Environment and Forestry (KLHK), Bogor 16118, Indonesia
9 Research Center for Biomass and Bioproducts, Research Organization for Life Sciences and Environment, National Research and Innovation Agency (BRIN), Bogor 16911, Indonesia
* Correspondence: retn030@brin.go.id or retno.agustarini@gmail.com; Tel.: +62-811-848-4080

Citation: Andadari, L.; Yuniati, D.; Supriyanto, B.; Murniati; Suharti, S.; Widarti, A.; Steven, E.; Sadapotto, A.; Winarno, B.; Minarningsih; et al. Lens on Tropical Sericulture Development in Indonesia: Recent Status and Future Directions for Industry and Social Forestry. *Insects* **2022**, *13*, 913. https://doi.org/10.3390/insects13100913

Academic Editors: Silvia Cappellozza, Morena Casartelli, Federica Sandrelli, Alessio Saviane and Gianluca Tettamanti

Received: 20 September 2022
Accepted: 4 October 2022
Published: 8 October 2022

Simple Summary: Sericulture is a labor-intensive agro-industry business that can increase the community's welfare and support environmental improvement. In Indonesia, silk, as the final product of sericulture, is a potential non-timber forest product (NTFP) that provides benefits to livelihoods and the forest ecosystem. Silk is a fiber produced by the domestic silkworm or mulberry silk moth, *Bombyx mori* L., belonging to the Lepidopteran order, Bombycidae family, probably providing more than 99% of the world's silk. However, there are many challenges to its development at both upstream and downstream levels, including the availability of quality eggs, optimal and efficient cultivation, pest and disease control, a lack of policy support, unsustainable production, low product quality, and competition with imported products. This paper discusses the recent status and future directions of sericulture development in Indonesia. Improvements in technical and social-economic aspects can support the development of sericulture in Indonesia through increasing productivity in the upstream sector along with conducive downstream policies and governance.

Abstract: The domestic silkworm or mulberry silk moth, *B. mori* L., provides more than 99% of the world's silk. Silk, as a sericulture product, was first introduced in Indonesia through a trade mechanism and began to develop in 1953. Several factors (economic, ecological, market, and cultural) support sericulture and make it become one of the non-timber forest product priorities. However, the competitive advantages alone have not encouraged the development of prospective sericulture industry in Indonesia yet. This paper is a review of tropical sericulture development in Indonesia. The literature on the development of sericulture in Indonesia between 1989 and 2022 is used to describe conditions related to mulberry cultivation (moriculture), and silkworm rearing (sericulture), as well as the state of socio-economic development, culture, and institutions. Moriculture and sericulture techniques, socio-economic aspects, institutional arrangements, and community motivations are intertwined, creating a challenging atmosphere for sericulture development. There are potential resources, such as exploring quality mulberry production and quality silkworm production through research and development, valuable cultural aspects, and potential stakeholders to build network engagement. Commitment, cooperation, and action from all stakeholders are needed to enhance the development of sericulture in Indonesia. In this context, the central government can play an important role in facilitating multi-stakeholder partnerships in the development of integrated sericulture in Indonesia.

Keywords: *Bombyx mori* L.; sericulture; non-timber forest product; livelihood; Indonesia

1. Introduction

The COVID-19 pandemic has caused the global agenda—e.g., the Sustainable Development Goals (SDGs) for 2030—to come under pressure. This crisis has plunged millions of people into unemployment and dire health conditions resulting in poverty [1–3]. In developing countries, including Indonesia, the impact of the pandemic has been more severe, especially in areas where community dependency on forest resources is high due to the limited availability of both on-farm and off-farm jobs [4–6]. The situation has been exacerbated by reduced access to inputs, labor, and agricultural land, resulting in a decrease in production, household income, and nutrition [7].

Many attempts have been made by the government to prevent the most severe negative impacts and to boost the local economy by developing several alternative forest-based agribusinesses [4,8]. In Indonesia, an economic recovery program, including social forestry, has been implemented, with the double objective of enhancing consumption and economic productivity [9]. Forests, as natural resource systems, have the potential to provide multiple benefits. In addition to wood products, forests can provide non-timber forest outputs and environmental benefits. NTPs are biological products other than wood of high value, generally obtained from wild biodiversity in natural or human-modified environments [10,11]. Many studies have revealed that non-timber forest products, including ecotourism, small-scale timber enterprises, and environmental services, could play a significant role in social and economic recovery during and after a pandemic [8,12,13]. Non-timber forest products contribute 80% of the social forestry business model in Indonesia [14].

One of the non-timber forest products with the potential to resolve economic challenges during and after a crisis is the natural silk produced by sericulture agribusiness. Sericulture is a prospective and potential activity that can regularly generate higher income [15,16]. Apart from its ability to provide gainful employment and economic improvement for people in rural areas due to the high selling price of its products [15,16], sericulture also plays an important role in preventing the migration of rural people to urban areas in search of employment [17,18]. The silk industry is labor-intensive [19] and so provides employment for 7.9 million people in India and 20,000 weaver families in Thailand [20]. In the context of Indonesia, there are 1200 silk farmers [21] and 4900 weavers [22] involved in silk agro-industry activities. Sericulture agribusiness has been determined to be one of the five NTFP priorities, having the potential to contribute to the economy of the country and tackle poverty [23–25].

Sericulture was first introduced to Indonesia through trade in the 10th century; since 1953 [26], it has developed rapidly as it is suitable to the agroclimatic conditions as well as the local culture. Silk yarn production reached its peak in 1971 with a production of 140 tons but then tended to decrease due to several factors [27], such as the low quality of silkworm seeds, pebrine disease attacks, and mulberry plants as feed have begun to be shifted to other horticultural commodities, low prices for silk products, and a lack of support from silkworm production programs and organization.

The silk agro-industry chain is a sequence of activities starting with mulberry cultivation (moriculture) and silkworm rearing. These stages are in the upstream segment and are commonly known as sericulture. The next phase is reeling raw silk from the cocoons and producing the yarn and its derivative products. These products then deliver to the weaving industry for further processing and marketing. Yarn processing, referred to as the manufacturing sector, is the downstream sector of the agro silk industry [21,24].

Several factors support sericulture and make it become one of the NTFP priorities. Economically, the silk industry is an important way of fulfilling domestic as well as export needs, either in the form of cocoons, yarns, or finished goods [19,28]. It is a potential commodity that can significantly contribute to foreign exchange earnings, raise living standards, and reduce poverty, hunger, gender discrimination, and disease risk. Sericulture is mostly developed in rural and suburban areas and hence provides job opportunities for rural communities [29]. In addition, the sericulture and silk industry in Indonesia represents an economically viable rural enterprise for inclusive development, which can effectively have fall-outs in various industrial sectors [20].

Ecologically, the silk industry involves an environmentally friendly production process. The mulberry tree ensures green land covering, soil conservation, and erosion protection and allows for the use of land that is not suitable for other crop cultivation. The silkworm rearing does not cause pollution, CO_2 emissions are very low, and waste can be easily degraded [30], so it can coexist well with already inhabited areas [31,32]. Silk produced from sericulture is a non-synthetic, renewable, and biodegradable fiber [15]. Furthermore, sericulture also contributes to the improvement of the microclimate of the area of development, which will eventually improve the surrounding environmental conditions [25]. In addition, mulberry cultivation can only be developed in tropical countries, including Indonesia, which has agroclimatic conditions that are suitable for mulberry and silkworm cultivation [33]. Because it has moderate dry and rainy seasons, this country allows mulberry plants to serve as the main food for silkworms to grow and be cultivated throughout the year [34–36].

In terms of market prospects, there is still a gap between the annual world demand for natural silk (92,743 tons per year) and the supply of yarn (83,393 tons) [24,25]. Although it is only ranked twelfth among silk yarn-producing countries in the world, Indonesia has a comparative advantage as it produces better quality yarn than other counties [37] and so has the potential to fill the gap. However, filling the market gap is not easy since there are high-quality standards for the international market. At a domestic level, the demand for silk has not yet been met, and hence must be supplied from imported products. The annual need for silk yarn is 500–800 tons, while the national production in 2015 was only about 12.13 tons, 8.95 tons of which were from South Sulawesi [21]. In the current pandemic situation, the need for cocoons as a raw material for domestically produced silk, which was previously imported from other silk-producing countries, such as China, cannot be fulfilled. The above-mentioned characteristics of the silk industry indicate that sericulture has several social, economic, and ecological advantages [25].

However, until now, the competitive advantage of sericulture has not encouraged the establishment of a strong and independent national sericulture industry in Indonesia. During the period 2011–2015, Indonesia continued to experience a decline in the production of both cocoons and silk yarn, correlated with a decline in the natural silk fabric industry in various regions in Indonesia. Some factors causing the decline of the silk agro-industry in Indonesia include: the limited availability of quality silkworm eggs and feed (mulberry leaves), limited skilled labor, the uncertainty of the product selling price [19,38,39], and pest and disease attacks [21]. Apart from those challenges, Sadapotto [27] stated that ineffective policy formulation for sericulture is actually the main factor behind the failure of sericulture in Indonesia due to neglecting institutional factors and focusing more on physical and technical factors in the analysis of problems regarding sericulture. Sericulture industry development involves lots of stakeholders from upstream to downstream. At the upstream level, the development of silk is the responsibility of the Ministry of Environment and Forestry [40]. At the downstream level, activities are under the authority of several relevant stakeholders, including the Ministry of Industry and Trade and the Ministry of Small–Medium Enterprise and Cooperatives. Strong coordination and collaboration among all stakeholders involved in natural silk production are essential for the development of the sericulture industry in Indonesia [21,27].

At the farm level, although it is not a capital-intensive business, sericulture requires sufficient capital to start by buying silkworm eggs, preparing the land to plant mulberries as silkworm feed, and setting up rearing rooms; not all farmers can afford the costs related to these investments. These challenges have led natural silk agribusiness activities to be increasingly abandoned by farmers who tend to choose other farm activities [41].

Various efforts have been made to overcome this situation by developing good-quality silkworm eggs and mulberry seeds and providing technical assistance to silk farmers; unfortunately, these have not provided significant results yet. The key to the successful development of natural silk lies mainly in the support from government policies, the availability of good quality silkworm eggs and feed, the availability of land resources and easy access to adequate funding, as well as training of workers who have skills and expertise in silk cultivation [42].

Many studies have been conducted regarding the development of the silk industry in Indonesia, but most are still partial. An analysis of the gap between existing practices and the strategies that should be developed is needed to determine how to optimize sericulture agribusiness through conducive upstream to downstream policies and governance suited to Indonesian conditions. This paper aims to describe in detail the various development potentials and challenges that should be faced in order to optimally develop sericulture agribusiness.

2. Examining Sericulture Development in Indonesia

Indonesia's sericulture production centers are in several provinces, including West Java, Central Java, and South Sulawesi. South Sulawesi contributes 70–80% of national silk yarn production [21,43,44]. Silk production has been developed in Indonesia since 1953 [26], and mulberry trees began to be cultivated in South Sulawesi in the early 1960s [19].

The sericulture in Indonesia has decreased significantly due to the pebrine disease epidemic [21]. Before 2012, the development areas for sericulture were located on the islands of Java, Sulawesi, Kalimantan, and Sumatra, regions that cover 11 provinces. After 2012, the sericultural activity continued in two provinces only, South Sulawesi and West Java (Figure 1).

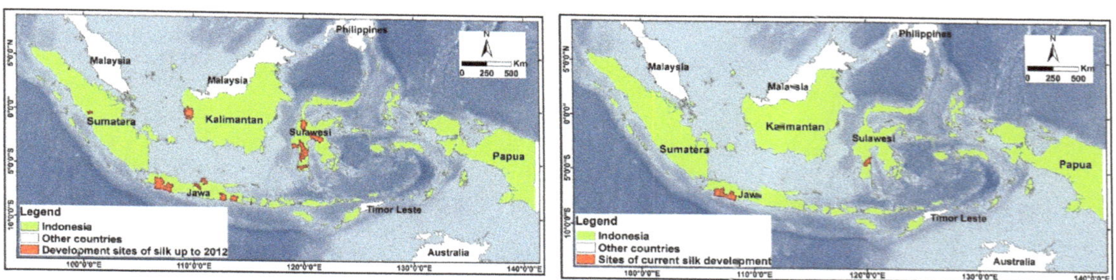

(a) (b)

Figure 1. Distribution of Indonesian sericulture development areas. (**a**) before 2012; (**b**) after 2012.

Various internal and external factors have influenced the decline in the distribution of sericulture development areas; therefore, it is crucial to study the dynamics of sericulture in Indonesia. This section will review the dynamics of sericulture in Indonesia, including the cultivation of mulberry plants (moriculture), rearing of silkworms, products, and social, economic, cultural, and institutional factors.

2.1. Cultivation of Mulberry Plant (Moriculture)

Mulberry plants, especially the leaves, are the sole food of the silkworm (*B. Mori* L.) [45,46]. Recently, more attention has been paid to mulberry plant production improvement (quality or quantity)—more than 60% of the total cost of cocoon production goes to mulberry cultivation [45].

Mulberry cultivation in Indonesia uses *Morus cathayana* Hemsl. [47–50], *M.s alba* L. [47–50], *M. nigra* L., *M. multicaulis* (Perr.) Loudon, *M. australis* Poir, and *M. bombycis* Koidz [47–49]. The mulberry species most widely cultivated to date is *M. cathayana* [50]. This species is preferred because it has comparative advantages such as being relatively easy to grow, adaptive to any location, and producing a higher number of leaves compared with other species of *Morus* developed in Indonesia [50]. The scientific name reference of the mulberry species or variety refers to https://powo.science.kew.org/.

In Indonesia, productivity is relatively low, ranging from 8 to 23 tons/ha/year [51,52] compared with in India (the majority of which are *M. alba* var. *multicaulis Loud* and *M. alba* var. *arthoputpurea* (Shahtut)), which reaches 10–30 tons/ha/year in tropical conditions and 25–30 tons/ha/year in temperate regions [53]. China, a major global silk producer, has mulberry productivity ranging from 15 to 46.5 tons/ha/year, depending on the species of mulberry and the conditions of the area of development [54], while production in India ranges from 18.40–23.06 tons/ha/year [55,56].

Mulberry leaves determine the productivity and quality of silkworm products, so we must consider their nutritional quality. The main factors affecting the quality and quantity of mulberry leaves are the genetic factors of each mulberry species, the environment, the cultivation methods, and pest and disease control [57].

Techniques to increase the productivity and quality of mulberry leaves via genetic factors are through type/variety selection and hybridization [47,58,59]. Several studies conducted from 1999 to 2015 in Indonesia have created several mulberry hybrids with 40–53 tons of leaves/ha/year that are resistant to drought, pests, and diseases and are easy to cultivate [60,61]. The crosses came from several sources, including *M. cathayana* Hemsl., *M. australis* Poir, *M. indica* L., *M. nigra* L., *M. amakusaguwa (Hybrid M. bombycis* Koidz *x M. acidosa* Griff.*)*, *M. acidosa* Griff. and *M. latifolia* Poir. [62,63]. Information on the productivity of mulberry hybridization in Indonesia can be seen in Table 1.

Table 1. The productivity of mulberry hybridization in Indonesia.

The Origin of the Cross	Leaf Production (ton/ha/year)
M. australis Poir *x M. indica* L.	23.2 *
M. nigra L. *x M. indica* L.	23 *
M. cathayana Hemsl. x *M. amakusaguwa (Hybrid M. bombycis* Koidz *x M.s acidosa* Griff.*)*	52.35 **
M. acidosa Griff.*x M. latifolia* Poir.	35.42 **
M. australis Poir *x M. indica* L.	29.56 **

Source: * [62], ** [63].

The result of hybridization between *M. cathayana* Hemsl. x *M. amakusaguwa (Hybrid M. bombycis* Koidz *x M. acidosa* Griff.*)*, having the highest leaf productivity, was launched by the Ministry of Forestry in 2013 and named SULI 01 [16]. The SULI 01 mulberry hybrid is suitable to be planted in lowlands and highlands. This accession can produce better quality cocoons than other mulberry varieties and has greater resistance to pests and diseases [64]. Unfortunately, this type of SULI 01 is not well known by the public, so its use as a silkworm feed is still limited. So far, silk farmers have not used hybrid types that have higher productivity than conventional ones.

The environment that affects the productivity of mulberry plants must also be considered. Some of the requirements for mulberry growth are soil conditions that are not loamy, not too sandy, and not acidic; optimal and balanced fertilization; and maintenance according to Standard Operational Procedures (SOPs) at each growth stage and water availability [65]. In Indonesia, the mulberry is found from the lowlands to the highlands (over 3600 m above sea level—m asl). It can grow well even if the leaf production is lower than that of mulberry plants cultivated in areas with an optimum altitude of 400–700 m asl and an optimum temperature of 23.9–26.6 °C [61]. Problems arise because, in extreme areas, silkworm cultivation experiences pose many challenges.

Although many areas of Indonesia are suitable for the growing of mulberry plants, the tropical climate in Indonesia causes the productivity and supply of mulberry leaves to be discontinuous [28,66]. This is related to the differences in the dry and rainy seasons. Mulberry plants require a lot of water in their growth process, so that in the rainy season, mulberry plants generally grow faster and produce more leaves. Apart from this, the decrease in leaf production in the dry season also depends on the genetic characteristics of the cultivated mulberry varieties [63]. Most of the mulberry plants in Indonesia are propagated vegetatively using stem cuttings, although they can be reproduced by gamic propagation (through seeds). Some mulberry plants in Indonesia can be seen in Figure 2.

(a)

(b)

(c)

Figure 2. Cultivation of mulberry plants in Indonesia. (**a**) South Sulawesi; (**b**) West Java; (**c**) Hybrid SULI 01.

Another challenge related to mulberry cultivation is pest and disease attacks. Mulberry plants have some diseases and pests that affect the quality and productivity of the leaves and interfere with the growth and nutrient content in the leaves [67]. Eight types of pests and four types of diseases are commonly found in Indonesia, especially in the central part of South Sulawesi [68], such as grasshoppers (*Valanga nigricornis* (Burmeister), leaf webber (*Glyphodes pulverulentalis* Hampson), snails (*Achatina fulica*), mealybug (*Maconellicoccus hirsutus* Green), white peach scale (*Pseudaulacapsis pentagona* Targioni-Tozzetti), stem borers (*Epepeotes plorator* Newman), powdery mildew (*Phyllactinia corylea* (Pers.) P.Karst), red rust fungus (*Aecidium mori* (Barclay) Barclay), and leaf spot (*Cercospora moricola* Cooke and *Sirosporium mori* (Sydow et P. Sydow) M.B. Ellis). It was stated that these pests and diseases affect the rate of leaf loss up to 15.32% and are influenced by factors related to where the mulberry plants grow. Whitefly (*Bemisia tabaci*) is the most common pest on mulberry plants in Indonesia and can cause damage of up to 80% or crop failure [69]. In India, whitefly attacks cause a leaf yield loss of 4500 kg/ha/year [70].

The use of insecticides and fungicides needs to be controlled because mulberry leaves are used as feed for silkworms in addition to other uses for humans. Due to the application of pesticides over the years, sericulture faces various problems, such as silkworm sensitivity and death. The use of pesticides can increase the mortality of silkworms in the 5th instar. *B. mori* L. silkworms are sensitive to the smell of pesticides, even though they are used for other plants close to mulberry fields or where silkworms are kept [40,71,72]. Pesticides containing dimethoate in mulberry leaf at rearing facilities silkworms will affect cocoon quality, the weight of instar fifth silkworms, and are contained as chemical residues; for this reason, harvest losses are recorded [73].

The area of the mulberry plantation will greatly affect cocoon production. This can be seen in the trend of cocoon production in South Sulawesi Province. Currently, the area of mulberry production in South Sulawesi is 65.2 ha, with cocoon production in two districts (Soppeng and Wajo) where are produced 14,075 kg [40]; in 2014, in the same two districts, there was an area of 238.35 ha of mulberries with cocoon production of 29,249 kg [74]. However, on average, cocoon productivity per hectare of mulberry plants is currently better than in 2014. This is possible because the use of superior eggs of silkworms and mulberry is more intensive and because SOPs for silkworm rearing with supervision extension workers in the area. Other challenges need to be anticipated. This relates to the trend of farmers converting their mulberry plantations into other commodity crops [40,41]. However, they still grow mulberries on part of their land and continue silkworm rearing.

2.2. Silkworm Rearing (Sericulture)

Nowadays, the domestic silkworm or mulberry silk moth, *B. mori* L., belonging to the Bombycidae family, provides more than 99% of the silk in the world. Therefore, we can divide silkworm accessions into Chinese, Japanese, European, and Tropical races [75]. However, there are only three accessions of silkworms in Indonesia, Chinese, Japanese and Tropical races.

The Chinese race has greenish eggs, yellow eggshells, plain silkworm patterns, blue belly color between segments, shorter silkworm life, oval/oval cocoon larvae shape, and high silk percentage. The Japanese race has gray eggs, white eggshells, heavier eggs, skin spots, pink color between segments of the abdomen, longer silkworm life, stronger silkworm, and peanut-shaped cocoons. Tropical races have white eggshell color, plain silkworm pattern, strong/disease resistant silkworm, short silkworm life, small cocoons, difficult cocoon shells to spin, poor productivity and fiber quality, but high adaptability in Indonesia. The differences in eggs, silkworms, and cocoons of each race in Indonesia can be seen in Figure 3.

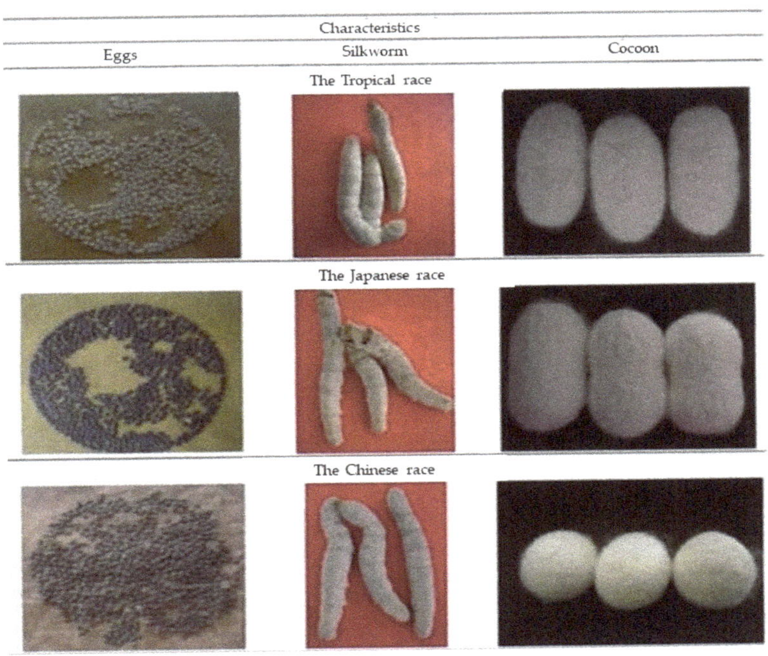

Figure 3. The characteristics of each group of races.

In Indonesia, what has been developed are the Chinese race silkworms, the Japanese race, and the result of crossing between the Japanese strains with the Chinese strains [26]. The quality of hybrid silkworm cocoons was better than that of the Chinese and Japanese silkworms [76].

Problems in the rearing of silkworms in Indonesia include: (1) the low quality of silkworm eggs [19]; (2) limited number and supply of silkworm eggs [77]; (3) incompatibility of the type of silkworm with the location of cultivation [78]; (4) inadequate conditions of facilities and infrastructure [78]; (5) lack of farmers' compliance with good silkworm-rearing standards (SOPs) [79]; and (6) silkworm pests and disease attacks [21]. The limited supply and low quality of silkworm eggs have caused farmers to use the same silkworm eggs for various geographic conditions [78], resulting in varying productivity [77]. Infrastructure and compliance with silkworm rearing standards (SOPs) are important points in silkworm rearing. Silkworm diseases are the most significant factor that impacts cocoon production [21,80,81].

2.2.1. Silkworm Breeding

In general, the silkworm eggs used are of three accessions, coming from three subtropical regions (China, Japan, and Europe), and are usually kept under optimum conditions. The silkworm eggs currently circulating are the result of crossing parents from subtropical areas (bivoltine), apparently not able to adapt well to tropical conditions such as Indonesia [82,83]. For tropical conditions such as those in Indonesia, where the agroclimate fluctuates, the mulberry leaf quality is low, and the silkworm rearing ability is limited, stronger silkworms are needed.

The success of silkworm rearing is highly dependent on the genetic quality of the eggs and their suitability for the environment. Therefore, it is recommended to obtain and use eggs suitable to the conditions of the rearing environment and with the desired properties as targets [84]. Hybridization in Indonesia is mostly carried out by the Sericulture Laboratory under the Ministry of Environment and Forestry and Perum Perhutani (a state forest enterprise). There are five silkworm hybridization products that are developed and used in Indonesia: Hybrid C-301 [52,75,77,85], BS 08 [86], BS 09 [86], PS 01 [87] and SINAR [88]. Information about silkworm hybridization products can be found in Supplementary Table S1.

2.2.2. Silkworm Nursery

After obtaining high-quality silkworm eggs, nursery development is needed so they can be commercialized. Until 2012, Indonesia had only two producers of silkworm eggs, namely Pusat Produksi Ulat Sutera/PPUS (Center for Silkworm Production) Candiroto and Kesatuan Pengusahaan Sutera Alam/KPSA (the Natural Silk Concession Unit) Soppeng [35,86,89,90]. Both institutions are under the management of Perum Perhutani Unit I Central Java (PPUS Candiroto) and Perum Pehutani Unit II East Java (KSPA Soppeng). The production of these two nurseries results from the silkworm hybridization that has been carried out, namely C301 [52,75,77,85] and BS 09 [86]. The following conveys the productivity of silkworm eggs in Indonesia and the institutions that produce them (Table 2).

Table 2. The productivity of silkworm eggs in Indonesia.

Egg Type	Producer	Hatching Percentage (%)	Characteristics	Cocoon Quality	Filament Quality
C301 *	KSPA Soppeng and PPUS Candiroto	81.65–81.95	- egg color: gray - eggshell color: white - silkworm pattern: spots - silkworm segment color: gray - silkworm life: 24–26 days - silkworms cannot stand being kept in standard conditions - The cocoon is white and oval shape	- Normal cocoons: 86.75–93.25% - Weight of cocoon: 1.7–2.04 g - Weight of cocoon shell: 0.41–0.46 g - Cocoon shell ratio: 22.3– 23.97%	- Filament length: 1026–1127 m, - Filament percentage: 21% - Filament denier: 2.71–3.63 - Decomposition: 76–89%
BS 09 **	KSPA Soppeng dan PPUS Candiroto	>90	- egg color: gray - eggshell color: white - silkworm pattern: spots - color between segments: blue - silkworms are somewhat sensitive to minimum conditions - cocoon shape: oval - cocoon color: white	- Normal cocoons: 90.00–96.00% - Cocoon shell ratio: 21.28–23.49%	- Filament length: 1060–1216 m, - Filament percentage: 16.64–19.09% - Filament denier: 2.3–3.26
PS 01 ***	Bina Mandiri Forest Farmers Group, Sukabumi	90.74–96.99	- egg color: gray - eggshell color: white - silkworm pattern: spots - silkworm color between segments: slightly bluish gray - silkworm growth is more uniform - resistant to disease - the life of the silkworm is 25–27 days - cocoon color: white - stable cocoon shape	- Normal cocoons: 97.5–100% - Weight of cocoon: 1.7–1.9 g - Weight of cocoon shell: 0.38–0.44 g - Cocoon shell ratio: 22.29–24.82%	- Filament length: 808–1003 m, - Rolling power: 95–100% - Filament denier: 2.3–2.43 - Spun yield: 13.29–15.31

Source: * [77], ** [91], *** [87].

Hybrid C301 is a silkworm strain widely used in various environmental conditions [39]. Silkworm eggs from two different nurseries reared in the same place sometimes resulted in different productivity [35,86]. Rahmathulla [92] confirmed that seasonal differences in the environmental components significantly affect the genotypic expression in the form of the phenotypic output of silkworm crops, such as cocoon weight, shell weight, and cocoon shell ratio.

The average rearing carried out by each silk farmer is one box containing 25,000 eggs. The average cocoon production in Indonesia from the rearing of one box is 25.03 kg, while the cocoon production in China, the highest producer in the world, can reach 39.97 kg/box [19]. The productivity of the C301 egg hybrid is 25–35 kg/box [74,84], that of the BS 09 hybrid is 27–30 kg/box [74], and that of the PS 01 hybrid is 35–40 kg/box [74,84]. The productivity of the PS 01 hybrid silkworm was better than that of the BS-09 hybrid, which is reared at an altitude of 250 m above sea level; as can be observed according to the body features, mortality of the instar first–four instar silkworms, and the quality of the silkworm cocoons [93].

The number and quality of eggs have experienced ups and downs, thus encouraging importing [94]. In general, silkworm eggs are imported from China. This can be seen from the source of silkworm eggs used by silkworm farmers in South Sulawesi Province, which come from the production of Perum Perhutani (KPSA Soppeng) and are also imported egg seeds from China [19,95].

The price of eggs from Perhutani is more affordable for silk farmers with relatively good productivity, although the production quality is unstable due to changes in weather conditions. The production of silkworm eggs from China is higher than that of *Perhutani* eggs [94,96], but it is only available in certain months [19]. Chinese eggs have higher adaptability and steadiness in yield than hybrid C301, but hybrid C301 showed better cocoon and yarn quality in the observed parameters. Importing silkworm eggs must be balanced with the need for adequate storage equipment, considering that the import of eggs is carried out in large quantities and is not directly absorbed by the market. Poor egg storage results in many eggs dying, hatching not being uniform, and the larvae of the following generation being weak [77,97].

Another problem with silkworm eggs is related to the limitations of the center and breeders. Up until now, there has been only one breeder operating (KSPA Soppeng) with a limited capacity. The limited number of breeders impacts the use of silkworm eggs in rearing locations that are far from the nursery location. With the lack of activity of the Candiroto Silkworm Breeding Center (PPUS), silkworm farmers in West Java and surrounding areas obtain silkworm eggs from Soppeng's KPSA in South Sulawesi, thus requiring a lengthy shipping process. The time the eggs are in transit increases the risk of egg damage.

The PS 01 hybrid production is a solution for the existence of superior silkworm eggs. Unfortunately, PS 01 silkworm eggs are not yet on the market [98]. The Directorate of Social Forestry and Customary Forest Business Development (Bina Usaha Perhutanan Sosial & Hutan Adat/BUPSHA), part of the General Directorate of Social Forestry and Environmental Partnership (PSKL) in collaboration with the Center for Forest Research and Development of Forest, has created Bina Mandiri Farmer Group (BMFG) to become the PS 01 breeder, functioning as a supplier of silkworm eggs for the Sukabumi Regency and surrounding areas [98]. However, until now, it has not been produced optimally.

2.2.3. Silkworm Rearing

Silkworm rearing is a series of activities to produce cocoons using silkworm eggs with certain cultivation standards. The success of cocoon production is highly dependent on compliance with its SOPs, temperature, and humidity, and the condition of the administered leaves [99]. Sericulture farmers in Sukabumi have good knowledge of silkworm rearing standards [100], but many do not apply them, resulting in low or even negative income [79]. Support from Sericulture Laboratory technicians of the Forest Research and Development Center for applying good silk-rearing standards in Forest Management Unit/FMU (Kesatuan Pemangkuan Hutan/KPH) Boalemo has resulted in higher cocoon productivity than that of FMU Yogyakarta without assistance [101]. Therefore, efforts are needed to increase farmers' adoption of good rearing standards by providing consistent, accessible, relevant, and practical information with more personal means of communication and stakeholder collaboration, using an interdisciplinary approach including social scientists, communication specialists, and marketing experts [102].

People who still cultivate silkworms at home can potentially interfere with silkworm cultivation activities [41]. *B. mori* L. is very delicate, highly sensitive to environmental fluctuations, and unable to survive extreme conditions [103]. Thus, traditional rearing activities carried out at home will result in low cocoon productivity.

Complete and appropriate infrastructure facilities (separate rearing houses for first and last instar silkworms, mulberry fields) are important for silkworm rearing. They will affect the number of rearing cycles in a year. Silk cultivation in FMU Boalemo can be done 12 times per year with appropriate facilities and infrastructures in one location. This differs from the BMFG in Sukabumi, which can only cultivate eight cycles in one year and keeps young silkworms and grown silkworms separate. More rearing cycles will increase yield productivity. Silkworm rearing in Indonesia can be seen in Figure 4.

(a) **(b)**

Figure 4. Silkworm rearing in Indonesia. (**a**) South Sulawesi Province; (**b**) West Java Province.

2.2.4. Pests and Diseases

Sericulture in Indonesia has experienced two waves of epidemics of pebrine disease and Nuclear Polyhedrosis Virus (NPV) infection, causing a drastic decline in sericulture production [21]. Both mulberries and silkworms are infested by a number of pests, which affect the cocoon quality and productivity, resulting in economic losses for farmers [104]. Among the silkworm diseases that cause economic damage, the highest crop loss is attributed to viral diseases, accounting for 70% of the total loss of the crop every year [105].

Diseases that often attack silkworms in Indonesia are NPV, Infectious Flacherie (FV), Aspergillus, Muscardine, pebrine, bacteria, and poisoning from agricultural chemicals [21,106,107]. The causes of these diseases include bacteria, fungi, viruses, and physical and chemical factors as well as environmental factors that result in abnormal conditions and death [108].

2.3. Silkworm Products

2.3.1. Main Products

The main products of sericulture are cocoons, yarns, and material fabrics (Figure 5). The trend of cocoon production [72] and yarn production [109] continues to decline. In the 1990s, Indonesia's silk yarn production reached 120–140 tons/year. Nowadays, it is only around 2.5 tons/year [20]. The current national demand for silk yarn is 900 tons per year [110], and it is predicted to increase by 12.2% annually [72]. The solution to the scarcity of silk yarn supply is generally importing. As stated by Antaranews [110], 95% of Indonesia's silk yarn is still imported from China, India, and Japan. Therefore, Indonesia has changed its sericulture strategy for the export market to meet domestic needs [111].

(**a**) (**b**) (**c**)

Figure 5. The main products of sericulture. (**a**) cocoons; (**b**) yarns; (**c**) material fabrics.

- Cocoons

Cocoons are the final product of silkworm rearing. The quality of the cocoons is determined by the hereditary nature of the silkworm strains and environmental conditions during rearing, spinning stage, etc. [112]. The requirements of a perfect cocoon are healthy (no defects), clean (clean white), the inside (pupa) not damaged or crushed, and a hard silk shell (layer of silk fibers) resistant to finger pressure. Several parameters of cocoon quality affect the quality of the raw materials for silk fiber and determine the quality, quantity, and efficiency of the process [113]. Cocoon characteristics, such as cocoon weight, cocoon shell weight, and cocoon shell ratio, are the most important economic determinants of silkworm rearing [114]. Different shapes and sizes of cocoons will produce various sizes of silk fiber and yarn quality.

The quality of the C301 hybrid showed higher yields compared with hybrids from China with respect to these parameters: percentage of normal cocoons, the weight of cocoons, weight of cocoon shells, and proportion of cocoon shells [77]. China's normal hybrid cocoons are 80.08–90.08%, while the C301 hybrid is 86.75–93.25%. The weight of the Chinese hybrid cocoon is 1.44–1.57 g, while that of the C301 hybrid is 1.7–2.04 g. The weight of the Chinese hybrid cocoon shell is 0.3–0.32 g, while the C301 hybrid is 0.41–0.46 g. The percentage of Chinese hybrid cocoon skin is 20.58–20.96%, while that of the C301 hybrid is 22.94–23.97% [77].

Hybrid PS 01 (Chinese race male and Japanese race female—(804 × 927)) has an egg hatching percentage of 97.00% compared with hybrid C301, which is only 96.01%. Hybrid PS 01 produces a cocoon weight of 1.6 g, cocoon shell weight of 0.29 g, and a cocoon shell ratio of 20.35–21.36%, the hybrid C301 cocoon weighs 1.27 g, the cocoons shell weighs 0.2 g, and the cocoon shell ratio is 17.9–19.54% [39].

The cocoon shell percentage is one of the benchmark measures for determining the selling price of cocoons. The highest level (grade of cocoon quality) is a percentage of cocoon shells of 22–25% [115].

High competition with producers in other countries, low government support for this industry, lagging technology, and poor handling of pests and diseases are thought to hinder the development of silkworm cultivation in Indonesia. In 2020, globally, Indonesia's silk production was only twelfth behind China, India, Brazil, Thailand, and Japan [20]. From year to year, Indonesia has experienced a decline in cocoon production.

- Silk Yarn

The quality of silk fiber is very important in the reeling process because it will affect the silk yarn obtained. These qualities include the dry cocoon weight, number of fiber breaks, remaining cocoon, percentage of fiber weight, and reelability (fiber unwinding). Reelability is the ability of silk fibers to unravel and coil when the cocoon is spun. Cocoon is highly dependent on the silkworm variety, temperature, and humidity during spinning. The cocoon mountages also affect the cocoon unraveling capacity [60].

Silkworm eggs from KSPA Soppeng and PPUS Candiroto, when maintained in conditions of high temperature and low humidity, did not show any difference in a produced cocoon. Still, there was a difference in the filament yield. Longer filaments were produced by silkworm eggs from KSPA Soppeng compared with the eggs from PPUS Candiroto [35].

The quality of the C301 hybrid filament is better than that of the Chinese hybrid for the parameters of filament length, percentage of filament, denier thickness, and reelability. The China hybrid filament length is 889.08–994.84 m; the filament percentage is 17.91–19.14%, the denier thickness is 2.37–2.97, the reelability is 85.57 –95.56, and the yarn weight per 50 cocoons is 8.87–10.7. The C301 hybrid produces a filament length of 1026.05–1127.83 m, filament percentage of 21.24–21.49%, denier thickness of 2.71–3.63, reelability of 76.11–89.58, and yarn weight per 50 cocoons of 9.51–13.66 [77]. Hybrid PS 01 (Chinese race male and Japanese race female—(804 $\times$ 927)) produced a filament length of 1003.67–1160 m, 32% longer than the C301 hybrid, which only had a filament of 755.67–883.33 m [39].

To encourage the development of the national silkworm industry, the Government of Indonesia (GOI), through the Ministry of Forestry, Ministry of Industry, and Ministry of Cooperatives and Small and Medium Enterprises—issued Joint Regulation No. P47/Menhut-II/2006, No. 29/M-IND/PER/6/2006 and No. 07/PER/M.KUKM/VI/2006 concerning the Guidance and Development of National Natural Silk with a Cluster Approach [27]. Through this joint regulation, the government plans to build an integrated silk industry cluster from upstream to downstream, including cultivating mulberries, silkworms rearing, and cocoons for the yarn and silk reeling industry. Furthermore, to encourage the above efforts, the Ministry of Forestry issued Ministerial Regulation no. P.56/Menhut-II/2007 concerning the Procurement and Circulation of Silkworm Eggs.

Ironically, when the cluster effort was launched (2006), the annual national silk yarn production fell to less than 32 tons and the yearly demand for domestic industry to around 700 tons [116]. On the other hand, at the global level, the market for silk yarn in 2005 continued to increase, reaching the demand of 100,000 tons. Meanwhile, to fill the shortage of domestic silk yarn, the government allowed imports of silk yarn of more than 600 tons from China, India, Thailand, and other ASEAN countries [116,117]. The imports from silk yarn-producing countries turned out to harm the national silk yarn industry. Domestic artisans prefer imported yarn, especially from China, because the price is lower. For silk entrepreneurs, the presence of spikes yarn or synthetic yarn at a low price affects the selling power of natural silk fabrics. Thus, lower-class people especially prefer it because the price is low [118].

- Material Fabrics

Spinning and weaving will produce material fabric. Spinning is the process of extruding cocoon fibers into spun strands. There are two categories of spinning, modern spinning and people spinning. Modern spinning is spinning that uses machines, both fully automatic and semi-automatic. People's spinning machines in South Sulawesi have three stages in the spinning process, and all stages affect the quality of the yarn. The three stages are the process of cocoon handling, spinning, and yarn handling spinning result [40]. The patterns of woven fabrics are produced based on regional characters. The results of woven fabrics from two areas of silk development in Indonesia can be seen in Figure 6.

(a) (b) (c) (d)

Figure 6. Process and material fabrics. (**a**) spinning; (**b**) weaving; (**c**) Sulawesi woven fabrics; (**d**) West Java woven fabric.

2.3.2. Diversification Product

Despite the aforementioned challenges, the untapped potential of silk-based products makes us optimistic. Beyond conventional textiles, there are many opportunities for product diversification. This is due to the fact that silkworm cocoons and pupae can be exploited as sericin [119], regenerated silk fibroin [120], and chitin/chitosan [121]. These are high-value biomaterials with potential applications in many fields. In the last couple of decades, progress around the science of silk has enabled the utilization of silk-based materials in exciting areas such as the pharmaceutical/medical field [121–124], cosmetics [125,126], electronics [127–129], the energy sector [130,131]. Although these are niche markets, product diversification efforts may lead to renewed interest from investors and developers in the sericulture industry. To advance in this direction, it is important for the scientific community to collaborate with farmers in the hope that high-value niche products may give some additional competitive advantage against other commodities at both national and global scales.

2.4. Social, Economic, Cultural, and Institutional Factors

2.4.1. Social, Economic, and Cultural Factors

Sericulture plays an important role, especially in rural communities [109,132]. The sericulture industry involves minimal investments, a short cultivation period, high employment potential, and very profitable returns that are suitable for the economy of a rural agrarian community [133,134].

Natural silk is a labor-intensive industry and provides jobs for many people [135]. The proportion of labor costs for silk cultivation reaches 73.14% [85]. Sericulture business actors, especially the upstream sector in Indonesia, are still dominated by people with low levels of education, mostly at an elementary school level [17,44,79,135]. However, some of them have a high school or undergraduate education [41,136], and the majority are aged between 15 and 64 years [137]. Even though education is not the main requirement [44], the sustainability of silk needs to be supported by an adequate level of education to update technical knowledge, management, and mindset [138] in the adoption of technology for a commercial business. In its implementation, the upstream sector involves a lot of family labor and does not use wage labor, so it is more of commercial household business.

Of 9223 Business Farmer Groups in the Social Forestry unit, only 27 focus on sericulture development (West Java, North of Kalimantan, and South Sulawesi); 23 focus on mulberry and cocoon production, and 4 on silk processing [139]. Farmers carry out sericulture cultivation as a primary or secondary job [44]. The average income of cocoon farmers is Rp. 2,383,969 (USD 156.56) per head of family/year, while the average income of yarn farmers is Rp. 8,004,500 (USD 525.68) per family head per year [140]. The income showed the shrink of the business and led these activities to be part of side occupations which contribute to the additional income for the family.

There is a decreasing number of silk farmers as they switch to other commodities [40]. An insufficient number of eggs was one of the main problems. On the other hand, if enough eggs are produced, the quality is sometimes not so good. Hence, the production will fail and cause uncertainty in the cocoon yields. The minimal number of cocoons produced affected the availability of the raw materials for weavers [141].

Women play an important role in natural silk cultivation [142,143]. According to [17,143,144]. silk activities involve distributed gender roles through internal agreements [145,146], and this condition is different in each region [40]. Men are more focused on mulberry cultivation and leaf harvesting, which require physical energy, while women are dominant in rearing, feeding, harvesting cocoons, weaving, and reeling, which require patience [142,147]. Some rural women consider silk activities as their primary or secondary livelihood. The distribution of internally agreed roles involves substantial social capital in supporting the future development of labor-intensive sericulture [147].

Processing cocoons into yarn and silk is characterized as a traditional household business [148]. The craft of silk weaving was initially to meet the family's needs and turned into a commercial enterprise with a cluster of handicraft industries. However, traditional characteristics remain attached to simple reeling (non-machine looms). In South Sulawesi, there are around 4900 silk woven fabric artisans that produce 99,000 sarongs per year; 227 of them are non-machine loom craftsmen with a fabric production of 1,580,000 m per year [22]. Thus, the silk business is a cultural asset of the Indonesian nation that must be maintained and preserved.

The foremost opportunity for the development of Indonesian silk is the culture of the people in the use and management of silk production [136]. Silk cloth became a traditional costume and a symbol of social status. Silk is used at traditional, religious, and official national events [40,44] and is a mainstay commodity and souvenir typical of South Sulawesi [41]. Silk motifs and creations also represent the wearer's position, happiness, and safety [149].

2.4.2. Sericulture Development Policy

The policy of developing silk in Indonesia aims to encourage agribusiness to produce silk products of good quality in large quantities to meet the demands and standards of the domestic and export markets [24,25]. Sericulture was first developed in South Sulawesi Province through the cooperation of the Directorate General of Forestry, Ministry of Agriculture, with the Japan International Cooperation Agency (JICA).

In 2002 through Forestry Minister Decree No. 664/Kpts-II/2002, a Natural Silk Center was formed to support the development of silk in Indonesia. Balai Persuteraan Alam (BPA) has a vast working area, covering almost the entire territory of Indonesia with natural silk development activities, starting from institutions arrangement and skills development through training to data collection from upstream to downstream. Silk development refers to the Joint Regulation of the Minister of Forestry Number P.47/MENHUT-II/2006; the Ministry of Industry Number 29/M-IND/PER/6/2006, and the State Ministry of Cooperatives and Small-Medium Enterprises Number 07/PER/M.UMKM/VI/2006 concerning the Guidance and Development of National Natural Silk through a cluster approach. Business development with a cluster system (upstream-downstream) was developed through a partnership scheme. In this scheme, business groups engaged in the sericulture business (mulberry farmers, silkworm cultivators, and silk yarn industries) and related institutions are geographically close together.

At the provincial level, the government of South Sulawesi launched the "Restore the Glory of Silk" program to recreate the level of Indonesian silk production achieved in the 1970s. In order to support this agenda, the South Sulawesi Regional Government has developed various programs to strengthen the upstream and downstream sectors, including technical assistance, equipment, marketing, farmer empowerment, and capital [41]. To improve the quality and quantity of silkworm eggs and ensure the quality and availability of silk cocoons, in 2017, the Ministry of Forestry launched MoF number P.37/MenLHK/Setjen/Kum.1/6/2017, which regulates the procurement and circulation of silkworm eggs.

In 2016, there was an institutional change at BPA to become the Center for Social Forestry and Environmental Partnerships (Balai Perhutanan Sosial dan Kemitraan Lingkungan/BPSKL) through the Regulation of the Minister of Environment and Forestry of the Republic of Indonesia Number P. 14 /Menlhk/Setjen/OTL.O/I/2016.

The development of the silk agro-industry in Indonesia is influenced by changes in policies related to institutions. Starting in 2016, the silk agro-industry development activities became part of the social forestry (PS) activities supported by the Decree of the Minister of Environment and Forestry No. 83/2016 concerning Social Forestry, which provides access rights for communities in forest management areas. However, silk activities have become limited because they are only a small part of social forestry. They are, therefore, only supported by 1.2–1.5% of the total budget of BPSKL.

Fortunately, in 2020 the GOI stipulated Law Number 11 of 2020 concerning Job Creation and Government Regulation (PP) and its derivatives Number 23 of 2021 concerning forestry administration which opened up opportunities for forest areas to permit holders to develop multiple forestry businesses, including the development of silk cultivation through Forest Utilization Business Permits (PBPH). The dynamics of government policies that affect the development of the silk agro-industry in Indonesia can be seen in Supplementary Table S2.

2.4.3. Institutional Factors

Sericulture agribusiness is carried out by three leading players, namely the public sector (government institutions, both central and regional), the private sector (legal entity organizations), and farmer groups [150]. The public sector has a role in setting regulations [150] to ensure mulberry availability and quality, the provision of silkworm technical guidance, evaluating facilities and infrastructure, increasing human resources, and updating databases [40]. Private sector groups play a role in terms of providing capital, distributing various inputs needed in mulberry and silkworm cultivation (fertilizer, mulberry, and silkworm planting material), the yarn spinning industry, and marketing the products. The farmer group has a role as a mulberry and silkworm cultivation actor in producing cocoons.

These three stakeholders are involved in the upstream and downstream activities of sericulture. The parties involved are as follows: (a) Upstream Sector: public groups (BPA/BPSKL, Provincial Forestry Service, District Forestry Service), private groups (Perum Perhutani and BMFG, which produces silkworm), mulberry farmer groups and silkworm farmers, (b) Downstream Sector: public groups (Provincial Industry and Trade Office, Regency Industry and Trade Office), Business Groups (PT. Kokon Sutera Sulawesi*);* and (c) Supporters (financial institutions, Forest Research and Development Agency, Industry & Trade Research Institute, colleges).

The parties involved in silk activities have different organizational structures and rules [151]. From 2006 until 2012, natural silk still relied on cluster policies [27]. During this period, institutional development was carried out by the Ministry of Environment and Forestry, the Ministry of Industry, and the Ministry of Cooperatives and Small Enterprises. However, in reality, these three institutions are still running independently [27]. This shows that there is a fragmentation between upstream and downstream in the management of Indonesia's silk, which has the potential to weaken efforts toward improvement.

The motivation for establishing silk institutions at the upstream level was different in the silk centers of Indonesia. The establishment of silk institutions in West Java and Central Java was generally initiated by the national reforestation program. On the other hand, the silk farmers 'institutions in South Sulawesi were formed on the motivation of the socio-cultural attachment of the community to the use of silk fabrics. This has induced Wajo farmers to be more independent. This condition is different from silk farmers in West Java, who still depend heavily on assistance from other parties [100].

The institutional capacity of the sericulture business in Indonesia is generally at a small scale, even at the household level, and only reaches the silkworm cocoon stage.

Small-scale development, only up to the cocoon production stage, causes the bargaining position of farmers to be relatively weak, with prices determined by buyers [85]. Widiarti et al., [152] found that the basic problems of small businesses such as sericulture are a lack of management skills and professionalism, as well as limited access to capital, technological innovation, and marketing networks. These challenges have hampered institutional performance, especially at the farmer level, and often led to the reluctance of large companies to cooperate.

The change in the centralization of the forestry sector from regencies to provinces has also disrupted the partners of natural silk farmer groups in the field [40]. The ongoing changes in government institutions hamper the sustainability of the silk program, such as the lack of control in the development of techniques, infrastructure, human resource development, and a database that supports decision-making. In addition, public sector institutions that handle silk often change their functions. The Natural Silk Center, as one of the technical implementing units of the Ministry of Environment and Forestry that is specialized in handling natural silk, has turned into a Social Forestry and Environmental Partnership Center (BPSKL), whose main task is not only dealing with silk. This change aims to simplify the structure but is affected functionality. However, the problem of silk is very complex (upstream and downstream, from mulberry to yarn), so organizational changes are not sufficient. One of the consequences of these institutional changes is the loss of information (in the term of database) on silk production, such as production potential, demand, the number of parties involved, the area used for mulberry cultivation, etc.

3. The Way Forward

3.1. Moriculture

Based on the existing conditions of natural silk related to the fulfillment of silkworm (*B. mori* L.) feed sources, the main problem was the low and discontinuous productivity of mulberries. The food availability factor is important since silkworms are monophagous; hence their entire life cycle depends on mulberry leaves [153]. However, the tropical climate in Indonesia causes the productivity and supply of mulberry leaves to be discontinuous [28,66].

Efforts that can be made to overcome the problems with mulberry productivity as a source of silkworm feed include:

1. Enhance the commitment and compliance of all related stakeholders to apply standard operating procedures in cultivating mulberries (selection of suitable locations for mulberry cultivation, cultivation techniques, intensive rearing, appropriate fertilization, and preventive measures to anticipate plant pests and diseases).
2. Superior mulberry seeds with high productivity that can be planted under various growing conditions. Indonesia, through the Ministry of Environment and Forestry, has initiated and produced a mulberry hybrid with productivity of 52.35 tons/ha/year [39].
3. Integrating mulberry agroforestry with other plants to increase land productivity and mulberry production in Indonesia. Agroforestry that integrates mulberry plantations with agricultural crops provides multidimensional benefits, including the efficient use of land area, increased production and total leaf yield, improved nutrition, efficient use of light and water from the mulberry plant, and its inherent ability to combat pests and diseases [46]. Expanding the mulberry plantation to the potential of the Social Forestry areas of 5.03 million Ha [139].
4. Managing mulberry blocks to ensure the continuity of mulberry production. A good field division will optimize feed production so that, in addition to optimal leaf quality and productivity, it can also increase the intensity of silkworm rearing per year. The separation of young and grown silkworms into different facilities with the different dedicated mulberry fields has the advantages of intensive maintenance, pest control, and feed productivity prediction, considering that young silkworms need one-month-old mulberry feed. In contrast, grown silkworms require two months old mulberry feed. Otherwise, it will cause losses due to separation, which will result in high

maintenance operational costs and will require a wider allocation of the mulberry plantation area.

3.2. Silkworm Cultivation (Sericulture)

One of the main obstacles to sericulture development in Indonesia is the absence of quality silkworm eggs and the application of silkworm rearing that does not follow the appropriate procedures. Silkworm egg producers are still limited with low production. Some of the efforts that can be made include:

1. Creation of silkworm breeding centers as silkworm egg producers with superior strains of silkworm eggs. The scheme for establishing a silkworm nursery center in collaboration with related stakeholders in Sukabumi can be an example to be implemented in other areas, especially in South Sulawesi.
2. Develop superior silkworm hybrids. The Ministry of Environment and Forestry has initiated the distribution of the superior silkworm hybrid strains PS 01, which has 35–40 kg/box cocoon production. This strain can be developed more widely to increase the productivity of Indonesian sericulture.
3. Enforcement in implementing SOP for silkworm rearing. The SOP for maintaining silkworms that are not carried out properly has an impact on the productivity of cocoons and yarns. Adherence to selecting a good and suitable location (effectiveness of feed sources and minimal air pollution), carrying out room and silkworm disinfection procedures, and paying attention to the timing and rearing procedures correctly will be steps to ensure optimal cocoons and yarns.
4. The separation of young and grown silkworm farmers can increase productivity. Young silkworms require intensive care and are susceptible to disease. Productivity will increase if skilled farmers carry out the maintenance of young silkworms. Only then can the maintenance of grown silkworms be carried out by novice farmers. The advantages of this separation are the effective use of the rearing room, and the productivity is maintained because young silkworms are susceptible to disease, so they will only be kept by skilled farmers. Good cooperative arrangements can increase the intensity of maintenance. Disadvantages will arise if communication is established between young and grown silkworm farmers; high mobility is also required because of the distance between the two.

3.3. Products

The main products of sericulture are cocoons and yarns. So far, the quality of the cocoons produced does not match the market or consumer standard, which has an impact on yarn imports and causes competition with synthetic yarns such as rayon and cupro. A strategic step to overcome this problem is the selection of cocoons at the time of silkworm rearing so that the quality is optimal and the price can be higher. Supposing the requirement for quality silk yarn can be appropriately fulfilled. The problem of competition with synthetic yarns can be avoided, considering that the existence of synthetic yarns is due to the unavailability of raw materials for silk yarns. Silk's loyal consumers have a different market share; hence, they will not be affected by synthetic yarns. Moreover, the cultural factor in South Sulawesi, which requires the use of silk, means the silk market can be maintained.

3.4. Social, Economic, and Cultural Factors

Sericulture can increase people's income, absorb labor, and be applied in rural areas. Livelihoods in the sericulture sector are discontinuous due to the absence of eggs causing many people to turn to other commodities. Some efforts can be made to tackle the situation:

1. Build production networks and cooperate with stakeholders, facilitated by the government to ensure the sustainability of the production of superior eggs and the propagation of silkworm egg producers.

2. Internalization of silkworm rearing (sericulture) through both formal and informal education involving various stakeholders, including business stakeholders, to maintain socio-cultural values and develop the value of silk entrepreneurship as part of the nation's identity.
3. Increasing the community's capacity for silkworm rearing in terms of technical aspects, farm management, farmer group organization management, and improved access to markets. The yield of silk products is low due to a lack of assistance regarding proper rearing procedures. Efforts are needed to increase farmers' adoption of good rearing standards by providing consistent, accessible, relevant, and practical information via more personal means of communication and coupling with stakeholders by a transdisciplinary approach, including social scientists, communication specialists, entrepreneurs, and marketing experts. The role of the Forest Management Unit/FMU, which is the implementation of management at the site level, needs to be more intensive for facilitating sericulture development.

3.5. Institutional Factors

One important factor that needs to be considered in the institutional system is the arrangement between rights and obligations contained in a mutually binding agreement. One example is the collaboration between silk farmer groups in Sukabumi with PT. Begawan Sutera Nusantara (PT. BSN) in terms of joint utilization of production activities that can increase competitive advantage through capacity utilization and product diversification [85,152]. Shared use arises from the opportunity to share activities (PT BSN, which carries out the egg production process and raises young silkworms, is a producer of eggs for young silkworms, while farmers are rearing grown silkworms and harvesting cocoons that are taken and paid for by PT BSN), similarities in technology (assistance is provided by the company), distribution or other factors. The joint use of production activities can increase utilization capacity and product diversification and, in turn, achieve economies of scale. This will increase the productivity and competitiveness of sericulture. Improving the ability to build networks, increasing social capital within the group, and improving the organizational management and negotiation skills of farmers are important factors in supporting silk farming institutions. In addition, local governments can provide formal support through local regulations, policies, and silk development programs at the local level.

4. Conclusions

Sericulture in Indonesia involves the production of non-timber forest products and has been practiced for generations by the local community for their livelihood. However, there has been a decrease in this activity over the last 5–10 years. There are various and complex problems from upstream to downstream that have led to this unfortunate development. Moriculture and sericulture techniques, the quality and quantity of the products, the social-economic aspect, institutional arrangements, and community motivation are interconnected. They have resulted in a challenging atmosphere for sericulture development. On the other hand, there are potential resources such as exploring qualified mulberry production and quality silkworm production through research and development, valuable cultural aspects, and potential stakeholders for building network engagement. The commitment, cooperation, and action of all stakeholders are needed to improve sericulture development in Indonesia. In this context, the central government can play an important role in facilitating multi-stakeholder partnerships in integrated sericulture development in Indonesia.

Supplementary Materials: The following supporting information can be downloaded at: https://www.mdpi.com/article/10.3390/insects13100913/s1, Table S1: Silkworm hybridization in Indonesia, Table S2: The dynamics of government policy affecting silk development in Indonesia.

Author Contributions: L.A., D.Y., B.S., M. (Murniati), S.S., A.W., E.S., A.S., B.W., M. (Minarningsih), R.A., N.M., W.I., Y.H., Y.A., I.Y., R.D., A.N., S.D.R., K.E.M., L.M.N. and B.H.N. contributed equally as main contributors to the conception and design of the project, collecting data, performing the analysis, providing constructive feedback on each section, and writing and editing the manuscript. All authors have read and agreed to the published version of the manuscript.

Funding: This research received no external funding.

Data Availability Statement: Not applicable.

Conflicts of Interest: The authors declare no conflict of interest.

References

1. Sparrow, R.; Dartanto, T.; Hartwig, R. Indonesia Under the New Normal: Challenges and the Way Ahead. *Bull. Indones. Econ. Stud.* **2020**, *56*, 269–299. [CrossRef]
2. Halimatussadiah, A.; Cesarina, A.; Siregar, A.A.; Hanum, C.; Wisana, D.; Rahardi, F.; Bintara, H.; Rezki, J.F.; Husna, M.; Azar, M.S.; et al. *Thinking Ahead: Indonesia's Agenda on Sustainable Recovery from COVID-19 Pandemic*; LPEM FEB UI dan BAPPENAS: Jakarta, Indonesia, 2020.
3. United Nations. Ending Poverty. Available online: https://www.un.org/en/global-issues/ending-poverty (accessed on 15 November 2020).
4. Golar, G.; Malik, A.; Muis, H.; Herman, A.; Nurudin, N.; Lukman, L. The social-economic impact of COVID-19 pandemic: Implications for potential forest degradation. *Heliyon* **2020**, *6*, e05354. [CrossRef] [PubMed]
5. Djalante, R.; Lassa, J.; Setiamarga, D.; Sudjatma, A.; Indrawan, M.; Haryanto, B.; Mahfud, C.; Sinapoy, M.S.; Djalante, S.; Rafliana, I.; et al. Review and analysis of current responses to COVID-19 in Indonesia: Period of January to March 2020. *Progress in Disaster Science* **2020**, *6*, 100091. [CrossRef]
6. Suryahadi, A.; Al Izzati, R.; Suryadarma, D. *The Impact of COVID-19 Outbreak on Poverty: An Estimation for Indonesia (Draft)*; SMERU Working Paper; The SMERU Research Institute: Jakarta, Indonesia, 2020; pp. 1–20.
7. FAO. *The impact of disasters and crises on agriculture and food security: 2021*; Food and Agriculture Organization of The United Nations: Rome, Italy, 2021.
8. Cahyadi, H.S.; Newsome, D. The post COVID-19 tourism dilemma for geoparks in Indonesia. *Int. J. Geoheritage Park.* **2021**, *9*, 199–211. [CrossRef]
9. Supriyanto, B. *Cara Belajar Baru Petani Dalam Buku Salam Lima Jari*; Direktorat Penyiapan Kawasan Perhutanan Sosial: Jakarta, Indonesia, 2020.
10. Shackleton, C.; Delang, C.O.; Shackleton, S.; Shanley, P. Non-timber Forest Products: Concept and Definitions. In *Non-Timber Forest Products in the Global Context*; Shackleton, S., Shackleton, C., Shanley, P., Eds.; Springer: Berlin/Heidelberg, Germany, 2011; pp. 3–21.
11. Sardeshpande, M.; Shackleton, C. Wild Edible Fruits: A Systematic Review of an Under-Researched Multifunctional NTFP (Non-Timber Forest Product). *Forests* **2019**, *10*, 467. [CrossRef]
12. Simangunsong, B.C.H.; Manurung, E.G.T.; Elias, E.; Hutagaol, M.P.; Tarigan, J.; Prabawa, S.B. Tangible economic value of non-timber forest products from peat swamp forest in Kampar, Indonesia. *Biodiversitas* **2020**, *21*, 5954–5960. [CrossRef]
13. Wahyudi, W. Non-Timber Forest Product (Ntfp) Commodities Harvested and Marketed By Local People At the Local Markets in Manokwari—West Papua. *Indones. J. For. Res.* **2017**, *4*, 27–35. [CrossRef]
14. Kartodihardjo, H.; Suwito. *Survei dan Indeks Perhutanan Sosial: Jalan Menuju Kesejahteraan Rakyat dan Kelestarian Hutan*; Katadata Insight Center: Jakarta, Indonesia, 2020.
15. Barcelos, S.M.B.D.; Salvador, R.; Guedes, M.D.G.; de Francisco, A.C. Opportunities for improving the environmental profile of silk cocoon production under Brazilian conditions. *Sustainability* **2020**, *12*, 3214. [CrossRef]
16. Agustarini, R.; Andadari, L.; Minarningsih; Dewi, R. Conservation and breeding of natural silkworm (*Bombyx mori* L.) in Indonesia. *IOP Conf. Ser. Earth Environ. Sci.* **2020**, *533*, 012004. [CrossRef]
17. Kasi, E. Role of Women in Sericulture and Community Development. *SAGE Open* **2013**, *3*(3), 2158244013502984. [CrossRef]
18. Hogeboom, R.J.; Hoekstra, A.Y. Water and land footprints and economic productivity as factors in local crop choice: The case of silk in Malawi. *Water (Switzerland)* **2017**, *9*, 802. [CrossRef]
19. Harbi, J.; Nurrochmat, D.R.; Kusharto, C.M. Pengembangan Usaha Persuteraan Alam Kabupaten Wajo, Sulawesi Selatan. *Risal. Kebijak. Pertan. Dan Lingkung. Rumusan Kaji. Strateg. Bid. Pertan. Dan Lingkung.* **2015**, *2*, 128. [CrossRef]
20. ISC. Global Silk Industry. Available online: https://inserco.org/en/statistics (accessed on 21 June 2021).
21. Nuraeni, S. Gaps in the thread: Disease, production, and opportunity in the failing silk industry of south sulawesi. *For. Soc.* **2017**, *1*, 110–120. [CrossRef]
22. Rusniati. *Analisis Pengendalian Proses Produksi LIPA SABBE' (Sarung Sutera) Sengkang di Kecamatan Tanasitolo Kabupaten Wajo*; Universitas Muhammadiyah Makassar: Makassar, Indonesia, 2018.

23. Minister of Forestry. *Peraturan Menteri Kehutanan Republik Indonesia No.P.21/Menhut-II/2009 tentang Kriteria dan Indikator Penetapan Jenis Hasil Hutan Bukan Kayu Unggulan*; Ministry of Environment and Forestry of The Republic of Indonesia: Jakarta, Indonesia, 2009.
24. Mutiara, F.; Asnani, D. Strategi pengembangan agribisnis Ulat Sutera Pemakan Daun Singkong di Kabupaten Malang. *J. Ilmu-Ilmu Peternak.* **2017**, *27*, 24–38. [CrossRef]
25. Tenriawaru, A.N.; Fudjaja, L.; Jamil, M.H.; Rukka, R.M.; Anisa, A.; Halil. Natural silk agroindustry in Wajo Regency. *IOP Conf. Ser. Earth Environ. Sci.* **2021**, *807*, 012037. [CrossRef]
26. Hartati. *Analisis Fenotip Ulat Sutera (bombyx mori L.) Hasil Persilangan Ras Jepang, Chna, dan Rumania*; Global Research and Consulting Institute (Global-RCI): Makassar, Indonesia, 2015; pp. 21–23.
27. Sadapotto, A. Proses Kebijakan Persuteraan Alam di Sulawesi Selatan. *J. Parennial* **2012**, *8*, 1–5.
28. Nurjayanti, E.D. Budidaya Ulat Sutera Dan Produksi Benang Sutera Melalui Sistem Kemitraan Pada Pengusaha Sutera Alam (PSA) Regaloh Kabupaten Pati. *Mediagro* **2011**, *7*, 1–10.
29. Zambrano-Gonzalez, G.; Ramirez-Gonzalez, G.; Almanza-P, M.I. The evolution of knowledge in sericultural research as observed through a science mapping approach. *F1000Research* **2017**, *6*, 2075. [CrossRef]
30. Sathyanarayana, K.; Saratchandra, B. *Carbon trading through afforestation programmes with host plants of tasar silkworm (Antheraea mylitta Drury) in private waste lands: A Value Add for Tribals*; Council of Research and Experiments in Agriculture, Apiculture and Sericulture Unit of Bologna: Padua, Italy, 2013.
31. Rohela, G.K.; Shukla, P.; Muttanna; Kumar, R.; Chowdhury, S.R. Mulberry (*Morus* spp.): An ideal plant for sustainable development. *Trees For. People* **2020**, *2*, 100011. [CrossRef]
32. Deepa, K.B.; Vishaka, G.V.; Kumar, N. Mulberry as a avenue plant. *J. Pharmacogn. Phytochem.* **2020**, *9(4S)*, 135–137.
33. McNair, P.D. Reviving the Sericulture Industry. In Proceedings of the 5th BACSA International Conference "Sericulture for Multi Products—New Prospects for Development" SERIPRODEV, Bucharest, Romania, 11–15 April 2011.
34. Tarigan, D.; Fauzi, A.M.; Suryani, A.; Kaomini, M.; Tarigan, D.; Fauzi, A.M.; Suryani, A.; Kaomini, M. Strategi pengembangan agroindustri sutera alam melalui pendekatan klaster a strategy for the development of silk agroindustry using cluster approach. *J. Teknol. Ind. Pertan.* **2019**, *20*, 39–47.
35. Nursita, I.W. Perbandingan produktifitas ulat Sutra dari dua tempat pembibitan yang berbeda pada kondisi lingkungan pemeliharaan panas. *J. Ilmu Peternak.* **2011**, *3*, 10–17.
36. Muin, N.; Suryanto, H.; Minarningsih. Ujicoba Hibrid Morus Khunpai dan Morus Indica Sebagai Pakan Ulat Sutera (*Bombyx Mori.* Linn). *J. Penelit. Kehutan. Wallaceae* **2015**, *4*, 137–145. [CrossRef]
37. Pudjiono, S. *Ulat Sutra*; Badan Litbang dan Inovasi: Jakarta, Indonesia, 2015; pp. 42–59.
38. Shantibala, T.; Lokeshwari, R.K.; Gusheinzed, W.; Agarwala, B.K. *Documentation of Ethno-Entomophagy Practices in Ethnic Communities of Manipur, North-East India. Ancestral Knowledge in Agri-Allied Science of India*; India Publishing Agency: New Delhi, India, 2013.
39. Andadari, L. Pemilihan jenis hibrid ulat sutera yang optimal untuk dikembangkan di dataran tinggi dan/atau dataran rendah. *J. Penelit. Hutan Tanam.* **2016**, *13*, 13–21. [CrossRef]
40. Sadapotto, A.; Palulungan, L.; Riwu, M.; Sahide, M.A.K.; Sirimorok, N.; Parenreng, S.M.; Salindeho, Y.M.; Yahya, A.F.; Mulyani, A.; Said, R.D.; et al. *Laporan Kajian Final Rantai Nilai Komoditas Sutra Sulawesi Selatan*; Bappelitbangda Provinsi Sumatera Selatan: Makassar, Indonesia, 2021.
41. Muin, N.; Bisjoe, A.R.H. Budidaya Ulat Sutera Di Desa Sudu, Kecamatan Alla, Kabupaten Enrekang, Sulawesi Selatan. *J. Penelit. Hutan Tanam.* **2013**, *10*, 229–239. [CrossRef]
42. Hadi, P.; Rustiono, D. Silkworm Agribusiness in Bejen Village Temanggung. *Int. J. Agric. Innov. Res.* **2015**, *3*, 1592–1594.
43. Maturidy, A.F. Preferensi pelaku usaha dalam pengembangan persuteraan alam di Provinsi Sulawesi Selatan. *J. Pasca Sarj. Univ. Hasanuddin* **2011**, 1–11.
44. Isnan, W. Perception and Motivation of Farmers in the Development of Natural Silk Business in Soppeng Regency Sulawesi Selatan. *J. Wasian* **2019**, *6*, 1–10. [CrossRef]
45. Sharma, A.; Krishna, V.; Kaur, P.; Rayal, R. Characterization and Screening of Various Mulberry Varieties Throgh Morpho-Biochemical Characteristics. *J. Glob. Biosci.* **2015**, *4*, 1186–1192.
46. Mir, M.R.; Khan, I.L.; Baqual, M.F.; Banday, M.; Raja, R.; Khursheed, S. *Mulberry-Based Agroforestry System: An Effective Way of Maintaining Livelihood Security and Climate Change Mitigation*; Apple Academic Press: Palm Bay, FL, USA, 2019.
47. Andadari, L.; Minarningsih; Dewi, R. The Effect of Mulberry Types on the Productivity of Cocoon of Two *Bombyx mori* L. Silkworm Hybrids. *Widyariset* **2017**, *3*, 119. [CrossRef]
48. Sasmita, N.; Komara, L.L. Species Test of Morus alba and Morus cathayana in Indonesia. *Adv. Biol. Sci. Res.* **2021**, *11*, 246–252. [CrossRef]
49. Daulay, N.S. *Pengaruh Kualitas Daun Murbei Moruscathayana Terhadap Indeks Nutrisi Ulat Sutera Bombyx mori L. (Lepidoptera: Bombicidae)*; Repositori Institusi Universitas Sumatera Utara: Sumatera Utara, Indonesia, 2013.
50. Triandi, A.; Mubin, N.; Nurulalia, L. Identification of Mulberry Pests and Its Natural Enemies at "Rumah Sutera Alam" Ciapus, West Java. *Akta Agrosia* **2021**, *24*, 9–18. [CrossRef]
51. Sulthoni, A. *Peranan Entomologi Hutan dan Tantangannya Menghadapi Pembangunan Hutan Tanaman Industri. Pidato Pengukuhan Guru Besar dalam Ilmu Hama Tanaman*; Universitas Gadjah Mada: Yogyakarta, Indonesia, 1991.

52. Pudjiono, S.; Na'iem, M. Pengaruh Pemberian Pakan Murbei Hibrid Terhadap Produktivitas dan Kualitas Kokon. *J. Pemuliaan Tanam. Hutan* **2007**, *1*, 81–87. [CrossRef]
53. Singh, B.; Makkar, H.P.S. The Potential of Mulberry Foliage as Feed Supplement in India. In Proceedings of the FAO Electronic Conference on Mulberry for Animal Production (Morus I-L), Rome, Italy, August 2020; pp. 1–20.
54. Yongkang, H. Mulberry cultivation and utilization in China. In Proceedings of the Electronic Conference, 2000.
55. Doss, S.G.; Chakraborti, S.P.; Roychowdhuri, S.; Das, N.K.; Vijayan, K.; Ghosh, P.D. Development of mulberry varieties for sustainable growth and leaf yield in temperate and subtropical regions of India. *Euphytica* **2012**, *185*, 215–225. [CrossRef]
56. Bhatia Narendra, K.; Yousuf, M.; Nautiyal, R. Yield Gap Analysis of Mulberry Sericulture in Northwest India. *Int. J. Ind. Entomol.* **2013**, *27*, 131–141. [CrossRef]
57. Sori, W.; Gebreselassie, W. Evaluation of mulberry (*Morus* spp.) genotypes for growth, leaf yield and quality traits under Southwest Ethiopian condition. *J. Agron.* **2016**, *15*, 173–178. [CrossRef]
58. Nursyamsi. Propagasi Tiga Varietas Murbei Melalui Teknik Kultur Jaringan. *J. Penelit. Hutan Tanam.* **2012**, *9*, 75–82. [CrossRef]
59. Pudjiono, S.; Septina, S. Morfologi Tanaman Hibrid Murbei di Purwobinangun Yogyakarta. *J. Pemuliaan Tanam. Hutan* **2008**, *2*(1), 163–171. [CrossRef]
60. Atmosoedarjo, S.; Kartasubrata, J.; Kaomini, M.; Saleh, W.; Murdoko, W. *Sutera Alam Indonesia*; Yayasan Sarana Wana Jaya: Jakarta, Indonesia, 2000.
61. Andadari, L.; Pudjiono, S.; Suwandi; Rahmawati, T. *Budidaya Murbei dan Ulat Sutra*; Forda Press: Bogor, Indonesia, 2013.
62. Santoso, B.; Budisantoso, H. Adaptasi Varietas Murbei Hasil Silangan. In Proceedings of the Prosiding Ekspose Hasil Penelitian dan Pengembangan Kehutanan Ujung Pandang, Makassar, Indonesia, 1999.
63. Andadari, L.; Dewi, R.; Pudjiono, S. Adaptation Test of Five New Hybrid Mulberry to Improve Natural Silk Productivity. *Widyariset* **2016**, *2*(2), 96–105. [CrossRef]
64. Pudjiono, S.; Andadari, L.; Darwo. Pemilihan Jenis Hibrid Murbei Untuk Dikembangkan Di Dataran Tinggi. *J. Penelit. Hutan Tanam.* **2016**, *13*, 133–138.
65. Gupta, S.K.; Dubey, R.K. Impact of heavily polluted air and acid rain on mulberry silkworm and cocoon production of Bombyx mori Linn. *Int. J. Fauna Biol. Stud.* **2021**, *8*, 26–30. [CrossRef]
66. Yulistiani, D. Tanaman murbei sebagai sumber protein hijauan pakan domba dan kambing. *Wartazoa* **2012**, *22*, 46–52.
67. Mahadeva, A. Insect Pest Infestation, an Obstacle in Quality Mulberry Leaves Production. *Asian J. Biol. Sci.* **2018**, *11*, 41–52. [CrossRef]
68. Prayudyaningsih, R.; Tikupadang, H.; Santoso, B. Hama Dan Penyakit Jenis Murbei Eksot Dan Tingkat Kehilangan Daunnya Pada Akhir Musim Kemarau. *J. Penelit. Hutan dan Konserv. Alam* **2006**, *3*, 429–435. [CrossRef]
69. Dewi, R.; Andadari, L.; Maharani, K.E. Tinjauan Bioekologi dan Pengendalian Hama Kutu Kebul (*Bemisia tabaci* Genn.). In Proceedings of the Prodising Seminar Nasional PEI, Jatinangor, Bandng, 25–26 October 2017; pp. 40–45.
70. Mukhopadhyay, S.K.; Trivedy, K. Intregated management and forecasting of insect pests of mulberry (*Morus Alba* L.) for Eastern and Northeastern regions of India. *J. Environ. Sociobiol.* **2016**, *13*, 137–145.
71. Soliman, N.H.; Sherif, D.F.E. rovided for non-commercial research and education use. Not for reproduction, distribution or commercial use. Hepatoprotective Agents in Rats. *Egypt. Acad. J. Biol. Sci.* **2019**, *11*, 91–95.
72. Maru, R.; Badwi, N.; Abbas, I.; Nurfadillah; Nur, M.M.; Basram, N.F. Opportunities and Challenges of SilkWorm Cultivation Development in Geography Perspectives. *La Geogr.* **2021**, *19*, 201–210.
73. Shi, M.; Li, G.; Zheng, X.; Liu, F.; Zhao, S.; Cai, M.; Zhu, Y. Effects of dimethoate on the gonads and oxidative stress response in the silkworm, Bombyx mori. *Acta Entomol. Sin.* **2019**, *62*, 1186–1196.
74. BPA. *Laporan Bulanan. Pelaksana Harian Balai Persuteraan Alam (BPA), Kabupaten Soppeng*; Balai Persuteraan Alam (BPA): Makassar, Indonesia, 2014.
75. Fambayun, R.A.; Agustarini, R.; Andadari, L. Cultivation and Breeding Techniques for Increase Silk Productivity in Indonesia. *IOP Conf. Ser. Earth Environ. Sci.* **2022**, *995*, 012055. [CrossRef]
76. Deni; Diba, F.; Tavita, G.E. Deni, Farah Diba, Gusti Eva Tavita. *J. Hutan Lestari* **2019**, *7*, 874–883.
77. Andadari, L.; Kuntadi. Comparison of Chinese-Origin and Local Silkworm (L.) Hybrids in South Sulawesi. *J. Penelit. Hutan Tanam.* **2014**, *11*, 173–183. [CrossRef]
78. Djabar, M.; Utiarahman, N. Kelayakan Usaha Budidaya Ulat Sutera (*Bombix Mori* L.) Berdasarkan Aspek Non Finanasial Kabupaten Boalemo. *J. Penelit. Kehutan. Bonita* **2019**, *1*, 15–22. [CrossRef]
79. Syam, D. *Analisis Pendapatan Pemelihara Ulat Sutera Pada Pemeliharaan Konvensional di Desa Sering, Kecamatan Donri-Donri, Kabupaten Sorong (Studi Kasus Kelompok Tani Batu Tungke'E)*; Universitas Hasanuddin: Makassar, Indonesia, 2017.
80. Watanabe, H. Genetic resistance of the silkworm, Bombyx mori to viral diseases. *Curr. Sci.* **2002**, *83*, 439–446.
81. Potrich, M.; Alves, L.F.A.; Brancalhão, R.C.; Dalcin, G. Entomopatógenos Associados a Lagartas De *Bombyx Mori* L. (Lepidoptera: Bombycidae) No Estado Do Paraná. *Arq. Do Inst. Biológico* **2007**, *74*, 363–367. [CrossRef]
82. Nuraeni, S. *Kuantitas dan kualitas produksi F1 ulat sutera (Bombyx mori L.) hasil persilangan antara Ras Rusia dengan Ras China*; Fakultas Pertanian universitas Hasanuddin: Makassar, Indonesia, 1993.
83. Budisantoso. Pengeringan Kokon dengan Alat Pengering Tenaga Matahari dan pengaruhnya Terhadap Kualitas Serat Sutra. *Bul. Penelit. Kehutan.* **2001**, *7*(1), 24–36.

84. Kasip, M. *Pembentukan Galur Baru Ulat Sutera (Bombyx mori L) Melalui Persilangan Ulat Sutera Bivoltin dan Polivoltin*; IPB University: Bogor, Indonesia, 2001.
85. Yuniati, D.; Suharti, S.; Widiarti, A.; Andadari, L.; Heryati, Y.; Agustarini, R. Business feasibility of several PS-01 hybrid silkworms (*Bombix mori* L.) cultivation scheme. *IOP Conf. Ser. Earth Environ. Sci.* **2021**, *917*, 012031. [CrossRef]
86. Estetika, Y.; Endrawati, Y.C. Produktivitas Ulat Sutera (*Bombyx mori* L.) Ras BS-09 di Daerah Tropis. *J. Ilmu Prod. Dan Teknol. Has. Peternak.* **2018**, *6*, 104–112. [CrossRef]
87. Minister of Forestry. *Keputusan Menteri Kehutanan Republik Indonesia No. SK. 794/Menhut-II/2013 Tentang Pelepasan Bibit Ulat Sutera (Bombyx mori L.) Hibrid PS-01*; Ministry of Environment and Forestry of The Republic of Indonesia: Jakarta, Indonesia, 2013.
88. Ministry of Environment and Forestry. *Decree of the Minister of Environment and Forestry of the Republic of Indonesia No: SK.300/Menlhk/Setjen/KUM.1/4/2019 Dated 24 April 2019*; Ministry of Environment and Forestry of The Republic of Indonesia: Jakarta, Indonesia, 2019.
89. Cholis, N. Studi Penampilan Produksi Ulat Sutera F1 Hibrid Hasil Persilangan RAS Jepang dan RAS Cina yang Berasal dari Pusat Pembibitan Soppeng dan Temanggung. *J. Ternak Trop.* **2014**, *15*, 72–76.
90. Cholis, N. Studi tentang polimorfisme ulat sutera F1 hibrid hasil persilangan ras Jepang dan ras Cina yang berasal dari pusat pembibitan Soppeng dan Temanggung dengan menggunakan enzim restriksi Pst1 dan EcoR1. *J. Ilmu-Ilmu Peternak.* **2015**, *25*, 61–65. [CrossRef]
91. Minister of Forestry. *Keputusan Menteri Kehutanan Indonesia No. SK. 369/Menhut-II/2004 Tentang Pelepasan Bibit Ulat Sutera (Bombyx mori L.) Hibrid BS-08 dan BS-09*; Ministry of Environment and Forestry of The Republic of Indonesia: Jakarta, Indonesia, 2004.
92. Rahmathulla, V.K. Management of Climatic Factors for Successful Silkworm (*Bombyx mori* L.) Crop and Higher Silk Production: A Review. *Psyche* **2012**, *11*, 1–12. [CrossRef]
93. Mubarak, H. *Produktivitas Ulat Sutera (Bombyx Mori L.) Candirotobs- 09dan Hibrid PS.01 Yang Diberi Pakandaun Murbei (Morus Multicaulis)*; Institut Pertanian Bogor: Bogor, Indonesia, 2017.
94. Nuraeni, S.; Putranto, B. Aspek Biologis Ulat Sutera (*Bombyx mori* L.) dari Dua Sumber Bibit di Sulawesi Selatan. *J. Perenn.* **2007**, *4*, 10–17. [CrossRef]
95. Sadapotto, A.; Kartodihardjo, H.; Triwidodo, H.; Darusman, D.; Sila, M. Penataan institusi untuk peningkatan kinerja persuteraan alam di Sulawesi Selatan. *J. Forum Pascasarj.* **2010**, *33*, 133–140.
96. Andadari, L.; Minarningsih, S. The effect of feeding various species of mulberry (*Morus* spp.) on the growth of silkworm and quality of cocoon hybrid BS 09. *IOP Conf. Ser. Earth Environ. Sci.* **2021**, *914*, 012017. [CrossRef]
97. San-Ming, W. *Silkworm Egg Production*; FAO: Rome, Italy, 1989; p. 64.
98. Yuniati, D.; Widiarti, A.; Suharti, S. Pengembangan Persutraan Alam di Kabupaten Sukabumi. In *Pembelajaran Dunia Riset untuk Pengembangan HHBK*; Widiarti, A., Irma, Y., Minarningsih, A.R., Eds.; IPB Press: Bogor, Indonesia, 2020.
99. Prihatin, J. Pemeliharaan dan Pengokonan Ulat Sutera Bombyx mori L. Menggunakan Seriframe Standar dan Seriframe Modifikasi. *Bioedukasi* **2009**, *7*, 31–38.
100. Fauziyah, E.; Wijayanto, N. Strategi pengembangan usaha persuteraan alam di kabupaten sukabumi, propinsi jawa barat (Development Strategy of Striving Nature Silk in Kabupaten Sukabumi, Propinsi Jawa Barat). *J. Manaj. Hutan Trop.* **2007**, *13*, 59–77.
101. Agustarini, R.; Andadari, L. Implementasi Persuteraan Alam dalam Mendukung Perhutanan Sosial. In Proceedings of the Prosiding Seminar Nasional Ekosistem Unggul Membangun Hutan Sebagai Ekosistem Unggul Berbasis Das: Jaminan Produksi, Pelestarian, dan Kesejahteraan, Yogyakarta, Indonesia, 23 August 2018; pp. 151–162.
102. Ritter, C.; Jansen, J.; Roche, S.; Kelton, D.F.; Adams, C.L.; Orsel, K.; Erskine, R.J.; Benedictus, G.; Lam, T.; Barkema, H.W. Invited review: Determinants of farmers' adoption of management-based strategies for infectious disease prevention and control. *J. Dairy Sci.* **2017**, *100*, 3329–3347. [CrossRef] [PubMed]
103. Sarkhel, S.; Shrivastava, S.; Pouranik, M. The Effective Influence of Temperature on the Varied Characteristic of Silkworm: A Review. *Asian J. Exp. Sci.* **2017**, *31*, 31–37.
104. Sakthivel, N.; Kumaresan, P.; Qadri, S.M.H.; Ravikumar, J.; Balakrishna, R. Adoption of integrated pest management practices in sericulture—A case study in Tamil Nadu. *J. Biopestic.* **2012**, *5*, 212–215.
105. Sharma, A.; Sharma, P.; Thakur, J.; Murali, S.; Bali, K. Viral diseases of Mulberry Silkworm, *Bombyx mori* L. —*A Review.* **2020**, *9*, 415–423.
106. Nuraeni, S.; Muin, N.; Sanusi, D. Respon Empat Ras Ulat Sutera Terhadap Bombyx mori Nuclear Polyhedrosis Virus (BmNPV). In Proceedings of the Prosiding Seminar Nasional Mikrobiologi Kesehatan dan Lingkungan, Makassar, Indonesia, 29 January 2015.
107. Muin, N. *Pengaruh Pakan Pada Resistensi Ulat Sutra (Bombyx mori L.) Terhadap Penyakit Grasserie*; IPB University: Bogor, Indonesia, 2009.
108. Guo-Ping, K.; Xi-Jie, G. Overview of silkworm pathology in China. *Afr. J. Biotechnol.* **2011**, *10*, 18046–18056. [CrossRef]
109. Nuraeni, S.; Baharuddin. The adaptability of the commercial seeds of silkworm (Bombyx Mori L) on different altitude preservation. *IOP Conf. Ser. Earth Environ. Sci.* **2019**, *270*, 012034. [CrossRef]
110. Antaranews. *Ulat "Sinar" Diharapkan Penuhi Kebutuhan Benang Sutera Lokal*; Antaranews: Jakarta, Indonesia, 2019.
111. ASSIA. World Sericulture in Industry 4.0. In Proceedings of the ASEAN Country Member National Brief on Silk and Sericulture: Indonesia, Bangkok, Thailand, 28–29 July 2019.

112. Hussain, M.; Khan, S.A.; Naeem, M.; Aqil, T.; Khursheed, R.; ul Mohsin, A. Evaluations of silkworm lines against variations in temperature and RH for various parameters of commercial cocoon production. *Psyche* **2011**, *2011*, 1–11. [CrossRef]
113. Gowda, B.N.; Reddy, N.M. Influence of different environmental conditions on cocoon parameters and their effects for reeling performance of bivoltine hybrids of silkworm, Bombyx mori L. *Int. J. Indust. Entomol* **2007**, *14*, 15–21.
114. Mirhosseini, S.Z.; Nematollahian, S.; Ghanipoor, M.; Seidav, A. Comparison of Phenotypic and Genetic Performance of Local Silkworm Groups and Two Commercial Lines. *Biol. Res.* **2010**, *43*, 411–416. [CrossRef]
115. Kim, S.E. Silkworm Breeding. In *Principle and Practices in Sericulture*; NSERI: City, Korea, 1998.
116. BPS. *Data dan Informasi Kemiskinan 2005−2006, Buku 1: Provinsi [Poverty Data and Information 2005−2006, Book 1: Province]*; BPS: Jakarta, Indonesia, 2007.
117. Rochmawati, R. *Kualitas Kokon Hasil Silangan Ulat Sutera (Bombyx mori L.) Ras Cina dengan Ras Jepang Secara Resiprocal*; IPB University: Bogor, Indonesia, 2011.
118. Sadapotto, A. Strategi Pengembangan Sutra Alam di Kabupaten Poliwali Mandar, Sulawesi Barat. In Proceedings of the Seminar Nasional MAPEKI XVII, Medan, Indonesia; 2014.
119. Martinez-Mora, C.; Mrowiec, A.; Garcia-Vizcaino, E.M.; Alcaraz, A.; Cenis, J.L.; Nicolas, F.J. Fibroin and sericin from Bombyx mori silk stimulate cell migration through upregulation and phosphorylation of c-Jun. *PLoS ONE* **2012**, *7*, e42271. [CrossRef]
120. Rockwood, D.N.; Preda, R.C.; Yucel, T.; Wang, X.; Lovett, M.L.; Kaplan, D.L. Materials fabrication from Bombyx mori silk fibroin. *Nat. Protoc.* **2011**, *6*, 1612–1631. [CrossRef]
121. Battampara, P.; Nimisha Sathish, T.; Reddy, R.; Guna, V.; Nagananda, G.S.; Reddy, N.; Ramesha, B.S.; Maharaddi, V.H.; Rao, A.P.; Ravikumar, H.N.; et al. Properties of chitin and chitosan extracted from silkworm pupae and egg shells. *Int. J. Biol. Macromol.* **2020**, *161*, 1296–1304. [CrossRef] [PubMed]
122. Lin, M.J.; Lu, M.C.; Chan, Y.C.; Huang, Y.F.; Chang, H.Y. An Insulin-like Growth Factor-1 Conjugated Bombyx mori Silk Fibroin Film for Diabetic Wound Healing: Fabrication, Physicochemical Property Characterization, and Dosage Optimization In Vitro and In Vivo. *Pharmaceutics* **2021**, *13*(9), 1459. [CrossRef]
123. Zhang, Y.; Liu, Y.; Jiang, Z.; Wang, J.; Xu, Z.; Meng, K.; Zhao, H. Poly(glyceryl sebacate)/silk fibroin small-diameter artificial blood vessels with good elasticity and compliance. *Smart Mater. Med.* **2021**, *2*, 74–86. [CrossRef]
124. Veiga, A.; Castro, F.; Rocha, F.; Oliveira, A.L. Protein-Based Hydroxyapatite Materials: Tuning Composition toward Biomedical Applications. *ACS Appl. Bio Mater.* **2020**, *3*, 3441–3455. [CrossRef] [PubMed]
125. Su, D.; Ding, S.; Shi, W.; Huang, X.; Jiang, L. *Bombyx mori* silk-based materials with implication in skin repair: Sericin versus regenerated silk fibroin. *J. Biomater. Appl.* **2019**, *34*, 36–46. [CrossRef]
126. Ulfa, M.; Yahya, P.T.; Angriani, P.D.; Nur, A.A. Formulasi dan Uji Efektivitas Krim Limbah Air Kokon Ulat Sutera (*Bombyx mori*) Asal Kabupaten Soppeng sebagai Pelembab Kulit. *J. Farm. UIN Alauddin Makassar* **2020**, *8*, pagination. [CrossRef]
127. Hwang, S.W.; Tao, H.; Kim, D.H.; Cheng, H.; Song, J.K.; Rill, E.; Brenckle, M.A.; Panilaitis, B.; Won, S.M.; Kim, Y.S.; et al. A physically transient form of silicon electronics. *Science* **2012**, *337*, 1640–1644. [CrossRef]
128. Kim, D.-H.; Viventi, J.; Amsden, J.J.; Xiao, J.; Vigeland, L.; Kim, Y.-S.; Blanco, J.A.; Panilaitis, B.; Frechette, E.S.; Contreras, D.; et al. Dissolvable films of silk fibroin for ultrathin conformal bio-integrated electronics. *Nat. Mater.* **2010**, *9*, 511–517. [CrossRef] [PubMed]
129. Aramwit, P.; Kanokpanont, S.; Nakpheng, T.; Srichana, T. The Effect of Sericin from Various Extraction Methods on Cell Viability and Collagen Production. *Int. J. Mol. Sci.* **2010**, *11*, 2200–2211. [CrossRef]
130. Xiang, M.; Wang, Y.; Wu, J.; Guo, Y.; Wu, H.; Zhang, Y.; Liu, H. Natural Silk Cocoon Derived Nitrogen-doped Porous Carbon Nanosheets for High Performance Lithium-Sulfur Batteries. *Electrochim. Acta* **2017**, *227*, 7–16. [CrossRef]
131. Ho, J.; Cao, C.; Idrees, F.; Ma, X. Hierarchical Porous Nitrogen-Doped Carbon Nanosheets Derived from Silk for Ultrahigh-Capacity Battery Anodes and Supercapacitors. *ACS Nano Publ.* **2015**, *9*, 2556–2564. [CrossRef]
132. Iwang, B.; Sudirman. Peranan Pemerintah dalam Memajukan Perusahaan Sutera di Sulawesi Selatan, Indonesia. *Southeast Asian Soc. Sci. Rev.* **2020**, *5*, 103–132.
133. Chanotra, S.; Bali, K.; Bali, R. Sericulture: An opportunity for the upliftment of rural livelihood. *J. Entomol. Zool. Stud.* **2019**, *7*, 1100–1103.
134. Pratama, A.G.; Supratman, S.; Makkarennu, M. Examining forest economies: A case study of silk value chain analysis in Wajo District. *For. Soc.* **2019**, *3*, 22–33. [CrossRef]
135. Dewangan, S.K.; Sahu, K.R.; Soni, S.K. Breaking of poverty through sericulture among the tribe- A Socio-Economic study of Dharamjaigarh block. *Res. J. Recent Sci.* **2012**, *1*, 371–374.
136. Ibrahim, H. *Pemberdayaan Pengrajin Ekonomi Kreatif Kerajinan Sutera Di Perdesaan Provinsi Sulawesi Selatan*; IPB Press: Bogor, Indonesia, 2014.
137. BPS. *Sulawesi Selatan Dalam Angka 2020*; Badan Pusat Statistik Provinsi Sulawesi Selatan: Makassar, Indonesia, 2020.
138. Sari, R.P.; Holilulloh; Yanzi, A. Faktor-faktor Yang Mempengaruhi Pola Pikir Masyarakat Terhadap Pentingnya Pendidikan di Desa Cugung. *J. Kult. Demokr.* **2015**, *3*, pagination.
139. goKUPS. Perhutanan Sosial dan Kemitraan Lingkungan. Available online: http://sinav.usahahutan.id/index.php/frontend/analisa/kups (accessed on 20 August 2022).
140. Haruna, A.; Ridwan Sadapotto, A. Nilai Tambah Produk Persutraan Alam Dari Produk Intermediate Kokon ke Produk Akhir Benang di Kelurahan Walannae Kecamatan Sabbangparu Kabupaten Wajo. *J. Eboni* **2021**, *3*, 24–38.

141. Muin, N.; Hayati, N. *Penerapan Iptek Persuteraan Alam di Sulawesi Selatan [Laporan Hasil Penelitian]*; Balai Penerapan Standarisasi Instrumen Kementrian Lingkungan Hidup dan Kehutanan: Makassar, Indonesia, 2017.
142. Alam, A.M.K. *Peran Perempuan pada Usaha Persuteraan Alam di Desa Pising Kecamatan Donri-Donri Kabupaten Soppeng*; Universitas Hasanuddin: Makassar, Indonesia, 2016.
143. Bukhari, R.; Kour, H.; Aziz, A. Women and the Indian Sericulture Industry. *Int. J. Curr. Microbiol. Appl. Sci.* **2019**, *8*, 857–871. [CrossRef]
144. Geetha, G.S.; Indira, R. Silkworm Rearing by Rural Women in Karnataka: A Path to Empowerment. *Indian J. Gend. Stud.* **2011**, *18*, 89–102. [CrossRef]
145. Hanum, F. *Kajian dan Dinamika Gender*; Instrans Publishing: Malang, Indonesia, 2018.
146. Hubeis, A.V.S. *Pemberdayaan Perempuan dari Masa Ke Masa*; IPB Press: Bogor, Indonesia, 2010.
147. Muin, N.; Bisjoe, A.R.H. Strengthening Gender Role in Managing Private Forests in South Konawe, South East Sulawesi Province. *J. Penelit. Sos. Dan Ekon. Kehutan.* **2019**, *16*, 127–135. [CrossRef]
148. Harbi, J. Kajian teknis usaha persuteraan alam kelurahan walennae, kec. sabbangparu, kab. wajo, sulawesi selatan. *J. Penelit. Ilmu-Ilmu Kehutan.* **2016**, *5*, 45–53.
149. Muhammad, S.; Dharmawan, A.; Unito, S.; Damanhuri, D. Kearifan Lokal dalam Sistem Sosial Ekonomi Masyarakat Penenun Bugis-Wajo. *Mudra J. Seni Budaya* **2013**, *28*(2), 129–142.
150. Kumar, S.N. *Strategies for Adoption of Appropriate Innovations and Technologies for the Development of Sericulture in SAARC Countries*; SAARC Agriculture Center: Dhaka, Bangladesh, 2015.
151. Anantanyu, S. Kelembagaan Petani: Peran Dan Strategi Pengembangan Kapasitasnya. *SEPA* **2011**, *7*, 102–109.
152. Widiarti, A.; Andadari, L.; Suharti, S.; Heryati, Y.; Yuniati, D.; Agustarini, R. Partnership model for sericulture development to improve farmer's welfare (a case study at bina mandiri farmer group at Sukabumi Regency). *IOP Conf. Ser. Earth Environ. Sci.* **2021**, *917*, 012009. [CrossRef]
153. Kanafi, R.R.; Ebadi, R.; Mirhosseini, S.Z.; Seidavi, A.R.; Zolfaghari, M.; Etebari, K. A review on nutritive effect of mulberry leaves enrichment with vitamins on economic traits and biological parameters of silkworm Bombyx mori L. *Invertebr. Surviv. J.* **2007**, *4*, 86–91.

insects MDPI

Review

Bibliometric Analysis of Trends in Mulberry and Silkworm Research on the Production of Silk and Its By-Products

Domenico Giora [1,*], Giuditta Marchetti [1], Silvia Cappellozza [2], Alberto Assirelli [3], Alessio Saviane [2], Luigi Sartori [1] and Francesco Marinello [1]

1 Department of Land, Environment, Agriculture and Forestry, University of Padova, Agripolis, Legnaro, 35020 Padova, Italy; giuditta.marchetti@unipd.it (G.M.); luigi.sartori@unipd.it (L.S.); francesco.marinello@unipd.it (F.M.)
2 Council for Agricultural Research and Economics, Research Centre for Agriculture and Environment, Sericulture Laboratory, 35143 Padua, Italy; silvia.cappellozza@crea.gov.it (S.C.); alessio.saviane@crea.gov.it (A.S.)
3 Council for Agricultural Research and Economics, Research Centre for Engineering and Agro-Food Processing, Monterotondo, 00015 Rome, Italy; alberto.assirelli@crea.gov.it
* Correspondence: domenico.giora@phd.unipd.it

Simple Summary: Over the past two decades scientific research on sericulture, the agricultural activity of silk production, generated a great number of outputs in the form of articles reported and classified by one of the most well-known and used database dealing with scientific literature. This occurrence demonstrates an increasing interest in this sector especially starting from 2000s; results presented in relevant papers showed their applicability even in fields apparently not related to silk production as commonly meant, like medicine, cosmetics, and engineering. To understand how sericulture has been transcending its usual boundaries, which are its current "hotspots", and links with other fields of study, the authors propose a text-mining based analysis of the outputs of scientific research on sericulture and silk; the final goal is to establish "quantitative" indicators for researchers, entrepreneurs, and scholars.

Abstract: Traditionally, sericulture is meant as the agricultural activity of silk production, from mulberry (*Morus* sp.pl.) cultivation to silkworm (*Bombyx mori* L.) rearing. The aim of the present work is to analyze the trends and outputs of scientific research on sericulture-related topics during the last two decades, from 2000 to 2020. In this work the authors propose a text-mining analysis of the titles, abstracts and keywords of scientific articles focused on sericulture and available in the SCOPUS database considering the above-mentioned period of time; from this article collection, the 100 most recurrent terms were extracted and studied in detail. The number of publications per year in sericulture-related topics increased from 87 in 2000 to 363 in 2020 (+317%). The 100 most recurrent terms were then aggregated in clusters. The analysis shows how in the last period scientific research, besides the traditional themes of sericulture, also focused on alternative products obtainable from the sericultural practice, as fruits of mulberry trees (increment of +134% of the occurrences in the last five years) and chemical compounds as antioxidants (+233% of occurrences), phenolics (+330% of occurrences) and flavonoids (+274% of occurrences). From these considerations, the authors can state how sericulture is an active and multidisciplinary research field.

Keywords: silk; silkworm; mulberry; sericulture; bibliometric analysis

Citation: Giora, D.; Marchetti, G.; Cappellozza, S.; Assirelli, A.; Saviane, A.; Sartori, L.; Marinello, F. Bibliometric Analysis of Trends in Mulberry and Silkworm Research on the Production of Silk and Its By-Products. *Insects* **2022**, *13*, 568. https://doi.org/10.3390/insects13070568

Academic Editor: Toru Shimada

Received: 25 May 2022
Accepted: 21 June 2022
Published: 23 June 2022

1. Introduction

Sericulture is the agricultural activity that traditionally consists in the cultivation of mulberry trees (*Morus* sp.pl.) to yield leaves that are used for feeding silkworms (*Bombyx mori* L.), reared for silk production. In an ancient period of Human history, the trade of silk represented an important activity that allowed a first kind of globalization, connecting, since about 2000 years ago, Europe to Asia through the so-called "Silk Road" [1–4]. Nowadays, silk production represents about the 0.2% in value of total world textile production and it is spread over about 60 countries [5]. The most important countries for silk production are China, India, Uzbekistan, Vietnam, Thailand, and Brazil [5].

The term "silk" refers to a class of natural structural proteins produced and spun as long fibrous filament by different Arthropoda, for example Lepidoptera and spiders [6–9]. The most famous and studied type of silk is produced by the domestic silkworm *B. mori* L. [6].

From the beginning of silk manufacturing, silk has been used essentially for textile and medical applications, since efforts in medical research demonstrated how silk could be successfully applied to surgery due to its biocompatibility [10–12]. Over the last few years, there has been a growing interest in the engineering of biomaterials capable of mimicking the structure and characteristics of tissues and several studies have focused on the use of natural silk fibroin [10,13–15], due to its high biocompatibility when separated from sericin [13]. In recent years even sericin acquired a high commercial value [16] thanks to a series of properties that make it suitable for application in the pharmaceutical, cosmetic and food sector [17]. In fact, sericin has been shown to have several biological activities, such as antioxidant and anti-tyrosinase properties [17–20] and it is used for controlled drug-releasing biomaterials to promote stability and prolonged administration of drugs, enzymes and insulin [20–22].

To produce commercial silkworm eggs male and female larvae of pure strains are reared in a dedicated environment to create the parental lines that are then crossed among them [23]; the most common parental lines belong to Chinese and Japanese strains [24,25]. Hybrids and polyhybrids are in general more productive than pure strains and more resistant to biotic and abiotic stress; Nevertheless the management of environmental parameters, as air temperature and relative humidity, is fundamental for the optimal development of reared larvae since these factors heavily affects larval metabolism [26].

Traditionally, *B. mori* larvae feed on fresh leaves of the white mulberry (*Morus alba* L.) [27,28], although even leaves of other species belonging to *Morus* genus can be used [29]. Leaves of *M. alba* contain the optimal amount of nutrients and compounds that are fundamental for the growth and development of *B. mori* larvae and for their silk cocoon spinning. Besides being the only feeding source for silkworms, several studies have also been focused on alternative uses of by-products from *Morus*: leaves, roots, stems, and fruits of different species belonging to the *Morus* genus (in particular, *M. alba* and *nigra* L.), contain relatively high amounts of phenols [30–32], in particular flavonoids [32–34] with predominant quercetin and kaempferol glycosides as well as anthocyanins [35]. Different experiments showed how mulberry cultivation can represent an interesting and sustainable source of these compounds for nutraceuticals, which are usually contained in the mulberry leaf powder and mulberry fruit juice [36,37].

The previous paragraphs offer a brief overview about the whole compart of the silk production and show how the theme is wide and linked to several other topics, as engineering, medicine, and nutrition. In this complex framework, a text-mining analysis as proposed by Cogato et al. [38] and Ferrari et al. [39] seems to be appropriate to appreciate the interconnection among the hotspots of the research over the past two decades, from 2000 to 2020. This kind of analysis, in fact, allows to objectively identify and weigh the most important topics in a specific research field and to study how they interact.

The aim of the present analysis is to provide a general and comprehensive review of the state-of-the-art of literature about sericulture in the fields of agriculture and biological sciences. The specific objectives of this work were to: (i) provide a description of the temporal trend of publications over the past 20 years; (ii) identify the most important topics to which the research for sericulture field has been mainly directed; and (iii) analyze the most important links among topics; this aspect is very important since sericulture is based on the interactions of different production chains and processes. In order to obtain such results authors decided to perform a quantitative analysis that is the most effective and objective methodology to achieve the above-mentioned specific goals.

2. Materials and Methods

As first step for the bibliometric analysis, the authors downloaded a collection of documents indexed by the SCOPUS database, taking advantage of the *advanced search* to precisely define the topics. The analysis of scientific documents contained in the SCOPUS database allowed the authors to understand how the interests of researchers developed and changed in the past years.

The authors performed a *text mining* process of the selected documental collection, in order to derive significant numeric indices and information, by analyzing unstructured (textual) information. The statistical analysis of these indices provides a key to the text interpretation, obtaining high-quality information useful for content interpretation. The words appearing in the title, keywords, and abstract of all articles were first pre-processed in the Microsoft® pre-installed Notepad environment and used for the creation of basic graphs in Microsoft® Excel®; the more elaborated graphical representations were obtained using Gephi (Gephi® Consortium, Compiegne, France), an open-source software developed to perform network analysis.

2.1. Article Selection

The analysis was based on the words *"silkworm"* and *"mulberry"* both as parts of the process that leads to silk production. Moreover, to include articles containing the derived forms of those two nouns, the scripts "silkworm * " and "mulberry * " were used for the initial research in SCOPUS, since the asterisk "*" indicate any word-related declination: e.g., "mulberry", "mulberries", "mulberroside" and so on (Table 1). With the initial examination, the program selected the articles that contain the string "silkworm" and "mulberr" and their derived terms in the title, keywords, or abstract. The authors applied some filters (in Table 1, Filter application n.1) for a more pertinent selection of the articles: the field was limited to "Agricultural and Biological Sciences", the language was limited to "English" and the type of document was limited to "Scientific Article"; the time span considered was 2000–2020. According to Author's expectations, this analysis resulted in finding many articles (more than five thousand) even with the application of the described filters.

Table 1. Scripts for the extraction of research papers in the SCOPUS database.

Step	Script	Number of Papers [1]
Initial research	TITLE-ABS-KEY (mulberry *) OR TITLE-ABS-KEY (silkworm *)	23,580
Filter application n.1	TITLE-ABS-KEY (mulberry *) OR TITLE-ABS-KEY (silkworm *) AND (LIMIT-TO (DOCTYPE, "ar")) AND (LIMIT-TO (SUBJAREA, "AGRI")) AND (LIMIT-TO (LANGUAGE, "English"))	5184
Filter application n.2	TITLE-ABS-KEY (mulberry *) OR TITLE-ABS-KEY (silkworm *) AND NOT TITLE-ABS-KEY (Antheraea)) AND (LIMIT-TO (DOCTYPE, "ar")) AND (LIMIT-TO (SUBJAREA, "AGRI")) AND (LIMIT-TO (LANGUAGE, "English"))	4598

[1] Referred to years from 2000 to 2020.

The authors then applied a second series of filters to reduce misleading results. In particular, the word *"Antheraea"* was excluded from the research since it is linked to the topic of silk production but referred to a wild Lepidopteran rather than to *B. mori*. Articles collected by popular and dissemination journals were also excluded by the research. The new script is shown in Table 1 at Filter application n.2 row. With the applications of those new filters, the number of articles dropped to about 4600.

The download procedure from the SCOPUS database was repeated for each year between 2000 and 2020; the .csv format was chosen for the downloads.

2.2. Article Elaboration

The results of the research were converted and saved as .txt files to have all the data stored in a file format that could be easily processed. The first step of the data pre-processing was the so-called *tokenization*, a procedure through which the sentences are divided into their essential components (i.e., single words). All the other parts of the text as punctuation marks, hyphens, brackets, and others special characters were removed. The authors run a further elaboration to convert all capital letters into lowercase and to identify and convert singular/plural nouns (e.g., mulberry/mulberries or leaves/leaf) and synonyms (e.g., white mulberry and *M. alba*).

All .txt files obtained at the end of the previously described process were imported into Microsoft® Excel® (Microsoft 365 MSO Version 2111 Build 16.0.14701.20254 64-bit software, Microsoft Corporation, Redmond, WA, USA) to order all the single selected words and to count how many times each one appeared. This elaboration was necessary for the identification of the more frequent words in each considered year of the past two decades. Using Excel, the 100 most relevant words were identified and used for the subsequent analysis. Another interesting function provided by the Excel® software is the so-called *sparkline graphs* that creates a small graph into a single cell; the authors used this function to create time series graphs for all the analyzed nouns, thus recognizing in a fast way which words increased in importance during the analyzed time. As final step, the results were imported as .csv dedicated files in the Gephi software [40], which is a free open-source software that allows for the creation of complex graphical representation of word associations. To create a connection graph in Gephi, the user has to import into the software a .csv file containing information about the *nodes* (here, the most frequent nouns derived from the previous elaboration) and another .csv file containing information about the *edges*, the connection thanks to which each node is related to the others. Gephi considers the edges as vectors, directed or undirected, each one with its specific *weight*, which is generated as described in Section 2.2.1.

2.2.1. Combination Matrix

A word–word connection matrix was built as detailed by Ferrari et al. [39], based on the list of the 100 most recurrent words. The matrix had 100^2 cells and, that was the result of a total of 10,000 couples. Starting from this matrix, the authors built a dedicated connection matrix in which, for each couple of words (e.g., *mulberry-silkworm*) the number of articles that contained both the nouns was indicated; the connections of words are not directional, namely the order of the words forming the couple was not considered (e.g., *mulberry-silkworm* is equivalent to *silkworm-mulberry*) in order to avoid duplications. As a result of the matrix, 4950 couples of nouns were obtained.

2.2.2. Cluster Definition

Besides the analysis for the combination matrix and therefore for studying how the nouns are interconnected, the authors performed a so-called *cluster analysis*. The aim of the cluster analysis was to define a rule (or a feature, or a set of features) that allows to create groups containing a set of objects (the words in this case) and sharing one or more previously defined features: in this precise kind of cluster analysis the aim is grouping words according to the thematic area they belong to. When the main topic can be subdivided in several interrelated parts, as the one here-considered, cluster analysis allows to study and characterize the relationship among the different sections of the main topic: for example, how the cultivation of mulberry trees connects to silkworm rearing and to silkworm applications. As this analysis is only qualitative, by further differentiating results for the years or for the geographical area, it is possible to describe in a precise way the evolution of the relationship among clusters. It is worth noting that there are no fixed rules to define clusters, but the experience of the authors and the definition of precise rules, useful to minimize equivocations, allow the authors to discriminate the clusters. In the present work, the authors propose the different activities of the sericultural chain as criteria to define the clusters and, in this way, five clusters were identified: Cultivation, Silkworm, Rearing, Process and Product. All the words regarding the mulberry cultivation group in the cluster "cultivation": for example, words pertaining to the agronomic techniques and to the plant biology. The cluster "silkworm" collects all the words referring to the biology of silkworms (e.g., strains or different stages of their life). In the cluster named "rearing" the authors pooled all the words regarding the activity of silkworm rearing, including diseases or environmental factors (e.g., air temperature) that are fundamental for the insect development. The cluster "process" includes all terms that are linked to the industrial processing of products deriving from both silkworm rearing and mulberry cultivation. The cluster "product" groups all the words that describe the production obtainable by both mulberry (e.g., important metabolites) and silkworm. Although these clusters and rules were defined to avoid ambiguities and overlapping, the authors used the combination matrix and the specific research of documents on SCOPUS to define which words best fit to different cluster. Table 2 shows the cluster composition.

Table 2. Name of the considered clusters and words that compose them.

Cluster	Words
Cultivation	Crop, cultivar, fruit, germplasm, leaf, *Morus* sp. Pl, *M. alba*, *M. nigra*, mulberry, plant, root, seed, shoot, soil, species, variety, tree, water
Silkworm	Body, *Bombyx*, *B. mori*, breed, cocoon, diapause, diet, disease, DNA, egg, embryonic, expression, female, gene, gland, hemolymph, hormone, instar, larva, male, hybrid, immune, infection, instar, juvenile, larva, Lepidoptera, male, midgut, molting, moth, prothoracic, pupa, sex, shell, silkmoth, silkworm, spinning, stage, strain
Rearing	*B. bassiana*, feeding, heat, humidity, nutrition, *N. bombycis*, NucleoPolyhedroVirus, rearing, sericulture, stress, survival, temperature
Process	Drying, fermentation, industry, purification
Product	Acid, amino acid, anthocyanin, antioxidant, composition, diabetic, edible, enzyme, extract, fiber, fibroin, filament, flavonoid, food, juice, lipid, oil, peptide, phenol, polyphenol, powder, product, protein, quality, sericin, serine, silk, yield

The authors classified the most frequent words by assigning to them a score calculated as weighed average of the word in the considered years. The authors calculated, per word and per year, the ratio of the number of occurrences to the total number of publications in the sericultural field; in this way the authors could objectively characterize the relative importance of the selected word in a certain year and study its trend over the considered interval of time. The authors assigned the highest weight to the most recent years to pay more attention to the last trend of research in sericulture. The score for each word was calculated as proposed by Ferrari [41] (see Equation (1)).

$$W_i = \frac{\sum_{i=1}^{21} w_i \cdot \frac{o_i}{S_i}}{\sum_{i=1}^{21} w_i}, \tag{1}$$

In Equation (1), w_i is the weight of the i-th year (from 2000 to 2020), o_i is the absolute number of occurrences in the ith year and S_i is the total number of publications in the ith year.

The authors analyzed the impact of the five most frequent terms of each cluster, according to Equation (1). To this end, the impact of the papers related to such terms was quantified in terms of Hirsch's h-index. Thus, for a given term, the h-index has been calculated here by counting the number h of publications including that term and cited at least that same number h of times. The average was then computed for the five terms of each cluster.

The conceptual flux of the analysis is represented in Figure 1.

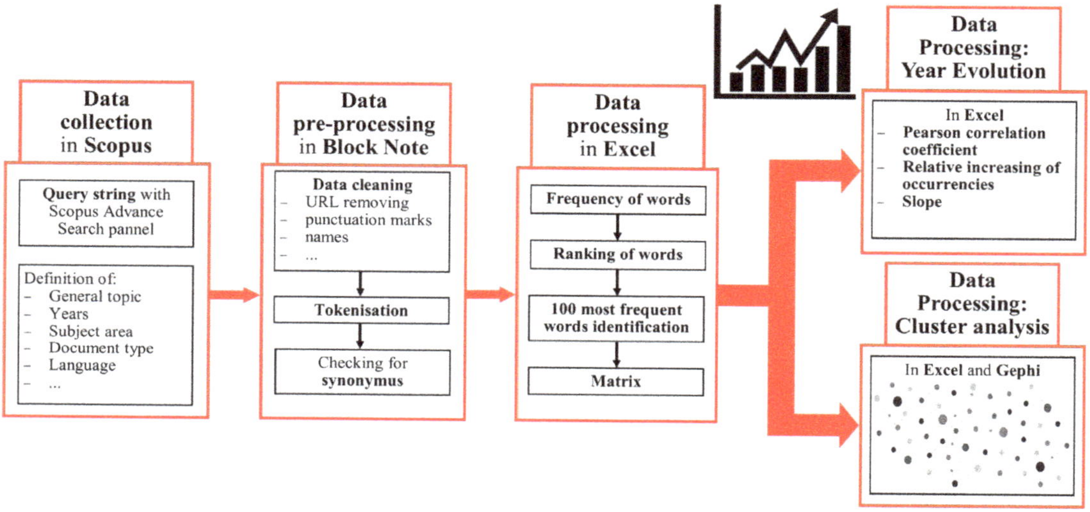

Figure 1. Scheme of the whole analysis process of the text data.

3. Results

3.1. Analysis of the Trends

As first analysis, the authors considered the number of articles published per year in the cumulative research on mulberry and silkworm and the ratio to the total number of publications in the agricultural and biological science field as an indicator of the general interest for the topic. As can be seen in Figure 2, the number of publications recorded a constant and quite rapid increase between 2000 and 2011, with an average increase of 20 articles per years. Between 2012 and 2015 a drop in the number of publications per year can be detected, reaching a relative minimum in 2015. Then, the number of publications per year increased until 2020.

On the basis of Figure 2, in the time span from 2000 to 2020 two main peaks of interest for the silkworm and mulberry topics can be detected, in 2005 and 2011, also considering the ratio of the number of publications regarding the mulberry and silkworm topics to the total number of publications in the agricultural and biological science field. From 2012 to 2018, the ratio of the selected topics to the total number of publications decreased, reaching its minimum in 2015, and then remained stable for almost five years, although the absolute number of publications in the silkworm and mulberry topics increased.

To characterize the geographic distribution of the research in the sericultural topic, an analogue analysis was performed considering the number of publications per country. Figure 3 shows the top five countries for number of publications in the 2000–2020 period.

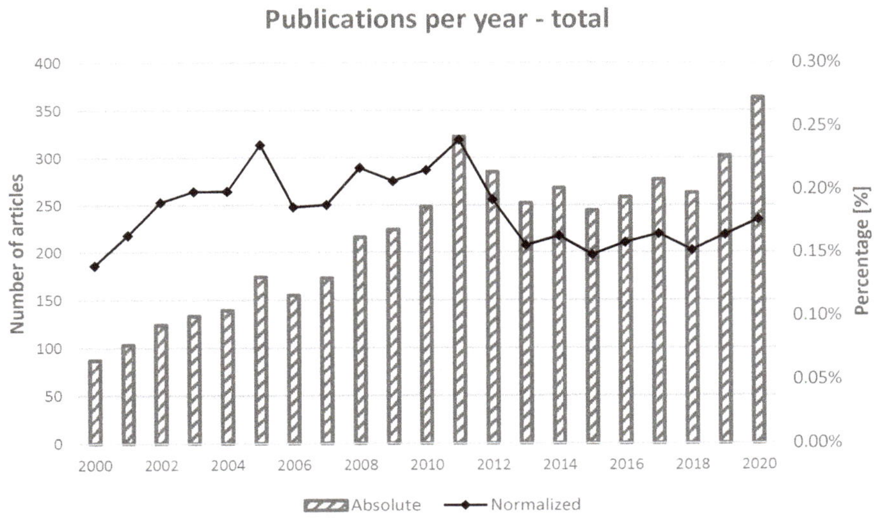

Figure 2. Publications per year (bars) and ratio between publications on the mulberry and silkworm topic in the sector "Agri" and total publications in Agricultural and Biological Science field (line).

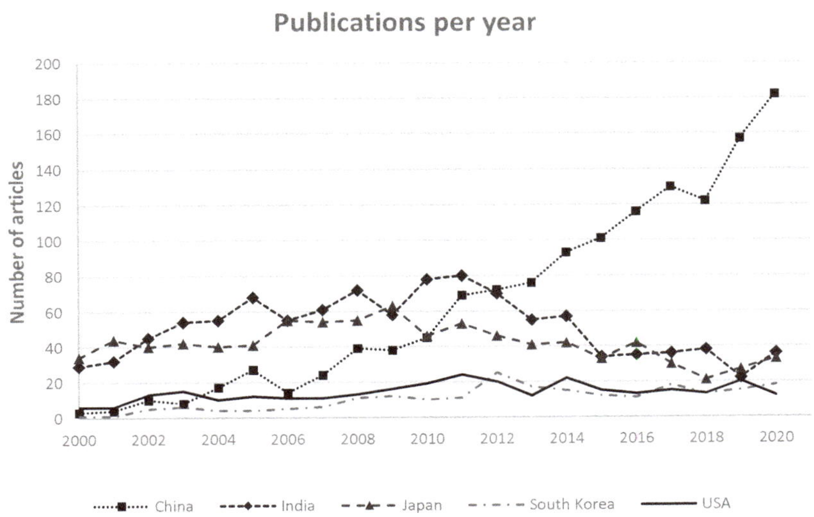

Figure 3. Geographic distribution of interest for the sericulture topic.

The three most important countries for the research in the sericultural topic are China, where the number of publications increased from 3 to 180 per years, India and Japan, where the interest increased more or less until 2010 and then decreased.

A more specific analysis about the geographical extent of the research is the one focused on collaborations among Universities and Research Institutes from different countries that aims to determine how the chosen topic can generate collaborations in the scientific community. Figure 4 shows in a graphic way how the web of collaborations for the sericultural topic is well distributed in the World among different countries. More detailed information about collaborations among different countries are presented in Table 3, where data about the first five countries for the total number of collaborations that led to articles publication between 2000 and 2020 are summarized; data of collaborations were derived from the affiliation metadata of the SCOPUS database.

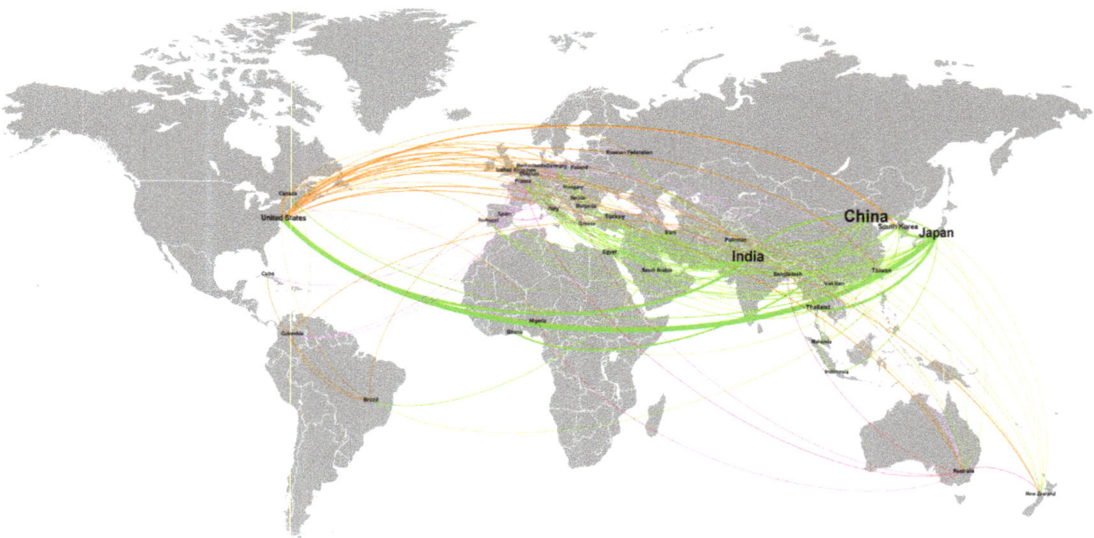

Figure 4. Geographic extent of collaborations for research in sericulture-related topics.

Table 3. Collaborations among countries.

Country	Total Number of Collaborations	Number of Partner Countries	Average Number of Collaborations per Country
Japan	142	37	3.9
USA	102	41	2.6
India	74	24	3.1
China	52	25	2.1
Italy	27	22	1.6

As it can be noticed, the first country for number of collaborations is Japan with a total of 142 articles derived from cooperation. Japan collaborated with 37 different countries, with an average of 3.9 articles derived from the collaboration with each one. The USA are the first country for number of partners of research, but with a lower total amount of published articles and average number of collaborations per country.

3.2. *Analysis of the Most Important Words*

The authors then analyzed the collection of the most recurrent words previously derived as described in Section 2.2. The authors started the analysis on the list of the most recurrent words both in a cumulative way and considering their evolution over time too; this study was performed to characterize in a better way the research outcomes during the last two decades.

The authors choose three different parameters to characterize the evolution over time of the selected 100 words objectively. For the computation of those parameters, the authors used Microsoft Excel. The first parameter is the *mean slope* of the regression line that fits the number of publications per year; this parameter is calculated by Microsoft Excel dividing the variation on y-axis by the variation on x-axis. This parameter allows the authors to understand the rate of increment or decrement of the selected word during the investigated span of time. The second parameter is the *Pearson correlation coefficient*. This parameter is adimensional and ranges between +1 and −1. It expresses the linearity of the trend (i.e., evaluates whether the increment or decrement of the occurrences, for the selected word, are constant over the years). The third parameter is the last *five years relative change*, calculated as ratio between the average number of occurrences per word from 2016 to 2020 and the average number of occurrences from 2000 to 2015. Table 4 resumes some of the most cited nouns and their associated parameters.

Table 4. The most important words considering their slope, Pearson correlation coefficient and their relative increment in the past five years.

Word	Slope	Pearson Correlation Coefficient	Five Years Relative Changing [%]	Graph
Gene	18.2	0.85	+63%	
Protein	12.6	0.90	+68%	
Acid	8.7	0.93	+172%	
Fruit	7.3	0.92	+134%	
Extract	9.6	0.93	+159%	
Morus alba	6.4	0.92	+118%	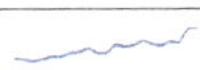
Morus nigra	3.3	0.91	+161%	
Antioxidant	8.6	0.97	+218%	
Silk	4.9	0.81	+47%	
Expression	14.3	0.93	+106%	
Anthocyanin	4.5	0.93	+270%	
Phenolic	3.4	0.87	+278%	
Polyphenol	1.9	0.91	+276%	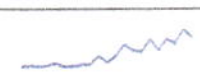
Flavonoid	3.0	0.90	+274%	
Immune	2.4	0.88	+307%	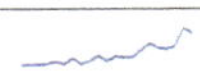
Juice	2.3	0.78	+234%	

Then, the same parameters used for characterizing words in the previous analysis were calculated out of the normalized number of occurrences per year; the normalized number of occurrences was calculated as ratio of the absolute value of the occurrences of the selected word and the absolute number of publications in the corresponding year. Results for the same nouns of Table 4 are reported in Table 5.

Table 5. Most important words normalized per topic considering their slope, Pearson regression coefficient and their relative increment in the past five years.

Word	Slope	Pearson Regression Coefficient	Five Years Relative Changing [%]	Graph
Gene	2.30	0.54	+15%	
Protein	1.32	0.47	+16%	
Acid	2.24	0.87	+101%	
Fruit	1.99	0.85	+77%	
Extract	2.64	0.87	+96%	
Morus alba	1.24	0.76	+50%	
Morus nigra	0.83	0.76	+94%	
Antioxidant	2.73	0.96	+155%	
Silk	0.42	0.26	+3%	
Expression	3.45	0.84	+53%	
Anthocyanin	1.41	0.88	+194%	
Phenol	1.50	0.85	+215%	
Polyphenol	0.64	0.89	+217%	
Flavonoid	0.94	0.88	+200%	
Immune	0.76	0.86	+212%	
Juice	0.71	0.66	+166%	

3.3. Cluster Analysis

The 100 top frequent words have been grouped into six clusters. With reference to Equation (1), the authors used the weight assigned to each term to define the relative weight of the cluster to which that specific word belongs. Therefore, the relative weight of each cluster was calculated on the basis of the total sum of the relative weights of all the n words contained in the cluster itself. Table 6 shows the constitution of each cluster, with the relative weights of all the considered words.

Table 6. Clusters composition with the relative weight of each word and the overall relative weight of each cluster.

Cluster	Words	Cluster Relative Weight
Cultivation	Crop (1.6%), cultivar (1.4%), fruit (6.5%), germplasm (0.6%), leaf (15.2%), *Morus* sp.pl. (3.4%), *Morus alba* (6.5%), *Morus nigra* (2.9%), mulberry (32.2%), plant (7.7%), root (2.4%), seed (1.0%), shoot (1.3%), soil (2.0%), species (6.5%), tree (2.9%), variety (2.6%), water (3.3%)	23.6%
Silkworm	Body (1.8%), *Bombyx* (0.6%), *Bombyx mori* (10.7%), breed (1.6%), cocoon (3.4%), diapause (1.1%), diet (1.3%), disease (1.7%), DNA (1.9%), egg (2.3%), embryonic (0.5%), expression (7.4%), female (1.5%), gene (12.6%), gland (2.0%), hemolymph (1.1%), hormone (1.3%), hybrid (1.4%), immune (0.8%), infection (2.3%), instar (1.5%), juvenile (0.5%), larva (5.8%), Lepidoptera (1.5%), male (0.9%), metabolism (0.9%), midgut (1.2%), moth (0.8%), prothoracic (0.3%), pupa (2.7%), sex (0.8%), shell (0.9%), silkmoth (0.3%), silkworm (15.3%), spinning (2.9%) stage (2.4%), strain (2.9%), voltinism (1.0%)	46.4%

Table 6. *Cont.*

Cluster	Words	Cluster Relative Weight
Rearing	*Beauveria bassiana* (5.1%), feeding (23.9%), heat (4.5%), humidity (2.7%), *Nosema bombycis* (5.1%), NucleoPolyhedroVirus (13.1%), rearing (8.0%), sericulture (7.4%), stress (9.5%), survival (7.2%), temperature (13.5%)	6.1%
Process	Drying (47.6%), fermentation (17.2%), industry (14.0%), purification (12.1%)	1.6%
Product	Acid (8.6%), aminoacid (2.4%) anthocyanin (3.4%), antioxidant (6.9%), composition (7.7%), diabetic (2.5%), edible (0.8%), enzyme (2.9%), extract (8.7%), lipid (4.3%), fiber (0.7%), fibroin (1.5%), filament (0.8%), flavonoid (2.3%), food (3.3%), juice (1.9%), oil (1.4%), peptides (1.9%), phenol (3.5%), polyphenol (1.4%), powder (1.4%), product (2.1%), protein (18.3%), quality (3.2%), sericin (1.5%), serine (1.1%), silk (8.2%), yield (3.5%),	22.2%

With reference to Table 4, the most important clusters are, in descending order, "silkworm", "cultivation", and "product". The "silkworm" cluster represents about one half (46.4%) of the total occurrences considered for this analysis. The authors grouped in this cluster all the nouns that refer to the biology of the silkworm *B. mori*. In this cluster, about one forth (26.0%) of occurrences are in general about the silkworm itself, respectively with the words "silkworm" (15.3%) and "*B. mori*" (10.7%). In absolute terms the occurrences of the word "gene" (12.6%) saw a rapid increase (slope = 18.2) that was constant and quite linear during the considered period (Pearson coefficient = 0.85); this trend is maintained considering the ratio of occurrences over the number of publications per year in sericulture. In relation to "gene", the word "expression" (7.4%) shows a significant increasing both in absolute (slope = 14.3) and relative terms (slope = 3.45) Another important theme in the biology of silkworm is represented by the sex identification, and in this analysis this topic represents the 3.3% of the weight of the cluster, respectively with words "female" (1.5%), "male" (0.9%), and "sex" (0.8%). Other important nouns belonging to this cluster are "larva", that represents 5.8% of the total weight of the cluster, "cocoon", with 3.4%, "shell", with 0.9% and "spinning", with 2.9%. The cluster named "cultivation" groups all the aspects linked to the mulberry cultivation and mulberry field management; its overall relative weight is 23.6%. The most recurrent word is "mulberry", with about one third of weight of the cluster (32.2%). In particular, the most present one is the white mulberry, *M. alba* (L.), whose weight in the cluster is 6.5%; then the second most studied species is the black mulberry, *M. nigra* (L.), with 2.9% of the total weight of the cluster. For both the species, the increment of interest in the last two decades in absolute terms showed a constant and quite linear increase, with a slope = 6.4 for *M. alba* and 3.3 for *M. nigra* and a Pearson coefficient = 0.92 and 0.91 respectively. *M. nigra* also shows the highest increase of interest in the last five years, trend confirmed both in absolute terms (+161% vs. +118%) and in relative terms (+94% vs. +50%). Other two important words of this cluster are "leaf" (15.2%) and "fruit" (6.5%). The noun "leaf" shows a good growth but not so constant in time in absolute terms (slope = 10.2 and Pearson coefficient = 0.73). However, in relation to the increasing number of publications in the sericulture field the interest for it decreased (slope and Pearson coefficient near to zero but negative and relative increment in the last five years ±0%). At the opposite, the interest of researchers for mulberry fruit increased both in absolute terms (slope = 7.3, Pearson coefficient = 0.92 and relative increment in the past five years = +134%) and in relative terms (slope = 1.99, Pearson coefficient = 0.85 and relative increment in the past five years = +77%). The third most relevant cluster is "Product", which is dedicated to the products obtainable from both mulberry cultivation and silkworm rearing. The relative weight of this cluster is 22.2%. In this cluster, the most important word is "protein". The importance and interest for the word "protein" arose in a rapid and linear way during the last two decades (slope = 12.6 and Pearson coefficient = 0.90) in absolute terms; in relative terms this increase is not so high, with a peak of interest during the beginning of 2000s (see graph in Table 5). An analogue consideration could be made for the word "silk" that shows a quite high interest in absolute terms, but this increase in interest is biased by the increasing number of publications per year in the sericultural field. In fact, when we perform the normalization of the trend on the number of publications per year, we can see that this interest has been constant for researchers. Linked to the word "silk" there are nouns referring to its principal components, "fibroin" (1.5%) and "sericin" (1.5%). In this cluster, another important category of words is represented by beneficial compounds that could be extracted from mulberry leaves. These words are "antioxidant" (6.9%), "anthocyanin" (3.4%), "phenol" (3.5%), "flavonoid" (2.3%) and "polyphenol" (1.4%) and their sum weighs about 17% of the whole cluster. In particular, the interest for this group of words rapidly increased in the past five years, both in absolute and relative terms, as attested by the parameter of relative increase reported in Tables 4 and 5.

As last analysis, the authors evaluated the temporal trend of the aggregated number of occurrences per cluster. Furthermore, the trend of the normalized total occurrences as a ratio to the number of publications was analyzed. The results are reported in Tables 7 and 8.

Table 7. Analysis of cluster considering their slope, Pearson regression coefficient, and their relative increment in the past five years.

Cluster	Slope	Pearson Regression Coefficient	Five Years Relative Changing [%]	Graph
Cultivation	83.4	0.93	+80%	
Process	7.4	0.89	+144%	
Product	90.0	0.98	+117%	
Rearing	18.9	0.87	+71%	
Silkworm	105.7	0.79	+33%	

Table 8. Analysis of clusters normalized per cluster by considering their slope, Pearson regression coefficient, and their relative increment in the past five years.

Cluster	Slope	Pearson Regression Coefficient	Five Years Relative Changing [%]	Graph
Cultivation	11.57	0.79	+25%	
Process	1.64	0.82	+73%	
Product	18.53	0.93	+55%	
Rearing	1.55	0.40	+16%	
Silkworm	−6.67	−0.35	−10%	

From the analysis of Tables 5 and 6 some interesting considerations can be derived. In particular, the "silkworm" cluster, in absolute terms, has the highest value of slope, followed by "product" and "cultivation"; the cluster "process" is the one with the major relative increase of publications in the last five years. On the other hand, when we normalize the number of publications per year, the silkworm cluster evidences a negative slope, which although linked to the relative increment of the last five years, testifies a slow decreasing interest in this topic. "Cultivation", "process" and "product" are the cluster that maintain a high relative interest in the field of sericultural research.

3.4. Impact Factor Analysis

The objective of this analysis was to derive sound information about the impact that the most frequent terms generated on scientific research. The proposed impact factors of considered word are reported in Table 9.

Table 9. Calculated impact factor of the most frequent words for the five clusters. The words are sorted in descending order considering their weights according to Equation (1); h-indices in brackets.

Cluster	Words	Cluster Mean
Cultivation	Mulberry (89), leaf (77), plant (94), *Morus alba* (75), species (84)	84
Silkworm	Silkworm (78), gene (95), *Bombyx mori* (77), expression (73), larva (63)	77
Rearing	Feeding (62), temperature (55), nucleopolyhedroviral (35), stress (60), rearing (61)	51
Process	Drying (38), fermentation (39), industry (42), purified (66)	46
Product	Protein (82), extract (82), acid (97), silk (56), antioxidant (77)	79

As reported in Table 9, the "cultivation" and "product" clusters are the ones exhibiting the highest impact, highlighting a vivid interest by the scientific community on these topics. Also, the "silkworm" cluster gives evidence of a significant impact: in this case, the high h-index should also be related to the highest number of papers published on this topic since 2000. A lower interest is apparently arising from research related to the process or to entomological aspects. On the other hand, the specific terms, "acid" (97) and "gene" (95) gained the highest impact, having collected more than 35,000 citing documents each.

3.5. Interrelationships among Words

The objective of this analysis was to derive sound information about the interrelationships among the analyzed words. To achieve this aim, each of the 100 most frequent words previously mentioned was coupled with each of the remaining 99 words, thus generating a table of 4950 possible combinations (avoiding repetitions). Each noun of this collection of paired words was studied in terms of total number of occurrences for the last considered 20 years.

As reported in Table 10, "silkworm" is the cluster with the highest number of co-occurrences, both inside the cluster itself and with other clusters; the maximum number of occurrences was reached by the interaction of "silkworm" with itself (61,864), with the cluster "cultivation", (20,626) and with the cluster "product" (8112), attesting the great attention of the scientific community to themes focused on the biology and life cycle of silkworms and to the production that could be derived from them. The cluster "cultivation", as expected, is the second one for number of co-occurrences; as for the cluster "silkworm", the co-interactions are mainly concentrated on the cluster "cultivation" itself and on the cluster "product". In this case, the interaction with the cluster "process" is also high.

Table 10. Number of interrelationships among clusters.

	Cultivation	Process	Product	Rearing	Silkworm
Cultivation	15,203	-	-	-	-
Process	17,010	25	-	-	-
Product	19,756	2156	13,028	-	-
Rearing	5415	520	6186	924	-
Silkworm	20,626	2396	50,422	12,135	61,864

Then, the authors proposed a brief analysis of the most recurrent couples of nouns, independently on the original cluster, to stress the importance of the most relevant topics. The couple of words with the highest number of co-occurrences is silk-silkworm (2435) *Bombyx-B. mori* (2097), and *B. mori*-Silk (1953). The first couple for co-occurrences related to the theme of mulberry cultivation is the couple leaf-mulberry (924) followed by *Morus-M. alba* (666); several couples of terms regard the area of genetic studies, as Silk-Gene (1366), Silkworm-Gene (1297), *Bombyx*-Gene (1245) and *B. mori*-Gene (1243).

Taking advantages of the Open-Source software Gephi, the authors generated the map of the interrelationship between the 100 most frequent terms showed in Figure 5.

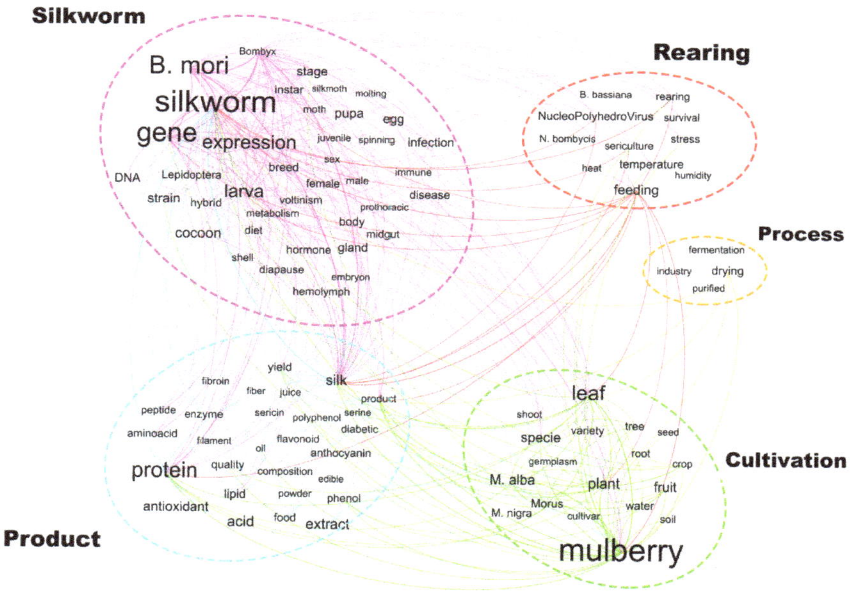

Figure 5. Interrelations among the top 100 most frequent words. In this picture, the curved lines represent relevant interrelations (number of co-occurrences is greater than 100).

4. Discussion

The main core of this research is represented by a text-mining analysis focused on the words in the title, abstract, and keywords of articles related to silk production. Different considerations can be drawn by the analysis of the most frequent nouns, which were previously identified and evaluated. Taking advantage of a procedure algorithm already tested by Cogato et al. and Ferrari et al. [38,39], the most significant relationships among nouns were recognized. Additionally, by including time parametrization, it was also possible to characterize the evolution of the research interests in sericulture. Thanks to the large number of documents considered in this study, a statistically robust analysis of research was achieved, supporting hypotheses on influences and trends in sericulture

From the characterization of the number of released publications in sericulture (see Figure 2), it is worth noting that the absolute number of articles published per year has been increasing since 2000; such trend reflects the great interest in the topic and the interdisciplinarity of the matter. By considering this trend, two main periods can be identified: from 2000 to 2011 the trend was constantly increasing, both in absolute number and in relative percentage (considering the total number of publications indexed by the SCOPUS database), reaching a peak in 2011; the second period began in 2012 and exhibited a decrease followed by an increase until 2017. Currently, the interest is still apparently growing in absolute terms. In relative terms, the interest appears quite stable; this is due to a rather great relative increase in the total number of publications per year (from about 62 k in 2000 to more than 200 k in 2020, documents indexed by the SCOPUS database). The next decades should be monitored to determine whether these new trends are stable, or they are not. From a geopolitical and temporal point of view, such a trend should be explained by two main reasons: the decreasing number of publications per year from India, and the increasing number of publications per year from China, which is now the leading country for the research in the sericultural field.

From the analysis of the trends of single words and clusters, interesting and more specific information have been extracted. From an aggregate point of view, considering the clusters, the main arguments of investigation of the research in the past twenty years were the study of the *Morus* cultivation, industrial processes, and products that could derive from sericulture, from both *Morus* cultivation and *B. mori* rearing.

From the analysis of the impact of the most important words of each cluster, it is interesting to note how those words are able to generate a considerable impact in the frame of sericulture, with an overall average impact factor of about 67.

The cluster "silkworm", related to the silkworm biology, is the largest one in terms of number of included words (according to the data of Table 6, 46.4%) but during the last decade, and in particular in the last five years, the interest of the scientific research decreased relatively to this sericultural main topic (Table 8). In particular, the most important words of the cluster are "silkworm" (15.3%) and "*B. mori*" (10.7%), also characterized by considerable high values of the impact factor (respectively 78 and 77); this is easily understandable since the work on the living organism is the main focus of the research on this topic. The silkworm has always been an important laboratory tool for genetics and physiology of insects; in particular the independent genome sequencing of 2004 by a Chinese [42] and a Japanese team [43] and the integrated silkworm database of the end of 2009 [44], gave probably place to the peak of publication on sericulture recorded in 2011. However, from 2012 (the year of FAO experts' consultation in Rome about entomophagy "Insects to feed the world") new insects started to be studied in relationship with the new trend of edible insects as perspective feed and food for the future; for example, *Tenebrio molitor* genome was sequenced in 2020 [45]. Exploitation of edible insects gave rise to a new impulse to research on the physiology of new and less known insects as *T. molitor* and *Hermetia illucens*, probably resulting in a diminished interest for traditional insects used as laboratory tools like silkworms.

However, silkworm genetics, still accounts for about 22% of the weight of this cluster. The Nucleopolyhedrosis virus (NPV) has also been used for genetic studies, and in particular as a vector for the recombinant protein technology [46–48]. NPV research could, therefore, be considered for its importance in determining both methods for biotechnologies and silkworm physiology studies or in establishing correct rearing techniques for disinfection, disease prevention and rearing facility planning. For this reason, NPV research and other studies on silkworm diseases like pebrine (*Nosema bombycis*) or white muscardine (*Beauveria bassiana*) could be regarded as horizontal literature common to the cluster "Silkworm" and "Rearing".

Another important theme of the silkworm biology is represented by the sex identification. Indeed, in order to obtain a highly-productive polyhybrid, the best male and best female individuals of different strains are reared and bred. As stated by Raj et al. [23] different solutions for non-destructive sex-sorting methods for silkworms have been proposed [23,41,49]. In fact, an efficient sexing method has not been found yet. Three other important terms of the cluster "Silkworm", counting about 10% of the weight of the cluster are "larva", "cocoon" and "shell", the raw materials for farmers who rear the larval stage of silkworm and obtain, at the end of their fifth larval instar, fresh silk cocoons.

The cluster "cultivation", related to mulberry cultivation, is the second most important one in terms of the relative weight of the included words (according to the data of Table 6 about one fourth). Since mulberry is the general word for the identification of *Morus* sp. pl. plants, the most interesting and reliable results are about specific species. As stated before, silkworms feed mainly on *M. alba* leaves [27,28], and the term "*Morus alba*" showed a high increases in the past twenty years. The term "*Morus nigra*" recorded a lower increase located in particular in the 2016–2020 period. The term "leaf" is the second most important word in this cluster, and this is related to the fact that mulberry leaves represent the only source of food for silkworms [27,28]. It is worth noting how "leaf" is a word that has not exhibited an increase, as for others considered terms (Table 5). The explanation of this trend could be related to the fact that *Morus* is currently studied much more for fruits and active ingredients for pharmaceuticals than as feed for the silkworm, for which artificial diets are an innovative and attractive matter of research.

In the "product" cluster, the most recurrent word is "protein", which represents the most important class of nutrients in mulberry leaves, the only source of nitrogen for silkworms. Thus, silkworms use the proteins obtained through dietary ingestion in order to extract the amino acids necessary to synthesize other required proteins both for their growth and for spinning cocoons [50]. Proteins are, therefore, also important products obtained from cocoons (fibroin and sericin) and from silkworm pupae. Recombinant proteins achieved from engineered larvae also represent a "hot spot" of sericulture [46–48,51]. It is worth noticing how the nouns "acid" (97), "extract" (82) and "protein" (82) are characterized by very high values of the impact factor here proposed, attesting a great interest of the scientific research about the topics related to these nouns.

The "process" cluster is the one with the lowest number of occurrences and thus with the lowest weight (1.6%). Nonetheless, an actual interest on the topic can be recognized, especially in the last five years (2016–2020), when the cluster gave evidence of the highest increment. New industrial processes are currently required for recombinant protein purification, for fibroin and sericin dissolution and reconstitution, to build 3D scaffolds, biomedical products, biocompatible implantable devices and biofilms. Furthermore, to purify mulberry organic compounds, fat and proteins from pupae, new methods and processing technologies are necessary and this explains the increase of the last years linked to the emerging fields of sericultural biotechnologies, pharmaceuticals from mulberry, and edible insects [52–54].

In the "rearing" cluster the two most relevant themes are those related to the words "feeding" (already discussed above) and "temperature". The word "temperature" is the one with highest weight in the cluster (13.5%) and this trend is somehow confirmed also by the high attention paid by farmers as well as by researchers to the temperature control for the sericultural practice [26,55,56]; in fact, some authors proposed sensor-based systems that allow the automatic control of temperature (concurrently with other environmental parameters) [57]. For the same reason, a clear interest is also focused on other two strictly related nouns namely "heat" (4.5%) and "humidity" (2.7%). We should also highlight that climatic changes have impacted on all the agricultural activities and sericulture was also much affected by the extreme environmental conditions recorded in the last years.

5. Conclusions

Due to the renovated high interest in natural fibers (on the SCOPUS database, from 367 document published in 2000 to 1768 document published in 2020), in this work the authors propose a detailed analysis of the outcomes of the scientific research on silk. The results of the analysis showed a high number of documents related to classical themes of silk production, as the quality of silk, the biology of *B. mori* and the cultivation of *M. alba*.

The analysis has highlighted some gaps of knowledge in sericulture, in particular a low amount of research is related to the themes of automation of processes. The automation applied to time-consuming tasks as feeding or environmental control in rearing rooms could be successfully carried out through new possibilities offered by the application of Machine Learning and Artificial Intelligence [58] and constitute an important research field for the future sericulture.

Author Contributions: Conceptualization, D.G. and F.M.; methodology, D.G. and F.M.; software, D.G.; validation, S.C., F.M. and A.S.; formal analysis, D.G.; investigation, D.G. and G.M.; resources, A.A.; data curation, D.G.; writing—original draft preparation, D.G.; writing—review and editing, S.C., A.S., A.A., F.M. and L.S.; visualization, D.G., F.M. and G.M. supervision, F.M.; project administration, L.S.; funding acquisition, L.S, A.A. and S.C. All authors have read and agreed to the published version of the manuscript.

Funding: This research was funded by Veneto Region, Measure 16.1-2 Programme of Rural Development for the Veneto Region, 2014-2020-DGR 2175 del 23 December 2016, grant number:55-04/12/2017 SERINNOVATION.

Institutional Review Board Statement: Not applicable.

Informed Consent Statement: Not applicable.

Data Availability Statement: Data presented in this study are available on request from the corresponding author. The data are not publicly available because they can be used by the Operational Group "Serinnovation" for future economic applications.

Acknowledgments: S.C. and A.S. acknowledge the support of the Operational Group "Serinnovation".

Conflicts of Interest: The authors declare no conflict of interest. The funders had no role in the design of the study; in the collection, analyses, or interpretation of data; in the writing of the manuscript, or in the decision to publish the results.

References

1. Goldsmith, M.R.; Shimada, T.; Abe, H. The Genetics and Genomics of the Silkworm, Bombyx Mori. *Annu. Rev. Entomol.* **2005**, *50*, 71–100. [CrossRef] [PubMed]
2. Kurin, R.; Borden, C.M. The silk road: Connecting cultures, creating trust. *Talk Story* **2002**, *21*, 1–11.
3. Datta, R.K.; Nanavaty, M. *Global Silk Industry: A Complete Source Book*; Universal-Publishers: Irvine, CA, USA, 2005; ISBN 1581124937.
4. Lopez, R.S. China silk in Europe in the Yuan period. *J. Am. Orient. Soc.* **1952**, *72*, 72–76. [CrossRef]
5. International Sericultural Commission. Statistics about Global Silk Producton. Available online: https://www.inserco.org/en/statistics (accessed on 24 May 2022).
6. Meinel, L.; Betz, O.; Fajardo, R.; Hofmann, S.; Nazarian, A.; Cory, E.; Hilbe, M.; McCool, J.; Langer, R.; Vunjak-Novakovic, G.; et al. Silk based biomaterials to heal critical sized femur defects. *Bone* **2006**, *39*, 922–931. [CrossRef]
7. Zhou, Q.Z.; Fu, P.; Li, S.S.; Zhang, C.J.; Yu, Q.Y.; Qiu, C.Z.; Zhang, H.B.; Zhang, Z. A Comparison of Co-expression Networks in Silk Gland Reveals the Causes of Silk Yield Increase During Silkworm Domestication. *Front. Genet.* **2020**, *11*, 225. [CrossRef]
8. Porter, D.; Guan, J.; Vollrath, F. Spider silk: Super material or thin fibre? *Adv. Mater.* **2013**, *25*, 1275–1279. [CrossRef]
9. Mortimer, B.; Gordon, S.D.; Holland, C.; Siviour, C.R.; Vollrath, F.; Windmill, J.F.C. The speed of sound in silk: Linking material performance to biological function. *Adv. Mater.* **2014**, *26*, 5179–5183. [CrossRef]
10. Altman, G.H.; Diaz, F.; Jakuba, C.; Calabro, T.; Horan, R.L.; Chen, J.; Lu, H.; Richmond, J.; Kaplan, D.L. Silk-based biomaterials. *Biomaterials* **2003**, *24*, 401–416. [CrossRef]
11. Jewell, M.; Daunch, W.; Bengtson, B.; Mortarino, E. The development of SERI® Surgical Scaffold, an engineered biological scaffold. *Ann. N. Y. Acad. Sci.* **2015**, *1358*, 44–55. [CrossRef]

12. van Turnhout, A.A.W.M.; Franke, C.J.J.; Vriens-Nieuwenhuis, E.J.C.; van der Sluis, W.B. The use of SERI™ Surgical Scaffolds in direct-to-implant reconstruction after skin-sparing mastectomy: A retrospective study on surgical outcomes and a systematic review of current literature. *J. Plast. Reconstr. Aesthetic Surg.* **2018**, *71*, 644–650. [CrossRef]
13. Holland, C.; Numata, K.; Rnjak-Kovacina, J.; Seib, F.P. The Biomedical Use of Silk: Past, Present, Future. *Adv. Healthc. Mater.* **2019**, *8*, 1800465. [CrossRef] [PubMed]
14. Melke, J.; Midha, S.; Ghosh, S.; Ito, K.; Hofmann, S. Silk fibroin as biomaterial for bone tissue engineering. *Acta Biomater.* **2016**, *31*, 1–16. [CrossRef] [PubMed]
15. Kasoju, N.; Bora, U.; Kasoju, N.; Bora, U. Silk Fibroin in Tissue Engineering. *Adv. Healthc. Mater.* **2012**, *1*, 393–412. [CrossRef] [PubMed]
16. Gupta, D.; Agrawal, A.; Chaudhary, H.; Gulrajani, M.; Gupta, C. Cleaner process for extraction of sericin using infrared. *J. Clean. Prod.* **2013**, *52*, 488–494. [CrossRef]
17. Chlapanidas, T.; Faragò, S.; Lucconi, G.; Perteghella, S.; Galuzzi, M.; Mantelli, M.; Avanzini, M.A.; Tosca, M.C.; Marazzi, M.; Vigo, D.; et al. Sericins exhibit ROS-scavenging, anti-tyrosinase, anti-elastase, and in vitro immunomodulatory activities. *Int. J. Biol. Macromol.* **2013**, *58*, 47–56. [CrossRef]
18. Dong, X.; Zhao, S.X.; Yin, X.L.; Wang, H.Y.; Wei, Z.G.; Zhang, Y.Q. Silk sericin has significantly hypoglycaemic effect in type 2 diabetic mice via anti-oxidation and anti-inflammation. *Int. J. Biol. Macromol.* **2020**, *150*, 1061–1071. [CrossRef]
19. Aramwit, P.; Siritientong, T.; Srichana, T. Potential applications of silk sericin, a natural protein from textile industry by-products. *Waste Manag. Res.* **2012**, *30*, 217–224. [CrossRef]
20. Zhang, Y.Q.; Tao, M.L.; Shen, W.D.; Zhou, Y.Z.; Ding, Y.; Ma, Y.; Zhou, W.L. Immobilization of L-asparaginase on the microparticles of the natural silk sericin protein and its characters. *Biomaterials* **2004**, *25*, 3751–3759. [CrossRef]
21. Rocha, L.K.H.; Favaro, L.I.L.; Rios, A.C.; Silva, E.C.; Silva, W.F.; Stigliani, T.P.; Guilger, M.; Lima, R.; Oliveira, J.M.; Aranha, N.; et al. Sericin from *Bombyx mori* cocoons. Part I: Extraction and physicochemical-biological characterization for biopharmaceutical applications. *Process Biochem.* **2017**, *61*, 163–177. [CrossRef]
22. Zhang, Y.Q.; Ma, Y.; Xia, Y.Y.; Shen, W.D.; Mao, J.P.; Xue, R.Y. Silk sericin-insulin bioconjugates: Synthesis, characterization and biological activity. *J. Control. Release* **2006**, *115*, 307–315. [CrossRef]
23. Raj, A.N.J.; Sundaram, R.; Mahesh, V.G.V.; Zhuang, Z.; Simeone, A. A Multi-Sensor System for Silkworm Cocoon Gender Classification via Image Processing and Support Vector Machine. *Sensors* **2019**, *19*, 2656. [CrossRef]
24. Grekov, D.; Kipriotis, E.; Tzenov, P. *Sericulture Training Manual*; National Agricultural Research Foundation: Komotini, Greece, 2005.
25. Cappellozza, S.; Saviane, A. Aspetti attuali dell'allevamento del baco da seta. In *Atti Seminario Insetti Utili*; MULSA (Museo Lombardo di Storia dell'Agricoltura: Milano, Italy, 2014; pp. 71–78.
26. Rahmathulla, V.K. Management of climatic factors for successful silkworm (*Bombyx mori* L.) crop and higher silk production: A review. *Psyche* **2012**, *2012*, 121234. [CrossRef]
27. Lim, S.-H.; Kim, Y.-T.; Lee, S.-P.; Rhee, I.-J.; Lim, J.-S.; Lim, B.-H. *Sericulture Training Manual*; Food & Agriculture Org.: Rome, Italy, 1990; ISBN 9251029040.
28. ITO, T. Nutritional requirements of the silkworm, *Bombyx mori* L. *Proc. Jpn. Acad.* **1967**, *43*, 57–61. [CrossRef]
29. Ayandokun, A.E.; Alamu, O.T. Cocoon production efficiency of silkworm (*Bombyx mori* L.) in response to host shift between two selected mulberry varieties. *Int. J. Trop. Insect Sci.* **2020**, *40*, 49–52. [CrossRef]
30. Zhang, M.; Chen, M.; Zhang, H.Q.; Sun, S.; Xia, B.; Wu, F.H. In vivo hypoglycemic effects of phenolics from the root bark of *Morus alba*. *Fitoterapia* **2009**, *80*, 475–477. [CrossRef] [PubMed]
31. Imran, M.; Khan, H.; Shah, M.; Khan, R.; Khan, F. Chemical composition and antioxidant activity of certain *Morus* species. *J. Zhejiang Univ. Sci. B* **2010**, *11*, 973–980. [CrossRef]
32. Lin, J.Y.; Tang, C.Y. Determination of total phenolic and flavonoid contents in selected fruits and vegetables, as well as their stimulatory effects on mouse splenocyte proliferation. *Food Chem.* **2007**, *101*, 140–147. [CrossRef]
33. Kim, S.Y.; Gao, J.J.; Lee, W.C.; Ryu, K.S.; Lee, K.R.; Kim, Y.C. Antioxidative flavonoids from the leaves of *Morus alba*. *Arch. Pharmacal Res.* **1999**, *22*, 81–85. [CrossRef]
34. Enkhmaa, B.; Shiwaku, K.; Katsube, T.; Kitajima, K.; Anuurad, E.; Yamasaki, M.; Yamane, Y. Mulberry (*Morus alba* L.) Leaves and Their Major Flavonol Quercetin 3-(6-Malonylglucoside) Attenuate Atherosclerotic Lesion Development in LDL Receptor-Deficient Mice. *J. Nutr.* **2005**, *135*, 729–734. [CrossRef]
35. Chen, Z.; Zhu, C.; Han, Z. Effects of aqueous chlorine dioxide treatment on nutritional components and shelf-life of mulberry fruit (*Morus alba* L.). *J. Biosci. Bioeng.* **2011**, *111*, 675–681. [CrossRef]
36. Kimura, T.; Nakagawa, K.; Kubota, H.; Kojima, Y.; Goto, Y.; Yamagishi, K.; Oita, S.; Oikawa, S.; Miyazawa, T. Food-Grade Mulberry Powder Enriched with 1-Deoxynojirimycin Suppresses the Elevation of Postprandial Blood Glucose in Humans. *J. Agric. Food Chem.* **2007**, *55*, 5869–5874. [CrossRef] [PubMed]
37. Awad, N.E.; Seida, A.A.; Hamed, M.A.; Mahmoud, A.H.; Elbatanony, M.M. Phytochemical and in vitro screening of some *Ficus* and *Morus* spp. for hypolipidaemic and antioxidant activities and in vivo assessment of *Ficus mysorensis* (Roth). *Nat. Prod. Res.* **2012**, *26*, 1101–1111. [CrossRef] [PubMed]
38. Cogato, A.; Meggio, F.; Migliorati, M.D.A.; Marinello, F. Extreme Weather Events in Agriculture: A Systematic Review. *Sustainability* **2019**, *11*, 2547. [CrossRef]

39. Ferrari, G.; Pezzuolo, A.; Nizami, A.S.; Marinello, F. Bibliometric Analysis of Trends in Biomass for Bioenergy Research. *Energies* **2020**, *13*, 3714. [CrossRef]
40. Bastian, M.; Heymann, S.; Jacomy, M. Gephi: An Open Source Software for Exploring and Manipulating Networks. *Proc. Int. AAAI Conf. Web Soc. Media* **2009**, *3*, 361–362.
41. Mahesh, V.G.V.; Raj, A.N.J.; Celik, T. Silkworm cocoon classification using fusion of Zernike moments-based shape descriptors and physical parameters for quality egg production. *Int. J. Intell. Syst. Technol. Appl.* **2017**, *16*, 246–268. [CrossRef]
42. Xia, Q.; Zhou, Z.; Lu, C.; Cheng, D.; Dai, F.; Li, B.; Zhao, P.; Zha, X.; Cheng, T.; Chai, C.; et al. A Draft Sequence for the Genome of the Domesticated Silkworm (*Bombyx mori*). *Science* **2004**, *306*, 1937–1940. [CrossRef]
43. Mita, K.; Kasahara, M.; Sasaki, S.; Nagayasu, Y.; Yamada, T.; Kanamori, H.; Namiki, N.; Kitagawa, M.; Yamashita, H.; Yasukochi, Y. The genome sequence of silkworm, *Bombyx mori*. *DNA Res.* **2004**, *11*, 27–35. [CrossRef]
44. Shimomura, M.; Minami, H.; Suetsugu, Y.; Ohyanagi, H.; Satoh, C.; Antonio, B.; Nagamura, Y.; Kadono-Okuda, K.; Kajiwara, H.; Sezutsu, H.; et al. KAIKObase: An integrated silkworm genome database and data mining tool. *BMC Genom.* **2009**, *10*, 486. [CrossRef]
45. Eriksson, T.; Andere, A.A.; Kelstrup, H.; Emery, V.J.; Picard, C.J. The yellow mealworm (*Tenebrio molitor*) genome: A resource for the emerging insects as food and feed industry. *J. Insects Food Feed.* **2020**, *6*, 445–455. [CrossRef]
46. Maeda, S.; Kawai, T.; Obinata, M.; Fujiwara, H.; Horiuchi, T.; Saeki, Y.; Sato, Y.; Furusawa, M. Production of human α-interferon in silkworm using a baculovirus vecto. *Nature* **1985**, *315*, 592–594. [CrossRef] [PubMed]
47. Ogawa, S.; Tomita, M.; Shimizu, K.; Yoshizato, K. Generation of a transgenic silkworm that secretes recombinant proteins in the sericin layer of cocoon: Production of recombinant human serum albumin. *J. Biotechnol.* **2007**, *128*, 531–544. [CrossRef] [PubMed]
48. Wang, F.; Xu, H.; Yuan, L.; Ma, S.; Wang, Y.; Duan, X.; Duan, J.; Xiang, Z.; Xia, Q. An optimized sericin-1 expression system for mass-producing recombinant proteins in the middle silk glands of transgenic silkworms. *Transgenic Res.* **2013**, *22*, 925–938. [CrossRef] [PubMed]
49. Liu, C.; Ren, Z.H.; Wang, H.Z.; Yang, P.Q.; Zhang, X.L. Analysis on gender of silkworms by MRI technology. In Proceedings of the BioMedical Engineering and Informatics: New Development and the Future—Proceedings of the 1st International Conference on BioMedical Engineering and Informatics, BMEI 2008, Sanya, China, 27–30 May 2008; Volume 2, pp. 8–12. [CrossRef]
50. Yogananda Murthy, V.N.; Ramesh, H.L.; Lokesh, G.; Munirajappa, D.Y.B.R.; Yadav, B.R. Leaf quality evaluation of ten mulberry (*Morus*) germplasm varieties through phytochemical analysis. *Int. J. Pharm. Sci. Rev. Res.* **2013**, *21*, 182–189.
51. Tomital, M.; Munetsuna, H.; Sato, T.; Adachi, T.; Hino, R.; Hayashi, M.; Shimizu, K.; Nakamura, N.; Tamura, T.; Yoshizato, K. Transgenic silkworms produce recombinant human type III procollagen in cocoons. *Nat. Biotechnol.* **2003**, *21*, 52–56. [CrossRef] [PubMed]
52. Nongonierma, A.B.; FitzGerald, R.J. Unlocking the biological potential of proteins from edible insects through enzymatic hydrolysis: A review. *Innov. Food Sci. Emerg. Technol.* **2017**, *43*, 239–252. [CrossRef]
53. Kim, H.W.; Setyabrata, D.; Lee, Y.J.; Jones, O.G.; Kim, Y.H.B. Pre-treated mealworm larvae and *silkworm pupae* as a novel protein ingredient in emulsion sausages. *Innov. Food Sci. Emerg. Technol.* **2016**, *38*, 116–123. [CrossRef]
54. Grabowski, N.T.; Klein, G. Microbiology of processed edible insect products—Results of a preliminary survey. *Int. J. Food Microbiol.* **2017**, *243*, 103–107. [CrossRef]
55. Hsieh, F.K.; Yu, S.J.; Su, S.Y.; Peng, S.J. Studies on the thermotolerance of the *silkworm, Bombyx mori*. *Chin. J. Entomol.* **1995**, *15*, 91–101.
56. Ueda, S.; Lizuka, H. Studies on the effects of rearing temperature affecting the health of silkworm larvae and upon the quality of cocoons-1 Effect of temperature in each instar. *Acta Sericologia Jpn.* **1962**, *41*, 6–21.
57. Dixit, M.A.; Kulkarni, A.; Raste, N.; Bhandari, G. Intelligent control system for sericulture. In Proceedings of the 2015 International Conference on Pervasive Computing (ICPC), Pune, India, 8–10 January 2015. [CrossRef]
58. Yang, J.; Guo, X.; Li, Y.; Marinello, F.; Ercisli, S.; Zhang, Z. A survey of few-shot learning in smart agriculture: Developments, applications, and challenges. *Plant Methods* **2022**, *18*, 28. [CrossRef] [PubMed]

insects

MDPI

Review

The Contribution of Silk Fibroin in Biomedical Engineering

Cristian Lujerdean *, Gabriela-Maria Baci *, Alexandra-Antonia Cucu and Daniel Severus Dezmirean

Faculty of Animal Science and Biotechnology, University of Animal Sciences and Veterinary Medicine Cluj-Napoca, 400372 Cluj-Napoca, Romania; antonia.cucu@usamvcluj.ro (A.-A.C.); ddezmirean@usamvcluj.ro (D.S.D.)

* Correspondence: cristian.lujerdean@usamvcluj.ro (C.L.); gabriela-maria.baci@usamvcluj.ro (G.-M.B.)

Simple Summary: In the medical area and beyond one of the most important biomaterials is the silk fibroin (SF) produced by the *Bombyx mori* L. silkworm. This outstanding biopolymer has received great attention from researchers due to its unique properties. Among them, the most important characteristic of SF is the high level of biocompatibility with the human organism. The biocompatibility, high mechanical strength, biodegradability and the biologically active properties have put SF in the spotlight, and thus numerous biomaterials have been developed. Furthermore, by using genetic engineering, biomaterials have been obtained that exhibit enhanced properties. In a wide range of studies, SF was used in order to develop sponges, hydrogels, nanospheres and films. By using SF-based biomaterials, tremendous progress has been made in tissue engineering and cancer therapy. In the specialized literature, various methods have been described regarding both extraction and processing of SF as a functional material. Moreover, SF-based biomaterials have been successfully obtained by using ecological methods of processing. Therefore, SF is considered to be the foremost green material.

Abstract: Silk fibroin (SF) is a natural protein (biopolymer) extracted from the cocoons of *Bombyx mori* L. (silkworm). It has many properties of interest in the field of biotechnology, the most important being biodegradability, biocompatibility and robust mechanical strength with high tensile strength. SF is usually dissolved in water-based solvents and can be easily reconstructed into a variety of material formats, including films, mats, hydrogels, and sponges, by various fabrication techniques (spin coating, electrospinning, freeze-drying, and physical or chemical crosslinking). Furthermore, SF is a feasible material used in many biomedical applications, including tissue engineering (3D scaffolds, wounds dressing), cancer therapy (mimicking the tumor microenvironment), controlled drug delivery (SF-based complexes), and bone, eye and skin regeneration. In this review, we describe the structure, composition, general properties, and structure–properties relationship of SF. In addition, the main methods used for ecological extraction and processing of SF that make it a green material are discussed. Lastly, technological advances in the use of SF-based materials are addressed, especially in healthcare applications such as tissue engineering and cancer therapeutics.

Keywords: silk fibroin; biocompatibility; green material; biomaterial; 3D scaffolds; cancer therapy

Citation: Lujerdean, C.; Baci, G.-M.; Cucu, A.-A.; Dezmirean, D.S. The Contribution of Silk Fibroin in Biomedical Engineering. *Insects* **2022**, *13*, 286. https://doi.org/10.3390/insects13030286

Academic Editors: Silvia Cappellozza, Morena Casartelli, Federica Sandrelli, Alessio Saviane and Gianluca Tettamanti

Received: 4 February 2022
Accepted: 10 March 2022
Published: 14 March 2022

1. Introduction

In the past years, research in the biomaterials field has expanded considerably through unprecedented progress and through the introduction of a significant number of innovative materials [1]. As a result, biomaterials currently have a widespread application in medicine due to their specific combination of resistance and biological compatibility. Their ability to induce tissue and organ regeneration has made it an extraordinary option for therapeutic treatments [2,3]. Moreover, modern medicine has been increasingly using various synthetic or natural materials (generally called "biomaterials") for improving the quality of life and longevity of human beings. A biomaterial is intended to interact with biological systems to evaluate, treat, enlarge, or replace any tissue, organ, or function of the body.

Numerous biomaterials are used in the human body, including metals, ceramics, and synthetic and natural polymers [4]. Of these, natural polymers (biopolymers) form the largest class of biomaterials used in medicine, possessing a number of unique properties covering a wide variety of clinical applications, especially in regenerative medicine and tissue engineering [5]. At the same time, they are able to mimic the structure of human tissue due to their physical and chemical resemblance, and most importantly, they are less toxic to native healthy tissue and more biocompatible compared to most synthetic materials [6,7].

Among biopolymers, silk fibroin (SF) has attracted more and more attention from life scientists all over the world from various fields such as chemistry, physics, engineering, biology, and last but not least, medicine. This protein is regarded as a remarkable bioactive material for tissue engineering applications due to its biocompatibility, biodegradability, and tunability [8], and it is studied in the production of 3D scaffolds mixed with various materials (natural or synthetic polymers, drugs, nanoparticles, growth factors, bioactive molecules, etc.) that together promote wound healing [9–14]. In addition, it can be processed in a wide range of formats, including sponges, wafers, gauze, particles, hydrogels, fibers, and films [15], being feasible for many other applications. SF as a biomaterial displays remarkable biological and mechanical properties such as controlled biodegradability, high biocompatibility, optical transparency, flexibility, mechanical resistance and processability. It is one of the main natural fibers produced by the *Bombyx mori* L. silkworm, and the purification and further processing of this protein as regenerated silk fibroin (RSF) can be obtained by eco-friendly methods that reduce or remove the need for chemical solvents and energy-intensive equipment. These eco-properties combined with a water-based extraction and purification process make SF a promising material for sustainable manufacturing, enabling it to partially replace synthetic, plastic-based and non-biodegradable material [16,17].

In recent years, numerous studies have been conducted focusing on the widespread use of SF as a functional biomaterial in the biomedical field. This review is expected to provide information on superior mechanical and biological properties of SF, the processing methods as a green material, recent advances in tissue engineering where SF plays a major role in mimicking the microenvironment of cells, and the importance of SF-based biomaterials used in drug delivery and cancer therapy. Figure 1 presents a schematic image of the main SF application.

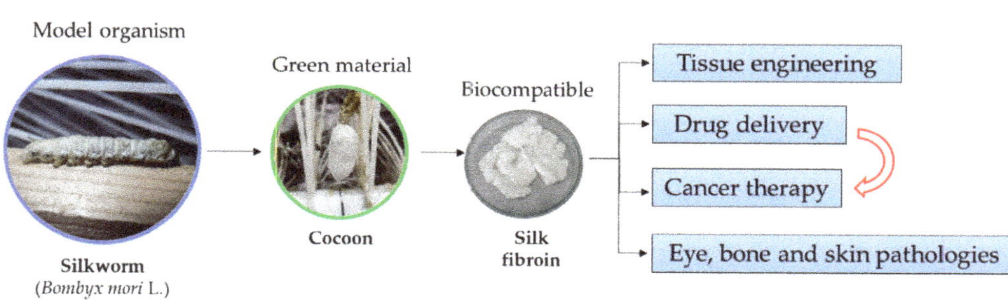

Figure 1. Schematic representation of SF main applications.

2. SF—Overview

Silks, as natural polymers, play a pivotal role for insects, being involved in the processes of survival and reproduction. Owning remarkable properties, silk has been the subject of various studies for more than a century [18,19], and it is appreciated by the scientific community in the field of biology, biotechnology and materials [20]. The main sources for this outstanding natural protein biopolymer are the spider (*Nephila clavipes*) and the silkworm (*B. mori*). *N. clavipes* exhibits certain behaviors that do not allow for the spider's large-scale harvesting [21]. Conversely, for more than 5000 years, silkworms have been domesticated and are currently being reared in numerous countries, such as China, India, and Japan, etc. As anticipated, the silk obtained from *B. mori* has been the most explored. Keeping this in mind, silkworms are the most feasible biological system in the context of silk production [22–24]. In the past, the silkworms were reared in order to obtain silk for textile applications; however, currently, silk receives considerable attention for its various purposes in life science areas [25].

The silk produced by *B. mori* is composed of two proteins, fibroin and sericin. Of the two proteins, fibroin (~400 kDa) is the core peptide. The silk fibers involve two fibroin filaments banded together by sericin, which has a glue-like activity [26,27]. Every fibroin molecule involves two chains, the fibroin heavy chain (FibH) and the fibroin light chain (FibL). Between the two chains, there is a disulfide bond through which the two proteins are linked. Besides FibH and FibL, SF also includes one more protein, namely fibrohexamerin/P25. The FibH chain contains multiple repetitive sequences that are involved in the development of anti-parallel structures. However, the FibL chain contains non-repetitive sequences that are responsible for the chain's hydrophilicity and fibroin's elasticity [28]. The molar ratio of FibH, FibL and P25 in fibroin molecules is 6:6:1 [29]. In a recent study, it was shown that the genes that encode the three fibroin proteins are directly regulated by the promoter-interacting proteins. Particularly, their data revealed that there are numerous proteins of this type that impact the ribosome and metabolism pathways [30]. The architecture and mechanical strength of SF assembled by *B. mori* are defined by the protein's amino acid structure and sequence. With regard to the fibroin's primary structure, three main amino acids are described, namely glycine, alanine and serine [31]. However, tyrosine, valine, histidine and tryptophan are also involved in fibroin's structure, but in smaller amounts. SF possesses crystalline domains, but it also contains amorphous domains [32]. This outstanding protein, SF, displays three distinct polymorphs (I, II, III) [26,33]. Silk I and silk II are the crystalline forms of fibroin. The first one represents the structure of fibroin before spinning and belongs to the orthorhombic crystal system, and the last one is the protein's solid state obtained after the spinning process, as part of the monoclinic crystal system. However, by treating silk I with potassium phosphate, it will be converted to the second polymorph [28].

The silk gland is the silk producing unit, with fibroin being assembled in the posterior silk gland (PSG) and sericin synthesized in the middle silk gland. After being synthetized, the proteins form an aqueous solution and are deposited into the silk gland lumen. When the process of spinning occurs, the proteins proceed through the anterior silk gland (ASG) duct [34].

3. SF Key Properties

SF exhibits numerous properties that convert this outstanding polymer into one of the most appreciated biomaterials (Figure 2). Currently, there are various processing forms for SF: scaffolds, sponges, or films, etc. [35–40]. It has been shown that the biocompatibility feature is influenced by the purification method and the purification process. Numerous studies have reported that SF's biocompatibility is maximized after the removal of the sericin [41,42]. However, by using sericin as a biomaterial, minimal inflammatory responses have been reported, indicating that the immune system is active by the combination of the two proteins [43].

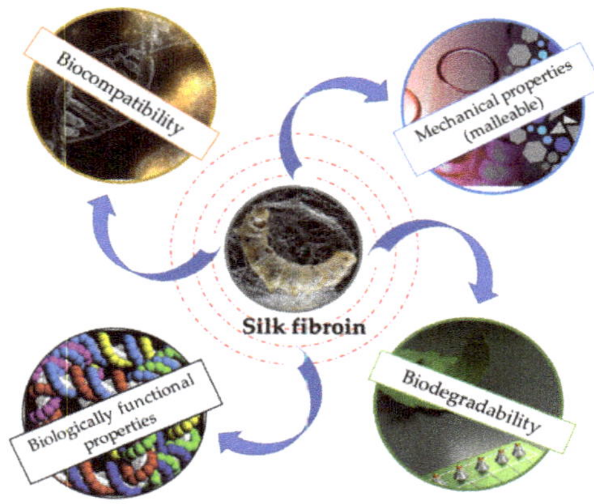

Figure 2. Schematic representation of SF properties.

3.1. Biocompatibility

The applicability of SF in the medical field dates to the 19th century. It has been intensively used as sutures due to its high biocompatibility with the human organism. There are numerous reports that have confirmed this extraordinary property of SF [44–51].

Burn wounds represent one of the most severe categories of injuries; they cause massive loss of life worldwide. When a burn injury occurs, the patient exhibits great vulnerability to numerous photogenic bacteria [52]. Through reoccurring bacterial infections, the healing process can be severely delayed. In order to prevent bacterial infections, the scientific community is currently focused on developing advanced dressings that can exhibit great protection against certain pathogens [53]. Yin et al. (2022) developed SF-based hydrogels and incorporated rhein into its structure, due to the fact that rhein possesses certain antibacterial and anti-inflammatory properties. After developing the target biomaterial, the authors determined the hydrogel biocompatibility by evaluating the hemolysis ratio observed after the blood was exposed to the hydrogel for 1 h. Their results showed favorable biocompatibility. Furthermore, the authors used mice as an animal model to demonstrate in vivo the protective activity against infections, thus highlighting the feasibility of using SF-based dressings for accelerated tissue healing [54].

One of the most common neurodegenerative disorders is Alzheimer's disease. In Alzheimer's disease, the hippocampus is the most vulnerable are of the brain [55,56]. Tang et al. (2008) [57] performed an in vitro study in order to investigate the biocompatibility of hippocampal neurons with SF biomaterials. The authors removed the sericin from the silk fibers and developed an SF-based substrate in order to use it to culture the hippocampal neurons. With regard to central nervous system therapy, one of the most used biomaterials is polyglycolic acid (PGA). Their findings showed that by using SF as a biomaterial, similar results with PGA fibers were observed, indicating that, in vitro, fibroin displays the essential conditions for a biomaterial [57]. Another in vitro study evaluated the biocompatibility of SF fibers and extract fluid, respectively, with rat dorsal root ganglia and Schwann cells derived from sciatic nerves. The data indicated that SF did not exhibit any cytotoxicity against either biological specimens. This progress is extremely important for the development of artificial nerve grafts [46].

Skin represents the largest and one of the most important organs of humans. It has numerous functions; thus, the wound healing process plays a pivotal role in human health [58]. In this direction, Zhang et al. (2017) [59] explored the effectiveness of using SF-based films in the skin repair process by using rabbits and porcine as animal models. This study highlighted the high biocompatibility and the effectiveness of SF as a biomaterial [59].

Macrophages play a key role in the regulation of the host's immune response. An ideal biomaterial should exhibit great biocompatibility; therefore, it should not induce appreciable macrophage response. The macrophage response had to be evaluated in order to confirm the biocompatibility of each biomaterial. In this direction, a research group investigated the macrophage response by adhering SF films to L929 murine fibroblasts. By performing this experiment, extremely low macrophage activation was observed, suggesting the high biocompatibility of SF as a biomaterial [60].

Moreover, in a recent study, SF produced by *B. mori* was used to develop bioink for three-dimensional (3D) bioprinting. Bioink is in the spotlight due to its applicability in the area of tissue and organ engineering. In this study, the authors used glycidyl methacrylate to chemically modify the SF in order to obtain the bioink. They acquired SF-based hydrogels by employing 3D bioprinting and demonstrated its great biocompatibility by using NIH/3T3 fibroblasts. Furthermore, in order to investigate the long-term biocompatibility of this specific biomaterial, the SF hydrogels were used for printing various complex biological structures, such as cartilaginous trachea or lungs. This research highlighted the feasibility of SF-based hydrogels for 3D bioprinting [61].

3.2. Biodegradability

Biodegradation is the process of disruption of natural polymers into numerous small-scale compounds. Regarding the applicability of biomaterials in the medical area, one of the most key requirements is biodegradability. Due to SF's biodegradable behavior, this natural biopolymer, together with chitosan or collagen, represents one of the most used biomaterials for tissue engineering applications and beyond. There are three main proteolytic enzymes that degrade SF, namely chymotrypsin, carboxylase and actinase. After the process of biodegradation occurs, both the structure and the molecular mass of SF suffer notable changes [62].

SF's degradation behavior is impacted by several factors. The most important factor that influences the degradation rate of SF-based biomaterials is pore size [63–65]. By comparing the degradation rate of SF-based scaffolds with large and small pore sizes, it has been demonstrated that the biomaterials that had large pores degraded faster than the ones with smaller pores. Conversely, scaffolds with higher pore density deteriorated slowly compared to the ones with lower pore density [66].

In a recent study, Sun et al. (2022) [67] developed SF sponges by using the ice-cold-induced phase separation-based method. Moreover, the cryo-sponges contained exosomes in order to obtain controlled delivery of the specific extracellular vesicles. Their data showed that the encapsulated exosomes maintained their characteristics, and for two months, the vesicles remained undigested. Furthermore, in vivo assessments showed that the sustained release promoted cell migration activity, but also stimulated the development of novel blood vessels. Moreover, their results showed that the SF sponges' biodegradability rate was correlated with the fibroin concentrations [67]. This study highlights the great potential of SF as a biodegradable biomaterial for sustained drug delivery.

Catto et al. (2015) [68] successfully designed SF-based tubular vascular grafts. They tested, in vitro, the degradation behavior of SF tubes and confirmed the fact that the enzymatic activity impacted the SF's crystallinity [68]. In another study, SF was associated with chitin in order to develop scaffolds that harbor silver nanoparticles that exhibit antimicrobial activity for wound dressings. The authors demonstrated the antimicrobial effect of nanocomposite scaffolds against several bacteria. The biodegradability of the obtained biomaterial was tested in vitro by using lysozyme. Their data showed that scaffolds exhibit great biodegradable behavior; moreover, the degradation products did not exhibit cytotoxicity [69].

Fan et al. (2009) [70] used pigs as an experimental model in order to evaluate the potential of employing SF scaffolds for clinical applications. By using an SF scaffold in pigs, the authors aimed to regenerate one of the key players for the strength of the knee joint, namely the anterior cruciate ligament. Furthermore, they developed mesenchymal stem cells-seeded SF-based scaffolds and implanted them in the experimental animal model. After 24 weeks, the mesenchymal stem cells displayed fibroblast morphology; moreover, great scaffold degradation was perceived. Conversely, after 24 weeks, the ultimate tensile load of the repaired anterior cruciate ligament was sustained [70].

In another study, 3D porous SF scaffolds were obtained by using two different processing methods, aqueous or organic solvent-based techniques. Additionally, the impact of several processing variables on SF scaffold behavior was investigated; specifically, the authors explored how the SF concentration and scaffolds' pore size influenced short- and long-term in vivo behavior of this biomaterial. As an animal experimental model, two types of rats were used, nude and Lewis. The aqueous-derived scaffolds compared with the organic solvent (hexafluoroisopropanol)-derived, exhibited more homogeneous degradability. However, the hexafluoroisopropanol-derived scaffolds were more impacted by the SF concentration and pore size; more specifically, a slower degradation rate was observed by using a higher SF concentration and smaller pore size [71].

3.3. Mechanical Properties

Silkworm SF has a toughness of 80–78 MJ m-3 and exhibits an ultimate tensile strength (UTS) of 300–740 MPa. The breaking strain of this natural polymer is 4–26%, and the Young's modulus of SF is 10–17 GPa. The presence of β-sheet crystallites determines the strength and stiffness of SF; thus, these structures impact the stability of this polymer [33]. Conversely, by removing the sericin, the UTS is impacted. In the biomaterial area, the mechanical strength and the elasticity of SF are remarkable properties. However, the properties of SF are impacted by several parameters, for instance the processing techniques and β-sheet patterns [72]. Nevertheless, it was reported that by influencing the molecular weight of SF, and its solvent composition, the stiffness of this polymer is impacted [73]. There are several studies that explored the impact of temperature on SF properties. By applying heat treatment on this natural polymer, its mechanical features changed, specifically the β-sheet patterns became predominant [32]. Conversely, numerous studies described complex strategies to improve the properties of SF by blending it with other polymers [74–76].

The extraordinary mechanical strength, elasticity and resistance of SF produced by *B. mori* are essential, especially for regenerative medicine and tissue regeneration applications [77]. Keeping this in mind, Chen et al. (2018) [78] developed 3D SF-based scaffolds in order to explore the potential of this biomaterial for tracheal defect repairing. They tested in vitro and in vivo the performance of SF scaffolds by analyzing the regeneration process of tracheal epithelium. Their data showed that SF scaffolds sustained tracheal mucosa regeneration within four weeks [78]. Another research group developed SF scaffolds by using high concentrations of SF aqueous solutions. Two different methods were combined for preparing the scaffolds, namely the salt-leaching and freeze-drying techniques. It was observed that the concentration of SF directly impacted the mechanical properties of the scaffolds. The authors revealed that after 30 days the biomaterials maintained their initial structure, suggesting that SF scaffolds are an appropriate choice for cartilage regeneration [79]. Table 1 points out the mechanical differences between several polymers.

Table 1. Comparison between the mechanical properties of several polymers.

Polymer	Source	UTS (MPa)	Modulus (GPa)	Breaking Strain (%)	References
SF	*B. mori*	300–740	10–17	4–26	[33]
Silk	*B. mori*	740	10	20	[80]
Silk	N. clavipes	875	10.9	16.7	[33]
Polylactide	Corn	69.8 ± 3.2	1777 ± 42	5.7 ± 0.3	[81]
Polyethylene-terephthalate	Synthetic	56	2.2	7300	[82]
Polypropylene	Synthetic	34.5	1.7	400	[82]
Cellulose	Bacteria	11.6 ± 0.8	180.3 ± 10.6	8.2 ± 0.6	[83,84]

3.4. Biologically Functional Properties

One of the most complex processes in the human body is wound healing. This process consists of four different phases, specifically hemostasis, inflammation, cell proliferation and resolution. However, there are numerous factors that affect this process [85]. In consequence, major efforts are being made for the development of biomaterials that promote wound healing, such as oxygenation, infections, or age and gender [86]. SF's extraordinary properties are continually being studied for numerous applications in the medical area, including the process of wound healing. Mrowiec et al. (2012) [87] investigated the impact of fibroin and sericin in the wound healing process. Furthermore, they explored the molecular basis of biological functional properties of the two proteins. For this purpose, the authors used breast cancer (MDA-MB-231) and epithelial (Mv1Lu) cells and performed the wound healing scratch assessment. Their results confirmed that the silk proteins induced cell migration. Moreover, they determined that fibroin and sericin are involved in two signaling pathways, namely ERK and JNK, which are members of the family of mitogen-activated protein kinases. By phosphorylating the c-Jun, the two kinases activate and upregulate it [87].

Besides the fact that SF promotes cell migration, it has been shown that this protein improves cell adhesion. Nikam et al. (2020) [88] investigated the impact of SF nanofibers on fibroblast-like cells. After culturing the target cells on silk nanofibers for one week, their data showed that SF significantly improved cell adhesion [88].

Due to SF's great characteristics, it has been shown that it is a promising biomaterial as a coating agent in terms of drug delivery [89]. In a recent study, SF was used by Kwon et al. (2021) [90] for coating probiotic strains. The aim of this study was to improve the stability of probiotics in the human body. By performing this experiment, the authors observed that the survival rate of the target probiotics that were SF coated was improved. Furthermore, by using SF as a biomaterial, cell adhesion was significantly increased [90].

3.5. Enhanced SF by Genetic Engineering

Aiming to meet the high demand of the medical field of biomaterials, numerous efforts have been made by the scientific community. For this purpose, in the last decades, genetic engineering has made extraordinary progress. In this direction, various studies have reported the development of enhanced SF fibers (Table 2) [91–93].

Spider silk exhibits superior mechanical properties that could be extraordinary in the biomaterials branch. It has been shown that the great toughness and elasticity of spider silk are determined by the presence of major ampullate silk protein (MaSp). The MaSp protein has a large and highly repetitive sequence, and due to this fact, it could not be obtained by using bacteria, yeast or plants [94]. Conversely, *B. mori* possesses the ability of spinning highly repetitive proteins. Xu et al. (2018) [24] obtained SF with enhanced mechanical properties by genetically manipulating the *B. mori* genome. The authors simultaneously knocked-out the *FibH* gene and knocked-in the major ampullate spidroin-1 gene from N. clavipes. On the same page, Kuwana et al. (2014) [94] obtained transgenic *B. mori* that expressed an altered dragline spider protein from *Araneus ventricosus*. By performing this, they obtained a tougher silk; furthermore, this mechanical property was improved by 53% [94].

Table 2. Advances in genetically manipulated *B. mori* for obtaining enhanced SF.

Exogenous Gene	Enhanced SF	Reference
Insulin-like Growth factor-1	Displays improved strength, elongation and tenacity	[95]
Human acidic fibroblast growth factor	Promotes cell proliferation	[96]
Human basic fibroblast growth factor and transforming growth factor-b1	Promotes cell proliferation and exhibits anti-inflammatory activity	[97]
Cecropin B and moricin	Antimicrobial activity	[98]
Green fluorescent protein and cecropin	Antibacterial activity and fluorescence	[99]
Enhanced green fluorescent protein, DsRed monomer fluorescent protein and monomeric Kusabira orange	Exhibits fluorescence	[100]
Laminin and fibronectin peptide adhesive fragments	Exhibits increased adhesive activity	[101]
Polyalanine motifs	Displays improved mechanical properties	[102]
Collagen and fibronectin	Improves cell adhesive properties	[103]

4. SF as a Green Material

SF is a protein produced by the silkworm *B. mori*, and it is an extensively researched material with applications in various fields, including the cosmetics industry, biomedicine and biomaterials [104,105]. Despite having a natural origin and being a biocompatible mechanically superior, biodegradable and functionalizable material [33], the processing methods of SF oftentimes include highly complex processes that can have a damaging impact on the environment. This disadvantage comes in contradiction with one of the main motivators of researching SF's potential as a biomaterial, which is its eco-friendly origin and properties (Figure 3). Moreover, the way that processing techniques can affec its microstructure and properties is still not fully understood.

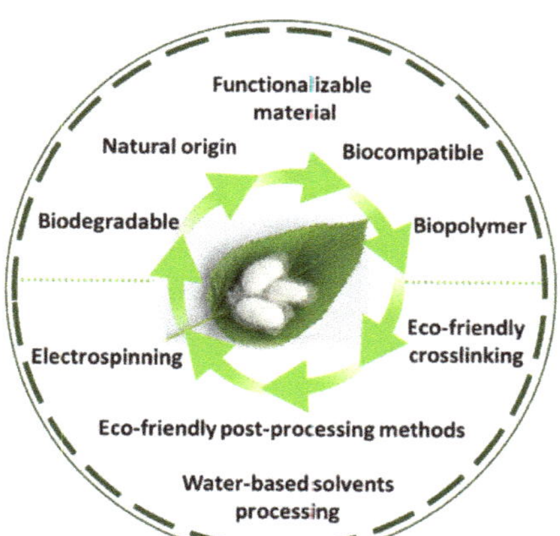

Figure 3. The main properties and methods of processing SF as a green material.

The classical regeneration of SF largely includes the use of toxic organic solvents including calcium chloride/formic acid [106–109], hexafluoroisopropanol (HFIP) [110–112] lithium salt solutions such as lithium thiocyanate (LiSCN) [113] and lithium bromide (9.3 M $LiBr$-H_2O) [114] and calcium nitrate/methanol ($Ca(NO_3)_2$)/CH_3OH) mixtures [115]. Unfortunately, these solvents are hazardous and tend to be environmentally unfriendly [116 Furthermore, they can directly affect the characteristics of SF and particularly impact the mechanical properties of the resulting silk biomaterials in subsequent processing [117,118

Along with legislation and evolving attitudes toward environmental issues, new (ecological) alternative solvents that are used for SF processing have appeared. These include 0.02 M sodium carbonate (Na_2CO_3) [114], N-methyl morpholine-N-oxide (NMMO) [119], ionic liquids [120–122] and the so-called Ajisawa's reagent consisting of calcium chloride/water/ethanol ($CaCl_2/H_2O/C_2H_5OH$) [123].

Water-based solvents have a lower environmental impact, and interest in their potential use in the processing of SF has gradually increased. Reizabal et al. (2021) [124] analyzed the effect of two dissolving techniques ($CaCl_2/H_2O$/EtOH ternary and $LiBr/H_2O$ binary solutions), three regeneration approaches (gas foaming, lyophilization, gelation), and one post-processing method (ethanol—EtOH) on porosity, macro- and microstructure, molecular and structural conformation, thermal behavior, water uptake and stability, and last but not least, heavy metal adsorption. Their results were promising, showing that it is possible to control SF properties through processing and that the green processing methods have the potential of expanding, even further, the applicability of SF-based biomaterials [124].

Bae et al. (2020) [125] prepared a water-soluble SF derivative by the methacrylation (introduction of glycidyl methacrylate—GMA) of SF's reactive side chains ($-NH_2$, $-OH$, $-COOH$). Through a green, eco-friendly electrospinning process using distilled water as a spinning solvent, GMA-modified SF water-soluble ultrathin fibers were obtained. Next, a structural change from random coils to b-sheets was facilitated by the treatment with an aqueous ethanol solution in order to insolubilize the fibers, and chemical crosslinking improved the water resistance. Electrospinning is a suitable alternative for the green processing of SF without the use of toxic substances [125]. Another study by Yang et al. (2017) [10] reported the feasibility of developing natural green composite matrices based on SF and MN (Manuka honey), used as an antimicrobial dressing in wound healing (ulcer). The SF was dissolved in a ternary solvent system of a $CaCl_2/H_2O/CH_3CH_2OH$ solution (1:8:2 in molar ratio) and was mixed with various amounts of MH over 12 h to obtain stable spinning solutions. Subsequently, the mixtures loading with different amounts of MH were manufactured by a process of green electrospinning, where the ecological damage associated with solvents used is often overlooked [126]. In this regard, SF played a key role in indicating permeability of these membranes and in biocompatibility.

Freezing-induced silk I crystallization is another entirely green alternative for obtaining a water-insoluble SF material with the silk I crystalline structure from an SF aqueous solution. Crystallization induction allows for the obtainment of a water-insoluble silk I structure without any highly processing approaches or the use of organic solvents. Through a freezing-generated concentration and thermodynamically driven assembly, the silk I crystallization process can be controlled and can be environmentally friendly [127].

Yang et al. (2016) [128] improved the strain performance, up to 120%, of a green electrospun SF nanofiber through an eco-friendly 24 h process of hyaluronic acid (HA)/1-ethyl-3-(3-dimethylaminopropyl) carbodiimide (EDC)/N-hydroxysuccinimide (NHS)-crosslinking. Not only did their SF nanofibrous matrices have better tensile performance than the matrices treated conventionally (ethanol soaking or fumigating), but the hydrophilicity was higher, and the cytocompatibility was good. Thus, HA/EDC/NHS-crosslinking might be an eco-friendly option to improve the mechanical properties of SF nanofibers, which are the main restriction in their use as tissue engineering scaffolds [128].

Fei et al. (2013) [129] used SF as a biotemplate to produce silver nanoparticles in situ under light (both incandescent light and sunlight) at room temperature. The composite solution (RSF-AgNPs) was prepared by an ecological method where the $AgNO_3$ powder was added to RSF solution and mixed until a transparent RSF-$AgNO_3$ solution formed. The mixture solution obtained was then exposed under the light with an incandescent bulb and was incubated at room temperature for 24 h to produce an RSF-AgNPs composite solution. In this case, SF serves as the reducing agent of silver and the dispersing and stabilizing agent of the resulting silver nanoparticles. As the reaction does not need any other chemicals and only uses light as a power source, the synthetic route of silver nanoparticles reported here is environmentally friendly and energy saving.

Raho et al. (2020) [130] used a green synthesis approach to create a composite hydrogel (CoHy) of carboxymethylcellulose-Na (CMC-Na) stabilized and loaded with AgNPs. An RSF solution was obtained by dissolving the fibroin with a solution of $CaCl_2$ ($CaCl_2/CH_3CH_2OH/H_2O$) instead of LiBr and mixing it with a CMC-Na aqueous solution. The silver nitrate ($AgNO_3$) was weighed and slowly added to the solution. For a great crosslinking, the composite solution was sonicated and exposed to UV radiation (for green synthesis of AgNPs). In this work, the synthesis of the hydrogel (CoHy) was based on a green process with ecological methods and natural components such as SF that served as a reducing agent in the green synthesis of AgNPs [129,131].

Switching toward eco-friendlier processing techniques of SF has gained popularity in materials science research and has raised the awareness of the negative impact of more environmentally toxic approaches. More research is necessary in order to find the ideal methods for maximizing SF's potential as a biomaterial, with the lowest impact on the environment.

5. SF and Tissue Engineering

The purpose of tissue engineering is to combine cells with scaffold materials and growth factors in order to help the regeneration of or even replace damaged tissue and/or organs. The need for a biomaterial matrix that could aid in the development of viable and biologically active tissue, either in vivo or in vitro, has increased in the last few years [132]. It has been proven that SF stimulates the attachment and growth of human cells [133]; thus, it is not surprising that it became a Food and Drug Administration (FDA)-approved biomaterial that can successfully be used as a scaffold in tissue engineering due to its unique properties and complex structure [2].

In order for it to be successfully used in tissue engineering, SF has to undertake a dissolution process, with different morphologies of the regenerated SF observed according to the different solvent systems used [134]. SF has a tightly packed structure; thus, it is insoluble in most of the solvents used for the dissolution of polymers for biotechnology applications, including water. Obtaining an SF solution requires the use of near-saturated bromides or thiocyanates and dialysis, after which the solution is unstable and hard to store for a longer time [135].

Additionally, multiple techniques have been used to process SF, including chemical (surface modifications only, entire material modification, and sonochemical methods) and enzymatic approaches (surface/bulk material modifications) [135]. From gas plasma treatments [136] and UV irradiation to more complex systems where SF films were first aminated by ammonia, NH_3, and then covalently sulfonated by sulfur dioxide [137], chemical modifications of SF are still of interest, despite their disadvantages. Biochemical enzymatic treatments of SF that have been studied include protease XIV and a-chymotrypsin treatment of SF for corneal regeneration [138], proteinase K for a porous network by a co-precipitation process with hydroxyapatite solution [139], and collagenase IA [140]. Furthermore, the enzymatic grafting of biomolecules such as lactoferrin [141] and chitosan was also described [142], and they are beginning to reveal their potential in tissue engineering.

Among the multiple and complex applications of SF in tissue engineering, SF dressings are a viable and promising option for the improvement of wound healing strategies through tissue engineering techniques [14]. Through the use of SF in developing sponges [143], hydrogels [144,145], SF micro- and nanoparticles [146], nanofibrous matrices [147], scaffolds, and composite films [148], SF and its potential applications in tissue engineering have been intensely studied in the last few years.

SF dressings can also improve the management of chronic wounds, as shown by Li et al. (2017) [149]. They prepared a functionalized SF dressing with topical bioactive insulin release by coaxial electrospraying of aqueous SF solution under mild processing conditions. This was based on the ability of insulin to stimulate cell migration and certain wound recovery methods [150]. The encapsulated insulin in the inner layer of the SF microparticles could thus be gradually released up to 28 days. The therapeutic effects of these microparticles loaded into an SF sponge (obtained by multilayer loading and freeze-drying method) was evaluated in vivo on dorsal full-thickness wounds of diabetic Sprague–Dawley rats. The degree of wound healing by measuring the area of wound closure showed the promotion of wound healing through accelerated wound closure and stimulation of vascularization and collagen deposition [149].

Moreover, processed SF after extraction of silk–sericin was also studied as a potential material for tissue engineering of the anterior cruciate ligament (ACL). A six-cord silk wire-rope matrix was designed that, based on ultimate tensile strength, linear stiffness, yield point, and percent elongation at break, had similar mechanical properties with the human ACL cells. This silk matrix supported both the attachment, as well as the expansion and differentiation, of adult human progenitor bone marrow stromal cells [132].

A highly porous 3D silk scaffold was designed by Wang et al. (2005) [151] and was combined with autologous adult mesenchymal stem cells (MSCs) in an attempt to perform in vitro cartilage tissue engineering. After 3 weeks of cell cultivation, RT-PCR analysis for cartilage-specific ECM gene markers and histological and immunohistochemical evaluations of cartilage-specific ECM components showed that the MSCs took the chondrogenic pathway with zonal architecture, including spatial cell arrangement and type II collagen distribution similar to native cartilage tissue [151]. Since cartilage has a limited self-repair capacity, and the current approaches for treating cartilage damage are not good enough for restoring the lost functionality, the potential of silk scaffolds in cartilage tissue engineering is of great importance and should be further evaluated in vivo. More recently, a 4D bioprinting system including SF was designed by Kim et al. (2020) [152]. Their system was based on digital light processing 3D bioprinting technology and photopolymerizable SF bioink. After obtaining a 4D-bioprinted silk hydrogel and using it to make trachea mimetic tissue, they successfully implanted it in the damaged trachea of an 8-week-old rabbit and proved the applicability of this new bioprinting system in vivo [152].

In another study, Mallepally et al. (2015) [153] used CO_2-assisted acidification for the synthesis of SF hydrogels that were subsequently converted to SF aerogels using the salt-leaching method [154]. Moreover, these SF aerogels were structurally, mechanically and in vitro biologically characterized using human foreskin fibroblast cells. The results showed that the aqueous SF concentration plays an important role in tuning the morphology and textural properties of the SF aerogels. These presented a significantly higher surface area and pore volume compared to the freeze-dried SF scaffolds. Furthermore, the SF-aerogels presented are highly porous with an interconnected network of nanofibrils, contributing to the success of cytocompatibility with human foreskin fibroblasts cells and their propagation. The SF aerogel scaffolds obtained by supercritical CO_2 have the potential for applications in tissue engineering, and the method of their synthesis can be used as an alternative method for 3D scaffold preparation.

Pacheco et al. (2020) [155] reported the use of several biopolymers in the development of SF/CS/SA multilayer membranes as a system for controlled drug release in wound healing. In this regard, the aim of the paper was to combine, in a single-composite material, the mechanical and biocompatibility properties of SF, the antimicrobial action of chitosan (CS) and the ideal exuded absorption of alginate (AS), thus achieving a synergistic effect. The membranes were prepared by pouring, where diclofenac sodium was incorporated as a model drug, into the chitosan solution before the solvent evaporated, which was stored in the middle layer of the membrane. The results showed that the incorporation of the drug did not affect the mechanical, thermal or barrier properties. Drug release was evaluated in vitro using a simulated solution of a wound exudate at 37 °C, where Fickian diffusion behavior was shown to be the dominant release mechanism. The results supported the idea that these multilayer membranes (SF/CS/SA multilayer membranes) could serve in biomedical applications as high-performance wound dressings.

Among the varieties of materials tested in the field of tissue engineering, SF continues to be of increasing interest in the medical field, this being a promising material in scaffold (3D) fabrication, mimicking the natural extracellular matrix (Figure 4). Furthermore, the ease of processing, the excellent biocompatibility, the remarkable mechanical properties and the personalized degradability of SF make it a competitive material in tissue engineering and regenerative medicine.

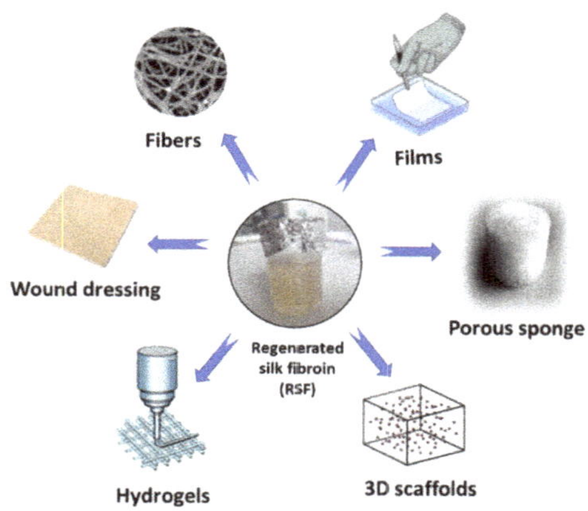

Figure 4. SF as a functional biomaterial for the tissue engineering field.

6. SF Involved in the Treatment of the Eye, Bone and Skin Regeneration

6.1. SF Involved in the Treatment of the Eye

The cornea is a half-millimeter thick structure with three types of tissues: corneal epithelium (anterior to an acellular area—Bowmann's layer), avascular corneal stroma with mesenchymal cell types (keratocyte), and an innermost single layer of corneal endothelial cells (posterior to an acellular zone—Descemet's membrane).

Harkin et al. (2011) [156] hypothesized that, among the complex eye structures that SF could be used as for a bioengineered tissue replacement would be the corneoscleral limbus, the corneal stroma, the corneal endothelium, and Ruysch's complex (the outer blood-retinal barrier). For the corneoscleral limbus, they believe that a composite material such as a prosthetic basement membrane could be used to grow epithelium on, which would support the exchange of nutrients and regulatory substances that must be present for the maintenance of the stem cells. Overlying this prosthetic basement membrane, they proposed a 3D scaffold that is more biomimetic for the stromal cells [156]. Although, for the corneoscleral limbus, transplant from a deceased donor has been associated with a high rejection rate [157], this type of procedure seems to be highly successful for corneal stroma reparation. The risk of transmission of different diseases through a transplant makes it desirable that viable alternatives from bioengineered tissues should be available. In order for a biomaterial to contribute to corneal stroma regeneration, it should support the growth of an epithelial layer and provide a vehicle for the transplantation of keratocytes. However, they should also maintain a high level of transparency, unlike corneoscleral limbus reconstruction.

Two of the methods currently used for obtaining SF films for corneal applications are the centrifugal casting technique and the dry casting method. Lee et al. (2016) [158] used both of these methods and compared the results. Their results showed that the centrifugal method offers more benefits, with less surface roughness of the films, increased growth of corneal fibroblasts (primary human corneal keratocytes) and better tensile strength and transparency [158]. Thus, the centrifugal casting technique is a promising method in tissue engineering of SF films for corneal regeneration.

The potential of SF as a biomaterial for corneal epithelial cell sheet generation has further been evaluated by Liu et al. (2012) [159], who compared the biological cell behavior of human and rabbit primary and immortalized corneal epithelial cells on SF and denuded human amniotic membrane (AM). The cells adhered to and proliferated on both SF and AM and expression of DNp63a, a progenitor cell marker [160], and keratin 3/12, a differentiation marker [161], existed on both cell culture types. Moreover, after being cultured at the air–liquid interface for 7 days, cells on the SF formed a stratified graft with a 2- to 3-cell layering with compact columnar cells on the basal layer and squamous cells present on the apical layers [159]. Since no cytotoxic response or inhibition of cell growth was observed in the SF cultures, their study strongly supports the high biocompatibility of SF biomaterial to corneal epithelial cells and their potential for corneal regeneration.

Lawrence et al. (2009) [162] designed 2-micromillimeter thick, surface-patterned silk films with 0.5–5.0-micromillimeter pores and demonstrated human and rabbit corneal fibroblast proliferation, alignment and corneal extracellular matrix expression in both 2D and 3D cultures on these films. The thickness was carefully chosen to imitate corneal collagen lamellae dimensions, whereas the surface pattern was meant to guide cell alignment. Pores were used for the enhancement of translamellar diffusion of nutrients and promotion of cell–cell interactions [162]. Their results further strengthen the potential of SF in bioengineered tissues for corneal injuries and pathologies.

The eye is an extremely complex organ, and its functionality in normal parameters highly influences the quality of life. Multiple research branches are currently performing joint work with the sole purpose of improving the regeneration of eye components. In this context, SF has proven to be a suitable candidate for a biomaterial that can be used for tissue engineering of eye components, especially for the cornea. Despite the enthusiasm that SF has seeded, more research is needed in order to better understand the interaction of its properties as a biomaterial and the human living organism, especially within the eye.

6.2. SF Used in the Treatment of the Bone

Bone is a rigid tissue with the essential propriety of providing structural support and protection of certain internal organs while supporting muscular contraction and serving as a deposit for certain minerals [163]. While it heals and remodels without leaving gross scars, efficient therapeutic strategies are needed for the efficient management of bone injuries. Bone grafts were the main option for a while, but their multiple disadvantages [164,165] have encouraged further research for viable alternatives.

For this purpose, in a recent study, Zhao et al. (2022) [166] developed SF-based scaffolds. One of the key players when it comes to bone repair and healing processes is hydroxyapatite. Being the foremost mineral element in bones, the authors combined SF with hydroxyapatite and obtained porous scaffolds. Moreover, they developed poly(lactic-co-glycolic acid) (PLGA)-based microspheres for controlled release of naringin, which is known for its ability to promote stem cell differentiation into osteoblasts. The microspheres were attached to the developed scaffold. First, the authors used human bone mesenchymal stem cell cultures in order to investigate the impact of the developed biomaterial on stem cell differentiation potential. Their data showed that the microspheres inlaid into the developed scaffolds have great potential in promoting osteogenic differentiation of target cells. Second, they used rabbits as a model organism to evaluate the bone repair process; more specifically, they surgically induced significant lesions of distal femoral epiphysis. The results indicated that the obtained biomaterial has great potential in the bone regeneration area by promoting bone defect healing [166].

A study conducted by Meinel et al. (2005) [167] used SF to form porous scaffolds on which human mesenchymal stem cells (hMSC) were cultured for 5 weeks. Afterward, they used the tissue as an implant for mouse critical calvarial defects. After another 5 weeks, multiple analyses including gene expression analysis, biochemical assays, and X-ray diffractometry demonstrated that SF successfully induced advanced bone formation [167].

The properties of SF of providing an osteogenic environment for bone-related outcomes was also proven by Zhao et al. (2009) [168]. They used apatite-coated SF scaffolds seeded with osteogenically induced autologous bone marrow stromal cells (bMSCs). The graft was used to repair inferior mandibular border defects in a canine model in comparison with bMSCs and SF scaffolds (no apatite-coating), apatite-coated SF scaffolds alone, SF scaffolds alone, autologous mandibular grafts, and untreated blank defects. Four weeks after the operation, new bone formation was detected through histological and radiographic examinations, with the defects being completely repaired for the bMSCs together with the apatite-coated SF group [168]. Their favorable results suggest that SF scaffolds could successfully be used with an apatite coating in combination with osteogenically induced autologous bone marrow stromal cells for the management of bone defects through tissue engineering.

Hydroxyapatite-coated SF composites were proven through fluorescence microscopy and scanning electron microscopy (SEM) to sustain cell adhesion and proliferation of mesenchymal cells derived from EGFP-expressing transgenic rat bone marrow in 10-day cultures. The results were comparable to the adhesion and growth of the cells seeded on tissue culture polystyrene dishes, which are standard scaffolds for the culture of cells, and which strongly support the use of SF scaffolds for regenerative medicine with applications in bone-related pathologies [169].

Another novel nanocomposite containing SF that might be used as tissue engineering scaffold or bone replacement was obtained by Cui et al. (2007) [170]. During its synthesis in an aqueous solution containing SF, hydroxyapatite (HA) nanocrystallites nucleated and grew preferentially along the plane (002) of the HA crystal, suggesting that SF might have an essential role in the mineralization of HA nanocrystallites. Moreover, the resulting nanocomposite had a compression strength of 97.6 MPa, which is higher than that of woven bone [170]. This compression strength was much higher than the one found in the case of the SF-chitosan/nano-hydroxyapatite porous scaffolds by Wen et al. (2007) [171]. Although their scaffold had good bioactivity, proven in the simulated body fluid experiments, when a layer of bone-like apatite crystals that were randomly oriented formed on the scaffold surface, their compressive strength varied only from 0.26 to 1.96 MPa. Both the porosity and the compressive strength seemed to change in relation to organic phase concentration and pre-freezing temperatures, and the authors concluded that a pre-freezing temperature of -80 °C and 5% organic concentration would be the best choices for bone tissue engineering [171].

The multiple properties and functions that bones have in the body make it necessary to have more than one viable option for bone regeneration, especially in the context of the high frequency of accidents involving bony structures. Considering the disadvantages of bone grafts and the impact that post-traumatic bone lesions can have on the quality of life through the impairment it causes, SF research as a potential tissue engineering biomaterial for bone-replacement structure synthesis has recently gained popularity. Although this potential still has to be tested in vivo, the properties of SF in current research, as well as the in vitro experimental work that has been performed, suggest that this is the proper direction to go.

6.3. *SF Involved in the Treatment of Skin Regeneration*

Skin has a complex function, including protection, thermo-regulation and a barrier between the outside of the body and the underlying tissues. Because skin lesions are frequent, over time, multiple approaches have been suggested for the proper treatment of injuries and pathologies developed at this level. Skin grafts have multiple classifications (based on thickness, donor origin, etc.), and each has different advantages and disadvantages. Their limitations of size, donor site availability and tolerability make skin grafting an imperfect option, and they have encouraged the search for a bioengineered skin equivalent [172].

A complex SF-based formic acid-crosslinked 3D nonwoven scaffold was studied by Dal Pra et al. (2006) [173] and was proven to be a great candidate for human tissue engineering. Among the methods they used was scanning electron microscopy (SEM) and differential scanning calorimetry, as well as thermogravimetric and tensile strength studies. These showed that the scaffolds are a composite material with anisotropic SF fibers enclosed within an isotropic matrix of SF in film form. They also obtained fibers, which as well as films, were firmly crosslinked and water insoluble. The crosslinking was hypothesized to have been induced by formic acid treatment and the lack of water solubility to their ß-sheet crystalline structure. Both human embryonic kidney cells and human dermal fibroblasts were attached to the scaffolds within 3 h and started proliferating and colonizing the scaffolds after 24 h. The cultures kept growing, with an active metabolism and no sign of protein catabolism, for at least 15 weeks [173].

A chitosan/SF blend film was evaluated as a biomaterial for skin tissue engineering by Luangbudnark et al. (2012) [174]. Chitosan appears to have enhanced β-sheet conformation of fibroin as well as increased flexibility, enzyme degradation, and swelling index. Fourier transform infrared spectroscopy and differential scanning calorimetry analysis showed intermolecular interactions between chitosan and SF. Furthermore, XTT assay showed no toxicity of the chitosan/SF blend film on fibroblast cell cultures, which showed proliferation. The blend films had a percentage of elongation at the break between 6.4% and 14.4%, which seemed to increase proportionally with the chitosan content, while the tensile strength was between 52.8 and 58.3 MPa, independent of the chitosan proportion. Conversely, increasing the fibroin content of the CS/SF blend films led to lower flexibility [174].

Another type of recently studied blended scaffold of SF was in combination with human hair keratin with a potential use as a dermal substitute [175]. Three-dimensional (3D) blended scaffolds were fabricated by freeze-drying and were tested on the L929 mouse-fibroblastic cell line. Functional fibroblasts showed good biocompatibility and proliferation on the 3D scaffolds, with a greater expression of collagen type I proven by immunohistochemical staining methods. Cellular compatibility evaluated through MTT assay showed good cell viability for 14 days of culture. Further, these composite scaffolds supported cell attachment and proliferation along with intact extracellular matrix (ECM) deposition, specifically for Col I. These results demonstrate that SF–keratin blended scaffolds are promising substrates for dermal reconstruction, wound healing, and other biomedical applications [175].

Although, for the treatment of skin injuries, skin grafting has been studied and used for a long time, they have a limitation of size, are dependent on donor size availability, and have the potential to be rejected by the host organism. These limitations cannot be overlooked, especially in complex and extended burn-cases that require additional support and special care. The biocompatibility of SF, as well as its natural origin and abundance, give it the potential to overcome certain issues of skin grafts (Figure 5). Besides its potential use for bioactive wound dressings, SF might become an important resource for in vivo regeneration of the skin.

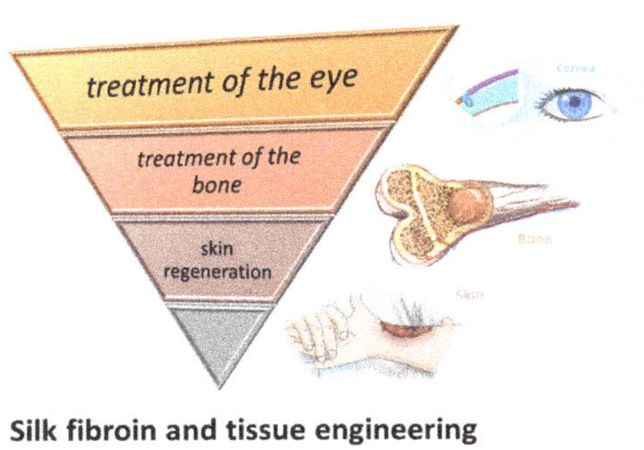

Figure 5. SF involved in the treatment of the eye, bone, and skin regeneration.

7. SF in Cancer Therapy

Globally, cancer is the leading cause of mortality [176]. Currently, chemotherapy is the standard therapy for numerous types of cancer. However, there are numerous disadvantages when it comes to chemotherapy, the most important being its high toxicity [177].

Keeping these aspects in mind, there is a major need for the development of drugs that exhibit reduced toxicity. Due to its extraordinary properties, SF has great potential as a drug delivery system. Another great advantage of using SF is its great feasibility of modification and the ease of processing [178,179]. Being extremely versatile, SF has been used to deliver numerous compounds, such as vaccines, proteins, therapeutic compounds and so on. Sun et al. (2019) [180] encapsulated doxorubicin in SF-based nanoparticles for exploring its feasibility in target cancer therapy. In addition, on the nanoparticles' surface, the authors covalently grafted folic acid in order to target the tumor cells. Their results showed that the prepared SF nanoparticles did not affect cell viability or proliferation. This approach displayed promising results, as the particles were able to target the tumor cells. Moreover the doxorubicin was controllably released [180].

In a recent study, Saqr et al. (2021) [181] developed SF nanoparticles that were loaded with docetaxel to target breast cancer cells. Their data revealed that the developed nanoparticles exhibited an enhanced cytotoxic effect against the targeted cancer cells [181]. Mottaghitalab et al. (2017) [182] encapsulated gemcitabine in SF nanoparticles in order to target lung cancer cells. As an experimental model animal, the authors used mice. The results of this approach showed that SF nanoparticles were able to reduce the side effect of gemcitabine. Furthermore, the therapeutic activity resulting from this formulation was higher than other approaches [182]. In the same year, Moin et al. (2021) [183] developed both nano- and micro-fibroin-based particles as drug delivery systems to encapsulate doxorubicin. The authors used two types of cancer cell lines, specifically MCF-7 and SAOS2, and one normal human cell line, namely HFF. By using the MTT assessment, they evaluated the toxicity exhibited by micro- and nanoparticles on cell lines. Their data showed that cell growth was inhibited. Furthermore, the impact of SF-based particles on *p53* expression was evaluated. Numerous studies have highlighted that *p53* overexpression plays a key role when administering doxorubicin to cancer patients; specifically, the overexpression if this gene is linked to higher sensitivity of malignant cell sensitivity to this chemotherapy agent. By using the real-time PCR method to evaluate *p53* expression, the doxorubicin-loaded SF nanoparticles induced higher *p53* expression in the malignant cell lines and decreased its expression in the normal cell line. Conversely, the doxorubicin-loaded SF microparticles decreased *p53* expression in the SAOS cell line and significantly increased it in the MCF-7 and HFF cell lines. Moreover, the authors investigated the reactive oxygen species levels after treating the cell lines with the developed biomaterials and observed that the reactive oxygen species levels were lower in two cell lines, namely MCF-7 and HFF. These findings highlight the great contribution of SF-based biomaterials in cancer therapy [183].

In 2022, a group of researchers successfully developed SF-based double-layer microneedles. The authors developed the delivery devices in order to encapsulate and administer triptorelin in a controlled released manner. Triptorelin is widely used in the treatment of prostate gland carcinomas due to its inhibitory effect on testosterone; however, it is an injectable suspension that exhibits certain disadvantages such as cold-chain storage or the administration method. After developing the SF-based microneedles and successfully encapsulating triptorelin in the microneedles' tip, the authors subcutaneously implanted the devices in rats in order to evaluate its therapeutic effects. The authors showed that by encapsulating triptorelin into SF microneedles, its half-life was extended compared with the half-time obtained after directly subcutaneously injecting the drug. In order to confirm the inhibitory effect of encapsulated triptorelin on testosterone concentration, as controls castrated and healthy rats were used. Their results showed that by using the developed microneedles to administer triptorelin, the testosterone levels were the same as the levels observed in the castrated rats, and the inhibitory effect was maintained for more than seven days. Conversely, by subcutaneously injecting the target drug, the repressive action of triptorelin was maintained for less than one day. These findings suggest that SF-based microneedles are a promising approach for delivering triptorelin [184].

Furthermore, SF is currently being used as a tumor model to investigate cancer biology. By using a tumor model to mimic the tumor microenvironment, valuable insights could be observed. Dondajewska et al. (2018) [185] used an SF-based scaffold to construct a breast cancer model. The model provided insights into the interactions between the stromal fibroblasts and cancer cells. This complex model represents a promising approach for expanding the understanding of tumor biology [185]. In another study, scaffolds were developed that contained both SF and chitosan to mimic a tumor's microenvironment. These scaffolds allow for the investigation of tumor drug resistance mechanisms and provide insights for the evaluation of therapeutic agents [186]. Table 3 shows numerous studies that successfully used SF in cancer therapy, and Figure 6 illustrates the use of SF in cancer therapy.

Table 3. SF-based biomaterials used in cancer therapy.

Biomaterial	Type of Cancer	Reference
SF hydrogels	Hepatocellular carcinoma	[187]
SF–Thelebolan matrix	Soft tissue carcinoma	[188]
Doxorubicin loaded SF nanoparticles	Brain cancer	[189]
SF–Sodium alginate nanocarriers	Colorectal cancer	[190]
SF-based metastasis model	Breast cancer	[191]
Triptolide–Celastrol-loaded SF nanoparticles	Pancreatic cancer	[192]
Alpha-mangostin loaded SF nanoparticles	Colon cancer; Breast cancer	[193]
SF rods	Breast cancer	[194]
Quercetin loaded SF nanoparticles	Breast cancer; Lung metastasis	[195]
Biliverdin–SF hydrogel	Glioma	[196]
Floxuridine-loaded SF nanospheres	Digestive tract cancer; Lung cancer	[197]
Curcumin-loaded SF nanoparticles	Breast cancer	[198]

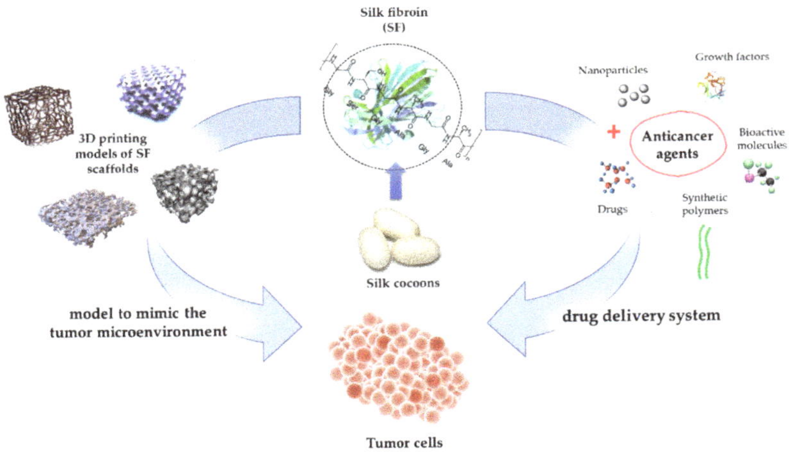

Figure 6. SF-based drug delivery and the production of 3D SF scaffolds for the growth of cancer cells.

8. Conclusions

SF represents the major structural element of the silk produced by *B. mori*. Numerous studies have demonstrated the unique properties of this protein. One of the most important characteristics is the great level of biocompatibility with human organisms. Moreover, SF has great mechanical properties and biodegradable behavior. Due to its unique properties, this protein has an extraordinary potential in the medical field and beyond. It has been used for a long time as a suture in surgeries, but currently, it is the main component of numerous biomaterials. Great advances have been made in the medical area due to SF's ability to generate numerous biomaterials, such as scaffolds or nanoparticles. Moreover, by applying different techniques, SF's features can be controlled and remodeled. In recent years, SF has gained attention due to the fact that it can be considered a green material.

SF is currently being used in various areas, and a wide range of techniques has been employed for obtaining different types of SF-based biomaterials. There have been numerous types of scaffolds obtained by using this natural polymer, such as hydrogels, films, nanospheres, sponges and coatings. With the current progress that is being made for a better understanding of SF's structure, properties, and processing methods, there are new opportunities for its applicability. Moreover, SF has been associated with a wide range of compounds in order to maximize its mechanical and biological functional properties. There have been various studies that have used SF for the cornea, bone, and its applications. Beside SF's extraordinary contribution to tissue engineering, this biomaterial has a great input in cancer therapy. It is currently being used as a coating agent in the treatment of several types of cancers, for instance in lung and breast cancer. Furthermore, molecular engineering has been applied in order to obtain enhanced SF for biomedical applications.

Author Contributions: Conceptualization, G.-M.B. and C.L.; resources, A.-A.C., C.L. and D.S.D.; writing—original draft preparation, C.L., G.-M.B., A.-A.C. and D.S.D.; writing—review and editing, A.-A.C., C.L. and G.-M.B.; supervision, G.-M.B. and C.L. All authors have read and agreed to the published version of the manuscript.

Funding: The publication was supported by funds from the National Research Development Projects to finance excellence (PFE)-14/2022-2024 granted by the Romanian Ministry of Research and Innovation.

Institutional Review Board Statement: Not applicable.

Data Availability Statement: Not applicable.

Conflicts of Interest: The authors declare no conflict of interest.

References

1. Ozkale, B.; Selman, M.; Mooney, D.J. Active Biomaterials for Mechanobiology. *Biomaterials* **2021**, *267*, 120497. [CrossRef] [PubMed]
2. Sun, W.; Gregory, D.A.; Tomeh, M.A.; Zhao, X. Silk Fibroin as a Functional Biomaterial for Tissue Engineering. *Int. J. Mol. Sci.* **2021**, *22*, 1499. [CrossRef] [PubMed]
3. Holland, C.; Numata, K.; Rnjak-Kovacina, J.; Seib, F.P. The Biomedical Use of Silk: Past, Present, Future. *Adv. Healthc. Mater.* **2019**, *8*, 800465. [CrossRef] [PubMed]
4. Sunija, A.J. Biomaterials and Biotechnological Schemes Utilizing TiO_2 Nanotube Arrays—A Review. In *Fundamental Biomaterials: Metals*; Elsevier Ltd: Amsterdam, The Netherlands, 2018; pp. 175–196. [CrossRef]
5. Shanmugam, K.; Sahadevan, R. *Bioceramics—An Introductory Overview*; Elsevier: Amsterdam, The Netherlands, 2018. [CrossRef]
6. Sionkowska, A. Current Research on the Blends of Natural and Synthetic Polymers as New Biomaterials: Review. *Prog. Polym. Sci.* **2011**, *36*, 1254–1276. [CrossRef]
7. Ullah, S.; Chen, X. Fabrication, Applications and Challenges of Natural Biomaterials in Tissue Engineering. *Appl. Mater. Today* **2020**, *20*, 100656. [CrossRef]
8. Wu, R.; Li, H.; Yang, Y.; Zheng, Q.; Li, S.; Chen, Y. Bioactive Silk Fibroin-Based Hybrid Biomaterials for Musculoskeletal Engineering: Recent Progress and Perspectives. *ACS Appl. Bio Mater.* **2021**, *4*, 6630–6646. [CrossRef]
9. Kostag, M.; Jedvert, K.; El, O.A. Engineering of Sustainable Biomaterial Composites from Cellulose and Silk Fibroin: Fundamentals and Applications. *Int. J. Biol. Macromol.* **2021**, *167*, 687–718. [CrossRef]
10. Yang, X.; Fan, L.; Ma, L.; Wang, Y.; Lin, S.; Yu, F.; Pan, X.; Luo, G.; Zhang, D.; Wang, H. Green Electrospun Manuka Honey/Silk Fibroin Fibrous Matrices as Potential Wound Dressing. *Mater. Des.* **2017**, *119*, 76–84. [CrossRef]
11. Tao, G.; Cai, R.; Wang, Y.; Liu, L.; Zuo, H.; Zhao, P.; Umar, A.; Mao, C.; Xia, Q.; He, H. Bioinspired Design of AgNPs Embedded Silk Sericin-Based Sponges for Efficiently Combating Bacteria and Promoting Wound Healing. *Mater. Des.* **2019**, *180*, 107940. [CrossRef]
12. Bakhsheshi-Rad, H.R.; Fauzi, A.; Aziz, M.; Akbari, M.; Hadisi, Z.; Omidi, M.; Chen, X. Development of the PVA/CS Nano Fibers Containing Silk Protein Sericin as a Wound Dressing: In Vitro and in Vivo Assessment. *Int. J. Biol. Macromol.* **2020**, *149*, 513–521. [CrossRef]

3. Zhou, L.; Yu, K.; Lu, F.; Lan, G.; Dai, F.; Shang, S.; Hu, E. Minimizing Antibiotic Dosage through in Situ Formation of Gold Nanoparticles across Antibacterial Wound Dressings: A Facile Approach Using Silk Fabric as the Base Substrate. *J. Clean. Prod.* **2020**, *243*, 118604. [CrossRef]
4. Patil, P.P.; Reagan, M.R.; Bohara, R.A. Silk Fibroin and Silk-Based Biomaterial Derivatives for Ideal Wound Dressings. *Int. J. Biol. Macromol.* **2020**, *164*, 4613–4627. [CrossRef]
5. Zakeri-Siavashani, A.; Chamanara, M.; Nassireslami, E.; Shiri, M.; Hoseini-Ahmadabadi, M.; Paknejad, B. Three Dimensional Spongy Fibroin Scaffolds Containing Keratin/Vanillin Particles as an Antibacterial Skin Tissue Engineering Scaffold. *Int. J. Polym. Mater. Polym. Biomater.* **2022**, *71*, 220–231. [CrossRef]
6. DeBari, M.L.; King, C.I.; Altgold, T.A.; Abbott, R.D. Silk Fibroin as a Green Material. *ACS Biomater. Sci. Eng.* **2021**, *7*, 3530–3544. [CrossRef]
7. Benfenati, V.; Toffanin, S.; Chieco, C.; Sagnella, A.; Virgilio, N. Di; Posati, T.; Varchi, G.; Natali, M.; Ruani, G.; Muccini, M.; et al. Silk Fibroin Based Technology for Industrial Biomanufacturing. In *Factories of the Future*; Springer International Publishing: Berlin/Heidelberg, Germany, 2019; pp. 409–430. [CrossRef]
8. Hu, W.; Lu, W.; Wei, L.; Zhang, Y.; Xia, Q. Molecular Nature of Dominant Naked Pupa Mutation Reveals Novel Insights into Silk Production in *Bombyx mori*. *Insect Biochem. Mol. Biol.* **2019**, *109*, 52–62. [CrossRef]
9. Song, J.; Chen, Z.; Liu, Z.; Yi, Y.; Tsigkou, O.; Li, J.; Li, Y. Controllable Release of Vascular Endothelial Growth Factor (VEGF) Bywheel Spinning Alginate/Silk Fibroin Fibers for Wound Healing. *Mater. Des.* **2021**, *212*, 110231. [CrossRef]
0. Pérez-Rigueiro, J.; Ruiz, V.; Cenis, J.L.; Elices, M.; Pugno, N.M. Lessons From Spider and Silkworm Silk Guts. *Front. Mater.* **2020**, *7*, 1–8. [CrossRef]
1. Xu, J.; Dong, Q.; Yu, Y.; Niu, B.; Ji, D.; Li, M.; Huang, Y.; Chen, X. Mass Spider Silk Production through Targeted Gene Replacement in *Bombyx mori*. *Proc. Natl. Acad. Sci. USA* **2018**, *115*, 8757–8762. [CrossRef]
2. Etebari, K.; Mirhoseini, S.Z.; Matindoost, L. A Study on Interaspecific Biodiversity of Eight Groups of Silkworm (*Bombyx mori*) by Biochemical Markers. *Insect Sci.* **2005**, *12*, 87–94. [CrossRef]
3. Abdelli, N.; Peng, L.; Keping, C. Silkworm, *Bombyx mori*, as an Alternative Model Organism in Toxicological Research. *Environ. Sci. Pollut. Res.* **2018**, *25*, 35048–35054. [CrossRef]
4. Xu, H.; O'Brochta, D.A. Advanced technologies for genetically manipulating the silkworm Bombyx mori, a model lepidopteran insect. *Proc. R. Soc. B Biol. Sci.* **2015**, *282*, 20150487. [CrossRef]
5. Zhu, Z.; Imada, T.; Asakura, T. Preparation and Characterization of Regenerated Fiber from the Aqueous Solution of *Bombyx mori* Cocoon Silk Fibroin. *Mater. Chem. Phys.* **2009**, *117*, 430–433. [CrossRef]
6. Marín, C.B.; Fitzpatrick, V.; Kaplan, D.L.; Landoulsi, J.; Guénin, E.; Egles, C. Silk Polymers and Nanoparticles: A Powerful Combination for the Design of Versatile Biomaterials. *Front. Chem.* **2020**, *8*, 1–22. [CrossRef]
7. Wang, F.; Guo, C.; Yang, Q.; Li, C.; Zhao, P.; Xia, Q.; Kaplan, D.L. Protein Composites from Silkworm Cocoons as Versatile Biomaterials. *Acta Biomater.* **2021**, *121*, 180–192. [CrossRef]
8. Qi, Y.; Wang, H.; Wei, K.; Yang, Y.; Zheng, R.Y.; Kim, I.S.; Zhang, K.Q. A Review of Structure Construction of Silk Fibroin Biomaterials from Single Structures to Multi-Level Structures. *Int. J. Mol. Sci.* **2017**, *18*, 237. [CrossRef]
9. Wang, J.; Yan, S.; Lu, C.; Bai, L. Biosynthesis and Characterization of Typical Fibroin Crystalline Polypeptides of Silkworm *Bombyx mori*. *Mater. Sci. Eng. C* **2009**, *29*, 1321–1325. [CrossRef]
0. Ma, Y.; Luo, Q.; Ou, Y.; Tang, Y.; Zeng, W.; Wang, H.; Hu, J. New Insights into the Proteins Interacting with the Promoters of Silkworm Fibroin Genes. *Sci. Rep.* **2021**, *11*, 1–12. [CrossRef]
1. Lotz, B.; Cesari, F.C. The Chemical Structure and the Crystalline Structures of *Bombyx mori* Silk Fibroin. *Biochimie* **1979**, *61*, 205–214. [CrossRef]
2. Vidya, M.; Rajagopal, S. Silk Fibroin: A Promising Tool for Wound Healing and Skin Regeneration. *Int. J. Polym. Sci.* **2021**, *2021*, 9069924. [CrossRef]
3. Koh, L.D.; Cheng, Y.; Teng, C.P.; Khin, Y.W.; Loh, X.J.; Tee, S.Y.; Low, M.; Ye, E.; Yu, H.D.; Zhang, Y.W.; et al. Structures, Mechanical Properties and Applications of Silk Fibroin Materials. *Prog. Polym. Sci.* **2015**, *46*, 86–110. [CrossRef]
4. Dong, Z.; Zhao, P.; Zhang, Y.; Song, Q.; Zhang, X.; Guo, P. Analysis of Proteome Dynamics inside the Silk Gland Lumen of *Bombyx mori*. *Sci. Rep.* **2016**, *6*, 1–10. [CrossRef] [PubMed]
5. Leem, J.W.; Fraser, M.J.; Kim, Y.L. Transgenic and Diet-Enhanced Silk Production for Reinforced Biomaterials: A Metamaterial Perspective. *Annu. Rev. Biomed. Eng.* **2020**, *22*, 79–102. [CrossRef] [PubMed]
6. Panda, D.; Konar, S.; Bajpai, S.K.; Arockiarajan, A. Thermodynamically-Consistent Constitutive Modeling of Aligned Silk Fibroin Sponges: Theory and Application to Uniaxial Compression. *Int. J. Solids Struct.* **2018**, *138*, 144–154. [CrossRef]
7. Michelle, G.; Agostini, M.; Moraes, D.; Cecília, A.; Rodas, D.; Zazuco, O.; Masumi, M. Hydrogels from Silk Fibroin Metastable Solution: Formation and Characterization from a Biomaterial Perspective. *Mater. Sci. Eng. C* **2011**, *31*, 997–1001. [CrossRef]
8. Bassani, G.A.; Vincoli, V.; Biagiotti, M.; Valsecchi, E.; Zucca, M.V.; Clavelli, C.; Alessandrino, A.; Freddi, G. A Route to Translate a Silk-Based Medical Device from Lab to Clinic: The Silk Biomaterials Srl Experience. *Insects* **2022**, *13*, 212. [CrossRef]
9. Ghalei, S.; Handa, H. A Review on Antibacterial Silk Fibroin-Based Biomaterials: Current State and Prospects. *Mater. Today Chem.* **2022**, *23*, 100673. [CrossRef]

40. Khademolqorani, S.; Tavanai, H.; Chronakis, I.S.; Boisen, A. The Determinant Role of Fabrication Technique in Final Characteristics of Scaffolds for Tissue Engineering Applications: A Focus on Silk Fibroin-Based Scaffolds. *Mater. Sci. Eng. C* **2021**, *122*, 111867 [CrossRef]
41. Jo, Y.; Kweon, H.; Kim, D.; Baek, K.; Chae, W.; Kang, Y.; Oh, J.; Kim, S.; Garagiola, U. Silk Sericin Application Increases Bone Morphogenic Protein-2/4 Expression via a Toll-like Receptor-Mediated Pathway. *Int. J. Biol. Macromol.* **2021**, *190*, 607–617 [CrossRef]
42. Teuschl, A.; Griensven, M. Van; Redl, H.; Teuschl, A.H.; Bioreactors, E. Sericin Removal from Raw *Bombyx mori* Silk Scaffolds of High Hierarchical Order. *Tissue Eng. Part C Methods* **2014**, *20*, 1–25. [CrossRef]
43. Gholipourmalekabadi, M.; Sapruc, S.; Samadikuchaksaraei, A.; Reis, R.L.; Kaplan, D.L.; Subhas, C.K. Silk Fibroin for Skin Injury Repair: Where Do Things Stand? *Adv. Drug Deliv. Rev.* **2019**, *153*, 28–53. [CrossRef]
44. Crakes, K.R.; Herrera, C.; Morgan, J.L.; Olstad, K.; Hessell, A.J.; Ziprin, P.; Liwang, P.J.; Dandekar, S. Efficacy of Silk Fibroin Biomaterial Vehicle for in Vivo Mucosal Delivery of Griffithsin and Protection against HIV and SHIV Infection Ex Vivo. *J. Int AIDS Soc.* **2020**, *23*, 1–12. [CrossRef]
45. Madden, P.W.; Klyubin, I.; Ahearne, M.J. Silk Fibroin Safety in the Eye: A Review That Highlights a Concern. *BMJ Open Ophtalmol* **2020**, *5*, 1–10. [CrossRef]
46. Yang, Y.; Chen, X.; Ding, F.; Zhang, P.; Liu, J.; Gu, X. Biocompatibility Evaluation of Silk Fibroin with Peripheral Nerve Tissues and Cells in Vitro. *Biomaterials* **2007**, *28*, 1643–1652. [CrossRef]
47. Zhang, J.; Huang, H.; Ju, R.; Chen, K.; Li, S.; Wang, W.; Yan, Y. In Vivo Biocompatibility and Hemocompatibility of a Polytetra Fl Uoroethylene Small Diameter Vascular Graft Modified with Sulfonated Silk Fibroin. *Am. J. Surg.* **2017**, *213*, 87–93. [CrossRef]
48. Tian, Y.; Wu, Q.; Li, F.; Zhou, Y.; Huang, D.; Xie, R.; Wang, X.; Zheng, Z.; Li, G. A Flexible and Biocompatible *Bombyx mori* Silk Fibroin/Wool Keratin Composite Scaffold with Interconnective Porous Structure. *Colloids Surf. B Biointerfaces* **2021**, *208*, 112080 [CrossRef]
49. Wang, D.; Wang, L.; Lou, Z.; Zheng, Y.; Wang, K.; Zhao, L.; Han, W.; Jiang, K.; Shen, G. Biomimetic, Biocompatible and Robust Silk Fibroin-MXene Film with STable 3D Cross-Link Structure for Flexible Pressure Sensors. *Nano Energy* **2020**, *78*, 105252. [CrossRef
50. Wang, S.; Zhang, Y.; Wang, H.; Dong, Z. Preparation, Characterization and Biocompatibility of Electrospinning Heparin-Modified Silk Fibroin Nanofibers. *Int. J. Biol. Macromol.* **2011**, *48*, 345–353. [CrossRef]
51. Choi, Y.; Cho, D.; Lee, H. Development of Silk Fibroin Scaffolds by Using Indirect 3D-Bioprinting Technology. *Micromachines* **2022** *13*, 43. [CrossRef]
52. Stoica, A.E.; Chircov, C.; Grumezescu, A.M. Hydrogel Dressings for the Treatment of Burn Wounds: An Up-To-Date Overview *Materials* **2020**, *13*, 2853. [CrossRef]
53. Zheng, L.; Li, S.; Luo, J.; Wang, X. Latest Advances on Bacterial Cellulose-Based Antibacterial Materials as Wound Dressings *Front. Bioeng. Biotechnol.* **2020**, *8*, 1–15. [CrossRef]
54. Yin, C.; Han, X.; Lu, Q.; Qi, X.; Guo, C.; Wu, X. Rhein Incorporated Silk Fibroin Hydrogels with Antibacterial and Anti Inflammatory Efficacy to Promote Healing of Bacteria-Infected Burn Wounds. *Int. J. Biol. Macromol.* **2022**, *201*, 14–19. [CrossRef
55. Mu, Y.; Gage, F.H. Adult Hippocampal Neurogenesis and Its Role in Alzheimer's Disease. *Mol. Neurodegener.* **2011**, *6*, 1–9 [CrossRef]
56. Zhong, S.; Wang, M.; Zhan, Y.; Zhang, J.; Yang, X.; Fu, S.; Bi, D.; Gao, F. Single-Nucleus RNA Sequencing Reveals Transcriptional Changes of Hippocampal Neurons in APP23 Mouse Model of Alzheimer's Disease. *Biosci. Biotechnol. Biochem.* **2020**, *84*, 919–926 [CrossRef]
57. Tang, X.; Ding, F.; Yang, Y.; Hu, N.; Wu, H.; Gu, X. Evaluation on in Vitro Biocompatibility of Silk Fibroin-Based Biomaterials with Primarily Cultured Hippocampal Neurons. *J. Biomed. Mater. Res. A* **2008**, *91*, 166–174. [CrossRef]
58. Reinke, J.; Sorg, H. Wound Repair and Regeneration. *Eur. Surg. Res.* **2012**, *49*, 35–43. [CrossRef]
59. Zhang, W.; Chen, L.; Chen, J.; Wang, L.; Gui, X.; Ran, J Silk Fibroin Biomaterial Shows Safe and Effective Wound Healing in Animal Models and a Randomized Controlled Clinical Trial. *Adv. Healthc. Mater.* **2017**, *6*, 1–16. [CrossRef]
60. Acharya, C.; Ghosh, S.K.; Kundu, S.C. Silk Fibroin Protein from Mulberry and Non-Mulberry Silkworms: Cytotoxicity, Biocompatibility and Kinetics of L929 Murine Fibroblast Adhesion. *J. Mater. Sci. Mater. Med.* **2008**, *19*, 2827–2836. [CrossRef]
61. Kim, S.H.; Yeon, Y.K.; Lee, J.M.; Chao, J.R.; Lee, Y.J.; Seo, Y.B.; Sultan, T.; Lee, O.J.; Lee, J.S.; Yoon, S.; et al. Precisely Printable and Biocompatible Silk Fibroin Bioink for Digital Light Processing 3D Printing. *Nat. Commun.* **2018**, *9*, 1–14. [CrossRef]
62. Cao, Y.; Wang, B. Biodegradation of Silk Biomaterials. *Int. J. Mol. Sci.* **2009**, *10*, 1514–1524. [CrossRef]
63. Zhang, L.; Liu, X.; Li, G.; Wang, P.; Yang, Y. Tailoring Degradation Rates of Silk Fibroin Scaffolds for Tissue Engineering. *J. Biomed Mater.* **2019**, *107*, 104–113. [CrossRef]
64. Wang, H.; Zhang, Y.; Wei, Z. Characterization of Undegraded and Degraded Silk Fibroin and Its Significant Impact on the Properties of the Resulting Silk Biomaterials. *Int. J. Biol. Macromol.* **2021**, *176*, 578–588. [CrossRef] [PubMed]
65. Wang, Y.; Fan, S.; Li, Y.; Niu, C.; Li, X.; Guo, Y.; Zhang, J.; Shi, J.; Wang, X. Silk Fibroin/Sodium Alginate Composite Porou Materials with Controllable Degradation. *Int. J. Biol. Macromol.* **2020**, *150*, 1314–1322. [CrossRef] [PubMed]
66. Luo, Z.; Zhang, Q.; Shi, M.; Zhang, Y.; Tao, W.; Li, M. Effect of Pore Size on the Biodegradation Rate of Silk Fibroin Scaffolds. *Adv Mater. Sci. Eng.* **2015**, *2015*, 1–8. [CrossRef] [PubMed]
67. Sun, M.; Li, Q.; Yu, H.; Cheng, J.; Wu, N.; Shi, W.; Zhao, F.; Shao, Z.; Meng, Q.; Chen, H.; et al. Cryo-Self-Assembled Silk Fibroin Sponge as a Biodegradable Platform for Enzyme-Responsive Delivery of Exosomes. *Bioact. Mater.* **2022**, *8*, 505–514. [CrossRef]

68. Catto, V.; Farè, S.; Cattaneo, I.; Figliuzzi, M.; Alessandrino, A.; Freddi, G.; Remuzzi, A.; Cristina, M. Small Diameter Electrospun Silk Fibroin Vascular Grafts: Mechanical Properties, in Vitro Biodegradability, and in Vivo Biocompatibility. *Mater. Sci. Eng. C* **2015**, *54*, 101–111. [CrossRef]
69. Mehrabani, M.G.; Karimian, R.; Mehramouz, B.; Rahimi, M.; Kafil, H.S. Preparation of Biocompatible and Biodegradable Silk Fibroin/Chitin/Silver Nanoparticles 3D Scaffolds as a Bandage for Antimicrobial Wound Dressing. *Biol. Macromol.* **2018**, *114*, 961–971. [CrossRef]
70. Fan, H.; Liu, H.; Toh, S.L.; Goh, J.C.H. Anterior Cruciate Ligament Regeneration Using Mesenchymal Stem Cells and Silk Scaffold in Large Animal Model. *Biomaterials* **2009**, *30*, 4967–4977. [CrossRef]
71. Wang, Y.; Rudym, D.D.; Walsh, A.; Abrahamsen, L.; Kim, H.; Kim, H.S.; Kirker-head, C.; Kaplan, D.L. In Vivo Degradation of Three-Dimensional Silk Fibroin Scaffolds. *Biomaterials* **2008**, *29*, 3415–3428. [CrossRef]
72. Melke, J.; Midha, S.; Ghosh, S.; Ito, K.; Hofmann, S. Silk Fibroin as Biomaterial for Bone Tissue Engineering. *Acta Biomater.* **2016**, *31*, 1–16. [CrossRef]
73. Johari, N.; Moroni, L.; Samadikuchaksaraei, A. Tuning the Conformation and Mechanical Properties of Silk Fibroin Hydrogels. *Eur. Polym. J.* **2020**, *134*, 109842. [CrossRef]
74. Grabska-Zielińska, S.; Sionkowska, A.; Coelho, C.C.; Grabska-zieli, S.; Monteiro, F.J. Silk Fibroin/Collagen/Chitosan Scaffolds Cross-Linked by a Glyoxal Solution as Biomaterials toward Bone Tissue Regeneration. *Materials* **2020**, *13*, 3433. [CrossRef]
75. Eivazzadeh-keihan, R.; Ahmadpour, F.; Aghamirza, H.; Aliabadi, M.; Radinekiyan, F.; Maleki, A.; Madanchi, H.; Mahdavi, M.; Esmail, A.; Lanceros-m, S. Pectin-Cellulose Hydrogel, Silk Fibroin and Magnesium Hydroxide Nanoparticles Hybrid Nanocomposites for Biomedical Applications. *Int. J. Biol. Macromol.* **2021**, *192*, 7–15. [CrossRef]
76. Chen, T.; Wen, T.; Dai, N.; Hsu, S. Cryogel Hydrogel Biomaterials and Acupuncture Combined to Promote Diabetic Skin Wound Healing through Immunomodulation. *Biomaterials* **2021**, *269*, 120608. [CrossRef]
77. Hoon, D.; Tripathy, N.; Hun, J.; Eun, J.; Geun, J.; Dan, K.; Hum, C.; Khang, G. Enhanced Osteogenesis of Β-Tricalcium Phosphate Reinforced Silk Fibroin Scaffold for Bone Tissue Biofabrication. *Int. J. Biol. Macromol.* **2017**, *95*, 14–23. [CrossRef]
78. Chen, Z.; Zhong, N.; Wen, J.; Jia, M.; Guo, Y.; Shao, Z.; Zhao, X.; Accepted, J. Porous Three-Dimensional Silk Fibroin Scaffolds for Tracheal Epithelial Regeneration in Vitro and in Vivo. *ACS Biomater. Sci. Eng.* **2018**, *4*, 2977–2985. [CrossRef]
79. Yan, L.-P.; Oliveira, J.M.; Oliveira, A.L.; Caridade, S.G.; Mano, J.F.; Reis, R.L. Macro/Microporous Silk Fibroin Scaffolds with Potential for Articular Cartilage and Meniscus Tissue Engineering Applications. *Acta Biomater.* **2012**, *8*, 289–301. [CrossRef]
80. Vepari, C.; Kaplan, D.L. Silk as a Biomaterial. *Prog. Polym. Sci.* **2007**, *32*, 991–1007. [CrossRef]
81. Zhang, H.; Fang, J.; Ge, H.; Han, L.; Wang, X.; Hao, Y.; Han, C.; Dong, L. Thermal, Mechanical, and Rheological Properties of Polylactide/Poly (1, 2-Propylene Glycol Adipate). *Polym. Eng. Sci.* **2013**, *53*, 112–118. [CrossRef]
82. Bunster, G.F. Polyhydroxyalkanoates: Production and Use in Medicine. In *Encyclopedia of Biomedical Polymers and Polymeric Biomaterials*; CRC Press: Boca Raton, FL, USA, 2016; pp. 6412–6421. [CrossRef]
83. Guhados, G.; Wan, W.; Hutter, J.L. Measurement of the Elastic Modulus of Single Bacterial Cellulose Fibers Using Atomic Force Microscopy. *Am. Chem. Soc.* **2005**, *21*, 6642–6646. [CrossRef]
84. Pogorelova, N.; Rogachev, E.; Dige, I.; Chernigova, S.; Nardin, D. Bacterial Cellulose Nanocomposites: Morphology. *Materials* **2020**, *13*, 2849. [CrossRef]
85. Guo, S.; Dipietro, L.A. Factors Affecting Wound Healing. *Crit. Rev. Oral Biol. Med.* **2010**, *89*, 219–229. [CrossRef]
86. Chen, Z.; Zhang, Y.; Zheng, L.; Zhang, H.; Shi, H.; Zhang, X.; Liu, B. Mineralized Self-Assembled Silk Fibroin/Cellulose Interpenetrating Network Aerogel for Bone Tissue Engineering. *Mater. Sci. Eng. C* **2021**, 112549. [CrossRef]
87. Martınez-Mora, C.; Mrowiec, A.; Marı, E.; Alcaraz, A.; Jose, F. Fibroin and Sericin from *Bombyx mori* Silk Stimulate Cell Migration through Upregulation and Phosphorylation Of. *PLoS ONE* **2012**, *7*, e42271. [CrossRef]
88. Nikam, V.S.; Punde, D.S.; Bhandari, R.S. Silk Fibroin Nanofibers Enhance Cell Adhesion of Blood-Derived Fibroblast-like Cells—A Potential Application for Wound Healing. *Indian J. Pharmacol.* **2020**, *52*, 306–312. [CrossRef]
89. Gharehnazifam, Z.; Dolatabadi, R.; Baniassadi, M.; Shahsavari, H.; Kajbafzadeh, A.; Abrinia, K.; Baghani, M. Computational Analysis of Vincristine Loaded Silk Fibroin Hydrogel for Sustained Drug Delivery Applications: Multiphysics Modeling and Experiments. *Int. J. Pharm.* **2021**, *609*, 121184. [CrossRef]
90. Kwon, G.; Heo, B.; Kwon, M.J.; Kim, I.; Chu, J.; Kim, B.; Kim, B.; Park, S.S. Effect of Silk Fibroin Biomaterial Coating on Cell Viability and Intestinal Adhesion of Probiotic Bacteria. *J. Microbiol. Biotechnol.* **2021**, *31*, 592–600. [CrossRef]
91. Lee, O.J.; Sultan, M.T.; Hong, H.; Lee, Y.J.; Lee, J.S.; Lee, H.; Kim, S.H.; Park, C.H. Recent Advances in Fluorescent Silk Fibroin. *Front. Mater.* **2020**, *7*, 1–12. [CrossRef]
92. Baci, G.-M.; Cucu, A.-A.; Giurgiu, A.-I.; Muscă, A.-S.; Rațiu, C.A.; Bagameri, L.; Moise, A.R.; Bobiș, O.; Dezmirean, D.S. Advances in Editing Silkworms (*Bombyx mori*) Genome by Using the CRISPR-Cas System. *Insects* **2022**, *13*, 28. [CrossRef] [PubMed]
93. Nagano, A.; Tanioka, Y.; Sakurai, N.; Sezutsu, H.; Kuboyama, N.; Kiba, H. Regeneration of the Femoral Epicondyle on Calcium-Binding Silk Scaffolds Developed Using Transgenic Silk Fibroin Produced by Transgenic Silkworm. *Acta Biomater.* **2011**, *7*, 1192–1201. [CrossRef] [PubMed]
94. Kuwana, Y.; Sezutsu, H.; Nakajima, K.I.; Tamada, Y.; Kojima, K. High-Toughness Silk Produced by a Transgenic Silkworm Expressing Spider (*Araneus Ventricosus*) Dragline Silk Protein. *PLoS ONE* **2014**, *9*, e105325. [CrossRef] [PubMed]

95. Fujinaga, D.; Kohmura, Y.; Okamoto, N.; Kataoka, H.; Mizoguchi, A. Insulin-like Growth Factor (IGF)-like Peptide and 20-Hydroxyecdysone Regulate the Growth and Development of the Male Genital Disk through Different Mechanisms in the Silkmoth, *Bombyx mori*. *Insect Biochem. Mol. Biol.* **2017**, *87*, 35–44. [CrossRef]
96. Wang, F.; Xu, H.; Wang, Y.; Wang, R.; Yuan, L.; Ding, H.; Song, C.; Ma, S.; Peng, Z.; Peng, Z.; et al. Advanced Silk Material Spun by a Transgenic Silkworm Promotes Cell Proliferation for Biomedical Application. *Acta Biomater.* **2014**, *10*, 4947–4955. [CrossRef]
97. Wang, M.; Du, Y.; Huang, H.; Zhu, Z.; Du, S.; Chen, S.; Zhao, H. Silk Fibroin Peptide Suppresses Proliferation and Induces Apoptosis and Cell Cycle Arrest in Human Lung Cancer Cells. *Acta Pharmacol. Sin.* **2019**, *40*, 522–529. [CrossRef]
98. Saviane, A.; Romoli, O.; Bozzato, A.; Freddi, G.; Cappelletti, C.; Rosini, E.; Cappellozza, S.; Tettamanti, G. Intrinsic Antimicrobial Properties of Silk Spun by Genetically Modified Silkworm Strains. *Transgenic Res.* **2018**, *27*, 87–101. [CrossRef]
99. Li, Z.; Cao, G.; Xue, R.; Chengliang, G. Construction of Transgenic Silkworm Spinning Antibacterial Silk with Fluorescence Construction of Transgenic Silkworm Spinning Antibacterial Silk with Fluorescence. *Mol. Biol. Rep.* **2014**, *42*, 19–25. [CrossRef]
100. Iizuka, T.; Sezutsu, H.; Tatematsu, K.; Kobayashi, I.; Yonemura, N.; Uchino, K.; Nakajima, K.; Kojima, K.; Takabayashi, C.; Machii, H.; et al. Colored Fluorescent Silk Made by Transgenic Silkworms. *Adv. J. Mater.* **2013**, *23*, 5232–5239. [CrossRef]
101. Asakura, T.; Isozaki, M.; Saotome, T.; Tatematsu, K.; Sezutsu, H.; Kuwabara, N.; Nakazawa, Y. Recombinant Silk Fibroin Incorporated Cell-Adhesive Sequences Produced by Transgenic Silkworm as a Possible Candidate for Use in Vascular Graft. *J. Mater. Chem. B Mater. Biol. Med.* **2014**, *2*, 7375–7383. [CrossRef]
102. Zhao, S.; Ye, X.; Wu, M.; Ruan, J.; Wang, X.; Tang, X.; Zhong, B. Recombinant Silk Proteins with Additional Polyalanine Have Excellent Mechanical Properties. *Int. J. Mol. Sci.* **2021**, *22*, 1513. [CrossRef]
103. Yanagisawa, S.; Zhu, Z.; Kobayashi, I.; Uchino, K.; Tamada, Y.; Tamura, T.; Asakura, T. Improving Cell-Adhesive Properties of Recombinanant *Bombyx mori* Silk by Incorporation of Collagen or Fibronectin Derived Peptides Produced by Transgenic Silkworms. *Biomacromolecules* **2007**, *8*, 3487–3492. [CrossRef]
104. Nguyen, T.P.; Nguyen, Q.V.; Nguyen, V.; Le, T.; Le, Q.V. Silk Fibroin-Based Biomaterials for Biomedical. *Polymers* **2019**, *11*, 1933. [CrossRef]
105. Santos, M.V.; Paula, K.T.; Andrade, M.B. De; Gomes, E.M.; Marques, L.F.; Ribeiro, S.J.L.; Mendonc, C.R. Direct Femtosecond Laser Printing of Silk Fibroin Microstructures. *Appl. Mater. Interfaces* **2020**, *12*, 50033–50038. [CrossRef]
106. Ho, W.; Jeong, L.; Il, D.; Hudson, S. Effect of Chitosan on Morphology and Conformation of Electrospun Silk Fibroin Nanofibers. *Polymer* **2004**, *45*, 7151–7157. [CrossRef]
107. Çalamak, S.; Erdo, C.; Özalp, M.; Ulubayram, K. Silk Fibroin Based Antibacterial Bionanotextiles as Wound Dressing Materials. *Mater. Sci. Eng. C* **2014**, *43*, 11–20. [CrossRef]
108. Ha, S.; Tonelli, A.E.; Hudson, S.M.; Carolina, N. Structural Studies of *Bombyx mori* Silk Fibroin during Regeneration from Solutions and Wet Fiber Spinning. *Biomacromolecules* **2005**, *6*, 1722–1731. [CrossRef]
109. Zhang, F.; Ming, J.; Dou, H.; Liu, Z. Silk Dissolution and Regeneration at the Nanofibril Scale. *J. Mater. Chem. B* **2014**, *2*, 3879–3885. [CrossRef]
110. Sato, M.; Nakazawa, Y.; Takahashi, R.; Tanaka, K.; Sata, M.; Aytemiz, D.; Asakura, T. Small-Diameter Vascular Grafts of *Bombyx mori* Silk Fibroin Prepared by a Combination of Electrospinning and Sponge Coating. *Mater. Lett.* **2010**, *64*, 1786–1788. [CrossRef]
111. Aznar-cervantes, S.; Roca, M.I.; Martinez, J.G.; Meseguer-olmo, L.; Cenis, J.L.; Moraleda, J.M.; Otero, T.F. Fabrication of Conductive Electrospun Silk Fibroin Scaffolds by Coating with Polypyrrole for Biomedical Applications. *Bioelectrochemistry* **2012**, *85*, 36–43. [CrossRef]
112. Aznar-cervantes, S.D.; Vicente-cervantes, D.; Meseguer-olmo, L.; Cenis, J.L.; Lozano-pérez, A.A. Influence of the Protocol Used for Fibroin Extraction on the Mechanical Properties and Fiber Sizes of Electrospun Silk Mats. *Mater. Sci. Eng. C* **2013**, *33*, 1945–1950. [CrossRef]
113. Yamada, H.; Nakao, H.; Takasu, Y.; Tsubouchi, K. Preparation of Undegraded Native Molecular Fibroin Solution from Silkworm Cocoons. *Mater. Sci. Eng. C* **2001**, *14*, 41–46. [CrossRef]
114. Sofia, S.; Mccarthy, M.B.; Gronowicz, G.; Kaplan, D.L.; Al, S.E.T. Functionalized Silk-Based Biomaterials for Bone Formation. *J. Biomed. Mater. Res.* **2000**, *54*, 139–148. [CrossRef]
115. Mathur, A.B.; Tonelli, A.; Rathke, T.; Hudson, S. The Dissolution and Characterization of *Bombyx mori* Silk Fibroin in Calcium Nitrate-methanol.Pdf. *Fiber Polym. Sci.* **1998**, *42*, 61–74.
116. Kunz, W.; Katharina, H. Some Aspects of Green Solvents. *Comptes Rendus Chim.* **2018**, *21*, 572–580. [CrossRef]
117. Wang, H.; Zhang, Y.; Wei, Z. Dissolution and Processing of Silk Fibroin for Materials Science. *Crit. Rev. Biotechnol.* **2021**, *41*, 406–424. [CrossRef] [PubMed]
118. Wöltje, M.; Kölbel, A.; Aibibu, D.; Cherif, C. A Fast and Reliable Process to Fabricate Regenerated Silk Fibroin Solution from Degummed Silk in 4 Hours. *Int. J. Mol. Sci.* **2021**, *22*, 565. [CrossRef] [PubMed]
119. Freddi, G.; Pessina, G.; Tsukada, M. Swelling and Dissolution of Silk Fibroin (*Bombyx mori*) in N -Methyl Morpholine N -Oxide. *Biol. Macromol.* **1999**, *24*, 251–263. [CrossRef]
120. Carissimi, G.; Baronio, C.M.; Montalbán, M.G.; Víllora, G.; Barth, A. On the Secondary Structure of Silk Fibroin Nanoparticles Obtained Using Ionic Liquids: An Infrared Spectroscopy Study. *Polymers* **2020**, *12*, 1294. [CrossRef] [PubMed]
121. Garc, M.; Aznar-Cervantes, S.D.; Lozano-Perez, A.A.; Cenis, L.; Gloria, V. Production of Silk Fibroin Nanoparticles Using Ionic Liquids and High-Power Ultrasounds. *J. Appl. Polym. Sci.* **2015**, *41702*, 1–8. [CrossRef]

22. Phillips, D.M.; Drummy, L.F.; Conrady, D.G.; Fox, D.M.; Naik, R.R.; Stone, M.O.; Trulove, P.C.; Long, H.C. De; Mantz, R.A. Dissolution and Regeneration of *Bombyx mori* Silk Fibroin Using Ionic Liquids. *J. Am. Chem. Soc.* **2004**, *126*, 14350–14351. [CrossRef]
23. Ajisawa, A. Dissolution Aqueous of Silk Fibroin with Calciumchloride/Ethanol Solution. *J. Seric. Sci. Jpn.* **1997**, *67*, 91–94.
24. Reizabal, A.; Costa, C.M.; Saiz, P.G.; Gonzalez, B.; Perez-Alvarez, L.; Luis, R.F. de; Garcia, A.; Vilas-Vilela, J.; Lanceros-Mendez, S. Processing Strategies to Obtain Highly Porous Silk Fibroin Structures with Tailored Microstructure and Molecular Characteristics and Their Applicability in Water Remediation. *J. Hazard. Mater.* **2021**, *403*, 123675. [CrossRef]
25. Bae, S. Bin; Kim, M.H.; Park, W.H. Electrospinning and Dual Crosslinking of Water-Soluble Silk Fibroin Modified with Glycidyl Methacrylate. *Polym. Degrad. Stab.* **2020**, *179*, 109304. [CrossRef]
26. Mosher, C.Z.; Brudnicki, P.A.P.; Gong, Z.; Childs, H.R.; Lee, S.W.; Antrobus, R.M.; Fang, E.C.; Schiros, T.N.; Lu, H.H. Green Electrospinning for Biomaterials and Biofabrication. *Biofabrication* **2021**, *13*, 035049. [CrossRef]
27. Li, X.; Fan, Q.; Zhang, Q.; Yan, S.; You, R. Freezing-Induced Silk I Crystallization of Silk Fibroin. *R. Soc. Chem.* **2020**, *22*, 3884–3890. [CrossRef]
28. Yang, X.; Wang, X.; Yu, F.; Ma, L.; Pan, X.; Luo, G.; Lin, S.; Mo, X.; Wang, H. Hyaluronic Acid/EDC/NHS-Crosslinked Green Electrospun Silk Fibroin Nanofibrous Scaffolds for Tissue Engineering. *R. Soc. Chem.* **2016**, *6*, 99720–99728. [CrossRef]
29. Fei, X.; Jia, M.; Du, X.; Yang, Y.; Zhang, R.; Shao, Z.; Zhao, X. Green Synthesis of Silk Fibroin-Silver Nanoparticle Composites with Effective Antibacterial and Biofilm-Disrupting Properties. *Biomacromolecules* **2013**, *14*, 4483–4488. [CrossRef]
30. Raho, R.; Nguyen, N.; Zhang, N.; Jiang, W.; Sannino, A.; Liu, H.; Pollini, M.; Paladini, F. Photo-Assisted Green Synthesis of Silver Doped Silk Fibroin/Carboxymethyl Cellulose Nanocomposite Hydrogels for Biomedical Applications. *Mater. Sci. Eng. C* **2020**, *107*, 110219. [CrossRef]
31. El-Sheikh, M.A.; El-Rafie, S.M.; Abdel-Halim, E.S.; El-Rafie, M.H. Green Synthesis of Hydroxyethyl Cellulose-Stabilized Silver Nanoparticles. *J. Polym.* **2013**, *2013*, 1–11. [CrossRef]
32. Altman, G.H.; Diaz, F.; Jakuba, C.; Calabro, T.; Horan, R.L.; Chen, J.; Lu, H.; Richmond, J.; Kaplan, D.L. Silk-Based Biomaterials. *Biomaterials* **2003**, *24*, 401–416. [CrossRef]
33. Kamalathevan, P.; Ooi, P.S.; Loo, Y.L. Silk-Based Biomaterials in Cutaneous Wound Healing: A Systematic Review. *Adv. Ski. Wound Care* **2018**, *31*, 565–573. [CrossRef]
34. Cheng, G.; Wang, X.; Tao, S.; Xia, J.; Xu, S. Differences in Regenerated Silk Fibroin Prepared with Different Solvent Systems: From Structures to Conformational Changes. *J. Appl. Polym. Sci.* **2015**, *132*, 1–8. [CrossRef]
35. Volkov, V.; Ferreira, A.V.; Cavaco-paulo, A. On the Routines of Wild-Type Silk Fibroin Processing Toward Silk-Inspired Materials: A Review. *Macromol. Mater. Eng.* **2015**, *300*, 1199–1216. [CrossRef]
36. Jin, S.C.; Baek, H.S.; Woo, Y.I.; Lee, M.H.; Kim, J.; Park, J.; Park, Y.H.; Lee, S.J. Beneficial Effects of Microwave-Induced Argon Plasma Treatment on Cellular Behaviors of Articular Chondrocytes onto Nanofibrous Silk Fibroin Mesh. *Macromol. Res.* **2009**, *17*, 703–708. [CrossRef]
37. Gu, J.; Yang, X.; Zhu, H. Surface Sulfonation of Silk Fibroin Film by Plasma Treatment and in Vitro Antithrombogenicity Study. *Mater. Sci. Eng. C* **2002**, *20*, 199–202. [CrossRef]
38. Shang, K.; Rnjak-kovacina, J.; Tao, H.; Kaplan, D.L.; Lin, Y.; Hayden, R.S. Accelerated In Vitro Degradation of Optically Clear Low b -Sheet Silk Films by Enzyme-Mediated Pretreatment. *Transl. Vis. Sci. Technol.* **2013**, *2*, 2. [CrossRef]
39. Wang, L.; Nemoto, R.; Senna, M. Three-Dimensional Porous Network Structure Developed in Hydroxyapatite-Based Nanocomposites Containing Enzyme Pretreated Silk Fibroin. *J. Nanopart. Res.* **2004**, *6*, 91–98. [CrossRef]
40. Li, M.; Ogiso, M.; Minoura, N. Enzymatic Degradation Behavior of Porous Silk Fibroin Sheets. *Biomaterials* **2003**, *24*, 357–365. [CrossRef]
41. Wang, P.; Zhou, Y.; Cui, L.; Yuan, J.; Wang, Q.; Fan, X.; Ding, Y. Enzymatic Grafting of Lactoferrin onto Silk Fibroins for Antibacterial Functionalization. *Fibers Polym.* **2014**, *15*, 2045–2050. [CrossRef]
42. Chi, R.; Cheung, F.; Ng, T.B.; Wong, J.H.; Chan, W.Y. Chitosan: An Update on Potential Biomedical and Pharmaceutical Applications. *Mar. Drugs* **2015**, *13*, 5156–5186. [CrossRef]
43. Li, X.; Li, B.; Wang, X.; Zhang, S. Development of a Silk Fibroin/HTCC/PVA Sponge for Chronic Wound Dressing. *J. Bioact. Compat. Polym. Biomed. Appl.* **2014**, *29*, 398–411. [CrossRef]
44. Karahaliloğlu, Z. Curcumin-Loaded Silk Fibroin e-Gel Scaffolds for Wound Healing Applications. *Mater. Technol.* **2018**, *7857*, 1–12. [CrossRef]
45. Thangavel, P.; Ramachandran, B.; Kannan, R.; Muthuvijayan, V. Biomimetic Hydrogel Loaded with Silk and L -Proline for Tissue Engineering and Wound Healing Applications. *Soc. Biomater.* **2016**, *105*, 1401–1408. [CrossRef]
46. Luo, Z.; Jiang, L.; Xu, Y.; Li, H.; Xu, W.; Wu, S.; Wang, Y.; Tang, Z.; Lv, Y.; Yang, L. Biomaterials Mechano Growth Factor (MGF) and Transforming Growth Factor (TGF)—B3 Functionalized Silk Scaffolds Enhance Articular Hyaline Cartilage Regeneration in Rabbit Model. *Biomaterials* **2015**, *52*, 463–475. [CrossRef]
47. Mehrabani, M.G.; Karimian, R.; Rakhshaei, R.; Pakdel, F.; Eslami, H.; Fakhrzadeh, V.; Rahimi, M.; Salehi, R.; Kafil, H.S. Chitin/Silk Fibroin/TiO_2 Bio-Nanocomposite as a Biocompatible Wound Dressing Bandage with Strong Antimicrobial Activity. *Biol. Macromol.* **2018**, *116*, 966–976. [CrossRef]
48. Patil, P.P.; Meshram, J.V.; Bohara, R.A. ZnO Nanoparticle-Embedded Silk Fibroin–Polyvinyl Alcohol Composite Film: A Potential Dressing Material for Infected Wounds. *R. Soc. Chem.* **2018**, *42*, 14620–14629. [CrossRef]

149. Li, X.; Liu, Y.; Zhang, J.; You, R.; Qu, J.; Li, M. Functionalized Silk Fibroin Dressing with Topical Bioactive Insulin Release for Accelerated Chronic Wound Healing. *Mater. Sci. Eng. C* **2017**, *72*, 394–404. [CrossRef]
150. Hrynyk, M.; Neufeld, R.J. Insulin and Wound Healing. *Burns* **2014**, *40*, 1433–1446. [CrossRef]
151. Wang, Y.; Kim, U.; Blasioli, D.J.; Kim, H.; Kaplan, D.L. In Vitro Cartilage Tissue Engineering with 3D Porous Aqueous-Derived Silk Scaffolds and Mesenchymal Stem Cells. *Biomaterials* **2005**, *26*, 7082–7094. [CrossRef]
152. Hee, S.; Been, Y.; Kyu, Y.; Jin, Y.; Sang, H.; Sultan, T.; Min, J.; Seung, J.; Joo, O.; Hong, H.; et al. Biomaterials 4D-Bioprinted Silk Hydrogels for Tissue Engineering. *Biomaterials* **2020**, *260*, 120281. [CrossRef]
153. Mallepally, R.R.; Marin, M.A.; Surampudi, V.; Subia, B.; Rao, R.R. Silk Fibroin Aerogels: Potential Scaffolds for Tissue Engineering Applications. *Biomed. Mater.* **2015**, *10*, 35002. [CrossRef]
154. Nazarov, R.; Jin, H.; Kaplan, D.L. Porous 3-D Scaffolds from Regenerated Silk Fibroin. *Biomacromolecules* **2004**, *5*, 718–726. [CrossRef]
155. Pacheco, S.M.; Eiji, G.; Almeida, L. De; Santos, P.; Agostini, M.; Moraes, D. Silk Fibroin/Chitosan/Alginate Multilayer Membranes as a System for Controlled Drug Release in Wound Healing. *Int. J. Biol. Macromol.* **2020**, *152*, 803–811. [CrossRef]
156. Harkin, D.G.; George, K.A.; Madden, P.W.; Schwab, I.R.; Hutmacher, D.W.; Chirila, T.V. Biomaterials Silk Fibroin in Ocular Tissue Reconstruction. *Biomaterials* **2011**, *32*, 2445–2458. [CrossRef]
157. Calonge, M.; Nieto-Miguel, T.; Mata, A. De; Galindo, S.; Herreras, J.M.; Marina, L. Goals and Challenges of Stem Cell-Based Therapy for Corneal Blindness Due to Limbal Deficiency. *Pharmaceutics* **2021**, *13*, 1483. [CrossRef]
158. Lee, M.C.; Kim, D.; Lee, O.J.; Kim, J.; Ju, H.W.; Lee, J.M.; Moon, B.M.; Park, H.J.; Kim, D.W.; Kim, S.H.; et al. Fabrication of Silk Fibroin Film Using Centrifugal Casting Technique for Corneal Tissue Engineering. *Soc. Biomater.* **2016**, *104*, 508–514. [CrossRef]
159. Liu, J.; Lawrence, B.D.; Liu, A.; Schwab, I.R.; Oliveira, L.A.; Rosenblatt, M.I. Silk Fibroin as a Biomaterial Substrate for Corneal Epithelial Cell Sheet Generation. *Investig. Ophthalmol. Vis. Sci.* **2012**, *53*, 4130–4138. [CrossRef]
160. Pellegrini, G.; Dellambra, E.; Golisano, O.; Martinelli, E.; Fantozzi, I.; Bondanza, S.; Ponzin, D.; Mckeon, F.; Luca, M. De. P63 Identifies Keratinocyte Stem Cells. *Proc. Natl. Acad. Sci. USA* **2001**, *98*, 357–394. [CrossRef]
161. Schermer, A.; Galvin, S. Differentiation-Related Expression of a Major 64K Corneal Keratin In Vivo and In Culture Suggests Limbal Location of Corneal Epithelial Stem Cells. *J. Cell Biol.* **1986**, *103*, 49–62. [CrossRef] [PubMed]
162. Lawrence, B.D.; Marchant, J.K.; Pindrus, M.A.; Omenetto, F.G.; Kaplan, D.L. Biomaterials Silk Film Biomaterials for Cornea Tissue Engineering. *Biomaterials* **2009**, *30*, 1299–1308. [CrossRef] [PubMed]
163. Malliappan, P.; Alp, A.; Burcu, S.; Demir, E.; Cetinel, S. Bone Tissue Engineering: Anionic Polysaccharides as Promising Scaffolds. *Carbohydr. Polym.* **2022**, *283*, 119142. [CrossRef]
164. Sk, S. Fracture Non-Union: A Review of Clinical Challenges and Future Research Needs. *Malays. Orthop. J.* **2019**, *13*, 1–10. [CrossRef] [PubMed]
165. Gillman, C.E.; Jayasuriya, A.C. FDA-Approved Bone Grafts and Bone Graft Substitute Devices in Bone Regeneration. *Mater. Sci. Eng. C* **2021**, *130*, 112466. [CrossRef]
166. Zhao, Z.; Ma, X.; Ma, J.; Kang, J.; Zhang, Y.; Guo, Y. Sustained Release of Naringin from Silk- Fibroin-Nanohydroxyapatite Scaffold for the Enhancement of Bone Regeneration. *Mater. Today Bio* **2022**, *13*, 100206. [CrossRef]
167. Meinel, L.; Fajardo, R.; Hofmann, S.; Langer, R.; Chen, J.; Snyder, B.; Vunjak-novakovic, G.; Kaplan, D. Silk Implants for the Healing of Critical Size Bone Defects. *Bone* **2005**, *37*, 688–698. [CrossRef]
168. Zhao, J.; Zhang, Z.; Wang, S.; Sun, X.; Zhang, X.; Chen, J.; Kaplan, D.L.; Jiang, X. Apatite-Coated Silk Fibroin Scaffolds to Healing Mandibular Border Defects in Canines. *Bone* **2009**, *45*, 517–527. [CrossRef]
169. Hirose, M.; Hamada, K.; Tanaka, T. Nano-Scaled Hydroxyapatite/Silk Fibroin Composites as Mesenchymal Cell Culture Scaffolds. *Key Eng. Mater.* **2006**, *311*, 923–926. [CrossRef]
170. Cui, B.; Liang, L.; Lu, X.; Weng, J. Fabricating HYDROXYAPATITE—SIlk Fibroin Nanocomposite by Bone Bionics. *Key Eng. Mater.* **2007**, *332*, 345–348. [CrossRef]
171. Wen, G.; Wang, J.; Li, M.; Meng, X. Study on Tissue Engineering Scaffolds of Silk Fibroin-Chitosan/Nano-Hydroxyapatite Composite. *Key Eng. Mater.* **2007**, *332*, 971–975. [CrossRef]
172. Sun, B.K.; Siprashvili, Z.; Khavari, P.A. Advances in Skin Grafting and Treatment of Cutaneous Wounds. *Science* **2014**, *346*, 941–945. [CrossRef]
173. Pra, I.D.A.L.; Chiarini, A.; Boschi, A.; Freddi, G.; Armato, U. Novel Dermo-Epidermal Equivalents on Silk Fibroin-Based Formic Acid-Crosslinked Three-Dimensional Nonwoven Devices with Prospective Applications in Human Tissue Engineering/Regeneration/Repair. *Int. J. Mol. Med.* **2006**, *18*, 241–247.
174. Luangbudnark, W.; Viyoch, J.; Laupattarakasem, W.; Surakunprapha, P.; Laupattarakasem, P. Properties and Biocompatibility of Chitosan and Silk Fibroin Blend Films for Application in Skin Tissue Engineering. *Sci. World J.* **2012**, *2012*, 697201. [CrossRef]
175. Bhardwaj, N.; Sow, W.T.; Devi, D.; Ng, K.W.; Mandal, B.B.; Cho, N.-J. Insight Statement for "Silk Fibroin-Keratin Based 3D Scaffolds as a Dermal Substitute for Skin Repair and Regeneration. *Integr. Biol.* **2014**, *7*, 53–63. [CrossRef]
176. Lyu, J.; Kaur, M.; Dibble, K.E.; Connor, A.E. A National Study of Alcohol Consumption Patterns among Population-Based U. S. Cancer Survivors Compared with Cancer-Free Individuals. *Cancer Epidemiol.* **2022**, *77*, 102101. [CrossRef]
177. Desforges, A.D.; Hebert, C.M.; Spence, A.L.; Reid, B.; Dhaibar, A.; Cruz-topete, D.; Cornett, E.M.; David, A.; Urits, I.; Viswanath, O. Treatment and Diagnosis of Chemotherapy-Induced Peripheral Neuropathy: An Update. *Biomed. Pharmacother.* **2022**, *147*, 112671. [CrossRef]

178. Ma, D.; Wang, Y.; Dai, W. Silk Fibroin-Based Biomaterials for Musculoskeletal Tissue Engineering. *Mater. Sci. Eng. C* **2018**, *89*, 456–469. [CrossRef]
179. Ma, Y.; Canup, B.S.B.; Tong, X.; Dai, F.; Xiao, B.; Wang, J. Multi-Responsive Silk Fibroin-Based Nanoparticles for Drug Delivery. *Front. Chem.* **2020**, *8*, 1–5. [CrossRef]
180. Sun, N.; Lei, R.; Xu, J.; Kundu, S.C. Fabricated Porous Silk Fibroin Particles for PH- Responsive Drug Delivery and Targeting of Tumor Cells. *J. Mater. Sci.* **2019**, *54*, 3319–3330. [CrossRef]
181. Saqr, A. Al; Ud, S.; Wani, D.; Gangadharappa, H.V.; Aldawsari, M.F.; Khafagy, E.; Lila, A.S.A. Enhanced Cytotoxic Activity of Docetaxel-Loaded Silk Fibroin Nanoparticles against Breast Cancer Cells. *Polymers* **2021**, *13*, 1–18.
182. Mottaghitalab, F.; Kiani, M.; Farokhi, M.; Kundu, S.C.; Reis, R.L.; Gholami, M.; Bardania, H.; Dinarvand, R.; Geramifar, P.; Beiki, D.; et al. Targeted Delivery System Based on Gemcitabine Loaded Silk Fibroin Nanoparticles for Lung Cancer Therapy Targeted Delivery System Based on Gemcitabine Loaded Silk Fibroin Nanoparticles for Lung Cancer Therapy. *Appl. Mater. Interfaces* **2017**, *9*, 31600–31611. [CrossRef]
183. Moin, A.; Wani, S.U.D.; Osmadi, R.A.; Lila, A.S.A.; Khafagy, E.-S.; Arab, H.H.; Gangaharappa, H.V.; Allam, A.N. Formulation, Characterization, and Cellular Toxicity Assessment of Tamoxifen-Loaded Silk Fibroin Nanoparticles in Breast Cancer. *Drug Deliv.* **2021**, *28*, 1626–1636. [CrossRef]
184. Lu, X.; Sun, Y.; Han, M.; Chen, D.; Wang, A.; Sun, K. Silk Fibroin Double-Layer Microneedles for the Encapsulation and Controlled Release of Triptorelin. *Int. J. Pharm.* **2022**, *613*, 121433. [CrossRef]
185. Dondajewska, E.; Juzwa, W.; Mackiewicz, A.; Dams-, H. Heterotypic Breast Cancer Model Based on a Silk Fibroin Scaffold to Study the Tumor Microenvironment. *Oncotarget* **2018**, *9*, 4935–4950. [CrossRef] [PubMed]
186. Li, J.; Zhou, Y.; Chen, W.; Yuan, Z.; You, B.; Liu, Y.; Yang, S.; Li, F.; Qu, C.; Zhang, X. A Novel 3D in Vitro Tumor Model Based on Silk Fibroin/Chitosan Sca Ff Olds to Mimic the Tumor Microenvironment. *Appl. Mater. Interfaces* **2018**, *10*, 36641–36651. [CrossRef] [PubMed]
187. Qian, K.; Song, Y.; Yan, X.; Dong, L.; Xue, J.; Xu, Y.; Wang, B.; Cao, B.; Hou, Q.; Peng, W.; et al. Biomaterials Injectable Ferrimagnetic Silk Fibroin Hydrogel for Magnetic Hyperthermia Ablation of Deep Tumor. *Biomaterials* **2020**, *259*, 120299. [CrossRef] [PubMed]
188. Mukhopadhyay, S.K.; Naskar, D.; Bhattacharjee, P.; Mishra, A.; Kundu, S.C.; Dey, S. Silk Fibroin-Thelebolan Matrix: A Promising Chemopreventive Scaffold for Soft Tissue Cancer. *Colloids Surf. B Biointerfaces* **2017**, *155*, 379–389. [CrossRef] [PubMed]
189. Pandey, V.; Haider, T.; Chandak, A.R.; Chakraborty, A.; Banerjee, S.; Soni, V. Technetium Labeled Doxorubicin Loaded Silk Fibroin Nanoparticles: Optimization, Characterization and in Vitro Evaluation. *J. Drug Deliv. Sci. Technol.* **2020**, *56*, 101539. [CrossRef]
190. Anas, M.; Hadianamrei, R.; Sun, W.; Xu, D.; Brown, S.; Zhao, X. Stiffness-Tuneable Nanocarriers for Controlled Delivery of ASC-J9 into Colorectal Cancer Cells. *J. Colloid Interface Sci.* **2021**, *594*, 513–521. [CrossRef]
191. Talukdar, S.; Kundu, S.C. Engineered 3D Silk-Based Metastasis Models: Interactions Between Human Breast Adenocarcinoma, Mesenchymal Stem Cells and Osteoblast-Like Cells. *Adv. Funct. Mater.* **2013**, *9*, 5249–5260. [CrossRef]
192. Ding, B.; Wahid, M.A.; Zhijun, W.; Chen, X.; Arvind, T.; Sunil, P.; Wang, J. Triptolide and Celastrol Loaded Silk Fibroin Nanoparticles Show Synergistic Effect against Human Pancreatic Cancer Cells. *Nanoscale* **2017**, *9*, 11739–11753. [CrossRef]
193. Toan, D.; Saelim, N.; Tiyaboonchai, W. Alpha Mangostin Loaded Crosslinked Silk Fibroin-Based Nanoparticles for Cancer Chemotherapy. *Colloids Surf. B Biointerfaces* **2019**, *181*, 705–713. [CrossRef]
194. Yucel, T.; Lovett, M.L.; Giangregorio, R.; Coonahan, E.; Kaplan, D.L. Silk Fibroin Rods for Sustained Delivery of Breast Cancer Therapeutics. *Biomaterials* **2014**, *35*, 8613–8620. [CrossRef]
195. Zhang, X.; Huang, Y.; Song, H.; Canup, B.S.B.; Gou, S. Inhibition of Growth and Lung Metastasis of Breast Cancer by Tumor-Homing Triple-Bioresponsive Nanotherapeutics. *J. Control. Release* **2020**, *328*, 454–469. [CrossRef]
196. Yao, Q.; Lan, Q.; Jiang, X.; Du, C.; Zhai, Y.; Shen, X.; Xu, H. Bioinspired Biliverdin/Silk Fibroin Hydrogel for Antiglioma Photothermal Therapy and Wound Healing. *Theranostics* **2020**, *10*, 11739–11753. [CrossRef]
197. Yu, S.; Yang, W.; Chen, S.; Chen, M.; Liu, Y.; Shao, Z.; Chen, X. Floxuridine-Loaded Silk Fibroin Nanospheres. *RSC Adv.* **2014**, *4*, 18171–18177. [CrossRef]
198. Mishra, D.; Iyyanki, T.S.; Hubenak, J.R.; Zhang, Q.; Mathur, A.B. Silk Fibroin Nanoparticles and Cancer Therapy. In *Nanotechnology in Cancer*; Elsevier: Amsterdam, The Netherlands, 2017; pp. 19–44. [CrossRef]

insects

MDPI

Review

Black, Caspian Seas and Central Asia Silk Association (BACSA) for the Future of Sericulture in Europe and Central Asia

Panomir Tzenov [1,*], Silvia Cappellozza [2] and Alessio Saviane [2]

1 Agricultural Academy, Scientific Center on Sericulture, 3000 Vratsa, Bulgaria
2 Council for Agricultural Research and Economics, Research Center for Agriculture and Environment, Padua Seat, 35143 Padua, Italy; silvia.cappellozza@crea.gov.it (S.C.); alessio.saviane@crea.gov.it (A.S.)
* Correspondence: panomir@yahoo.com

Simple Summary: This paper describes the 16-years long activity of the Black Caspian Seas and Central Asia Silk Association, which was founded in 2005 to revive the sericultural activity in the area indicated by its own denomination. The reasons why this Association was established are described as a direct consequence of the decline in the sericulture agroindustry following the collapse of the Soviet Union and the world cocoon/raw silk decrease of production (except for China and India) since the 90s of the 20th century. Therefore, the enlargement of its membership to countries outside of the boundaries of the geographical area is outlined as well as its internal organization and the actions performed to promote the interaction among the member countries, especially the biyearly conferences. The international scenario is depicted to explain the criticalities experienced in promoting sericultural activities in the region, as well as the opportunities offered by the new applications of the silk, silk proteins and mulberry derivatives.

Abstract: The history and recent activities of the Black Caspian Seas and Central Asia Silk Association are presented in this paper: the countries that participated in its foundation, the FAO's action to revitalize sericulture in Eastern Europe and Central Asia, the following widening of the Association geographical limits of to enclose other European countries, which were not well-represented in other similar organizations. Some statistical data are illustrated for a better description of the scenario in which the BACSA executive board acted: the world silk production quantity and the relative production of BACSA countries in respect to the Chinese expansion. The themes treated in the BACSA conferences are reported to explain which matters the Executive Board considered the most relevant for the relaunch of this activity in relationships to the international challenges in the subsequent years; the project proposals that were presented to international donors are summarized. A SWOT (Strengths, Weaknesses, Opportunities, Threats) analysis is shown, where key-factors in determining the strengths and weaknesses of this organization and its member countries for a successful re-establishment of sericulture, are considered. In addition, future trends of sericulture with regard to innovative productions and the Green Deal are examined.

Keywords: sericulture; agroindustry; Food and Agriculture Organization of the United Nations; mulberry; moriculture; silkworm; germplasm preservation

Citation: Tzenov, P.; Cappellozza, S.; Saviane, A. Black, Caspian Seas and Central Asia Silk Association (BACSA) for the Future of Sericulture in Europe and Central Asia. *Insects* 2022, *13*, 44. https://doi.org/10.3390/insects13010044

Academic Editor: Cheryl Y. Hayashi

Received: 1 December 2021
Accepted: 28 December 2021
Published: 30 December 2021

1. Introduction

The Central Asia region of the ex-Soviet Union includes 5 countries (Kazakhstan, Kyrgyzstan, Tajikistan, Turkmenistan, and Uzbekistan), while the South Caucasus region is composed of three (Armenia, Azerbaijan and Georgia). These 8 countries obtained their independence in 1991 after the breakdown of the Soviet Union, and since then they have experienced a dramatic economic and social crisis, resulting from the transition from a "centrally planned" economy to a "market economy". In fact, the fast and forced industrialization of the Soviet era was full of enormous structural distortions and microeconomic ineffectiveness [1]. After the dissolution of the Soviet Union, many industrial enterprises located in this region lost their previous markets and were unable to compete under the new market conditions. As sericulture is an agroindustry, the criticalities of the Central Asian and Caucasic sericulture activities, which were experienced from the '90s, were thus predictable.

Similar problems occurred in the Central and Eastern Europe after 1989 (Berlin Wall's collapse). Economies throughout the region fell into recession in the 1990s. Efforts to privatize and open markets caused unemployment and social inequality. Bulgaria's economy contracted each year from 1989 through 1993, while Romania's GDP (Gross Domestic Product) dropped nearly 13% in 1991 and nearly 9% in 1992 [2].

These political, social, and economic changes greatly impacted sericulture, perhaps more than any other economic activity, in the ex-communist countries. In fact, to manage the sericultural chain it is necessary to coordinate the agricultural production at the farm level, with silkworm eggs sourcing, logistic steps for drying, sorting, preservation, stocking and marketing of cocoons; furthermore, coordination is also necessary with the subsequent steps performed in the reeling and textile plants.

What happened in the ex-communist countries in Eastern Europe and Central Asia during the transitional period from centralized to market economic system was a sudden stop of the governmental support to sericulture, the breakage of the traditional economic relationships among the countries, thus the destruction of the already established system of integration of the different processes of the sericultural production.

Until the beginning of '90s of the last century the approximate annual fresh cocoon production in Europe, Caucasus and Central Asia was around 50,000 tons, by this occupying the third place in the world after China and India; nearly one million farmer's households were engaged with sericulture [3]. At the Soviet Union breakdown the Central Asian region including Uzbekistan, Tajikistan, Turkmenistan and Iran was still a remarkable cocoon and silk producer (even now this region is engaging in the industry about 450,000 farmers) [3]; at that time the Eastern part of Europe, which was also under the communist umbrella of the USSR (Union of Soviet Socialist Republics) and also Turkey, were still producing large quantities of silk, while the Western Europe and Greek sericulture activities had already considerably declined.

Around the same period of the above-mentioned changes, i.e., in the years between 1983 and 1995, the world production of raw silk increased rather constantly and almost doubled, thanks to the exclusive contribution of China, which, in the same period tripled its production. Therefore, China in 1995 had the world monopoly of silk production with 74% of the global quantity of this fiber [4]. In the long term the constant increase of the Chinese silk that was poured over the market caused an excess in the circulating quantity of this fiber, therefore, its international price fell down (about 24% only in 1997) reaching even USD 22/kg of raw silk [4]. In this scenario only India, and Brazil, remained comparatively competitive, while all the other countries in the world were forced out of the market, or remained on the very edge (Uzbekistan, Thailand, Vietnam). This tendency to the concentration of silk production in China was even more dramatic for countries acquiring silk from Celestial Empire because of the "double price" policy, on which basis the internal silk industry could have access to raw material at a price 20% lower than that for foreign buyers [4]. The monopoly of silk production, joint to the double price policy permitted China to obtain the absolute leadership in the cost of raw silk and even of silk tissues and clothes [4].

2. Establishment and Purpose of the Black and Caspian Seas and Central Asia Silk Association (BACSA)

As mentioned above sericulture is an agro-based industry and agriculture has always been of huge importance for the countries of the Black and Caspian Seas and Central Asia area, supporting the income and providing employment to a large majority of inhabitants of the rural territories, but also protecting the environment through the sustainable use of natural resources. For this reason, the collapse of sericulture following the USSR dissolution was regarded with much concern by NGOs (Non-Governmental Organizations). The case of Uzbekistan is emblematic: due to the combined effect of the economic turmoil after the independence, and the Chinese monopoly of silk prices, Uzbek cocoon production dropped from 33,000 t in 1990 to 21,000 t in 1997 [5].

In order to counteract this decline in sericulture for the countries of the former USSR or previously enclosed under its political umbrella, in April 2005, the Food and Agriculture Organization (FAO) of the United Nations and the Uzbek Government organized in Tashkent, Uzbekistan, the "International Workshop on Revival and Promotion of Sericultural Industries and Small Enterprise Development in the Black, Caspian Seas and Central Asia Region". Twelve countries including Azerbaijan, Bulgaria, Egypt, Georgia, Greece, Japan, Kazakhstan, Republic of Korea, Tajikistan, Turkey, Ukraine, and Uzbekistan participated in this meeting. During the workshop a regional association, named Black, Caspian Seas and Central Asia Silk Association (BACSA) was established [5].

The main targets of this association were to generate sericulture projects from external resources, including bilateral and multilateral cooperation, sensitize respective governments and prospective donors, promote local and regional joint efforts, which would have allowed the cooperation among the countries of the region of Europe, Caucasus and Central Asia, develop concrete actions to fortify the sustainable development of the sericulture in the region, promote agreements for international scientific-technical cooperation and business relations among the countries involved, encourage market studies, training, and spreading sericultural germplasm, and silkworm eggs. In fact, the BACSA member countries have many problems and issues in their sericulture revival and development, which could be partly or completely solved by the help of regional cooperation among them. Such problems are the access to the EU (European Union) funds for research, specializations and training of students and technical personnel, exchange of sericulture germplasm resources and improvement of the mulberry sapling and silkworm egg quality; other issues consist of saving the sericulture germplasm of the BACSA region, increasing silkworm egg production or favoring its revival in some of the member countries, solving criticalities of cocoon and silk marketing, improving the raw silk quality and rising the share of exported or locally processed raw silk, producing small niche textiles from cocoons and developing silk handcrafts [6].

At the time of its establishment, BACSA member states that were producing silk carpets, such as Iran and Turkey, were also comparatively big raw silk importers as the local silk production was not able to satisfy the silk carpet needs of raw material [7]. In that period, as even presently, some BACSA countries did not have any own silkworm egg production and in some of the Central Asian countries the quality of currently locally produced silkworm eggs did not meet the international standards, or their quantity could not satisfy the local needs [8].

Therefore, one of the basic aims of expansion of BACSA inter—regional cooperation was/is to transmit and share sericulture germplasms, silkworm eggs, advanced technologies, training, dry cocoons, raw silk, and silk allied products.

3. The BACSA Activity

3.1. The BACSA Composition

When established in 2005 BACSA included 9 countries—Azerbaijan, Bulgaria, Georgia, Greece, Kazakhstan, Tajikistan, Turkey, Ukraine and Uzbekistan. Later on, and gradually, BACSA attracted for membership new countries—Albania, Armenia, Iran, Poland, Romania, Switzerland, Italy, Spain, Germany, Portugal, Slovenia, Russia, UK so that currently the association includes 22 countries and has also 64 individual members and 4 institutional members. The individual members from countries, which are completely out of the enlarged BACSA region are mostly from India (43), but also from China (2), Egypt (2), Korea (1), Ghana (1), Syria (1) and Indonesia (1). Presently BACSA has 26 members of the Executive committee from 22 different countries. The composition of the Executive committee has been totally renewed at the end of 2009 [9] and can be periodically revised according to changes in the membership.

3.2. The BACSA Structure

The management of the association is composed of a President, two Vice-Presidents, one national coordinator for each member country, which are the most representative part of the Executive Committee; all of them are democratically elected. The members of the Executive Committee are the direct coordinators of all the current activities for each country, within the regional context. The Executive Committee is the bridge among each country, the national coordinator and the other countries of the association, to execute the actions defined in the region. The Committee gathers in person at least once every two years; its members maintain contacts among themselves regularly by e-mail/phone, to perform the following functions [9]:

- To evaluate the work made by each national coordinator in the respective countries in relationship to BACSA.
- To recommend people in the association who should receive training abroad.
- To evaluate and to watch the handling of the "Rotatory Funds" and "Research Funds" possibly created and to give recommendations on the best use of these resources.
- To present/display the research proposals that require financing from the "Research Fund" and to approve the necessary resources for this aim.
- To give recommendations and suggestions on all publications and written material that are created within the framework of the BACSA.
- To advise the association's President on the advances and progresses that take place in the development of the activities and give recommendations to her/him on possible modifications and corrections.
- To decide the main BACSA activities, including accepting new members etc.

3.3. The BACSA Conferences

Until the end of 2019 BACSA organized 9 international conferences. Each conference was on a specific subject, connected with the problems of regional sericulture development. The BACSA conferences are shown in Table 1.

Table 1. BACSA conferences (2005–2021).

	BACSA Conference Title	Year	Location
(1)	Revival and Promotion of Sericultural Industries and Small Enterprise Development	2005	Taskent, Uzbekistan
(2)	Silk Handicrafts Cottage Industries and Silk Enterprises Development	2006	Bursa, Turkey
(3)	Sericulture Challenges in the 21st Century	2007	Vratsa, Bulgaria
(4)	Possibilities for Using Silkworm and Mulberry for Non-Textile Purposes (First Balkan workshop)	2008	Plovdiv, Bulgaria
(5)	Sericulture for multi products—new prospects for development	2011	Bucharest, Romania
(6)	Building Value Chains in Sericulture	2013	Padova, Italy
(7)	Organic Sericulture—Now and the Future	2015	Sinaia, Romania
(8)	Climate changes and chemicals—the new sericulture challenges	2017	Sheki, Azerbaijan
(9)	Sericulture preservation and revival—problems and prospects	2019	Batumi, Georgia

The main findings, conclusions and recommendations of these international meetings are listed below:

(1) Revival and Promotion of Sericultural Industries and Small Enterprise Development (2005): As recalled before, this conference posed the basis for the development of a short/medium-term strategy for the sericulture revival at a regional level and preparation of projects for the sericultural industry rehabilitation in the region countries, ready for donor search [6].
(2) Silk Handicrafts Cottage Industries and Silk Enterprises Development (2006): This conference was particularly focused on the creation of a network collaborating in the different sectors of the supply chain to improve competences in silk reeling and silk industry management specially to establish ecofriendly processes of dyeing and finishing and to promote silk handcraft production and export to foreign countries even through the touristic channel [7].
(3) Sericulture Challenges in the 21st Century (2007): The most important theme of this meeting was the preservation and exploitation of germplasm resources (silkworm and mulberry) of the region, which are particularly rich but at risk as their preservation is costly for the concerned countries [7].
(4) Possibilities for Using Silkworm and Mulberry for Non-Textile Purposes (2008): The target of the discussion was the proper utilization of secondary and waste products of the sericultural industry, which can generate an extra income in addition to silk, which represents the main output. The by-products of sericulture are sericin, pupae, moths, silkworm frasses, silk waste, mulberry branches, fruits and roots. New commercial products can be obtained from these raw materials with a valuable destination for pharmaceuticals, cosmetics, feed and food, new materials [10].
(5) Sericulture for multi products—new prospects for development (2011): This conference enlarged the issues treated in the previous one of 2008, particularly stressing the criticalities of non-textile silk production, from the scientific and regulatory point of view [10–12].
(6) Building Value Chains in Sericulture (2013): Stakeholders of different BACSA countries met to plan possible cooperation programmes in the field of science and technology transfer, education and training with the aim of promoting bilateral research projects, financed by each participating country's government through grant competitions, announced periodically by the Ministries of education and sciences or bi and/or multilateral projects, financed by the EU, in addition to specialization and training of students and technical personnel in leading research centers and commercial companies, financed by EU/national programmes [13].
(7) Organic Sericulture—Now and the Future (2015): The theme of this meeting was about the new trend of organic production in agriculture that well adapts to sericulture, especially in Europe, where mulberry cultivation has always been "naturally organic". How to develop a certification chain for textile and non-textile organic silk production and how to exploit this certification on the market were the key-points treated in the internal discussion [14].
(8) Climate changes and chemicals—the new sericulture challenges (2017): In most of the 22 countries currently members of the Black, Caspian Seas and Central Asia Silk Association, wherever they are, Europe, Caucasus and Central Asia, sericulture has been very negatively affected by the use of chemicals in agriculture and by climatic changes. In fact, through fluctuations in temperature, water regimes and the increase in carbon dioxide levels, global climate changes directly influence mulberry, soil, pests, and silkworm rearing. The specific climatic conditions in the BACSA region countries require mulberry to have high cold and drought tolerance and the silkworm strains to possess a good tolerance to adverse rearing conditions like high temperature, daily temperature fluctuations and feeding with coarse mulberry leaves. On the other hand, the wide use of insecticides can easily harm silkworm rearing and even completely destroy the whole sericulture value chains in some regions or countries. A common

effort to develop new strategies is required to mitigate the joint effects of the chemical use and climatic changes [15].

(9) Sericulture preservation and revival—problems and prospects (2019): During this conference the BACSA members discussed about the world trends of silk production in particular the decrease of the cocoon quantity annually available in China and the increase of the international price. These phenomena give a prospective of decrease in the silk use by the textile industry and in coming back to consider this fiber as a very luxury good for a niche market. Sericulture may change from an industry for the poorest farmers, to an agribusiness, requiring more investments and production costs but having high revenues because of the high market price of the products. BACSA countries should be prepared for that [3,16,17].

The BACSA conference foreseen for 2021 was postponed because of the COVID-19 problems.

3.4. The BACSA Project Proposals

The BACSA activity to prepare project proposals is illustrated in Table 2. This activity was very intense between 2006 and 2010, then after many failures, these kinds of attempts ceased in terms of projects studied for the whole area and focused mostly on more limited projects or bilateral agreements between members of BACSA [18].

Table 2. BACSA project proposals to different donors (2006–2021).

Presented Projects	Year of Proposal or Realization	Possible Donor	Financed (Yes/No)
Improvement of Income-Generation Options Based on Revival of Sericultural Industries and Promotion of Small Silk Enterprise Development in Eastern Europe and Central Asia—Concept Note	2006	UNDP, ADA, CIDA, DCI, DEZA, DFID, FAO, GTZ, IFAD, ITC, UNCTAD, JICA, KOICA, NORAD, SIDA, UNIDO, USAID, EBRD, World bank	N
Comparative studies of silkworm hybrids performance for sericultural enterprise development in Black, Caspian seas and Central Asia region	2006–2007	FAO	Y (only partially)
Support for unlocking and developing the research potential of silkworm breeding and innovative management techniques in Bulgaria, Greece and Romania, targeting to the small silk enterprise development	2007	EU FP 7 "Capacities" Program	N
Improvement of Income-Generation Options Based on Revival of Sericultural Industries and Promotion of Small Silk Enterprise Development in Eastern Europe and Central Asia—Full Proposal	2008	FAO	N
Regional sericulture germplasm resources network for Africa, Middle East, Central Asia and Europe	2008	FAO	N
Proposal of a Regional workshop on "Utilization of mulberry and silkworm genetic resources for sericultural enterprises development in Africa, Middle East, Central Asia and Europe".	2008	FAO	N
TCP/ALB 3101 "Revival and Development of Sericulture in Albania"	2008–2009	FAO	Y
SERINNOV- SEE EoI/B/481/1.3/X "Sericulture and silk products in South-East Europe—stepping from the tradition to innovation and strengthening the economic profile of respective regions	2009	EU—SEE program	N
Exploitation of the heavy metal movement and removal pathways and mechanisms in the mulberry- silkworm chain for the establishment of model applications for further practical use in bioremediation of contaminated soils.	2009	FP7—KBBE program	N
TCP/GEO/3201 "Sericulture sector study in Georgia"	2009–2010	FAO	Y

3.5. The BACSA Network for Facilitating Commercial Contacts among Stakeholders

BACSA makes all the efforts to establish connections among the producers, sellers and buyers of different sericultural products such as mulberry saplings, silkworm eggs, dry cocoons, raw silk, silk yarn, fabrics, and garments. These activities are performed through regular updating of the section "sell/buy information" on the BACSA website as well as responding in real time to all the enquiries from possible sellers and buyers, connecting stakeholders together and giving a chance to exhibit sericultural products by organizing international workshops, conferences, etc. [19].

3.6. Results of BACSA Global Activity

BACSA carried out a series of activities that can be better understood when we compare the situation before 2005, when BACSA was established, and at the present time; from this comparison it is evident this association's role in the regional sericulture preservation, revival and development (Table 3).

Table 3. BACSA's activity results (2005–2021).

Item	Situation in 2005	Situation in 2021
Contacts among the key specialists and institutions engaged in the sericultural industries	Very few, well-developed among some of the Ex-Soviet Union and Eastern Europe countries only	Good contacts among all BACSA member countries, key sericulture specialists and most of the institutions
Regional sericulture database	Not available	Available, uploaded on the BACSA website and regularly updated
Number of countries, associated in BACSA	Only 8 Eastern Europe, Caucasus and Central Asia countries	22 countries from Europe, Caucasus and Central Asia; out of them 9 EU member states, which give possibilities to apply for projects financing from the EU funds.
Enquiry system for sericultural products marketing	Not available	Available: enquiries to BACSA → distribution of the enquiries to the national coordinators → distribution of the enquiries to interested stakeholders in each member country for direct contacts; uploading the sell/buy enquiries on the BACSA web site.
Exchange of sericulture germplasm resources among the BACSA member countries	Very few, well-developed among some of the Ex-Soviet Union and Eastern Europe countries only	At a much higher scale, based on bilateral scientific projects
Export of mulberry saplings and silkworm eggs	Very few, among some of the Ex-Soviet Union states only	There is, but still in a too small scale
Regional conferences and meetings	Very few, well-developed among some of the Ex-Soviet Union and Eastern Europe countries only	Regular BACSA international conferences, at intervals of 2 years, 9 conferences already organized.
Specializations and training of students and technical personnel in leading research centers and commercial companies in BACSA member countries	No	Present, but still in a too small scale.
Sensitizing the national governments about the regional BACSA executive meetings results, decisions and follow ups	No	Yes, operated by the BACSA national coordinators

Table 3. *Cont.*

Item	Situation in 2005	Situation in 2021
Responding to different enquiries, concerning the sericultural industries in the region	No	Yes, operated by BACSA president and national coordinators
Supporting the sericultural institutions and specialists in the BACSA member countries by providing, when necessary, letters of support, personal recommendations, reviews of scientific monographs and doctoral dissertations etc.	No	Yes, operated by BACSA president and selected experts
Popularization of Europe and Central Asia sericultural industries in the world through participation to international conferences and meetings	Scarce	Yes
Development of regional sericulture projects and looking for donors	No	Yes
Promoting bilateral and multilateral agreements for cooperation among the BACSA member states	No	Yes

4. BACSA Becomes Attractive for Western and Central Europe Countries

As mentioned before, the situation of sericulture of the countries of Western and Central Europe non-belonging to the Communist block was even worse than that of BACSA countries, because here the decline of sericulture (referred to as agricultural and reeling activity) began much earlier than in the '90s of the last century and precisely after the second World War with the maximum peak between the '60s and '70s. On the other hand, these countries retained a flourishing silk industry (weaving, dyeing, printing, manufacturing), especially in Italy, France, Switzerland and England [9]. The main feature of this silk industry has been its complete dependence for silk from external sourcing, especially from China, for a long time [9]. During the '90s the international price of silk was low and the quantity on the market abundant; this was the reason why the European silk industry was not stimulated to look for other sources than China. However, some signals of market sufferance were already very clear: double pricing of silk and difficulty in finding a constant and high quality of this fiber production in China [4].

In Western Europe some residual cocoon production could be preserved due to the EC (European Communities) subsidies to sericulture (Regulation EEC n. 922/72) establishing a contribution per each reared silkworm box from which a minimum quantity of at least 20 kg of fresh cocoons was obtained. This contribution was completely stopped in Europe in 2014, due to the Common Agricultural Policy (CAP) reform; the EC governments were allowed to support some strategic sectors only with direct payments; the majority of EC countries decided that sericulture was not a priority, while only the government of Greece continued to pay subsidies to its farmers for fresh cocoon production.

In Italy and France this lack of support from the Government was partly justified by the catastrophic situation of sericulture caused by external factors; in fact, from 1989 to 2010 France and Northern Italy (where some cocoon production had been preserved even in a small amount) were affected for a long time by the pollution due to an Insect Growth Regulator (active ingredient: fenoxycarb, commercial name: Insegar). The chemical was sprayed on the fruit orchards and its drift was transported by the winds onto mulberry leaf. It was so effective to be capable of acting on the silkworm larvae at doses of nanograms (1.0×10^{-9} g) or even lower [10]. Due to this phenomenon the larvae were unable to spin their cocoons ("non-spinning syndrome") [11,12] and French and Italian rearers were completely discouraged from continuing to practice sericulture. For almost 20 years, until the date of the retirement (which was operated by the same producing company) of the chemical from the market and exhaustion of the stocks, it was impossible to rear silkworms in the above-mentioned areas.

In 2009 the INRA (National Research Institute for Agriculture) Sericultural National Unit of Lyon (France) closed and only the Sericulture Laboratory of CREA (Council for Agricultural Research and Economics) in Padua remained responsible for the preservation of all the genetic resources of Western Europe [13]. Furthermore, the International Sericultural Commission based in France, transferred its headquarters to India, which was a signal of disengagement of the European stakeholders from sericulture [14].

In the same period the international price of raw silk began to rise very rapidly, reaching levels of about USD 55–60/kg mostly because of the industrialization of China in the last decades, in particular the Eastern area of the country, which was traditionally more apt to sericulture. Although the Chinese government made a big effort to transfer sericulture to other regions, less industrialized and more devoted to agriculture, the silk quantity and mostly its quality decreased significantly; as a consequence, the price of the best quality of silk in the international market began to increase quite abruptly. The COVID 19 pandemic has been a brief parenthesis in this trend.

Therefore, during the last 10 years a new interest was expressed by the European silk industry (especially from Italy and Switzerland) for countries alternative to China where silk might be produced; on this basis the silk industry may re-consider establishing part of the cocoon production they need in Europe, Caucasus and/or Central Asia. A sign of this interest was the BACSA conference of 2013, which was held in Italy with the economic support and commitment of the Italian silk industry in collaboration with CREA (the Research organ of the Italian Ministry of Agricultural, Food and Forestry Policies) [15]. A delegation of the executive board was hosted in Como and visited "Ratti", one of the most important silk Italian companies belonging to the Marzotto group. Another sign of interest was the progressive association to BACSA of countries from Western Europe: Italy, Germany, Spain, Portugal, Slovenia, Switzerland, UK which are currently members of BACSA [16].

However, the restoration of the sericulture chain in Europe and BACSA countries is a very large goal that cannot be sustained by the industry alone; a public-private partnership should be envisaged, and the governmental and EC support should be provided to make it a realistic goal, which, however, needs long term investments.

5. Problems and Prospects

In this section BACSA will be examined according to the problems and prospects that characterize this organization. An attempt to summarize the criticalities and "plus values" of BACSA is reported in Figure 1.

5.1. Problems Related to the BACSA Structure

BACSA is a quite open and very flexible organization; the membership of the Black, Caspian Seas and Central Asia Silk Association (BACSA) is possible for all individuals/institutes/countries from Europe and Central Asia which are willing to share common goals for an industrial and economic growth through sericultural activities. The expected members may be of two categories: "Internal" from the same geographical region countries, and "External" from other countries, who are possibly interested in technical cooperation or business with Europe and Central Asia region as well as from those countries that are possible donors for sericulture revival projects [20]. Even Chinese, Indian, Japanese, Korean, African scientists/stakeholders applied to this organization for membership. Chinese and Indian delegations are admitted as observers to BACSA meetings. No membership fee is required so far, and this fact permits to participate even to poor countries or individuals. However, what in principle appears to be very democratic and egalitarian might be a problem in the future in case some divisive themes are discussed, or the interest of some single country is concerned. At the moment, the BACSA community reaches a good agreement on any decision basically because of three factors: (1) members are few and many of them have been known each other for years; (2) there are no huge interests at stake because sericulture in the BACSA countries is mostly confined to a niche economy; (3) China's presence in the silk world is so outstanding that any other country should decide its strategy to develop its own silk production/industry reacting to the Chinese monopoly more than fighting against other competitors. Therefore, it is possible that this association should assume another structure in the future; in fact, some problems are the same for most of the countries others are typical of a more restricted geographical area. The three macro-areas that can be outlined in the BACSA region are Central Asia, the Caucasus and Europe. It might be advisable in the next period to divide BACSA in three different working groups, joining together only on themes of common interest. For example, the income level of farmers in these three macro-areas is completely different, being the minimum in Central Asia maximum in Western Europe. It appears very clear that Europe should make a big effort to mechanize and specialize its silk production, while countries like Uzbekistan (Central Asia) have the potential of producing high quantity of cocoons at a comparatively low price by exploiting local manpower and can be identified as competitor with China in the middle term. Other principles to form working groups may be assumed: for example silk industrial converters versus cocoon producing/reeling countries, low quantity silk producing countries versus high quantity silk producing countries, etc.

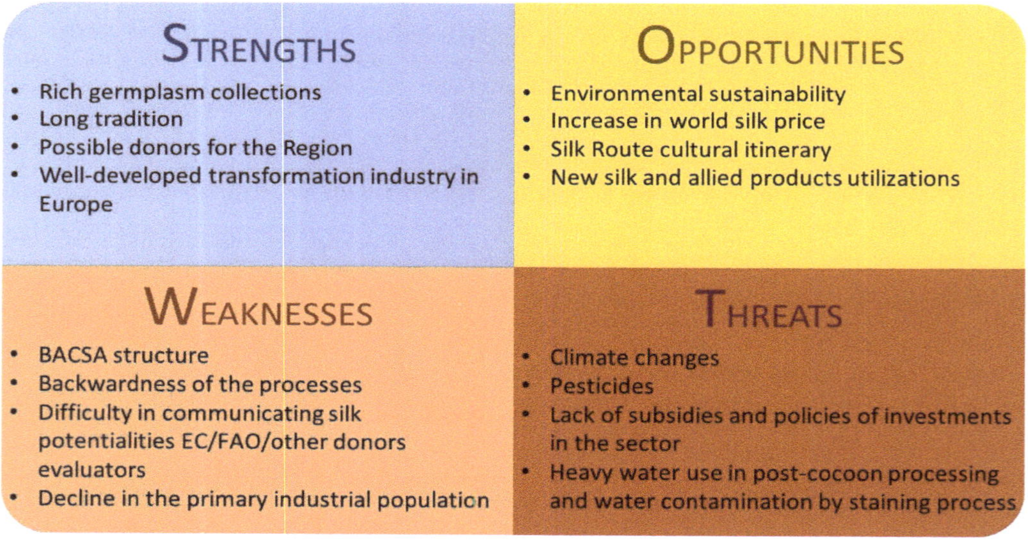

Figure 1. SWOT analysis representing criticalities and qualities of BACSA examined with respect to the possibilities of success in revitalizing the sericultural activity in the area.

5.2. Common Problems

5.2.1. Preservation of Genetic Resources

The global raw silk production was around 91,945 t in 2020, but out of them 53,359 t were produced by China and 33,770 t by India, while all the other countries produced only about 4816 t of raw silk [21] (see Figure 2).

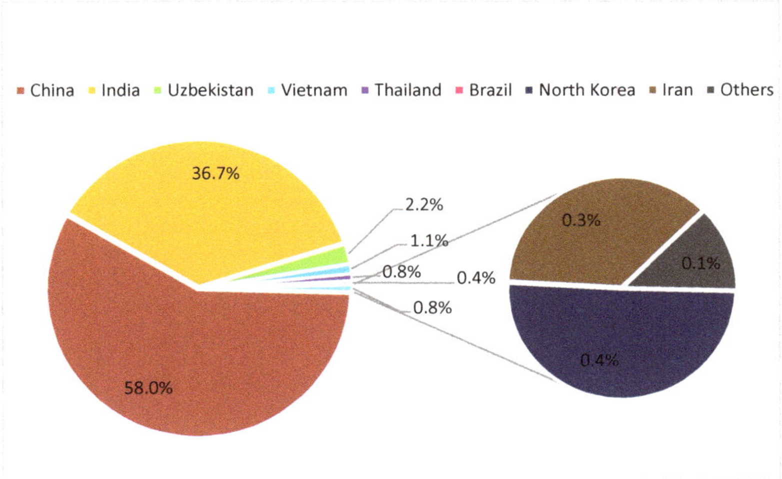

Figure 2. Pie-chart and relative shares of global silk production (data from https://inserco.org/en/statistics; accessed on 23 November 2021) in 2020; China and India alone cover more than 90% of the worldwide production (94.7%).

That means almost 95% of the total world silk production is from only two countries—China and India. On the other hand, although many countries in the World dealt with sericulture in the past or are still dealing nowadays, the majority of them only make efforts to preserve this activity and only few of them to revive the sericultural industry. Silk production volumes more than doubled from 1990 to 2019 but it saw a decrease over the last five years. Even in those countries that are the biggest world cocoon and silk producers, there are presently entire regions where the sericultural activities have been partly or even completely stopped and the sericultural expertise may be lost. Among these, BACSA associated countries, which still represents the third world producing area, face problems typical of nations with history of long tradition and scarce current production. The first problem for them is how to preserve their mulberry and silkworm germplasms, which usually constitute public genetic resources, mostly located in research institutes [8,22]; the preservation activity is quite expensive, governments, in consideration of the present scarce economical revenues from the sericultural agribusiness, tend to restrict funds destined for conservation. Any hypothesis to concentrate the germplasm in one center only, preserving accessions for all the region, is quite unrealistic and in addition dangerous for the possible losses of genetic material in case of diseases or unforeseen accidents. The best way to preserve genetic resources is their spreading ex-situ; however, this proved to be very difficult because of intellectual properties on the selected strains and varieties and could be done only for the genetic material that does not have any economic significance. Therefore, encouraging the use of these resources and restarting an economically viable sericulture is certainly the best way to guarantee their preservation. A contemporaneous rehabilitation of the activity in the whole BACSA region does not appear to be necessary; a partial restarting in those countries still endowed with well-preserved genetic material might also be useful for those countries that have lost their own sourcing, due to different accidental events, or have germplasm of inferior quality; in fact, even a partial sericultural revival will favor bilateral exchanges and commercial exploitation agreements among different countries, which might be regulated internally within the BACSA framework.

5.2.2. Backwardness of the Agroindustry Chain

Another problem for BACSA countries is the backwardness of the agricultural process (mulberry cultivation and silkworm rearing) and/or the silk industrial transformation. This is partially because the Chinese monopoly on silk has maintained too low the silk price to stimulate innovation and mechanization in countries with capitalistic economies; the only advanced part of the silk chain, in these countries, regards industrial processing after reeling, in which modernization and digitalization play a basic role on the final quality of manufacts for the exclusive fashion industry of the most important world brands. The communist governments highly promoted the mechanization and technical advances in the agriculture. In the communist countries after the '60s of the 20th century a big industrialization occurred; thus, moving of a huge mass of people from the rural areas to the cities resulted in the lack of agricultural workers. For this reason, in the '80s the mechanization in agriculture was a priority. Even in sericulture, where cocoon production, in some communist countries, was in the hands of the big agricultural cooperatives or state enterprises, they managed to organize large scale silkworm rearing, mechanized feeding of young instar larvae, using Japanese machineries, cardboard frame mountages, machines for cocoon deflossing. However, most of these efforts in sericulture appeared to be not economically effective due to the Chinese concurrence and were abandoned. To recover the technological gap of sericulture with respect to other agricultural crops or agro-industrial chains, more competitive in terms of economic revenues for farmers or investors, is not easy and requires many funds to develop innovations. This is one of the main reasons why sericulture is a niche production in several developed countries. This technological gap in sericulture became more evident in the last decades when the effect of climatic changes and environmental pollution began to seriously affect agriculture. For example, dramatic climatic changes began to negatively affect silkworm rearing with serious fluctuations in the average temperatures even in the seasons traditionally favorable to sericulture; the lack of climatic control of silkworm rearing facilities and of digital automatic control of temperature and humidity can cause big problems to larval development; heavy droughts, late frosts, excessive rains compromise mulberry leaf harvests; high temperatures and humidity favor spreading of insect pests, which are also more invasive due to the increased globalization of the transport of goods around the world, which works as an involuntary carrier. Fighting against these new insect pests with insecticides, in turn, affects sericulture. To solve these problems a lot of technology and research related to the environmental management of sericulture would be necessary; however, as previously mentioned, available funds to improve knowledge in the sericultural field are very limited [16].

5.2.3. Funding Access

With regard to structural funding for research, training, dissemination, demonstration and other activities related to the sericulture revival, as mentioned above, some BACSA associated countries are members of the EU; therefore, they should have access to EU funding for research institution and SMEs (Small and Medium Enterprises). However, until now the joint attempts for funding common projects have generally failed [18]. It is likely EC evaluators think that silk-focused projects have a minor impact on EC country development, with other agricultural/industrial activities considered more important than sericulture. Probably, the correct manner to attract financing is to enclose sericulture as a small part of wide projects focusing on other activities and where sericulture represents a case-study more than the central research or investment attractor. On the other hand, non-EC BACSA countries can have access to FAO funds or to those of other NGOs [18]. However, the lack of experts at the world level in the specific branch of sericulture often results in a minor attention to this theme. For example, two Korean experts gave a great personal contribution (Dr. Hoo Zoo Lea, FAO Senior officer and Dr. Jong Sung Lim, FAO consultant) to the BACSA creation, because of their knowledge and in-depth expertise in this sector [6]. Unfortunately, FAO now is missing these professional officers specialized in sericulture.

5.2.4. Governmental Subsidies

With regard to subsidies of governments to silk production, what occurred was that both in the EC and in the other BACSA countries, where they were applied, they did not prevent sericulture from the decline and, in some cases, they triggered fraud from farmer associations or other organizations. Therefore, it is clear enough that this instrument to guide the market might be useful only if coupled with a general policy of industrial and technological development of this sector.

5.3. Prospects for the Future

China, India, Brazil have probably already reached their maximum level of silk production and are not going to increase further their quote in the world market [23]. On the other hand, silk consumption, so far, has been a very small quote of total world fiber production (less than 1%) [24], but very stable over time, although with a diminishing trend in the last 5 years [21]. This stability of silk for the textile market is due to the general buyers' identification of silk with a luxury fiber, which is a part of human civilization; in many countries it is intrinsically connected with local culture, and it has a long-standing tradition. Silk is considered as "Queen of textiles"; it has some unique and important characteristics, such as its ability to keep the body warm when it is cool, and cool when it is warm, or being a healthy fiber because it breathes easily and naturally keeps away moisture from the skin or being actually soothing to skin diseases and itches. Therefore, according to the recent trend for a rising demand of natural fibers from the final consumers, who look for comfortable wearing and sustainability of clothes, an increase of the silk price is expected; this phenomenon might give opportunities to BACSA countries to be competitive with China, even considering that the top world quality silk fabrics and garment producing industries are concentrated in Europe (Italy, France, Switzerland, England).

Many BACSA countries might expand their artisanal and handicrafts production, especially by linking it to the tradition of their territories, countryside landscapes, culture. This idea gave birth to a project promoted by the Venice Municipality and the Council of Europe through the creation of a cultural itinerary "The European Silk Route"; it aims to be a local cultural network and infrastructure linking cities, regions, sites, museums and universities in order to enhance knowledge of a shared European cultural heritage, both tangible and intangible, and to promote new relationships within Europe and between Europe and the East through sharing of best practices and cultural tourism activities. The route would ideally be based on Marco Polo's travels eastward and include silk production and trade itineraries in Europe in the following centuries. The narrative to be developed will start from the commercial and religious exchanges that took place along the silk road, and specifically Marco Polo's trips. It will then analyze silk's impact in Europe through four main themes: the textile activity (from artisanal production to industrialization: innovations, technologies, the work world); silkworm rearing and its social, economic, agricultural and environmental consequences; the use of silk in paintings, fashion and design; and research and development in silk production [25]. The work undertaken to reach the certification of the cultural itinerary from the Institute of Luxembourg began in 2018 [25].

In addition to this exciting opportunity, the forecasts for the future represent that the demand of non-textile silk as constitutive proteins will increase at a steady rate, due to the new utilizations of silk as a versatile polymer for different aims (cosmetics, pharmaceuticals, biomedicals). Recently bio-technological sericulture has been developing [26,27]. For the first time in the world, in 2017, the legitimated rearing of genetically modified silkworms in conventional sericulture farms started in Japan. Functional silk is a promising material for medical applications. Using the methods of genetic engineering, absolutely new silks that have unprecedented functions were developed. These are transgenic spider silk [28], hyperfine silk of small diameter, artificial blood vessels, fluorescent silk [29]. Some of the BACSA countries are ready to face this biotechnological challenge. Silk regenerative medical materials like silk sponge, silk hernia mesh, wound dressing, silk surgical tape, hydrogel, films and 3D scaffolds for wound healing and tissue regeneration and reconstruction gels, powders, enzyme immobilization matrices were also created [30–34]. Transgenic sericin is used for several medical reagents, like blood test drugs, biomatrix for tissue engineering and cosmetics [35–38].

Therefore, new kinds of applications are likely to offer new opportunities; the interesting consideration is that, in this case, it is not necessary to produce silk in a huge quantity and it is not necessary to possess big reeling plants or transformation industries. These opportunities might allow BACSA countries to increase their production slowly and steadily. The region possesses some of the richest silkworm and mulberry germplasm collections. Several of the commercial silkworm hybrids, produced in the BACSA area manifest comparatively high productivity, namely single cocoon weight 2.2–2.5 g, shell ratio 23–24%, shell weight 0.500–0.600 g, filament length 1300–1500 m under laboratory conditions [39,40]. This might help in making the member countries attractive for this production. The level of sericultural science and technology in the region countries is comparatively high at a world level. This expertise might be particularly useful because new technological properties might be required for such a kind of silk production for innovative aims.

The EC green deal might also play a great role in promoting the development of sericulture in the BACSA countries: as mentioned before, sericulture might be an organic agricultural practice; the mulberry is environmentally useful to protect soil from erosion, to adsorb carbon dioxide, to prevent desertification in marginal areas; if it is exploited in a non-intensive way it requires limited fertilization and irrigation and no pesticides at all moriculture can also be practiced in polluted or salty soils to accelerate their recovery to production [41–43]. The silkworm is an environmental sentinel especially informative on the abuse of pesticides on agricultural crops cultivated nearby the rearing places. Furthermore sericulture and moriculture can be exploited for circular economies where by-products of some processes can become raw materials for others. Mulberry fruit can be consumed fresh dry or employed for production of juice, wine, jam and food additives. Pharmaceuticals can be extracted from mulberry branches, roots, leaves (for example, 1-deoxynojirimycin (DNJ) with antidiabetic aims [44]).

6. Conclusions

BACSA, being established only 16 years ago, is a rather young international organization, for example in comparison to another one in the same sector, the International Sericultural Commission (1960). BACSA is basically managed on a voluntary basis thanks to the work and support of individual members, mostly belonging to scientific institutions. According to its aims, it has been strengthening links and sharing knowledge among sericultural member countries, by giving a wide support to many actions dedicated to the revival of sericulture in Europe, Caucasus and Central Asia. Although this revival has not been possible yet on a large scale there are many hints about possible future developments so that the support action of this Association continues to be fundamental and would deserve more attention by the sector stakeholders.

Author Contributions: Writing—original draft preparation, P.T., S.C.; writing—reviewing and editing, A.S. All authors have read and agreed to the published version of the manuscript.

Funding: Project Serinnovation (Programma di sviluppo rurale per il Veneto 2014–2020–DGR N. 2175 del 23/12/2016 Misura 16); project RGV-FAO (V TRIENNIO DM 21076 del 25/07/2017).

Institutional Review Board Statement: Not applicable.

Informed Consent Statement: Not applicable.

Data Availability Statement: Data summarized in Figure 2 are available on the INSERCO web site at the following section/address: https://inserco.org/en/statistics (accessed on 23 November 2021).

Acknowledgments: We acknowledge the Veneto Region for the support to the research activity of CREA, through the project Serinnovation (Programma di sviluppo rurale per il Veneto 2014–2020–DGR N. 2175 del 23/12/2016 Misura 16) and the Italian Ministry of Agricultural, Food and Forestry Policies through the project RGV-FAO (V TRIENNIO DM 21076 del 25/07/2017).

Conflicts of Interest: The authors declare no conflict of interest.

References

1. Batsaikhan, U.; Dabrowski, M. Central Asia—Twenty-five years after the breakup of the USSR. *Russ. J. Econ.* **2017**, *3*, 296–320. [CrossRef]
2. Soergel, A. How the Fall of Communism in 1989 Reshaped Eastern Europe | Best Countries | US News. Available online: https://www.usnews.com/news/best-countries/articles/2019-11-08/how-the-fall-of-communism-in-1989-reshaped-eastern-europe (accessed on 23 November 2021).
3. Tzenov, P. Problems and prospects of sericulture preservation and revival in Europe, Causasus and Central Asia. In Proceedings of the SERIVIVAL, Batumi, Georgia, 7–12 April 2019; pp. 14–27.
4. Brunetti, G.; Visconti, F.; Marelli, M. *Euro e Distretti Industriali: Una Ricerca Nella Realtà Lombarda*; Franco Angeli—Fondazione CARIPLO per la Ricerca Scientifica: Milan, Italy, 2000; ISBN 8846419553.
5. Yamada, M.; Kawabata, Y.; Iikubo, M.; Vyacheslav, A.; Onwona-Agyeman, S. Revitalizing Silk-Road Silk Industry: A Case Study in Fergana Region, Uzbekistan. *J. Arid L. Stud.* **2012**, *22*, 179–182.
6. Tzenov, P.; Lea, H.Z. Regional strategies proposed for revival and promotion of sericultural industries and small enterprise development in the countries of Black, Caspian seas and Central Asia region. In Proceedings of the International Workshop on Revival and Promotion of Sericultural Industries and Small Silk Enterprise Development in the Black & Caspian Seas Region, Tashkent, Uzbekistan, 11–15 April 2005.
7. Tzenov, P.; Lea, H.Z. Silk Handcrafts Cottage Industries and Small Enterprises in Africa, Europe, Central Asia and the Near East. In Proceedings of the International Workshop on Silk Handcrafts Cottage Industries and Silk Enterprises Development in Africa, Europe, Central Asia and the Near East, Bursa, Turkey, 6–10 March 2006; pp. 19–61.
8. Tzenov, P. Present status and utilization of sericulture germplasm and comparative studies of different silkworm hybrids performance for sericultural enterprise development in the Black, Caspian seas and Central Asia (BACSA) region. In Proceedings of the Sericulture Challenges in the 21st Century—SERICHAL, Vratza, Bulgaria, 18–21 September 2007; pp. 13–35.
9. Bacsa Black, Caspian Seas and Central Asia Silk Association (BACSA) Proceedings. Available online: https://www.bacsa-silk.org/en/principles-of-work-organization/ (accessed on 23 November 2021).
10. Ichim, M.; Tanase, D.; Tzenov, P.; Grekov, D. Global trends in mulberry and silkworm use for non-textile purposes. In Proceedings of the Possibilities for Using Silkworm and Mulberry for Non-Textile Purposes, Plovdiv, Bulgaria, 23–26 September 2008; pp. 6–36.
11. Tzenov, P. Black, Caspian Seas and Central Asia Silk Association (BACSA) Activities for Revival and Development of Sericulture in Europe and Central Asia. In Proceedings of the Sericulture for Multi Products-New Prospects for Development—SERIPRODEV, Bucharest, Romania, 11–15 April 2011; pp. 13–18.
12. Tzenov, P.; Ichim, M.; Petkov, Z.; Vasileva, Y.; Arkova-Pantaleva, D. Identification and Possible Utilization of Some Silkworm Rearing Waste Products. In Proceedings of the Exploitation of Agricultural and Food Industry By-Products and Waste material Through Application of Modern Processing Techniques, Bucharest, Romania, 1–3 July 2008; pp. 42–49.
13. Tzenov, P.; Cappellozza, S. The role of regional cooperation between BACSA member countries in building value chains in sericulture. In Proceedings of the Building Value Chains in Sericulture—BISERICA, Padua, Italy, 7–12 April 2013; pp. 21–29.
14. Tzenov, P.; Ichim, M.; Mitova, D. The Organic Sericulture in the Context of Biological Agriculture and Organic Textile. In Proceedings of the Organic Sericulture—Now and the Future—ORGASERI, Sinaia, Romania, 19–24 April 2015; pp. 13–24.
15. Tzenov, P. Climate Changes Effect on Sericulture in Europe, Caucasus and Central Asia. In Proceedings of the Climate Changes and Chemicals—The New Sericulture Challenges—CLISERI, Sheki, Azerbaijan, 2–7 April 2017; pp. 9–16.
16. Ichim, M.; Tzenov, P.; Grekov, D.; Avramova, K. Mulberry Plantation Establishment Methods, Suitable for Europe, Caucasus and Central Asia. In Proceedings of the Sericulture Preservation and Revival—Problems and Prospects—SERIVIVAL, Batumi, Georgia, 7–12 April 2019; pp. 68–80.
17. Moorthy, M.; Sivaprasad, V.; Paramesh, B.; Bindhiya; Hukkeri, S.M.; Kumar, D.; Grekov, D.; Tzenov, P.; Teotia, R.S. Indo-Bulgarian Collaborative Research Project—A Way Forward in Silkworm Breed Improvement Programme in India and Bulgaria. In Proceedings of the Sericulture Preservation and Revival—Problems and Prospects—SERIVIVAL, Batumi, Georgia, 7–12 April 2019; pp. 158–159.
18. Bacsa Black, Caspian Seas and Central Asia Silk Association (BACSA) Proceedings. Available online: https://www.bacsa-silk.org/en/regional-project-proposal/ (accessed on 25 November 2021).
19. Bacsa Black, Caspian Seas and Central Asia Silk Association (BACSA) Proceedings. Available online: https://www.bacsa-silk.org/en/sellbuy-information/ (accessed on 25 November 2021).
20. BACSA Membership: The Black, Caspian Seas and Central Asia Silk Association (BACSA). Available online: https://www.bacsa-silk.org/en/membership/ (accessed on 25 November 2021).

21. International Sericultural Commission INSERCO Statistics. Available online: https://www.inserco.org/en/statistics (accessed on 23 November 2021).
22. Cappellozza, S.; Toso, L.; Saviane, A. La collezione di germoplasma di baco da seta (*Bombyx mori*) e gelso appartenente all'Unità di Apicoltura e Bachicoltura di Bologna, sede di Padova. In *Conservazione Biodiversità, Gestione Banche Dati e Miglioramento Genetico—BIODATI*; D'Andrea, F., Ed.; Edizioni Nuova Cultura: Roma, Italy, 2013; pp. 961–990. ISBN 9788868120986.
23. De Ponti, P. European Silk Forum—Per quanto tempo ancora la Cina fornirà la seta all'Europa? *La Seta* **2011**, *63*, 13–15.
24. Textile Exchange Preferred Fiber & Materials—Market Report 2020. *Prefer. Fiber Mater.—Mark. Rep. 2020* **2020**, 42–43. Available online: https://textileexchange.org/wp-content/uploads/2020/06/Textile-Exchange_Preferred-Fiber-Material-Market-Report_2020.pdf (accessed on 30 November 2021).
25. Council of Europe-Venice-European Silk Route: About the Project. Available online: https://www.coe.int/en/web/venice/about-the-project (accessed on 25 November 2021).
26. Tzenov, P.; Grekov, D. Research Achievements and Future Trends of the Silkworm *Bombyx mori* L. Breeding Work in Bulgaria. In Proceedings of the Recent Trends in Seribiotechnology—ICTS, Anantapur, India, 27–29 March 2008; pp. 43–54.
27. Tzenov, P.; Kipriotis, E. Product Diversification—An Alternative for Sericulture Development in the Black, Caspian Seas and Central Asia Region Countries. In Proceedings of the 21st Congress of the International Sericultural Commission, Athens, Greece, 3–6 November 2008; pp. 232–240.
28. Xu, J.; Dong, Q.; Yu, Y.; Niu, B.; Ji, D.; Li, M.; Huang, Y.; Chen, X.; Tan, A. Mass spider silk production through targeted gene replacement in *Bombyx mori*. *Proc. Natl. Acad. Sci. USA* **2018**, *115*, 8757–8762. [CrossRef] [PubMed]
29. Iizuka, T.; Sezutsu, H.; Tatematsu, K.I.; Kobayashi, I.; Yonemura, N.; Uchino, K.; Nakajima, K.; Kojima, K.; Takabayashi, C.; Machii, H.; et al. Colored fluorescent silk made by transgenic silkworms. *Adv. Funct. Mater.* **2013**, *23*, 5232–5239. [CrossRef]
30. Kapoor, S.; Kundu, S.C. Silk protein-based hydrogels: Promising advanced materials for biomedical applications. *Acta Biomater.* **2016**, *31*, 17–32. [CrossRef] [PubMed]
31. Huang, W.; Ling, S.; Li, C.; Omenetto, F.G.; Kaplan, D.L. Silkworm silk-based materials and devices generated using bio-nanotechnology. *Chem. Soc. Rev.* **2018**, *47*, 6486–6504. [CrossRef] [PubMed]
32. Chouhan, D.; Mandal, B.B. Silk biomaterials in wound healing and skin regeneration therapeutics: From bench to bedside. *Acta Biomater.* **2020**, *103*, 24–51. [CrossRef] [PubMed]
33. Nguyen, T.P.; Nguyen, Q.V.; Nguyen, V.; Le, T.; Le, Q. Van Silk Fibroin-Based Biomaterials for Biomedical. *Polymers* **2019**, *11*, 1933. [CrossRef] [PubMed]
34. Teramoto, H.; Iga, M.; Tsuboi, H.; Nakajima, K. Characterization and scaled-up production of azido-functionalized silk fiber produced by transgenic silkworms with an expanded genetic code. *Int. J. Mol. Sci.* **2019**, *20*, 616. [CrossRef] [PubMed]
35. Tomita, M. Transgenic silkworms that weave recombinant proteins into silk cocoons. *Biotechnol. Lett.* **2011**, *33*, 645–654. [CrossRef] [PubMed]
36. Arango, M.C.; Montoya, Y.; Peresin, M.S.; Bustamante, J.; Álvarez-López, C. Silk sericin as a biomaterial for tissue engineering: A review. *Int. J. Polym. Mater. Polym. Biomater.* **2021**, *70*, 1115–1129. [CrossRef]
37. Teramoto, H.; Iizuka, T.; Kameda, T.; Okada, E.; Sezutsu, H. Application of high molecular weight intact sericin as a cosmetic material. **2013**, 34–35. Available online: http://www.naro.affrc.go.jp/archive/nias/eng/research/h25/nias2013-16.pdf (accessed on 30 November 2021).
38. Teramoto, H.; Nakajima, K.I.; Takabayashi, C. Preparation of elastic silk sericin hydrogel. *Biosci. Biotechnol. Biochem.* **2005**, *69*, 845–847. [CrossRef] [PubMed]
39. Tzenov, P. Silkworm *Bombyx mori* L. Selection and Breeding in Bulgaria Recent Achievements. In Proceedings of the ASEAN Sericulture Conference 2010 and ASEAN Silk Fabric and Fashion Design Contest 2010, Nontaburi, Thailand, 22–27 August 2010.
40. Tzenov, P.; Grekov, D.; Vasileva, Y.; Moorthy, M.; Sivaprasad, V. Prospective hybrids of Bulgarian uni-bivolyine silkworm breeds and Indian polyvoltine breed, C.NICHI. *Sericologia* **2019**, *59*, 9–12.
41. Rohela, G.K.; Shukla, P.; Muttanna; Kumar, R.; Chowdhury, S.R. Mulberry (*Morus* spp.): An ideal plant for sustainable development. *Trees For. People* **2020**, *2*, 100011. [CrossRef]
42. Wan, X.; Lei, M.; Chen, T.; Tan, Y.; Yang, J. Safe utilization of heavy-metal-contaminated farmland by mulberry tree cultivation and silk production. *Sci. Total Environ.* **2017**, *599–600*, 1867–1873. [CrossRef] [PubMed]
43. Alahabadi, A.; Ehrampoush, M.H.; Miri, M.; Ebrahimi Aval, H.; Yousefzadeh, S.; Ghaffari, H.R.; Ahmadi, E.; Talebi, P.; Abaszadeh Fathabadi, Z.; Babai, F.; et al. A comparative study on capability of different tree species in accumulating heavy metals from soil and ambient air. *Chemosphere* **2017**, *172*, 459–467. [CrossRef] [PubMed]
44. Marchetti, L.; Saviane, A.; Montà, A.D.; Paglia, G.; Pellati, F.; Benvenuti, S.; Bertelli, D.; Cappellozza, S. Determination of 1-deoxynojirimycin (1-dnj) in leaves of italian or italy-adapted cultivars of mulberry (*Morus* sp.pl.) by hplc-ms. *Plants* **2021**, *10*, 1553. [CrossRef] [PubMed]

insects

MDPI

Review

Advances in Editing Silkworms (*Bombyx mori*) Genome by Using the CRISPR-Cas System

Gabriela-Maria Baci [1], Alexandra-Antonia Cucu [1], Alexandru-Ioan Giurgiu [1], Adriana-Sebastiana Muscă [1], Lilla Bagameri [1], Adela Ramona Moise [1,*], Otilia Bobiş [1], Attila Cristian Raţiu [2,*] and Daniel Severus Dezmirean [1]

1 Faculty of Animal Science and Biotechnology, University of Animal Sciences and Veterinary Medicine Cluj-Napoca, 400372 Cluj-Napoca, Romania; gabriela-maria.baci@usamvcluj.ro (G.-M.B.); antonia.cucu@usamvcluj.ro (A.-A.C.); alexandru.giurgiu@usamvcluj.ro (A.-I.G.); adriana-sebastiana.musca@usamvcluj.ro (A.-S.M.); lilla.bagameri@usamvcluj.ro (L.B.); obobis@usamvcluj.ro (O.B.); ddezmirean@usamvcluj.ro (D.S.D.)

2 Faculty of Biology, University of Bucharest, 050095 Bucharest, Romania

* Correspondence: adela.moise@usamvcluj.ro (A.R.M.); attila.ratiu@bio.unibuc.ro (A.C.R.)

Simple Summary: One of the most powerful gene editing approaches is the CRISPR (clustered regularly interspaced short palindromic repeats)-Cas (CRISPR-associated) tool. The silkworm (*Bombyx mori*) has a great impact on the global economy, playing a pivotal role in the sericulture industry. However, *B. mori* came into the spotlight by representing one of science's greatest contributors, being used to establish extraordinary bioreactors for the production of target proteins and illustrating a great experimental model organism. Herein, we focus on progress made in the field of *B. mori*'s genome manipulation by using CRISPR-Cas. In order to edit *B. mori*'s genome, remarkable advances were made, such as exposing gene functions and developing mutant lines that exhibit enhanced resistance against *B. mori* nucleopolyhedrovirus (BmNPV). We also discuss how CRISPR-Cas accelerated the fundamental investigation in *B. mori*, and beyond, thus highlighting the great potential of the insect's biotechnology in numerous scientific fields.

Abstract: CRISPR (clustered regularly interspaced short palindromic repeats)-Cas (CRISPR-associated) represents a powerful genome editing technology that revolutionized in a short period of time numerous natural sciences branches. Therefore, extraordinary progress was made in various fields, such as entomology or biotechnology. *Bombyx mori* is one of the most important insects, not only for the sericulture industry, but for numerous scientific areas. The silkworms play a key role as a model organism, but also as a bioreactor for the recombinant protein production. Nowadays, the CRISPR-Cas genome editing system is frequently used in order to perform gene analyses, to increase the resistance against certain pathogens or as an imaging tool in *B. mori*. Here, we provide an overview of various studies that made use of CRISPR-Cas for *B. mori* genome editing, with a focus on emphasizing the high applicability of this system in entomology and biological sciences.

Keywords: *Bombyx mori*; CRISPR-Cas; silkworms; genome engineering; insect biotechnology; entomology

Citation: Baci, G.-M.; Cucu, A.-A.; Giurgiu, A.-I.; Muscă, A.-S.; Bagameri, L.; Moise, A.R.; Bobiş, O.; Raţiu, A.C.; Dezmirean, D.S. Advances in Editing Silkworms (*Bombyx mori*) Genome by Using the CRISPR-Cas System. *Insects* **2022**, *13*, 28. https://doi.org/10.3390/insects13010028

Academic Editors: Silvia Cappellozza, Morena Casartelli, Federica Sandrelli, Alessio Saviane and Gianluca Tettamanti

Received: 19 November 2021
Accepted: 23 December 2021
Published: 27 December 2021

1. Introduction

The life sciences research fields were revolutionized by the outstanding development of various genome editing tools. By using specific techniques of genome editing, the genomic DNA of every living organism can be submitted to guided changes, such as deletions, insertions, and sequence substitutions [1].

In recent years, several genome editing tools have been in the spotlight. Among them, there are three remarkable technologies, namely those relying on programmable nucleases (i.e., the transcription activator like effector nucleases (TALENs)), zinc finger nucleases (ZFNs), and clustered regularly interspaced short palindromic repeat - associated nucleases (CRISPR-Cas) [2,3]. Currently, by using engineered nucleases, remarkable advances are being made regarding the correction of genetic mutations, gene expression regulation, and the development of therapeutic agents; these approaches are also used for a better understanding of gene functions and the mechanisms underlying the development of certain genetic disorders or various diseases [4].

When it comes to genome editing, it is crucial to avoid off-target effects, but overall, the CRISPR-Cas system exhibits reliable results, owing to a great degree of fidelity [4,5]. Since its discovery in bacteria, the CRISPR-Cas system has been continuously exploited, representing an extremely versatile tool for the scientific community due to its reprogrammable feature. Currently, this system is used to edit the genomes of various organisms, such as bacteria, insects, plants, or human cells [6].

For more than 5000 years, humans have been involved in the domestication of one of the most economically important insects, namely *Bombyx mori*. Historically, *B. mori* gained prominence due to silk production, but now it is one of the most valued model organisms for life science branches. Just to mention a few recent achievements, *B. mori* has been used as a bioreactor, for recombinant proteins production [7–10], while its silk is used to produce extraordinary silk-based biomaterials that exhibit great importance for the medical field [11]. The silk proteome contains two major proteins, namely fibroin and sericin. The silk gland consists of three anatomical areas, namely the posterior (PSG), the middle (MSG), and the anterior (ASG) silk gland. Fibroin is synthesized in the PSG and sericin is assembled in the MSG [12].

In 2018, Ma et al. [13] excellently summarized the advances of using genome editing tools in silkworms. However, herein, we mainly focus on the potential of CRISPR-Cas technology in editing *B. mori*'s genome [13]. We navigate through the recent progresses in using the outstanding CRISPR-Cas system in *B. mori* and discuss the latest studies that utilized this approach in order to investigate the genes function, to regulate the gene expression and to enhance the resistance against *B. mori* nucleopolyhedrovirus (BmNPV). Additionally, the key role of *B. mori* in the scientific fields will be discussed.

2. The CRISPR-Cas System

CRISPR-Cas is one of the key methods employed by many molecular biology scientific laboratories. Since its first description [14], genome editing focused research was implemented by countless research groups [15–17].

2.1. The CRISPR-Cas Complex Role in the Immunity System

When investigating the *iap* gene product in the opportunistic pathogen *Escherichia coli*, Ishino et al. (1987) [18] observed an atypical structure, specifically the repetition of several homologous sequences. Later, this type of structure was observed in various bacterial, as well as archaeal strains [19,20]. Subsequently, these repetitive sequences were linked with exogenous genetic material, and following several years, their assembling mechanism and function were elucidated [21]. This type of sequence can be placed on the chromosomal DNA, but it can also be found on the plasmid DNA [22].

The scientists demonstrated that CRISPR-Cas, which is present in one-third of bacteria and nearly in all archaea, has a key role in host's adaptive immunity. It protects the organism against various intruders, such as viruses, but it also offers protection against other mobile genetic elements, such as transposons or plasmids [23].

The CRISPR-Cas system structure includes three main components, i.e., the CRISPR arrays, the associated Cas proteins, and the leader nucleotide sequence. The first genetic component, the CRISPR locus, is characterized by identical repeats structures (21–37 bp) that are highly conserved and the spacer sequences that are acquired fragments of invader's nucleic acid material. The CRISPR array is located downstream from *cas* genes. The latter encodes for Cas proteins that are crucial to the immune reaction [24].

Initially, only four distinct Cas proteins (1–4) were reported, due to the rapid evolution of biological sciences; currently, there numerous Cas proteins have been described [25,26], Cas1 being the most analyzed [27].

CRISPR-Cas has a great adaptability, with host-related specificities; thus, it exhibits a significant diversity. The varying feature is defined by the CRISPR array and the *cas* gene sequences. The classification of these types of systems is based on the signature Cas proteins. Currently, there are two major classes of CRISPR-Cas systems, each also di-vided in several groups [28]. Regarding the leader nucleotide sequence, it has been shown that it has a key role by carrying the essential promoter sequences for the transcription of CRISPR loci. Besides the promoter, the leader contains specific signals that are crucial for the adaptation stage from the first phase of CRISPR-Cas activation [29].

The adaptation is the first functional stage of the CRISPR-Cas mechanism, during which the foreign nucleic acid is recognized by several Cas proteins [30] and consequently integrated next to a leader sequence. Through this mechanism, in evolution, the spacers are arranged chronologically, and this feature helps bacteria and archaea to enhance their protection against the genetic material of the latest foreign encounter [31]. Each new acquired spacer is accompanied by a repeat sequence; therefore, the CRISPR array expands with every invasion [32].

The CRISPR array is transcribed in the second step, specifically in the biogenesis phase [33]. First of all, it is being transcribed into a precursor CRISPR RNA (crRNA). At the end of this phase, there are numerous mature crRNAs molecules, resulting from the action of RNase III that process the precursor crRNA. Each crRNA includes a spacer and a repeat sequence [31,34,35].

The last step of the system's mechanism is the interference phase. It involves the degradation of the foreign nucleic acid, by targeting and cleaving it [36]. The products of the biogenesis phase, the crRNAs, act like guides for targeting the invader, which is then cleaved following a Cas proteins cascade that act like molecular scissors [37].

2.2. The CRISPR-Cas System as a Genome Editing Tool

When it comes to leading tools in genetic engineering, the CRISPR-Cas system can be considered the foremost instrument. After elucidating its function in various organisms, scientists aimed to exploit its versatility, in order to overcome the disadvantages of other available genome editing tools [38]. Even if scientific studies still report the use of ZFNs and TALENs techniques as editing tools, the CRISPR-Cas system is the most effective genome editing instrument, standing on top in regards to efficiency, cost-effectiveness, and the relative simplicity of use [39] (Table 1). Another considerable advantage of this system is represented by its capacity to simultaneously target multiple genes [40].

Table 1. Comparison between TALEN, ZFN, and CRISPR-Cas gene editing technologies.

Traits	TALEN	ZFN	CRISPR-Cas	References
Origin	Prokaryotic	Eukaryotic	Prokaryotic	[41]
Efficiency (%)	76	12	81	[42]
Specificity	Moderate	Low	High	[43–46]
Target site recognition	Any site	Any site	Pam motif (NGG) required	[43]
Multiplex potential	Low	Low	High	[43,45]
Processing time	Time consuming	Time consuming	Short	[45]
Methylation sensitive	Sensitive	Sensitive	Not sensitive	[42]
Engineering feasibility	Moderate/High	Moderate	Moderate/High	[42,45]
Dimerization required	Yes	Yes	No	[44]
Cost effectiveness	Moderate	No	Yes	[43,44]

Of the numerous CRISPR-Cas systems, CRISPR-Cas9 is currently the most used instrument in laboratories across the world [47]. The Cas9 nuclease is the signature protein of CRISPR-Cas II systems and it is responsible for double strand DNA breaks [27]. Three different methods to deliver the Cas9 endonuclease have been described. It can be directly delivered by microinjection into the embryos, while the other two delivery methods involve a plasmid that expresses the Cas9 enzyme, or a messenger RNA (mRNA) sequence that encodes it. Of the three techniques, in terms of genome engineering, the earliest mentioned is the best option due to its certain advantages. By directly delivering the protein, low immunogen effects were observed. Furthermore, the off-target activity is minimized compared with the other two methods [48]. The use of CRISPR-Cas9 is a simple but powerful genome editing tool, with various implementations, and their impact on new research trends has been reviewed elsewhere [49].

The CRISPR-Cas9 mechanism relies on the Cas9 nuclease and a guide sequence (gRNA). As the name implies, the gRNA has the role to guide the Cas9 nuclease to a target site in order to cleave the DNA. The key feature of gRNA is the extensive complementarity with the target sequence [50].The protospacer-adjacent motif (PAM) bordering the target complementary sequence has a key role, since in its absence, the CRISPR-Cas systems would degrade their own CRISPR loci. In order to perform a cleavage, the Cas9 protein scans for the PAM sequence. Even if the gRNA is complementary with the target sequence, the Cas9 endonuclease will not cleave it in the absence of PAM [51].

The central factor that influences the success of the gene editing process is the repair path of the double-strand breaks produced by Cas9. There are two main repair pathways: the homology-directed repair (HDR) and the nonhomologous end joining (NHEJ), respectively [52]. More often, NHEJ is exploited in order to acquire indels mutations, specially to obtain small deletions. These deletions are extremely useful for disclosing gene functions [53]. However, the HDR machinery is used not just to obtain knock-out or knock-down mutations, such as the expected output following NHEJ, but to generate target knock-ins. Therefore, by using HDR, exogenous sequences can be successfully integrated into the host's genome. Currently, major efforts are being made in order to enhance the sequence replacement by using the HDR mechanism [54].

CRISPR-Cas9 is currently used in multiple research fields, such as agriculture (editing of various agricultural plant genomes or pest insect's genome) [55–57], biotechnology, food industry, and medicine (modeling diseases using HeLa cells, deciphering HIV infection mechanisms, using various experimental models, such as *Danio rerio* to tackle cancer and neurological diseases, etc.) [58,59], just to mention a few (Figure 1) [60].

2.3. CRISPR-Cas9 in Entomology

Being the most diverse and numerous category of organisms for decades [61], insects have been intensively studied. Countless studies have been performed due to insects' key roles in ecology, agriculture, and medicine [62,63]. Considering this, numerous research groups aimed to use the CRISPR-Cas9 system to manipulate the insects' genome. The first application of CRISPR-Cas9 was performed in *Drosophila melanogaster* [64] due to its strategic importance as arguably the main experimental model organism for life sciences [65]. Besides *D. melanogaster*, the researchers also used the CRISPR-Cas9 applicability on *B. mori*, *Apis mellifera*, *Aedes aegypti*, and *Tribolium castaneum* [66,67].

Gratz et al. (2013) [68] programmed CRISPR-Cas9 to edit *Drosophila's* genome. The authors targeted the *yellow* gene, which is commonly used in various studies. First, they aimed to determine if this genome engineering tool could be efficient and could fulfill its role to induce breaks in the target sequence. By using the CRISPR-Cas9 system in *Drosophila*, not only was the *yellow* gene successfully knocked out, but the genome's alterations were also germline transmitted. Subsequent to the deletion of the target gene, a donor sequence was designed. This sequence provided the template for the HDR repair pathway and its use was to test the accuracy of specific replacement of *yellow* gene with an exogenous sequence. These sequence replacements were transmitted to descendants as well. Their data showed that there was no off-target activity and it highlighted the feasibility of using the CRISPR-Cas9 technology in eukaryotes [68].

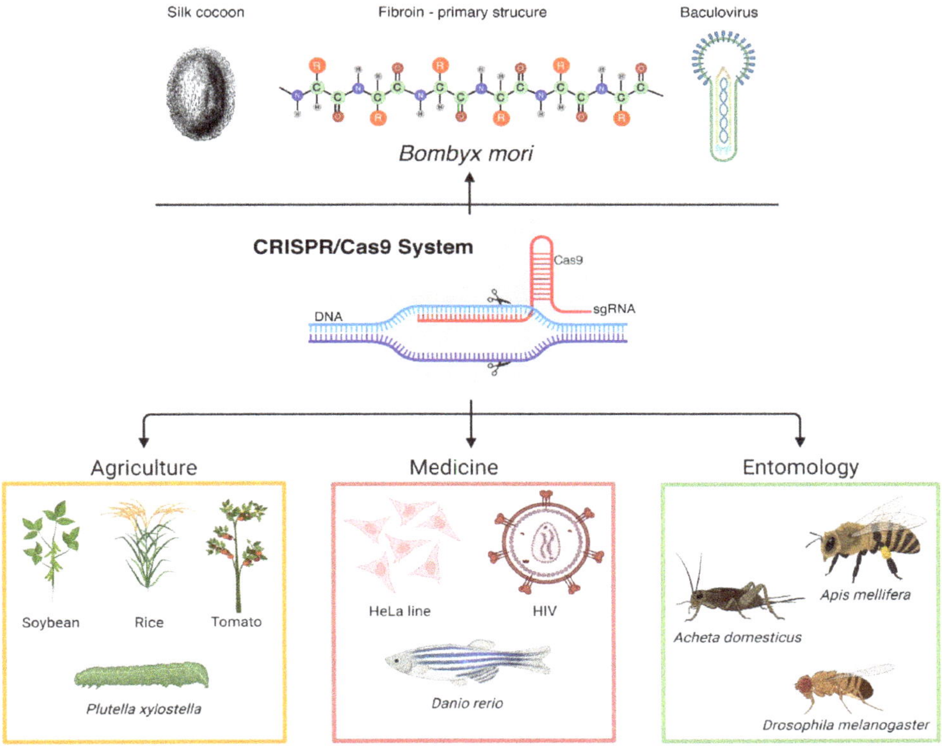

Figure 1. Schematic representation of the most important current applications of CRISPR-Cas9 in entomology, medicine, and agriculture. On top, a simplified description of CRISPR-Cas9 applicability in *B. mori* that is extensively described in the main text (created with BioRender.com, accessed on 2 December 2021).

Aiming to further highlight the feasibility of choosing this system to perform genome alteration in *Drosophila*, Yu et al. (2013) [69] designed two gRNAs to induce mutations in two regions of the *yellow* gene. In addition, they targeted other six sequences, both euchromatic and heterochromatic loci. Remarkably, a definite mutation in *ms(3)k81* was transmitted to descendants in a proportion of 100%. By successfully targeting heterochromatic loci, their result showed that the CRISPR-Cas9 system is efficient for altering the heterochromatin [69]. *Drosophila* have been used in numerous studies in order to examine the insecticide resistance [70–72]. In this direction, Douris et al. (2020) [56] notably summarized the progress in using CRISPR-Cas9 to explore the genetic basis of this mechanism.

The CRISPR-Cas9 technique was used to perform functional analysis concomitantly on two genes belonging to the cricket (*Gryllus bimaculatus)* [73]. *G. bimaculatus* is an important insect for experimental studies; for example, it plays an important role for evolutionary developmental studies and comparative biology, but it is also a relevant model organism for neurobiology and behavioral sciences [74]. The efficiency of inserting a donor sequence via a homology-independent technique was tested in two *hox* genes, namely *Gb-Ubx* and *Gb-abd-A*. After inserting the donor fragment into essential exons of both genes, their function was lost. Thus, functional investigations of *hox* genes could be carried out by using the knock-in/knock-out approaches [73].

Being one of the most important social insects [75] and as it plays a crucial role as a pollinator, the honeybee (*Apis mellifera*) has been intensively studied. It also plays a pivotal role in various therapeutic areas due to honey production. This natural product has extraordinary benefits for human health, exhibiting antioxidant, antiviral, and antibacterial effects [76]. Due to its special characteristics, the use of honey is not limited to humans, but this natural product is being used to improve certain features of other insects, such as silkworms [77]. There are numerous studies that detail functional analysis of *A. mellifera* genes by exploiting the CRISPR-Cas9 system. For instance, Hu et al. (2019) [67] reported the successful utilization of this system for knocking out the *mrjp 1* gene from the honeybee genome. The CRISPR-Cas9 complex was delivered through microinjection and they tested two specific regions of embryos, for identifying the most convenient structure for delivering the gRNA and the Cas9 endonuclease. By microinjecting the construct at the dorsal posterior side, there was a low rate of successful manipulation (11.8%); however, when choosing the ventral cephalic side, the results showed a great rate of gene editing (93.3%) Trying to validate the previous results, the authors also targeted *pax6*. Based on the previously obtained results, they microinjected the CRISPR-Cas9 construct at the ventral cephalic side. The results showed an editing rate of 100% [67]. Targeting the same gene, *mrjp 1*, similar results have been obtained in another study [78]. Thus, functional analysis of *A. mellifera* genes can be effectively performed by using the CRISPR-Cas9 system.

Considering the same topic of gene function research, Nie et al. (2021) [79] used the CRISPR-Cas9 technology to determine if the *yellow-y* gene plays a crucial role in the process of cuticular melanin synthesis in *A. mellifera*. They targeted this gene due to its great potential for mutants screening, being a selectable marker. By disrupting it, the phenotype of worker cuticle has changed, mainly due to the black pigment decreasing, thus confirming the *yellow-y* gene critical role in melanin pigmentation. However, as future prospects, this could be a great genetic marker for upcoming genomic research [79].

Referring to *A. mellifera* sex determination, it is controlled by the heterozygosity at a particular locus that harbors the key *complementary sex determiner* (*csd*) gene. The bees that are heterozygous at this specific locus are females, while the males are homozygous or hemizygous [80]. In a recent study, Wang et al. (2021) [81] used the CRISPR-Cas9 tool in order to knock out the *csd* gene and, thus, eliminated the genetic difference between females and males. Subsequently, they aimed to observe the transcriptome difference between the two sexes in this particular genetic background. They also successfully induced target mutations in mutant haploid individuals. It was observed that the expression level of several male-biased genes was higher in the mutant males. On the other hand, the expression level of several specific female-biased genes was lower. Their data also confirmed that *csd* interacts with certain genes, such as *fruitless*, *troponin T*, and *transformer-2* just to mention a few [81].

3. *Bombyx mori*

For numerous reasons, *B. mori* is one of the most studied insects, especially because it presents a real interest for the scientific community. It has been completely domesticated and plays the pivotal role in sericulture, being reared principally for silk production on a large scale [82]. Considering the great role of silk in the textile industry and its use as a biomaterial in medicine, major efforts are being made to enhance silk's quality, but also to increase its quantity [12]. Moreover, there are numerous studies that describe the process of obtaining enhanced silk with appreciable properties by genetically manipulating the silkworms [83–86]. *B. mori*'s genome was manipulated in order to enhance silk's properties. For instance, by knocking in the major ampullate silk protein from spiders into *B. mori*'s genome, a research group obtained silk with superior mechanical characteristics [84].

B. mori is an oligophagous insect and its main food source are the mulberry leaves, a nutritional preference that influences its biological and economical parameters. There is a major drawback when it comes to this source of nutrition, namely the limited availability of mulberry leaves. In order to be able to rear silkworms not only in spring and summer, artificial diets are currently being used for their nourishment [77,87].

Regardless of human medicine advances, certain microorganisms developed survival strategies and numerous infectious diseases remain a crucial problem worldwide [88]. In this context, *B. mori* is a reliable experimental model and exhibits the great advantage of a short development cycle that consists of four different stages. The first period of growth is the egg phase, followed by the larvae, pupa, and moth phases. The larvae phase plays a crucial role in silkworm's development, being the longest phase and including five different stages. Another significant advantage of using *B. mori* is the fact that in the last instar larvae, the body size has nearly 5 cm; hence, it is facile to manipulate or to exploit it for various purposes. Moreover, this size also facilitates the body's dissection; thus, multiple target tissues or organs can easily be obtained [89,90].

B. mori as a Model Organism

The level of attention received as a model experimental organism increased, since several groups made available *B. mori*'s genome data [91]. Various mutant strains have been described and the genetic analysis confirmed the numerous genetic traits. Another important aspect is that its manipulation is not associated with ethical concerns [90].

Being susceptible to various infectious agents, *B. mori* is one of the most used experimental models for drug screening, evaluation of different virulence factors, and identifying the pathogen genes responsible for its virulence [90]. Hitherto, several studies used *B. mori* to evaluate the effectiveness of antibiotics against certain human pathogens [92–94]. In a recent study, silkworms were used to examine the efficacy of three different glycopeptide antibiotics against *Staphylococcus aureus* infection. As a result, authors highlighted the great feasibility and efficacy of using *B. mori* to mimic bacterial infections in order to examine the therapeutic potential of antibiotics [95]. Moreover, recently, silkworms were used to evaluate the impact on several antibacterial compounds against *Cutibacterium acnes* [96]. As an experimental model organism, *B. mori* is currently being intensively used for various other purposes, as detailed in Table 2.

Table 2. Summary of various uses of *B. mori* as an experimental model organism.

Type of Model Organism	Brief Description	Purpose	References
Human disease model	Transgenic *B. mori* expressing *hIR* (human insulin receptor)	Drug evaluation for diabetes treatment	[97,98]
	Bacterial (*Listeria monocytogenes*) infection model in *B. mori*	Evaluating the interaction between host and pathogen; investigating the activity of vitamin A against microbial infections	[99]
	Fungal (*C. albicans*) infection model in silkworms	Assessing the *C. albicans* biofilm development	[100]
	Inducing deletions in the *BmSpr* gene, leading to sepiapterin reductase deficiency	Treatment options evaluations against sepiapterin reductase deficiency	[101]
Model for pesticide toxicity	Exposing silkworms to phoxim	Identifying specific biomarkers for phoxim stress; evaluating the toxicity reaction and the pretreatment with nanoparticulate titanium dioxide	[102]
	Inducing genotoxicity by feeding the silkworms with different doses of avermectin	Exploration of certain genes that are required for the DNA repairing mechanism	[103]
	Treating *B. mori* larvae with Fenvalerate-20EC	Evaluation of Fenvalerate-20EC impact on several digestive enzymes	[104]

Table 2. *Cont.*

Type of Model Organism	Brief Description	Purpose	References
Model for drugs toxicity	Injecting the silkworms with three different pharmacologically active agents (4-methyl umbelliferone, 7-ethoxycoumarine)	Evaluation of the metabolic pathway of these compounds	[92]
	Exposing the silkworms to fungal infections	Exploring pharmacokinetic parameters of an antifungal agent, Voriconazole	[105]
	Injecting cytotoxic drugs into *B. mori* larvae	Evaluation of cytotoxic drugs impact	[106]
Model for nanomaterials toxicity	Spreading silver nanoparticles on mulberry leaves	Toxicity evaluation of silver nanoparticles	[107–110]
	Injecting subcutaneously zinc oxide nanoparticles	Evaluation of zinc oxide nanoparticles toxicity, accumulation, and distribution	[111]
	Injecting in the dorsal vein different nanoparticles with great interest in various life science branches	Investigation of different silicon and carbon nanomaterials toxicity level against hemocytes	[112]

4. Applications of CRISPR-Cas in *B. mori*

The first communication of successful manipulation of *B. mori*'s genome by using the CRISPR-Cas9 tool, was reported by Wang et al. (2013) [113]. The authors targeted an essential gene [113], *BmBlos2*, that is orthologous to the *Blos2* human gene [114]. Two sgRNAs (23-bp) were designed to induce mutations leading to loss of target gene function. Each complex formed by one sgRNA and the Cas9 nuclease was injected in the preblastoderm embryonic stage. Ordinarily, the larval integument is opaque, but when the *BmBlos2* gene function is lost, the tegument becomes translucent. This effect could be viewed as a phenotypic marker for mutant's detection. Of all individuals, 94% and 95.6%, respectively were successfully edited by using the two sgRNAs. This study highlighted the feasibility of using CRISPR-Cas9 not only in *B. mori*, but also in other lepidopteran insects. These findings are of great interest by revealing CRISPR-Cas9 system's applicability in pest control approaches [113].

The multiplexable potential of CRISPR-Cas9 technology was highlighted by Liu et al. (2014) [115]. First, the *BmBlos2* gene was targeted for site specific mutagenesis, in order to confirm the feasibility of using the CRISPR-Cas9 system in *B. mori*. Following this, other six genes were targeted to confirm the system's multiplexable feature: *tyrosine hydroxylase*, *red egg*, *yellow-e*, *kynureninase*, *ebony*, and *flugellos*. Mutations were induced in each target gene, without evidence for the system's off-target activity [115]. The multiplexable feature of CRISPR-Cas9 is facilitating the process of genome engineering by simultaneously and precisely inducing mutations in different sites. This property of CRISPR-Cas9 enables researchers to perform precise and elaborate target mutagenesis in a time-effective manner [116].

Other approaches aimed to knock out the *BmKu70* gene by using CRISPR-Cas9 in order to target its second exon [117]. The *BmKu70* gene is coding a highly conserved protein, Ku70, which plays a key role in numerous mechanisms: cell adhesion, apoptosis, the maintenance of telomeres length, etc. In addition, numerous studies reported that by inactivating the *BmKu70* gene, the frequency of homologous repair is increased. In order to test this hypothesis, the authors performed a transient analysis in genetically manipulated embryos. They knocked in the *Em702* gene that is found on the Z chromosome. Their results confirmed that by knocking out *BmKu70* the homologous repair frequency is expanded. These promising results are of great interest for fundamental research in *B. mori*. [117].

It has been demonstrated by Fujinaga et al. (2017) [118] that in *B. mori*, the insulin-like growth factor-like peptide (IGFLP), is closely related to the genital disc, particularly involved in its growth. Due to the lack of studies performed in vivo on this topic, the same research group aimed to confirm the role of IGFLP by inactivating it [119]. For this purpose, they used the CRISPR-Cas9 genome editing tool. The absence of IGFLP leads to smaller ovaries and a lower number of laid eggs compared with the wild type. On the other hand, the size of laid eggs and its development were not affected. These findings indicated that this hormone has no impact on *B. mori*'s fertility. The ecdysteroids play a crucial role in IGFLPs production, by inducing gene expression. Furthermore, it has been shown previously that ecdysteroids have a key role in ovary development. Therefore, the authors initially appraised that a low ecdysteroids titre caused reduced ovary weights. However, by analyzing the transgenic females, their data showed that the ecdysteroids titre was the same. This study reveals insights on IGFLPs impact on ovarian development [119].

In a recent study [120], a research group explored the effect of activating the *BmFibH* gene in the *B. mori* embryonic cells. For this purpose, they constructed a complex that involved the inactive form of Cas9 nuclease (dCas9) and a VPR activating domain, driven by a certain promoter. This activating domain consists of several different activators: VP64, p65, and Rta, respectively. When it comes to sgRNAs, three specific constructs were designed to target the promoter of the gene of interest. In order to confirm the success of *BmFibH* activation, first they determined its expression in normal embryonic cells. Their results showed that the *BmFibH* gene is strongly downregulated in untransformed cells. On the other hand, the transfected cells exhibited a higher *BmFibH* expression level. Moreover, their data showed that the activation of the target gene impacted the cellular stress responses [120]. Cui et al. (2018) [121] targeted the *BmFibH* gene in order to explore its role in the development of the silk gland. After the CRISPR-Cas9 construct was designed, a total number of 630 eggs were microinjected, but only 12.5% hatched. After analyzing the unhatched eggs, they observed that all embryos were genetically edited. By knocking out the *BmFibH*, severe changes were observed, such as naked pupa or thin cocoons. All individuals that exhibited naked pupae died. Moreover, by inactivating this gene, several other genes involved in the processes of degradation, such as the autophagy, were upregulated. These findings offer a better understanding of FibH protein's role in silk gland development [121].

Keeping in mind the feasibility of expressing spider silk genes in *B. mori* [84], in order to obtain enhanced silk, Zhang et al. (2019) [122] used CRISPR-Cas9 to acquire high-performance fibers. By using this technique, the authors successfully knocked in spider silk genes in silkworms' genome. Accordingly, they designed two types of systems, FibL-CRISPR-Cas9 and FibH-CRISPR-Cas9. To avoid disrupting the protein production, spider silk genes were knocked in into one of the introns of *BmFibL* or *BmFibH*. They demonstrated the feasibility of employing CRISPR-Cas9 *in B. mori* to obtain silk with enhanced mechanical properties at industrial scale. The strategy described in this study can be further used to obtain numerous exogenous proteins that exhibit great interest for medical applications and beyond [122].

The microRNAs (miRNAs) are key regulators when it comes to gene expression. Notably, they recognize by complementarity specific mRNAs and inactivate them [123]. It has been revealed that *miR-2* is one of the most important miRNAs for wing morphogenesis in *D. melanogaster* and exhibits great influence on the Notch pathway. The *BmAwd* and *BmFng* genes are known positive regulators of this signaling pathway and are also potential *miR-2* targets. On this topic, Ling et al. (2015) [124] used the CRISPR-Cas9 technology to investigate the function of *miR-2* in *B. mori*. In the first phase of the study, the authors used the Gal4/UAS system to overexpress the *miR-2*, resulting in deformed adults' wings. However, in the second phase of the research, the CRISPR-Cas9 system was used to knock out the two *miR-2* target genes. The loss of function of *BmFng* and *BmAwd* also led to deformed wings. Both phases of the study confirmed that in silkworms, the *miR-2* plays a crucial role in wing development [124].

In another study, Liu et al. (2020) [125] used the CRISPR-Cas9 system to explore the function of *miR-34*, another miRNA that exhibits a great impact on insect development Firstly, they overexpressed *miR-34* in a transgenic line constructed by using a pBac plasmid. The *miR-34* overexpression negatively impacted the body size and wing morphology Secondly, the CRISPR-Cas9 system was used to inactivate the *miR-34* by using two different gRNAs. The *miR-34* ablation led to larvae developmental delay. Using several bioinformatic tools, they predicted that the *miR-34* target genes could be *BmE74*, *BmCpg4*, *BmLcp*, *BmWcp11*, and *BmBrc-z2*. Various analyses revealed that miR-34 target genes are just *BmE74* and *BmCpg4*. While it is well known that *BmE74* plays a key role in growth and morphogenesis, functional analysis of the second gene had to be performed. Given this, the CRISPR-Cas9 system was used to knock out *BmCpg4* resulting in affected wings, thus highlighting the gene's role in wing development [125].

Regarding the silkworms' innate defense mechanisms, there are two major activation pathways involved in the expression of numerous evolutionarily conserved antimicrobial peptides (AMPs) genes, and the Toll and the Imd pathways [126,127]. When Gram-negative bacterial or fungal contamination occurs, the Toll pathway is activated. One of the key genes that is involved in the Toll pathway is *BmCactus*. Being a negative regulator, once the infection with the mentioned pathogens occurs, this gene is being phosphorylated and inactivated. Considering this, Park et al. (2018) [128] used the CRISPR-Cas9 technology to perform site-target mutagenesis targeting the *BmCactus* gene in a specific *B. mori* cell line The authors designed six different gRNAs and transfected the CRISPR-Cas9 complex in the *B. mori* ovarian cell line by electroporation, but observed a very low survival rate of only 24%. Their data showed that all gRNAs determined site induced mutagenesis. By disrupting the *BmCactus* gene, the expression of several antimicrobial proteins (e.g., lysozyme and lebocin) was stimulated [128]. Keeping in mind the great importance of AMPs in clinical research [129], this study underlined the outstanding potential of *B. mori* usage in life science fields and the feasibility of using the advanced CRISPR-Cas9 genetic scissors to edit genomes for different purposes [128].

Ecdysteroids are steroid hormones that play a crucial role in the process of molting and metamorphosis in insects. The most important molting hormone is 20-hydroxyecdysone (20E) [130], but even if its biosynthesis processes have been intensively studied, the 20E metabolism has not been well-documented. However, there are several genes that are believed to be involved in the inactivation of 20E, but their biological functions have not been fully understood. Therefore, Li et al. (2015) [131] used *B. mori* to investigate the biological functions of one particular 20E inactivation enzyme, specifically the ecdysone oxidase (EO). Having a great impact on insects' key processes, it is crucial to regulate the ecdysteroids concentration. Accordingly, the EO participates in ecdysteroids' oxidation [132]. The CRISPR-Cas9 system was used to deplete the *BmEo* gene, and consequently, there was observed that the duration of the fifth instar larval phase was prolonged by 24 h [131].

Another key element that is involved in insects' development is the juvenile hormone (JH) [133]. The central role in the JH degradation process is played by the juvenile hormone esterase (JHE). Zhang et al. (2017) [134] used CRISPR-Cas9 in *B. mori* to deplete the encoding gene for JHE, specifically *BmJhe*, in order to investigate its function. Their data showed that by knocking out this gene, the fourth and fifth instar stages were prolonged due the fact that JH metabolism was delayed. These findings are not only important for functional analysis, but also for the sericulture industry. For silk production, it is of great impact to expand the larval stages, because this leads to larger larvae, and thus, larger cocoons are produced. These findings highlight the feasibility of using genome editing tools for economic purposes [134].

Moreover, CRISPR-Cas was used in *B. mori* to perform epigenetic modifications. In this context, Liu et al. (2019) [135] explored the impact of methylation on silkworms' development. This study is of great significance by providing a strategy for the investigation of DNA methylation importance in a target locus. Furthermore, their study represents a starting point for exploring the impact of DNA methylation status on different phenotypes in silkworms and beyond [135]. Furthermore, Xing et al. (2019) [136] used this technology for labeling endogenous regions in *B. mori* embryonic cell line and targeted the *BmFibH* gene. Using CRISPR-dCas9 as an imaging tool has extraordinary impact on performing fundamental research, but also on providing insights on insecticide resistance [136].

As for the detection of CRISPR-Cas induced mutations, there are a plethora of screening methods. On this topic, in a recent study, Brady et al. (2020) [137] described a new approach and provided a protocol for screening, characterizing, and stabilizing the mutant silkworm lines. The provided protocol involves several molecular methods that allow the recognition of the induced mutations on both autosomes and sex chromosomes [137].

In Table 3, several studies are described that used the CRISPR-Cas technology in order to genetically edit *B. mori*.

Applicability of CRISPR-Cas in Anti-BmNPV Therapy

Viruses represent a major threat to numerous hosts, including humans or insects, being one of the most rapidly mutating biological entities. However, there is an urgent need to develop new methods to combat these pathogens. Being one of the most valued molecular tools, the CRISPR-Cas system is currently being used for developing antiviral strategies in various organisms [155–157]. Of all nucleases described in the specialized literature, it has been shown that different variants of Cas9, Cas12, and Cas13 exhibit the most promising results regarding the antiviral approaches [155].

In respect to insect virology, BmNPV was the first discovered pathogen (1998) [158]. In the sericulture industry, this baculovirus causes extensive economic losses; therefore, it has been intensively studied. Being part of *Baculoviridae* family, its infection leads to the most severe disease and only a few silkworm strains are resistant to this virus [159]. Currently, there are several traditional methods that help to enhance silkworms' resistance to BmNPV, but they exhibit serious limitations [158,160,161].

However, CRISPR-Cas has been successfully used as an antiviral therapy in *B. mori*, especially against BmNPV. Chen et al. (2017) [162] selected two genes that are involved in baculovirus' replication and propagation processes: *immediate early-1 (ie-1)* and *me53*, respectively. The authors designed two gRNA for each target gene. The engineered plasmid contained three expression cassettes. One cassette contained the Cas9 nuclease, the second one included the gRNAs and the last one harbored a selecting marker, specifically the enhanced green fluorescent protein (EGFP). Even if the silkworms' viability and fecundity was not impacted, the transgenic homozygotes experienced a delay of the larval stage development. After performing viral inoculation in both wild type group and transgenic group, it was observed that the second category posed great viral resistance against BmNPV [162]. Likewise, another group of researchers targeted two other genes, *ie-0* and *ie-2*, by using CRISPR-Cas9. After being inoculated with occlusion bodies of BmNPV, the survival rate reached 65% for the transgenic silkworms [163].

The multiplexable feature of the CRISPR-Cas9 technology was underlined in a study performed by Dong et al. (2019) [164]. The researchers targeted in BmNPV three different genes essential for viral replication: *ie-1*, the *major envelope glycoprotein* and the *late expression factor-11*. The success of this study revealed one promising strategy in inhibiting different *B. mori* by using the powerful CRISPR-Cas method.

Table 3. CRISPR-Cas applications in *B. mori*.

Target Gene	Mutation Type	Delivery Approach	Objective	Gene Function	References
BmTim	Deletions	Plasmid	Functional gene analysis	Exhibit an impact on the embryo hatching process	[138]
BmApp	Deletions, insertions	mRNA	Functional gene analysis	Regulates wing development and cell mitosis	[139]
BmFoxo	Deletions	Plasmid	Analysis of *BmFoxo* and JH interaction	Involved in JH degradation	[140]
BmKmo	Deletions	Protein	Phenotypic analysis	Involved in the process of egg formation and eye coloring	[141]
BmYki	Deletions	Plasmid	Functional gene analysis	Involved in organ development and regeneration	[142]
BmTorso	Deletions	Plasmid	Functional gene analysis	Maintain the steroid hormones balance	[143]
BmEsp	Deletions	Plasmid	Functional gene analysis	Involved in a female's reproducibility	[144]
BmTudor	Deletions, insertions	Plasmid	Investigating the frequency of homologous recombination	Included in the stress granule formation	[145]
BmIdgf	Deletions	mRNA	Analyzing the pigmentation mechanism	Plays a key role in the melanization mechanism	[146]
BmBngr-a2	Deletions	Plasmid	Exploring functional studies of certain ion transport peptides	Involved in water homeostasis	[147]
BmTctp	Deletions	Plasmid	Functional analysis	Involved in different cell process, such as growth, development, and proliferation	[148]
BmGr66	Deletions	Plasmid	A better understanding of the specific feeding preference	Involved in silkworms' specific feeding preferences	[149]
BmOvo	Deletions	Plasmid	Functional analysis	Involved in germline sex determination and wing metamorphosis	[150]
BmPhyhd1	Deletions	Protein	Functional analysis	Exhibits a great impact on certain features of the epithelial cells	[151]
BmWnt1	Deletions	mRNA	Functional analysis	Involved in the embryogenesis	[152]
BmE75b	Deletions	mRNA	Functional analysis	Controls the developmental timing	[153]
BmOrco	Deletions	Plasmid	Exploration of adult mating behavior	Involved in silkworms' olfactory system, being an odorant receptor co-receptor	[154]

5. Conclusions

B. mori is one of the most important domesticated insects due to its great potential as a biotechnological platform to produce recombinant proteins, but also because of its great success as an experimental model organism [92,110]. Due to its extraordinary prospects, a wide range of studies have been performed that accelerated fundamental research and beyond in silkworms. Being the most feasible technology in terms of genome editing, the CRISPR-Cas system is currently used in many laboratories specialized in medicine, agriculture, alimentary industry, and entomology research [34,165]. A hallmark of CRISPR-Cas is the relative ease of designing CRISPR-based experiments. As reviewed elsewhere, there are numerous bioinformatics tools that facilitate the guide RNA design, as well as the prediction and evaluation of editing results [166]. In our experience, a thorough guide RNA design can also be achieved by using standard sequence alignment tools and manual inspection of potential target regions. Considering the aforementioned, the utilization of the CRISPR-Cas system as a gene editing tool augmented the research in *B. mori*. Most of the studies have been focused on using the CRISPR-Cas system to perform functional gene analysis, to elucidate certain mechanisms [135,148], or to enhance silkworms' resistance to BmNPV [164]. By reviewing the most remarkable work in this field, we provide deep insights that offer support for future research not only in *B. mori*, but also in other insect experimental models.

There are interesting future prospects when it comes to using the CRISPR-Cas technology in silkworms. Nowadays, by genetically manipulating the *B. mori* genome, major progress is being made for a better understanding of the process of fibroin and sericin synthesis, but also great advances are made to increase the understanding in the most important processes in silkworms. Notable great applications of CRISPR-Cas in *B. mori* refer to the development of enhanced silk fibers and the production of recombinant proteins that exhibit importance for various scientific fields. Even if compared with the other genome editing techniques, CRISPR-Cas9 exhibits lower off-target activity, although currently, it could not be confirmed that these unfavorable effects are completely eliminated. However, there is a major need to eliminate or at least reduce the off-target activity.

Author Contributions: Conceptualization, G.-M.B., A.C.R. and D.S.D.; resources, G.-M.B., A.-A.C. and D.S.D.; data curation, G.-M.B., A.C.R., O.B. and A.R.M.; writing—original draft preparation, G.-M.B., A.C.R. and A.-A.C.; writing—review and editing, G.-M.B., A.C.R., A.R.M. and O.B.; visualization, A.-I.G., O.B., A.-A.C., L.B., A.-S.M., A.R.M. and D.S.D.; supervision, D.S.D. and A.C.R. All authors have read and agreed to the published version of the manuscript.

Funding: This research received no external funding.

Institutional Review Board Statement: Not applicable.

Data Availability Statement: Not applicable.

Conflicts of Interest: The authors declare no conflict of interest.

References

1. Moon, S.B.; Kim, D.Y.; Ko, J.H.; Kim, Y.S. Recent advances in the CRISPR genome editing tool set. *Exp. Mol. Med.* **2019**, *51*, 1–11. [CrossRef] [PubMed]
2. Guha, T.K.; Wai, A.; Hausner, G. Programmable Genome Editing Tools and their Regulation for Efficient Genome Engineering. *Comput. Struct. Biotechnol. J.* **2017**, *15*, 146–160. [CrossRef] [PubMed]
3. Li, H.; Yang, Y.; Hong, W.; Huang, M.; Wu, M.; Zhao, X. Applications of genome editing technology in the targeted therapy of human diseases: Mechanisms, advances and prospects. *Signal Transduct. Target. Ther.* **2020**, *5*, 1–23. [CrossRef] [PubMed]
4. Schuijff, M.; De Jong, M.D.T.; Dijkstra, A.M. AQ methodology study on divergent perspectives on CRISPR-Cas9 in the Netherlands. *BMC Med. Ethics* **2021**, *22*, 48. [CrossRef]
5. Zhang, D.; Hussain, A.; Manghwar, H.; Xie, K.; Xie, S.; Zhao, S.; Larkin, R.M.; Qing, P.; Jin, S.; Ding, F. Genome editing with the CRISPR-Cas system: An art, ethics and global regulatory perspective. *Plant Biotechnol. J.* **2020**, *18*, 1651–1669. [CrossRef]
6. Li, P.; Wang, L.; Yang, J.; Di, L.; Li, J. Applications of the CRISPR-Cas system for infectious disease diagnostics. *Expert Rev. Mol. Diagn.* **2021**, *21*, 723–732. [CrossRef]
7. Sezutsu, H.; Sumitani, M.; Kondo, M.; Kobayashi, I.; Takasu, Y.; Suzuki, T.; Yonemura, N.; Iizuka, T.; Uchino, K.; Tamura, T.; et al. Construction of a Platform for the Development of Pharmaceutical and Medical Applications Using Transgenic Silkworms. *Yakugaku Zasshi J. Pharm. Soc. Jpn.* **2018**, *138*, 863–874. [CrossRef]

8. Kurihara, H.; Sezutsu, H.; Tamura, T.; Yamada, K. Production of an active feline interferon in the cocoon of transgenic silkworms using the fibroin H-chain expression system. *Biochem. Biophys. Res. Commun.* **2007**, *355*, 976–980. [CrossRef]
9. Nakaya, H.; Tatematsu, K.I.; Sezutsu, H.; Kuwabara, N.; Koibuchi, N.; Takeda, S. Secretory expression of thyroid hormone receptor using transgenic silkworms and its DNA binding activity. *Protein Expr. Purif.* **2020**, *176*, 105723. [CrossRef]
10. Itoh, K.; Kobayashi, I.; Nishioka, S.; Hidaka, T.; Tsuji, D.; Sezutsu, H. A transgenic silkworm overexpressing human lysosomal enzyme as a novel resource for producing recombinant glycobiologics and its application to development of enzyme replacement therapy for lysosomal diseases. *Mol. Genet. Metab.* **2016**, *120*, S69. [CrossRef]
11. Xu, H.; O'Brochta, D.A. Advanced technologies for genetically manipulating the silkworm *Bombyx mori*, a model lepidopteran insect. *Proc. R. Soc. B Biol. Sci.* **2015**, *282*, 20150487. [CrossRef]
12. Montali, A.; Romanelli, D.; Cappellozza, S.; Grimaldi, A.; De Eguileor, M.; Tettamanti, G. Timing of autophagy and apoptosis during posterior silk gland degeneration in *Bombyx mori*. *Arthropod Struct. Dev.* **2017**, *46*, 518–528. [CrossRef] [PubMed]
13. Ma, S.Y.; Smagghe, G.; Xia, Q.Y. Genome editing in *Bombyx mori*: New opportunities for silkworm functional genomics and the sericulture industry. *Insect Sci.* **2018**, *26*, 964–972. [CrossRef] [PubMed]
14. Jinek, M.; Chylinski, K.; Fonfara, I.; Hauer, M.; Doudna, J.A.; Charpentier, E. A Programmable Dual-RNA—Guided DNA endonuclease in adaptive bacterial immunity. *Science* **2012**, *337*, 816–822. [CrossRef] [PubMed]
15. Nidhi, S.; Anand, U.; Oleksak, P.; Tripathi, P.; Lal, J.A.; Thomas, G.; Kuca, K.; Tripathi, V. Novel CRISPR—Cas Systems: An Updated Review of the Current Achievements, Applications, and Future Research Perspectives. *Int. J. Mol. Sci.* **2021**, *22*, 3327 [CrossRef] [PubMed]
16. Hahn, F.; Loures, L.S.; Sparks, C.A.; Kanyuka, K.; Nekrasov, V. Efficient CRISPR/Cas-Mediated Targeted Mutagenesis in Spring and Winter Wheat Varieties. *Plants* **2021**, *10*, 1481. [CrossRef] [PubMed]
17. Hesami, M.; Yoosefzadeh Najafabadi, M.; Adamek, K.; Torkamaneh, D.; Jones, A.M.P. Synergizing off-target predictions for in silico insights of CENH3 Knockout in Cannabis through CRISPR/Cas. *Molecules* **2021**, *26*, 2053. [CrossRef]
18. Ishino, Y.; Shinagawa, H.; Makino, K.; Amemura, M.; Nakata, A. Nucleotide Sequence of the iap Gene, Responsible for Alkaline Phosphatase Isozyme Conversion in Escherichia coli, and Identification of the Gene Product. *J. Bacteriol.* **1987**, *169*, 5429–5433. [CrossRef]
19. Gophna, U.; Brodt, A. CRISPR/Cas systems in archaea. *Mob. Genet. Elem.* **2012**, *2*, 63–64. [CrossRef]
20. Horvath, P.; Barrangou, R. CRISPR/Cas, the immune system of Bacteria and Archaea. *Science* **2010**, *327*, 167–170. [CrossRef]
21. Shabbir, M.A.B.; Shabbir, M.Z.; Wu, Q.; Mahmood, S.; Sajid, A.; Maan, M.K.; Ahmed, S.; Naveed, U.; Hao, H.; Yuan, Z. CRISPR-cas system: Biological function in microbes and its use to treat antimicrobial resistant pathogens. *Ann. Clin. Microbiol. Antimicrob.* **2019**, *18*, 21. [CrossRef] [PubMed]
22. Rath, D.; Amlinger, L.; Rath, A.; Lundgren, M. The CRISPR-Cas immune system: Biology, mechanisms and applications. *Biochimie* **2015**, *117*, 119–128. [CrossRef] [PubMed]
23. Faure, G.; Shmakov, S.A.; Yan, W.X.; Cheng, D.R.; Scott, D.A.; Peters, J.E.; Makarova, K.S.; Koonin, E.V. CRISPR–Cas in mobile genetic elements: Counter-defence and beyond. *Nat. Rev. Microbiol.* **2019**, *17*, 513–525. [CrossRef] [PubMed]
24. Mcdonald, N.D.; Regmi, A.; Morreale, D.P.; Borowski, J.D ; Boyd, E.F. CRISPR-Cas systems are present predominantly on mobile genetic elements in Vibrio species. *BMC Genom.* **2019**, *20*, 105. [CrossRef] [PubMed]
25. Haft, D.H.; Selengut, J.; Mongodin, E.F.; Nelson, K.E. A Guild of 45 CRISPR-Associated (Cas) Protein Families and Multiple CRISPR/Cas Subtypes Exist in Prokaryotic Genomes. *PLoS Comput. Biol.* **2005**, *1*, e60. [CrossRef] [PubMed]
26. Makarova, K.S.; Haft, D.H.; Barrangou, R.; Brouns, S.J.J.; Charpentier, E.; Horvath, P.; Moineau, S.; Mojica, F.J.M.; Wolf, Y.I.; Yakunin, A.F.; et al. Evolution and classification of the CRISPR-Cas systems. *Nat. Rev. Microbiol.* **2011**, *9*, 467–477. [CrossRef] [PubMed]
27. Makarova, K.S.; Koonin, E.V. Annotation and Classification of CRISPR-Cas Systems. In *CRISPR*; Springer: Berlin/Heidelberg, Germany, 2015; pp. 1–27. [CrossRef]
28. Koonin, E.V.; Makarova, K.S. Origins and evolution of CRISPR-Cas systems. *Philos. Trans. R. Soc. B* **2019**, *374*, 20180087. [CrossRef]
29. Alkhnbashi, O.S.; Shah, S.A.; Garrett, R.A.; Saunders, S.J.; Costa, F.; Backofen, R. Characterizing leader sequences of CRISPR loci. *Bioinformatics* **2016**, *32*, i576–i585. [CrossRef]
30. Alkhnbashi, O.S.; Costa, F.; Shah, S.A.; Garrett, R.A.; Saunders, S.J.; Backofen, R. CRISPRstrand: Predicting repeat orientations to determine the crRNA-encoding strand at CRISPR loci. *Bioinformatics* **2014**, *30*, i489–i496. [CrossRef]
31. McGinn, J.; Marraffini, L.A. Molecular mechanisms of CRISPR–Cas spacer acquisition. *Nat. Rev. Microbiol.* **2018**, *17*, 7–12. [CrossRef]
32. Sorek, R.; Lawrence, C.M.; Wiedenheft, B. CRISPR-Mediated Adaptive Immune Systems in Bacteria and Archaea. *Annu. Rev. Biochem.* **2013**, *82*, 237–266. [CrossRef] [PubMed]
33. Roberts, A.; Barrangou, R. Applications of CRISPR-Cas systems in lactic acid bacteria. *FEMS Microbio. Rev.* **2020**, *44*, 523–537. [CrossRef] [PubMed]
34. Hryhorowicz, M.; Lipiński, D.; Zeyland, J.; Słomski, R. CRISPR/Cas9 Immune System as a Tool for Genome Engineering. *Arch. Immunol. Ther. Exp.* **2016**, *65*, 233–240. [CrossRef] [PubMed]
35. Terns, M.P.; Terns, R.M. CRISPR-based adaptive immune systems. *Curr. Opin. Microbiol.* **2011**, *14*, 321–327. [CrossRef] [PubMed]
36. Newsom, S.; Parameshwaran, H.P.; Martin, L.; Rajan, R. The CRISPR-Cas Mechanism for Adaptive Immunity and Alternate Bacterial Functions Fuels Diverse Biotechnologies. *Front. Cell. Infect. Microbiol.* **2021**, *10*, 1–10. [CrossRef]

37. Marraffini, L.A.; Sontheimer, E.J. CRISPR interference: RNA-directed adaptive immunity in bacteria and archaea. *Nat. Rev. Genet.* **2010**, *11*, 181–190. [CrossRef]
38. Ishino, Y.; Krupovic, M.; Forterre, P. History of CRISPR-Cas from Encounter with a Mysterious Repeated Sequence to Genome Editing Technology. *J. Bacteriol.* **2018**, *200*, e00580-17. [CrossRef]
39. Sontheimer, E.J.; Barrangou, R. The Bacterial Origins of the CRISPR Genome-Editing Revolution. *Hum. Gene Ther.* **2015**, *26*, 413–424. [CrossRef]
40. El-Mounadi, K.; Morales-Floriano, M.L.; Garcia-Ruiz, H. Principles, Applications, and Biosafety of Plant Genome Editing Using CRISPR-Cas9. *Front. Plant Sci.* **2020**, *11*, 56. [CrossRef]
41. Agustın-Pavon, C.; Isalan, M. Synthetic biology and therapeutic strategies for the degenerating brain. *Bioessays* **2014**, *36*, 979–990. [CrossRef]
42. Chen, L.; Tang, L.; Xiang, H.; Jin, L.; Li, Q.; Dong, Y.; Wang, W.; Zhang, G. Advances in genome editing technology and its promising application in evolutionary and ecological studies. *Gigascience* **2014**, *3*, 2047-217X. [CrossRef] [PubMed]
43. Tavakoli, K.; Pour-Aboughadareh, A.; Kianersi, F.; Poczai, P.; Etminan, A.; Shooshtari, L. Applications of CRISPR-Cas9 as an Advanced Genome Editing System in Life Sciences. *BioTech* **2021**, *10*, 14. [CrossRef]
44. Guha, T.K.; Edgell, D.R. Applications of Alternative Nucleases in the Age of CRISPR/Cas9. *Int. J. Mol. Sci.* **2017**, *18*, 2565. [CrossRef]
45. Khan, S.H. Genome-Editing Technologies: Concept, Pros, and Cons of Various Genome-Editing Techniques and Bioethical Concerns for Clinical Application. *Mol. Ther. Nucleic Acids* **2019**, *16*, 326–334. [CrossRef] [PubMed]
46. Chira, S.; Gulei, D.; Hajitou, A.; Zimta, A.A.; Cordelier, P.; Berindan-Neagoe, I. CRISPR/Cas9: Transcending the Reality of Genome Editing. *Mol. Ther. Nucleic Acids* **2017**, *7*, 211–222. [CrossRef]
47. Manghwar, H.; Lindsey, K.; Zhang, X.; Jin, S. CRISPR/Cas System: Recent Advances and Future Prospects for Genome Editing. *Trends Plant Sci.* **2019**, *24*, 1102–1125. [CrossRef] [PubMed]
48. Li, J.; Shi, Y.; Wu, J.; Li, H.; Smagghe, G.; Liu, T. CRISPR/Cas9 in lepidopteran insects: Progress, application and prospects. *J. Insect Physiol.* **2021**, *135*, 104325. [CrossRef] [PubMed]
49. Tyagi, S.; Kumar, R.; Das, A.; Won, S.Y.; Shukla, P. CRISPR-Cas9 system: A genome-editing tool with endless possibilities. *J. Biotechnol.* **2020**, *319*, 36–53. [CrossRef]
50. Zhang, Y.; Showalter, A.M. CRISPR/Cas9 Genome Editing Technology: A Valuable Tool for Understanding Plant Cell Wall Biosynthesis and Function. *Front. Plant Sci.* **2020**, *11*, 589517. [CrossRef] [PubMed]
51. Collias, D.; Beisel, C.L. CRISPR technologies and the search for the PAM-free nuclease. *Nat. Commun.* **2021**, *12*, 1–12. [CrossRef] [PubMed]
52. Yang, H.; Ren, S.; Yu, S.; Pan, H.; Li, T.; Ge, S.; Zhang, J.; Xia, N. Methods Favoring Homology-Directed Repair Choice in Response to CRISPR/Cas9 Induced-Double Strand Breaks. *Int. J. Mol. Sci.* **2020**, *21*, 6461. [CrossRef]
53. Bernheim, A.; Calvo-villamañán, A.; Basier, C.; Cui, L.; Rocha, E.; Touchon, M.; Bikard, D. Inhibition of NHEJ repair by type II-A CRISPR-Cas systems in bacteria. *Nat. Commun.* **2017**, *8*, 25–28. [CrossRef] [PubMed]
54. Di Stazio, M.; Foschi, N.; Athanasakis, E.; Gasparini, P.; d'Adamo, A.P. Systematic analysis of factors that improve homologous direct repair (HDR) efficiency in CRISPR/Cas9 technique. *PLoS ONE* **2021**, *16*, e0247603. [CrossRef]
55. Zhu, H.; Li, C.; Gao, C. Applications of CRISPR–Cas in agriculture and plant biotechnology. *Nat. Rev. Mol. Cell Biol.* **2020**, *21*, 661–677. [CrossRef] [PubMed]
56. Douris, V.; Denecke, S.; Van Leeuwen, T.; Bass, C.; Nauen, R.; Vontas, J. Using CRISPR/Cas9 genome modification to understand the genetic basis of insecticide resistance: Drosophila and beyond. *Pestic. Biochem. Physiol.* **2020**, *167*, 104595. [CrossRef] [PubMed]
57. Tyagi, S.; Kesiraju, K.; Saakre, M.; Rathinam, M.; Raman, V.; Pattanayak, D.; Sreevathsa, R. Genome Editing for Resistance to Insect Pests: An Emerging Tool for Crop Improvement. *ACS Omega* **2020**, *5*, 20674–20683. [CrossRef]
58. Yang, Y.; Xu, J.; Ge, S.; Lai, L. CRISPR/Cas: Advances, Limitations, and Applications for Precision Cancer Research. *Front. Med.* **2021**, *8*, 649896. [CrossRef]
59. Yang, Y.; Liu, X.; Li, S.; Chen, Y.; Zhao, Y.; Wei, Y.; Qiu, Y.; Liu, Y.; Zhou, Z.; Han, J.; et al. Genome-scale CRISPR screening for potential targets of ginsenoside compound K. *Cell Death Dis.* **2020**, *11*, 39. [CrossRef]
60. Xu, Y.; Li, Z. CRISPR-Cas systems: Overview, innovations and applications in human disease research and gene therapy. *Comput. Struct. Biotechnol. J.* **2020**, *18*, 2401–2415. [CrossRef]
61. Ma, X.; He, K.; Shi, Z.; Li, M.; Li, F.; Chen, X.-X. Large-Scale Annotation and Evolution Analysis of MiRNA in Insects. *Genome Biol. Evol.* **2021**, *13*, evab083. [CrossRef] [PubMed]
62. Brady, D.; Grapputo, A.; Romoli, O.; Sandrelli, F. Insect Cecropins, Antimicrobial Peptides with Potential Therapeutic Applications. *Int. J. Mol. Sci.* **2019**, *20*, 5862. [CrossRef] [PubMed]
63. Romoli, O.; Mukherjee, S.; Mohid, S.A.; Dutta, A.; Montali, A.; Franzolin, E.; Brady, D.; Zito, F.; Bergantino, E.; Rampazzo, C.; et al. Enhanced Silkworm Cecropin B Antimicrobial Activity against Pseudomonas aeruginosa from Single Amino Acid Variation. *ACS Infect. Dis.* **2019**, *5*, 1200–1213. [CrossRef]
64. Cui, Y.; Sun, J.; Yu, L. Application of the CRISPR gene-editing technique in insect functional genome studies—A review. *Entomol. Exp. Appl.* **2017**, *162*, 124–132. [CrossRef]
65. De Lazzari, F.; Sandrelli, F.; Whitworth, A.J.; Bisaglia, M. Antioxidant Therapy in Parkinson's Disease: Insights from Drosophila melanogaster. *Antioxidants* **2020**, *9*, 52. [CrossRef]

66. Taning, C.N.T.; Van Eynde, B.; Yu, N.; Ma, S.; Smagghe, G. CRISPR/Cas9 in insects: Applications, best practices and biosafety concerns. *J. Insect Physiol.* **2017**, *98*, 245–257. [CrossRef] [PubMed]
67. Hu, X.F.; Zhang, B.; Liao, C.H.; Zeng, Z.J. High-Efficiency CRISPR/Cas9-Mediated Gene Editing in Honeybee (*Apis mellifera*) Embryos. *G3 Genes Genomes Genet.* **2019**, *9*, 1759–1766. [CrossRef]
68. Gratz, S.J.; Cummings, A.M.; Nguyen, J.N.; Hamm, D.C.; Donohue, L.K.; Harrison, M.M.; Wildonger, J.; O'Connor-giles, K.M. Genome Engineering of Drosophila with the CRISPR RNA-guided Cas9 nuclease. *Genetics* **2013**, *194*, 1029–1035. [CrossRef]
69. Yu, Z.; Ren, M.; Wang, Z.; Zhang, B.; Rong, Y.S.; Jiao, R.; Gao, G. Highly Efficient Genome Modifications Mediated by CRISPR/Cas9 in Drosophila. *Genetics* **2013**, *195*, 289–291. [CrossRef] [PubMed]
70. Perry, T.; Batterham, P. Harnessing model organisms to study insecticide resistance. *Curr. Opin. Insect Sci.* **2018**, *27*, 61–67. [CrossRef]
71. Homem, R.A.; Davies, T.G.E. An overview of functional genomic tools in deciphering insecticide resistance. *Curr. Opin. Insect Sci.* **2018**, *27*, 103–110. [CrossRef]
72. Scott, J.G.; Buchon, N. Drosophila melanogaster as a powerful tool for studying insect toxicology. *Pestic. Biochem. Physiol.* **2019**, *161*, 95–103. [CrossRef] [PubMed]
73. Matsuoka, Y.; Nakamura, T.; Watanabe, T.; Barnett, A.A.; Noji, S.; Mito, T.; Extavour, C.G. Establishment of CRISPR/Cas9-based knock-in in a hemimetabolous insect: Targeted gene tagging in the cricket *Gryllus bimaculatus*. *bioRxiv* **2021**. [CrossRef]
74. Donoughe, S.; Extavour, C.G. Embryonic development of the cricket *Gryllus bimaculatus*. *Dev. Biol.* **2016**, *411*, 140–156. [CrossRef] [PubMed]
75. Evans, J.D.; Aronstein, K.; Chen, Y.P.; Hetru, C.; Imler, J.; Jiang, H.; Kanost, M.; Thompson, G.J.; Zou, Z.; Hultmark, D. Immune pathways and defence mechanisms in honey bees *Apis mellifera*. *Insect Mol. Biol.* **2006**, *15*, 645–656. [CrossRef] [PubMed]
76. Cucu, A.A.; Baci, G.M.; Moise, A.R.; Dezsi, S.; Marc, B.D.; Stângaciu, S.; Dezmiream, D.S. Towards a Better Understanding of Nutritional and Therapeutic Effects of Honey and Their Applications in Apitherapy. *Appl. Sci.* **2021**, *11*, 4190. [CrossRef]
77. Baci, G.; Cucu, A.; Moise, A.R.; Dezmirean, D.S. Applicability of Honey on Silkworms (*Bombyx mori*) and Quality Improvement of Its Biomaterials. *Appl. Sci.* **2021**, *11*, 4613. [CrossRef]
78. Kohno, H.; Suenami, S.; Takeuchi, H.; Sasaki, T.; Kubo, T. Production of Knockout Mutants by CRISPR/Cas9 in the European Honeybee, *Apis mellifera* L. *Zool. Sci.* **2016**, *33*, 505–512. [CrossRef] [PubMed]
79. Nie, H.; Liang, L.; Li, Q.; Li, Z.; Zhu, Y.; Guo, Y.; Zheng, Q.; Lin, Y.; Yang, D.; Li, Z.; et al. CRISPR/Cas9 mediated knockout of Amyellow-y gene results in melanization defect of the cuticle in adult *Apis mellifera*. *J. Insect Physiol.* **2021**, *132*, 104264. [CrossRef] [PubMed]
80. Gempe, T.; Hasselmann, M.; Schiøtt, M.; Hause, G.; Otte, M.; Beye, M. Sex Determination in Honeybees: Two Separate Mechanisms Induce and Maintain the Female Pathway. *PLoS Biol.* **2009**, *7*, e1000222. [CrossRef]
81. Wang, X.; Lin, Y.; Liang, L.; Geng, H.; Zhang, M.; Nie, H.; Su, S. Transcriptional Profiles of Diploid Mutant *Apis mellifera* Embryos after Knockout of csd by CRISPR/Cas9. *Insects* **2021**, *12*, 704. [CrossRef]
82. Horie, Y.; Watanabe, H. Recent advances in sericulture. *Annu. Rev. Entomol.* **1980**, *25*, 49–71. [CrossRef]
83. Saviane, A.; Romoli, O.; Bozzato, A.; Freddi, G.; Cappelletti, C.; Rosini, E.; Cappellozza, S.; Tettamanti, G.; Sandrelli, F. Intrinsic antimicrobial properties of silk spun by genetically modified silkworm strains. *Transgenic Res.* **2018**, *27*, 87–101. [CrossRef] [PubMed]
84. Xu, J.; Dong, Q.; Yu, Y.; Niu, B.; Ji, D.; Li, M.; Huang, Y.; Chen, X.; Tan, A. Mass spider silk production through targeted gene replacement in *Bombyx mori*. *Proc. Natl. Acad. Sci. USA* **2018**, *115*, 8757–8762. [CrossRef]
85. Tang, X.; Ye, X.; Wang, X.; Zhao, S.; Wu, M.; Ruan, J.; Zhong, B. High mechanical property silk produced by transgenic silkworms expressing the spidroins PySp1 and ASG1. *Sci. Rep.* **2021**, *11*, 1–9. [CrossRef]
86. Leem, J.W.; Fraser, M.J.; Kim, Y.L. Transgenic and Diet-Enhanced Silk Production for Reinforced Biomaterials: A Metamaterial Perspective. *Annu. Rev. Biomed. Eng.* **2020**, *22*, 79–102. [CrossRef]
87. Cappellozza, L.; Cappellozza, S.; Saviane, A.; Sbrenna, G. Artificial diet rearing system for the silkworm *Bombyx mori* (Lepidoptera: Bombycidae): Effect of vitamin C deprivation on larval growth and cocoon production. *Appl. Entomol. Zool.* **2005**, *40*, 405–412. [CrossRef]
88. Marrazzo, P.; Crupi, A.N.; Alviano, F.; Teodori, L.; Bonsi, L. Exploring the roles of MSCs in infections: Focus on bacterial diseases. *J. Mol. Med.* **2019**, *97*, 437–450. [CrossRef]
89. Abdelli, N.; Peng, L.; Keping, C. Silkworm, *Bombyx mori*, as an alternative model organism in toxicological research. *Environ. Sci. Pollut. Res.* **2018**, *25*, 35048–35054. [CrossRef]
90. Meng, X.; Zhu, F.; Chen, K. Silkworm: A promising model organism in life science. *J. Insect Sci.* **2017**, *17*. [CrossRef]
91. Yokoi, K.; Tsubota, T.; Jouraku, A.; Sezutsu, H.; Bono, H. Reference Transcriptome Data in Silkworm *Bombyx mori*. *Insects* **2021**, *12*, 519. [CrossRef]
92. Hamamoto, H.; Tonoike, A.; Narushima, K.; Horie, R.; Sekimizu, K. Silkworm as a model animal to evaluate drug candidate toxicity and metabolism. *Comp. Biochem. Physiol. Part C Toxicol. Pharmacol.* **2009**, *149*, 334–339. [CrossRef]
93. Hamamoto, H.; Kurokawa, K.; Kaito, C.; Kamura, K.; Razanajatovo, I.M.; Kusuhara, H.; Santa, T.; Sekimizu, K. Quantitative Evaluation of the Therapeutic Effects of Antibiotics Using Silkworms Infected with Human Pathogenic Microorganisms. *Antimicrob. Agents Chemother.* **2004**, *48*, 774–779. [CrossRef]
94. Kaito, C.; Sekimizu, K. A silkworm model of pathogenic bacterial infection. *Drug Discov. Ther.* **2007**, *1*, 89–93.

95. Montali, A.; Berini, F.; Brivio, M.F.; Mastore, M.; Saviane, A.; Cappellozza, S.; Marinelli, F.; Tettamanti, G. A Silkworm Infection Model for In Vivo Study of Glycopeptide Antibiotics. *Antibiotics* **2020**, *9*, 300. [CrossRef] [PubMed]
96. Matsumoto, Y.; Tateyama, Y.; Sugita, T. Evaluation of Antibacterial Drugs Using Silkworms Infected by Cutibacterium acnes. *Insects* **2021**, *12*, 619. [CrossRef] [PubMed]
97. Matsumoto, Y.; Ishii, M.; Ishii, K.; Miyaguchi, W.; Horie, R.; Inagaki, Y.; Hamamoto, H.; Tatematsu, K.-I.; Uchino, K.; Tamura, T.; et al. Transgenic silkworms expressing human insulin receptors for evaluation of therapeutically active insulin receptor agonists. *Biochem. Biophys. Res. Commun.* **2014**, *455*, 159–164. [CrossRef] [PubMed]
98. Matsumoto, Y.; Ishii, M.; Hayashi, Y.; Miyazaki, S.; Sugita, T.; Sumiya, E.; Sekimizu, K. Diabetic silkworms for evaluation of therapeutically effective drugs against type II diabetes. *Sci. Rep.* **2015**, *5*, 10722. [CrossRef] [PubMed]
99. Castillo, Y.; Suzuki, J.; Watanabe, K.; Shimizu, T.; Watarai, M. Effect of Vitamin A on Listeria monocytogenes Infection in a Silkworm Model. *PLoS ONE* **2016**, *11*, e0163747. [CrossRef]
100. Matsumoto, Y.; Kurakado, S.; Sugita, T. Evaluating Candida albicans biofilm formation in silkworms. *Med. Mycol.* **2021**, *59*, 201–205. [CrossRef]
101. Jiang, G.; Song, J.; Hu, H.; Tong, X.; Dai, F. Evaluation of the silkworm lemon mutant as an invertebrate animal model for human sepiapterin reductase deficiency. *R. Soc. Open Sci.* **2020**, *7*, 191888. [CrossRef]
102. Wang, L.; Su, M.; Zhao, X.; Hong, J.; Yu, X.; Xu, B.; Sheng, L.; Liu, N.; Shen, W.; Li, B.; et al. Nanoparticulate TiO 2 Protection of Midgut Damage in the Silkworm (*Bombyx mori*) Following Phoxim Exposure. *Arch. Environ. Contam. Toxicol.* **2015**, *68*, 534–542. [CrossRef] [PubMed]
103. Shen, W.; Zhao, X.; Wang, Q.; Niu, B.; Liu, Y.; He, L.; Weng, H.; Meng, Z.; Chen, Y. Genotoxicity evaluation of low doses of avermectin to hemocytes of silkworm (*Bombyx mori*) and response of gene expression to DNA damage. *Pestic. Biochem. Physiol.* **2011**, *101*, 159–164. [CrossRef]
104. Vyjayanthi, N.; Subramanyam, M.V.V. Effect of Fenvalerate-20EC on Sericigenous Insects: II. Digestive Enzymes in the Nutritive Physiology of Silkworm, *Bombyx mori* L. *Ecotoxicol. Environ. Saf.* **2002**, *53*, 212–220. [CrossRef]
105. Yasu, T.; Matsumoto, Y.; Sugita, T. Pharmacokinetics of voriconazole and its alteration by Candida albicans infection in silkworms. *J. Antibiot.* **2021**, *74*, 443–449. [CrossRef] [PubMed]
106. Inagaki, Y.; Matsumoto, Y.; Kataoka, K.; Matsuhashi, N.; Sekimizu, K. Evaluation of drug-induced tissue injury by measuring alanine aminotransferase (ALT) activity in silkworm hemolymph. *BMC Pharmacol. Toxicol.* **2012**, *13*, 13. [CrossRef]
107. Pandiarajan, J.; Jeyarani, V.; Balaji, S.; Krishnan, M. Silver Nanoparticles an Accumulative Hazard in Silkworm: *Bombyx mori. Austin J. Biotechnol. Bioeng.* **2016**, *3*, 1057.
108. Gad, A.A. Toxicity effect of Silver Nanoparticles to the Haemocytes and Antioxidant activity of Silkworm *Bombyx mori. Physiol. Entomol.* **2020**, *45*, 154–160. [CrossRef]
109. Meng, X.; Abdlli, N.; Wang, N.; Lü, P.; Nie, Z.; Dong, X.; Lu, S.; Chen, K. Effects of Ag Nanoparticles on Growth and Fat Body Proteins in Silkworms (*Bombyx mori*). *Biol. Trace Elem. Res.* **2017**, *180*, 327–337. [CrossRef]
110. Nouara, A.; Lü, P.; Chen, L.; Pan, Y.; Yang, Y.; Chen, K. Silver effects on silkworm, *Bombyx mori. J. Toxicol. Sci.* **2018**, *43*, 697–709. [CrossRef]
111. Xu, Y.; Wang, W.; Ma, L.; Cui, X.; Lynch, I.; Wu, G. Acute toxicity of Zinc Oxide nanoparticles to silkworm (*Bombyx mori* L.). *Chemosphere* **2020**, *259*, 127481. [CrossRef]
112. Li, K.L.; Zhang, Y.H.; Xing, R.; Zhou, Y.F.; Chen, X.D.; Wang, H.; Song, B.; Sima, Y.H.; Xu, S.Q. Different toxicity of cadmium telluride, silicon, and carbon nanomaterials against hemocytes in silkworm, *Bombyx mori*. *RSC Adv.* **2017**, *7*, 50317–50327. [CrossRef]
113. Wang, Y.; Li, Z.; Xu, J.; Zeng, B.; Ling, L.; You, L.; Chen, Y.; Huang, Y.; Tan, A. The CRISPR/Cas System mediates efficient genome engineering in *Bombyx mori*. *Cell Res.* **2013**, *23*, 1414–1416. [CrossRef] [PubMed]
114. Fujii, T.; Daimon, T.; Uchino, K.; Banno, Y.; Katsuma, S.; Sezutsu, H.; Tamura, T.; Shimada, T. Transgenic analysis of the BmBLOS2 gene that governs the translucency of the larval integument of the silkworm, *Bombyx mori*. *Insect Mol. Biol.* **2010**, *19*, 659–667. [CrossRef] [PubMed]
115. Liu, Y.; Ma, S.; Wang, X.; Chang, J.; Gao, J.; Shi, R.; Zhang, J.; Lu, W.; Liu, Y.; Zhao, P.; et al. Highly efficient multiplex targeted mutagenesis and genomic structure variation in *Bombyx mori* cells using CRISPR/Cas9. *Insect Biochem. Mol. Biol.* **2014**, *49*, 35–42. [CrossRef]
116. Sakuma, T.; Nishikawa, A.; Kume, S.; Chayama, K.; Yamamoto, T. Multiplex genome engineering in human cells using all-in-one CRISPR/Cas9 vector system. *Sci. Rep.* **2014**, *4*, 5400. [CrossRef]
117. Ma, S.; Chang, J.; Wang, X.; Liu, Y.; Zhang, J.; Lu, W.; Gao, J.; Shi, R.; Zhao, P.; Xia, Q. CRISPR/Cas9 mediated multiplex genome editing and heritable mutagenesis of BmKu70 in *Bombyx mori*. *Sci. Rep.* **2014**, *4*, 4489. [CrossRef]
118. Fujinaga, D.; Kohmura, Y.; Okamoto, N.; Kataoka, H.; Mizoguchi, A. Insulin-like growth factor (IGF)-like peptide and 20-hydroxyecdysone regulate the growth and development of the male genital disk through different mechanisms in the silkmoth, *Bombyx mori*. *Insect Biochem. Mol. Biol.* **2017**, *87*, 35–44. [CrossRef]
119. Fujinaga, D.; Shiomi, K.; Yagi, Y.; Kataoka, H.; Mizoguchi, A. An insulin-like growth factor-like peptide promotes ovarian development in the silkmoth *Bombyx mori*. *Sci. Rep.* **2019**, *9*, 1–12. [CrossRef]
120. Hu, W.; Wang, X.; Ma, S.; Peng, Z.; Cao, Y.; Xia, Q. CRISPR-Mediated Endogenous Activation of Fibroin Heavy Chain Gene Triggers Cellular Stress Responses in *Bombyx mori* Embryonic Cells. *Insects* **2021**, *12*, 552. [CrossRef]

121. Cui, Y.; Zhu, Y.; Lin, Y.; Chen, L.; Feng, Q.; Wang, W.; Xiang, H. New insight into the mechanism underlying the silk gland biological process by knocking out fibroin heavy chain in the silkworm. *BMC Genom.* **2018**, *19*, 215. [CrossRef]
122. Zhang, X.; Xia, L.; Day, B.A.; Harris, T.I.; Oliveira, P.; Knittel, C.; Licon, A.L.; Gong, C.; Dion, G.; Lewis, R.V.; et al. CRISPR/Cas9 Initiated Transgenic Silkworms as a Natural Spinner of Spider Silk. *Biomacromolecules* **2019**, *20*, 2252–2264. [CrossRef] [PubMed]
123. Michlewski, G.; Cáceres, J.F. Post-transcriptional control of miRNA biogenesis. *RNA* **2019**, *25*, 1–16. [CrossRef] [PubMed]
124. Ling, L.; Ge, X.; Li, Z.; Zeng, B.; Xu, J.; Chen, X.; Shang, P.; James, A.A.; Huang, Y.; Tan, A. MiR-2 family targets awd and fng to regulate wing morphogenesis in *Bombyx mori*. *RNA Biol.* **2015**, *12*, 742–748. [CrossRef] [PubMed]
125. Liu, Z.; Xu, J.; Ling, L.; Luo, X.; Yang, D.; Yang, X.; Zhang, X.; Huang, Y. miR-34 regulates larval growth and wing morphogenesis by directly modulating ecdysone signalling and cuticle protein in *Bombyx mori*. *RNA Biol.* **2020**, *17*, 1342–1351. [CrossRef]
126. Mahlapuu, M.; Håkansson, J.; Ringstad, L.; Björn, C. Antimicrobial Peptides: An Emerging Category of Therapeutic Agents. *Front. Cell. Infect. Microbiol.* **2016**, *6*, 194. [CrossRef]
127. Furukawa, S.; Tanaka, H.; Ishibashi, J.; Imanishi, S.; Yamakawa, M. Functional Characterization of a Cactus Homolog from the Silkworm *Bombyx mori*. *Biosci. Biotechnol. Biochem.* **2009**, *73*, 2665–2670. [CrossRef]
128. Park, J.W.; Yu, J.H.; Kim, S.W.; Kweon, H.Y.; Choi, K.H.; Kim, S.R. Enhancement of antimicrobial peptide genes expression in Cactus mutated *Bombyx mori* cells by CRISPR/Cas9. *Int. J. Ind. Entomol.* **2018**, *37*, 21–28.
129. Gan, B.H.; Gaynord, J.; Rowe, S.M.; Deingruber, T.; Spring, D.R. The multifaceted nature of antimicrobial peptides: Current synthetic chemistry approaches and future directions. *Chem. Soc. Rev.* **2021**, *50*, 7820–7880. [CrossRef]
130. Yang, H.; Wang, M.; Zhang, P.; Sabhat, A.; Malik, F.A.; Bhaskar, R.; Zhou, F.; Li, X.H.; Hu, J.-B.; Sun, C.G.; et al. Cloning and characterization of the *Bombyx mori* ecdysone oxidase. *Arch. Insect Biochem. Physiol.* **2011**, *78*, 17–29. [CrossRef]
131. Li, Z.; You, L.; Zeng, B.; Ling, L.; Xu, J.; Chen, X.; Zhang, Z.; Palli, S.R.; Huang, Y.; Tan, A. Ectopic expression of ecdysone oxidase impairs tissue degeneration in *Bombyx mori*. *Proc. R. Soc. B Biol. Sci.* **2015**, *282*, 20150513. [CrossRef]
132. Yamamoto, K.; Nagaoka, S. Identification and localization of a novel ecdysone oxidase in the silkworm, *Bombyx mori*. *J. Insect Biotechnol. Sericol.* **2017**, *53*, 49–53.
133. Tettamanti, G.; Casartelli, M. Cell death during complete metamorphosis. *Philos. Trans. R. Soc. B Biol. Sci.* **2019**, *374*, 20190065. [CrossRef] [PubMed]
134. Zhang, Z.; Liu, X.; Shiotsuki, T.; Wang, Z.; Xu, X.; Huang, Y.; Li, M.; Li, K.; Tan, A. Depletion of juvenile hormone esterase extends larval growth in *Bombyx mori*. *Insect Biochem. Mol. Biol.* **2017**, *81*, 72–79. [CrossRef] [PubMed]
135. Liu, Y.; Ma, S.; Chang, J.; Zhang, T.; Chen, X.; Liang, Y.; Xia, Q. Programmable targeted epigenetic editing using CRISPR system in *Bombyx mori*. *Insect Biochem. Mol. Biol.* **2019**, *110*, 105–111. [CrossRef]
136. Xing, W.; Ma, S.; Liu, Y.; Xia, Q. CRISPR/dCas9-mediated imaging of endogenous genomic loci in living *Bombyx mori* cells. *Insect Sci.* **2019**, *27*, 1360–1364. [CrossRef] [PubMed]
137. Brady, D.; Saviane, A.; Cappellozza, S.; Sandrelli, F. An Efficient Workflow for Screening and Stabilizing CRISPR/Cas9-Mediated Mutant Lines in *Bombyx mori*. *Methods Protoc.* **2020**, *4*, 4. [CrossRef]
138. Nartey, M.A.; Sun, X.; Qin, S.; Hou, C.X.; Li, M.W. CRISPR/Cas9-based knockout reveals that the clock gene timeless is indispensable for regulating circadian behavioral rhythms in *Bombyx mori*. *Insect Sci.* **2020**, *28*, 1414–1425. [CrossRef] [PubMed]
139. Yu, Y.; Liu, X.; Ma, X.; Zhang, Z.; Wang, T.; Sun, F.; Hou, C.; Li, M. A palmitoyltransferase Approximated gene Bm-app regulates wing development in *Bombyx mori*. *Insect Sci.* **2020**, *27*, 2–13. [CrossRef] [PubMed]
140. Zeng, B.; Huang, Y.; Xu, J.; Shiotsuki, T.; Bai, H.; Palli, S.R.; Huang, Y.; Tan, X.A. The FOXO transcription factor controls insect growth and development by regulating juvenile hormone degradation in the silkworm, *Bombyx mori*. *J. Biol. Chem.* **2017**, *292*, 11659–11669. [CrossRef]
141. Hong, J.W.; Jeong, C.Y.; Yu, J.H.; Kim, S.B.; Kang, S.K.; Kim, S.W.; Kim, N.S.; Kim, K.Y.; Park, J.W. *Bombyx mori* kynurenine 3-monooxygenase gene editing and insect molecular breeding using the clustered regularly interspaced short palindromic repeat/CRISPR associated protein 9 system. *Biotechnol. Prog.* **2020**, *36*, e3054. [CrossRef] [PubMed]
142. Xu, X.; Zhang, Z.; Yang, Y.; Huang, S.; Li, K.; He, L.; Zhou, X. Genome editing reveals the function of Yorkie during the embryonic and early larval development in silkworm, *Bombyx mori*. *Insect Mol. Biol.* **2018**, *27*, 675–685. [CrossRef]
143. Zhang, Z.J.; Liu, X.J.; Yu, Y.; Yang, F.Y.; Li, K. The receptor tyrosine kinase torso regulates ecdysone homeostasis to control developmental timing in *Bombyx mori*. *Insect Sci.* **2020**, *28*. 1582–1590. [CrossRef] [PubMed]
144. Xu, X.; Wang, Y.; Liu, Z.; Wang, Y.; He, L.; Li, K.; Huang, Y. Disruption of egg-specific protein causes female sterility in *Bombyx mori*. *Insect Sci.* **2021**, 1–11. [CrossRef]
145. Zhu, L.; Mon, H.; Xu, J.; Lee, J.M.; Kusakabe, T. CRISPR/Cas9-mediated knockout of factors in non-homologous end joining pathway enhances gene targeting in silkworm cells. *Sci. Rep.* **2015**, *5*, 18103. [CrossRef] [PubMed]
146. Gao, Y.; Liu, Y.C.; Jia, S.Z.; Liang, Y.T.; Tang, Y.; Xu, Y.S.; Kawasaki, H.; Wang, H.B. Imaginal disc growth factor maintains cuticle structure and controls melanization in the spot pattern formation of *Bombyx mori*. *PLoS Genet.* **2020**, *16*, e1008980. [CrossRef]
147. Sun, L.; Zhang, Z.; Zhang, R.; Yu, Y.; Yang, F.; Tan, A. Molecular Disruption of Ion Transport Peptide Receptor Results in Impaired Water Homeostasis and Developmental Defects in *Bombyx mori*. *Front. Physiol.* **2020**, *11*, 424. [CrossRef] [PubMed]
148. Liu, Z.L.; Xu, J.; Ling, L.; Zhang, R.; Shang, P.; Huang, Y.P. CRISPR disruption of TCTP gene impaired normal development in the silkworm *Bombyx mori*. *Insect Sci.* **2018**, *26*, 973–982. [CrossRef] [PubMed]
149. Zhang, Z.; Zhang, S.S.; Niu, B.L.; Ji, D.F.; Liu, X.J.; Li, M.W.; Bai, H.; Palli, S.R.; Wang, C.Z.; Tan, A.J. A determining factor for insect feeding preference in the silkworm, *Bombyx mori*. *PLoS Biol.* **2019**, *17*, e3000162. [CrossRef]

150. Bi, H.; Xu, X.; Li, X.; Zhang, Y.; Huang, Y.; Li, K.; Xu, J. CRISPR Disruption of BmOvo Resulted in the Failure of Emergence and Affected the Wing and Gonad Development in the Silkworm *Bombyx mori*. *Insects* **2019**, *10*, 254. [CrossRef]
151. Fujii, T.; Banno, Y. Enlargement of egg size by CRISPR/Cas9-mediated knockout of a sex-linked gene in the silkworm, *Bombyx mori*. *J. Insect Biotechnol. Sericol.* **2018**, *87*, 3_071–3_078. [CrossRef]
152. Zhang, Z.; Aslam, A.F.; Liu, X.; Li, M.; Huang, Y.; Tan, A. Functional analysis of Bombyx Wnt1 during embryogenesis using the CRISPR/Cas9 system. *J. Insect Physiol.* **2015**, *79*, 73–79. [CrossRef] [PubMed]
153. Li, K.; Tian, L.; Guo, Z.; Guo, S.; Zhang, J.; Gu, S.; Palli, S.R.; Cao, Y.; Li, S. 20-Hydroxyecdysone (20E) Primary Response Gene E75 Isoforms Mediate Steroidogenesis Autoregulation and Regulate Developmental Timing in *Bombyx*. *J. Biol. Chem.* **2016**, *291*, 18163–18175. [CrossRef] [PubMed]
154. Liu, Q.; Liu, W.; Zeng, B.; Wang, G.; Hao, D.; Huang, Y. Deletion of the *Bombyx mori* odorant receptor co-receptor (*BmOrco*) impairs olfactory sensitivity in silkworms. *Insect Biochem. Mol. Biol.* **2017**, *86*, 58–67. [CrossRef] [PubMed]
155. Baddeley, H.J.E.; Isalan, M. The Application of CRISPR/Cas Systems for Antiviral Therapy. *Front. Genome Ed.* **2021**, *3*, 745559. [CrossRef] [PubMed]
156. Murphy, B.G.; Wolf, T.; Vogel, H.; Castillo, D.; Woolard, K. An RNA-Directed Gene Editing Strategy for Attenuating the Infectious Potential of Feline Immunodeficiency Virus-Infected Cells: A Proof of Concept. *Viruses* **2020**, *12*, 511. [CrossRef] [PubMed]
157. Fareh, M.; Zhao, W.; Hu, W.; Casan, J.M.; Kumar, A.; Symons, J.; Zerbato, J.M.; Fong, D.; Voskoboinik, I.; Ekert, P.G.; et al. Reprogrammed CRISPR-Cas13b suppresses SARS-CoV-2 replication and circumvents its mutational escape through mismatch tolerance. *Nat. Commun.* **2021**, *12*, 1–16. [CrossRef]
158. Yao, Q.; Li, M.W.; Wang, Y.; Wang, W.B.; Lu, J.; Dong, Y.; Chen, K.P. Screening of molecular markers for NPV resistance in *Bombyx mori* L. (Lep., Bombycidae). *J. Appl. Entomol.* **2003**, *127*, 134–136. [CrossRef]
159. Guo, H.; Zhang, B.; Zheng, X.; Sun, J.; Guo, H.; Li, G.; Zhao, G.; Xu, A.; Qian, H. Pathogenicity Detection and Genome Analysis of Two Different Geographic Strains of BmNPV. *Insects* **2021**, *12*, 890. [CrossRef]
160. Isobe, R.; Kojima, K.; Matsuyama, T.; Quan, G.X.; Kanda, T.; Tamura, T.; Sahara, K.; Asano, S.I.; Bando, H. Use of RNAi technology to confer enhanced resistance to BmNPV on transgenic silkworms. *Arch. Virol.* **2004**, *149*, 1931–1940. [CrossRef]
161. Qiong, Y.; Dong Xu, X.; Qing Rong, L.; Yang, X.; Ming Qiang, Y. Standard method for detecting *Bombyx mori* nucleopolyhedrovirus disease-resistant silkworm varieties. *Rev. Bras. Entomol.* **2017**, *62*, 19–22. [CrossRef]
162. Chen, S.; Hou, C.; Bi, H.; Wang, Y.; Xu, J.; Li, M.; James, A.A.; Huang, Y.; Tan, A. Transgenic Clustered Regularly Interspaced Short Palindromic Repeat/Cas9-Mediated Viral Gene Targeting for Antiviral Therapy of *Bombyx mori* Nucleopolyhedrovirus. *J. Virol.* **2017**, *91*, e02465-16. [CrossRef] [PubMed]
163. Dong, Z.; Hu, Z.; Qin, Q.; Dong, F.; Huang, L.; Long, J.; Chen, P.; Lu, C.; Pan, M. CRISPR/Cas9 Mediated Disruption of the immediate early-0 and 2 as a Therapeutic Approach to *Bombyx mori* Nucleopolyhedrovirus in Transgenic Silkworm. *Insect Mol. Biol.* **2018**, *28*, 112–122. [CrossRef] [PubMed]
164. Dong, Z.; Qin, Q.; Hu, Z.; Chen, P.; Huang, L.; Zhang, X.; Tian, T.; Lu, C.; Pan, M. Construction of a One-Vector Multiplex CRISPR/Cas9 Editing System to Inhibit Nucleopolyhedrovirus Replication in Silkworms. *Virol. Sin.* **2019**, *34*, 444–453. [CrossRef] [PubMed]
165. Xu, X.; Bi, H.; Wang, Y.; Li, X.; Xu, J.; Liu, Z.; He, L.; Li, K.; Huang, Y. Disruption of the ovarian serine protease (Osp) gene causes female sterility in *Bombyx mori* and Spodoptera litura. *Pest Manag. Sci.* **2019**, *76*, 1245–1255. [CrossRef]
166. Sledzinski, P.; Nowaczyk, M.; Olejniczak, M. Computatonal Tools and Resources Supporting CRISPR-Cas Experiments. *Cells* **2020**, *9*, 1288. [CrossRef]

insects

MDPI

Systematic Review

Nutritional Composition of *Bombyx mori* Pupae: A Systematic Review

Luca Tassoni [1,2,*], Silvia Cappellozza [1], Antonella Dalle Zotte [2], Simone Belluco [3], Pietro Antonelli [3], Filippo Marzoli [3] and Alessio Saviane [1]

1 Consiglio per la Ricerca in Agricoltura e l'Analisi dell'Economia Agraria, Centro di Ricerca Agricoltura e Ambiente (CREA-AA), 35143 Padova, Italy; silvia.cappellozza@crea.gov.it (S.C.); alessio.saviane@crea.gov.it (A.S.)

2 Department of Animal Medicine, Production and Health, University of Padova, Agripolis, Viale dell'Università 16, 35020 Padova, Italy; antonella.dallezotte@unipd.it

3 IZSVe, Istituto Zooprofilattico Sperimentale delle Venezie, Viale dell'Università 10, 35020 Padova, Italy; sbelluco@izsvenezie.it (S.B.); pantonelli@izsvenezie.it (P.A.); fmarzoli@izsvenezie.it (F.M.)

* Correspondence: tassoni.luca@gmail.com

Simple Summary: The mulberry silkworm (*Bombyx mori*) is a domesticated insect traditionally reared to produce silk. Its pupae are historically eaten in Asian countries and are obtained as waste products from the silk reeling industry. Pupae are a promising novel food in Western countries as well as a source of proteins, lipids, and minerals. Several varied results are reported in the literature regarding the nutrient composition of silkworm pupa, and several factors must be considered when comparing the research. Some of the variables that could affect the pupal nutritional content include rearing techniques, diets, silkworm strains, killing, and drying techniques. This literature systematic review identifies the most important research areas and aids authorities and producers in the evaluation and development of silkworm pupae for novel uses.

Abstract: As insects have started to enter the eating habits of Western countries, an increasing amount of literature regarding the mulberry silkworm (*Bombyx mori*) prospective application as food has been published. Despite this growing interest, there is currently no systematic review of silkworm nutritional composition available. In this paper, we performed a systematic review of the recent available literature on the nutrient composition of mulberry silkworm pupae. After screening the titles and abstracts of 14,008 studies retrieved from three scientific databases, data about nutrients was extracted from 29 selected papers, together with their related variables. This systematic review provides an overview of the variety of data reported in the literature and highlights that many elements contribute to hindering a sound comparison of the different nutritional values reported for silkworm pupae. The observed variability of the composition data reported could be due to differences in diet, strains, pretreatments, and origin of the silkworm analyzed. However, all these variables were not always available and should be reported in future studies to simplify the data comparison.

Keywords: *Bombyx mori*; mulberry silkworm pupae; nutritional composition

Citation: Tassoni, L.; Cappellozza, S.; Dalle Zotte, A.; Belluco, S.; Antonelli, P.; Marzoli, F.; Saviane, A. Nutritional Composition of *Bombyx mori* Pupae: A Systematic Review. *Insects* **2022**, *13*, 644. https://doi.org/10.3390/insects13070644

Academic Editor: Toru Shimada

Received: 16 June 2022
Accepted: 13 July 2022
Published: 17 July 2022

1. Introduction

The only fully domesticated insect among those reared by humans is the mulberry silkworm (*Bombyx mori* L., 1758), which was first raised in captivity around 7500 years ago [1]. Its ability to generate silk is what gives it its significance, although it also has applications as food or feed, a pet, and a model organism in scientific research [2]. Furthermore, the silkworm has been used as a functional bioreactor to create a variety of molecules, mostly those with pharmaceutical implications [3].

Pupae are the most commonly employed life stage for food or feed since they make up around 60% of the dry cocoon weight [4] and are the principal by-product of the silk industry. They can be used as food or feed, which helps to support circular production chains and prevent the release of potentially harmful waste into the environment. The consumption of silkworm pupae is already widespread in Asia, particularly in India, China, Japan, South Korea, and Thailand [5–9], where they are valued as street food, or as a nutraceutical in traditional medicine [10–12].

Before consumption, it is important to weigh the dangers associated with allergens, microorganisms, or chemicals, in addition to the nutritional advantages of silkworm pupae. Regarding Europe, an EFSA (the European Food Safety Agency) opinion [13] explored the possibility of silkworms as a novel food, reflecting the growing interest in this topic in Western nations.

In this systematic review, we concentrated on analyzing and summarizing the literature on the nutritional composition of *B. mori* pupae (further called SP for "silkworm pupae"). There are reviews addressing silkworm composition [14], but a systematic review addressing silkworm pupa composition is lacking. The main distinction between this systematic review and other reviews is that it compares all the data that are currently available about SP published in the searched literature in the declared time interval, allowing us to understand the potential and restrictions of SP as a food. On the other hand, we may propose some hypotheses regarding the main aspects that may affect the nutritional value of SP based on a thorough comparison of the findings of various authors.

Authorities may find this systematic review useful when they evaluate SP and its derivative products as novel foods. Understanding the potential technology uses and the impact of the technological replacement of conventional nutrient sources in novel formulations could potentially be helpful for food producers. Additionally, this analysis could serve as a basis for future research looking into novel approaches to manipulate SP composition.

2. Materials and Methods

This systematic review seeks to provide a food-based characterization of the nutritional composition of mulberry SP (*B. mori*). All articles published in peer-reviewed journals in the languages of English, French, and Spanish were taken into consideration for this scope at the onset of our study without regard to publication dates.

Using the keywords listed in Table 1, we conducted searches in PUBMED, Web of Science Core Collection, and EMBASE (Title/Abstract, Topic (TS), and Title, Abstract, Author keywords, respectively). The keywords also contained terms pertaining to the allergic and security aspects of SP. The records pertaining to these subjects were located and used for a second systematic review with a focus on these specific concerns. The search was performed on 14 May 2020, and it was updated on 5 May 2021, with a time limit beginning on 1 January 2020, to include all the documents published in the meantime. We utilized the software EPPI-4 Reviewer [15] to examine the retrieved records

Table 1. Keywords employed to retrieve relevant records reporting data on silkworm food safety and composition. The first keyword refers to the species and any alternate names it may have. The second keyword refers to all the terms related to nutritional composition that have been investigated in this review, while the last keyword refers to food safety terms. The first keyword was searched in conjunction with the second or third, linking them with logical operators.

Keywords (Title/Abstract)				
Bombyx OR Silkworm OR "silk worms" OR silkmoth OR "silk moths"	AND	nutrition OR composition OR centesimal OR nutrient OR nutrients OR protein OR proteins peptide OR peptides OR aminoacid OR aminoacids OR "amino acid" OR "amino acids" OR acid OR acids OR polypeptide OR polypeptides OR fat OR fats OR lipid OR lipids OR "fatty acid" OR "fatty acids" OR "fatty alcohols" OR sugar OR sugars OR carbohydrate OR carbohydrates OR disaccharide OR disaccharides OR monosaccharide OR monosaccharides OR polysaccharide OR polysaccharides OR ash OR mineral OR minerals OR macronutrient OR macronutrients OR micronutrient OR micronutrients OR oligoelement OR oligoelements OR microelement OR microelements OR vitamin OR vitamins OR oil OR oils OR "trace element" OR "trace elements"	OR	safety OR hazard OR hazards OR risk OR risks OR microorganism OR microorganisms OR pathogen OR pathogens OR contaminant OR contaminants OR contamination OR contaminations OR chemical OR chemicals OR toxic OR toxics OR toxicity OR metal OR metals OR toxin OR toxins OR allergy OR allergies OR allergen OR allergens OR allergic OR allergenic OR sensitization OR sensitisation OR cross-reactivity OR anaphylactic OR anaphylaxis OR poisoning OR poison OR compound OR compounds OR pesticide OR pesticides OR residual OR residue OR residues OR antibiotic OR antibiotics OR antiparasitic OR antiparasitics OR mycotoxin OR mycotoxins OR dioxin OR dioxins OR polluting OR pollutant OR pollutants

There were two main phases to the selection process. Six researchers (S.B., F.M., A.P., L.T., A.S., and S.C.) conducted the first phase using a double-screening approach, in which two researchers separately classified each record to determine whether it related to the study topic "composition" or "safety" and to determine whether it was a review.

This step's inclusion criteria were as follows: (1) the language used had to be either English, French, or Spanish; (2) the data had to be taken directly from research publications and not from other reviews. All the records with unclear titles or abstracts were included for evaluation in the following stage. Based on the titles and abstracts, the selected studies were then split into two groups: those dealing with nutritional composition and those dealing with biological/chemical risk related to SP use. Three reviewers (S.B., F.M., and P.A.) did this classification.

The second phase concentrated on original papers addressing SP composition. The task involved classifying and evaluating each paper's data typology in full text for each record. Then, pertinent information was retrieved and listed in tables. We specifically found publications that discussed the following topics: (1) the overall chemical composition (macronutrients and ash), (2) the AA profile, (3) the FA profile, and (4) the composition of the macro and microelements.

The following additional exclusion criteria were applied:

1. Research involving pupae as feed;
2. Research describing the composition of the larval stage;
3. Research published before the year 2000.

As a result, the entire time frame of the systematic review is from 1 January 2000, through 5 May 2021. Additionally, we excluded data from food labeling when SP had been purchased from suppliers as food and only kept original data from the selected trials that were gathered in the research through an analytical method. Furthermore, we did not include information on SP fed a particular diet supplement. The information obtained from the investigations was arranged into pre-defined tables. Three researchers conducted this second stage (L.T., A.S., and S.C.). Information on silkworm strain, feeding (mulberry leaves or artificial diet), and sample processing was provided when available.

Insect-killing techniques, sample storage conditions, and drying technologies used before the analytical stages are among the sample processing variables that may have an impact on compositional results. Another potential source of variation was the analytical approach itself, which varied throughout the articles, particularly for fiber and ash. The existence of experimental replicates and the standard deviation, which were mentioned in this review if they were accessible in the original manuscripts, are additional factors to take into account. The relative standard deviation was estimated in cases when the data were required to be summed up or standardized using the principles of error propagation.

Data about chemical composition were extracted and normalized, if necessary, to be comparable among themselves. This included information about total nitrogen or protein content, ash, lipid or fat content, crude fiber, nitrogen-free extracts (NFE), amino acid (AA) profile, fatty acid (FA) profile, and oligo-elements. All these extracted data were reported in Table 2.

Only a small percentage of the selected papers stated the composition on a wet basis, necessitating the calculation of the composition on a dry matter basis using the moisture values. Therefore, the nutritional values were normalized according to the dry matter content and expressed as a percentage or "g per 100 g of dry matter". Protein content was also expressed using the nitrogen-to-protein conversion factor suggested by Janssen et al. [16].

To normalize and compare data given in articles included for the final review, amino-acid absolute values for the AA profile were transformed into percentage values (Table 3). The FA profile is shown in Table 4 as a percentage of all FAME (Fatty Acid Methyl Esters). To make data comparison easier, the isomeric forms of the same FA were combined together. The last step was to express the macro and microelements as mg/g on a dry matter basis (Table 5). According to the principles of error propagation, the standard deviation of the data produced by elaboration (addition or normalization of the dry matter) of the original values was determined.

3. Results

After removing duplicates and reading the abstracts of all 14,008 papers that were found using our search parameters and logical operators, 385 of them were retained for additional analysis. Among them, 185 of the chosen studies addressed safety issues, and 156 dealt with chemical composition. As previously said, the former were taken into consideration for a second evaluation of safety issues. Twenty-six papers out of 156 dealing with chemical composition met our inclusion criteria and were utilized to gather data on nutritional information (Figure 1).

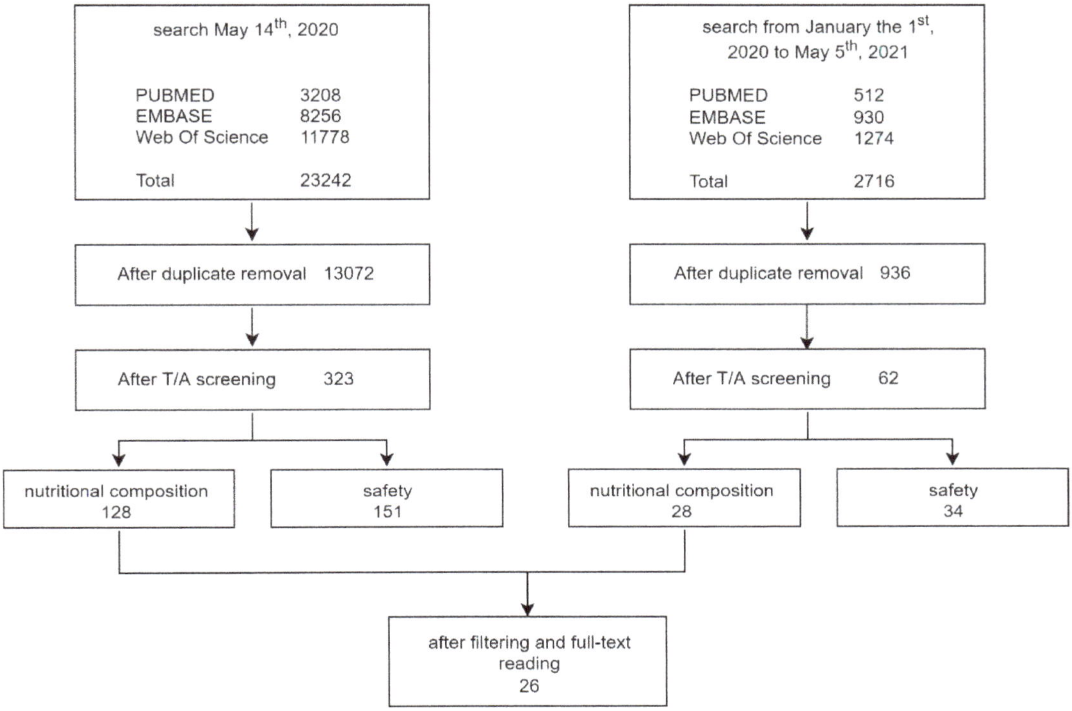

Figure 1. Representation of the procedure used to review and select the studies from which to extract the data for the current systematic review.

3.1. Macronutrients and Ash

A total of thirty-three data entries were retrieved from 16 studies reporting at least one nutrient value (Table 2). Two data entries referring to protein values on fresh weight [17] were not considered due to the absence of the moisture value that would have made the two values comparable. The two main types of reported data were "full-fat" SP and "defatted" SP. The values of protein (n = 33), lipids (n = 26), ash (n = 25), nitrogen-free extracts (n = 17), and crude fibers (n= 16) were reported in the 33 data entries.

An average nitrogen-to-protein conversion factor of 6.25 was used to calculate the protein values in each study. The corrected nitrogen-to-protein conversion factor of 4.76 proposed by Janssen et al. [16] was only adopted by Akande et al. [18] and Birman et al. [19]. The nitrogen-to-protein conversion factor has been "insect adjusted," and the protein values are as reported in the original studies shown in Table 2. The results discussed below, for simplicity's sake, apply to protein values obtained using the standard nitrogen conversion factor of 6.25. In three studies including defatted SP, protein levels ranged from 67.5 [20] to 82.9% [21], while residual lipids ranged from less than 0.2 [22] to 4.75% [20].

Table 2. Macro nutrient values and main processing variables (g/100 g) retrieved from selected articles. Data were normalized on the dry matter where necessary. * Protein values were originally calculated using Kp = 4.76 as a nitrogen conversion factor. ** the values were indicated in the original paper as "total fiber".

Reference	Treatment, Strain, Diet		Separation	Protein (Kp = 6.25)	Protein (Kp = 4.76)	Fat	Crude Fiber	NFE (Often Indicates as Carbohydrates in the Original Paper)	Ash
Mishra et al., 2003 [5]	Cocoons were boiled at 100 °C for 30 min. Pupae were taken out and dissected to remove the intestine and waste material. Composition was expressed on wet weight (100 g).		full-fat	34.38 ± 0.25	26.19 ± 0.19	57.64 ± 0.22	0.54 ± 0.00	5.16 ± 0.19	2.27 ± 0.17
Akande et al., 2020 [18]	Edible insects boiled without seasonings were prepared with 100 g of each of the edible insects with 150 mL of potable water and boiled for 15 min. Data are referred to as boiled silkworm pupae without seasoning (SWOS).		full-fat	47.17 ± 1.36	35.92 ± 1.04 *	32.16 ± 0.85	1.68 ± 0.02	28.15 ± 1.23	2.12 ± 0.03
David-Birman et al., 2019 [19]	Finely milled silkworm pupae flour (SWF) was purchased from JR Unique Foods Ltd. (Udon Thani, Thailand).		full-fat	53.05 ± 0.09	40.4 ± 0.07 *	29.56 ± 0.08	n/a	n/a	n/a
Kim et al., 2016 [20]	Dried, untreated SP flour. The whole insects were cleaned with distilled water, sieved, ground using a hammer mill, and passed through a 20-mesh sieve.		full-fat	47.87 ± 1.72	36.46 ± 1.31	30.85 ± 3.81	6.38 ± 0.32 **	n/a	6.96 ± 0.61
		The SP flour was then defatted using hexane	defatted	67.54 ± 0.97	51.44 ± 0.74	4.75 ± 1.79	9.54 ± 0.58 **	n/a	9.54 ± 1.22
		SP defatted using hexane. The insect flour was also acid-hydrolyzed and dried in the dry oven at 70 °C	defatted	73.51 ± 0.74	55.99 ± 0.56	2.05 ± 0.5	3.24 ± 0.19 **	n/a	8.04 ± 0.18
Anootthato et al., 2019 [21]	Frozen silkworm pupae (−20 °C) were thawed at 4 °C overnight and washed three times before blanching at 95 °C for 10 min.		full-fat	56.64 ± 0.42	43.14 ± 0.32	34.07 ± 0.62	n/a	n/a	n/a
		The sample was further dried at 60 °C for 18 h. Dried samples were ground and packed in an aluminum foil bag until use	full-fat	52.69 ± 0.46	40.13 ± 0.35	36.21 ± 0.23	n/a	n/a	n/a
		The sample was further dried at 60 °C for 18 h, ground, and packed in an aluminum foil bag until use. Dried samples were mixed and defatted using 99.8% ethanol (1:3 *w*/*v*) at 50 °C with stirring for 30 min. Then, each sample was filtered. Defatting was carried out twice before drying in the tray dryer at 50 °C for 3 h. Defatted samples were ground to 150 µm before hydrolysis	defatted	82.88 ± 0.5	63.12 ± 0.38	1.47 ± 0.44	n/a	n/a	n/a

Table 2. *Cont.*

Reference	Treatment, Strain, Diet		Separation	Protein (Kp = 6.25)	Protein (Kp = 4.76)	Fat	Crude Fiber	NFE (Often Indicates as Carbohydrates in the Original Paper)	Ash
Felix et al., 2020 [22]	The silkworm protein concentrate (SPC) used in this study was supplied by FeedStimulants (Amsterdam, The Netherlands).		full-fat	55.43 ± 1.10	42.22 ± 0.84	31.72 ± 0.22	n/a	5.60 ± 1.42	7.24 ± 0.88
		SPC was defatted using n-hexane	defatted	81.14 ± 2.06	61.80 ± 1.57	<0.2	n/a	8.23 ± 2.04	10.63 ± 1.27
Akande et al., 2020 [23]	Freshly harvested mulberry silkworm pupae (SWP) were stifled in an oven at 93 °C for 1 h, cut out of their cocoons and dried in a hot air oven (Gen. Lab. oven) at 40 °C for 8 h. The dried SWP was milled into flour using an electric blender (Binatone, Model No. 51-777) and kept airtight. The data are referred to as silkworm pupae powder (SWP).		full-fat	60.7 ± 0.2	46.23 ± 0.15	23.5 ± 0.21	1.1 ± 0.16	11.3 ± 0.42	0.9 ± 0.56
Kim et al., 2016 [24]	The dried edible insects were purchased from three vendors located in Chungnam, Korea. Nutrients were originally expressed as g/kg dry weight.		full-fat	52.58 ± 0.36	40.04 ± 0.27	19.21 ± 0.34	n/a	23.41 ± 0.01	4.8 ± 0.71
Kuntadi et al., 2018 [25]	The SP were oven dried at 60–70 °C for 12–24 h and then ground.		full-fat	60.03	45.72	29.47	n/a	0.92	5.79
Lamberti et al., 2019 [26]	SP were collected 7 days after reaching the cocoon stage. Pupae were lyophilized and stored at −80 °C.	male silkworm reared on an artificial diet	full-fat	61.45	46.80	n/a	n/a	n/a	5.68
		male silkworm reared on leaf	full-fat	55.02	41.90	n/a	n/a	n/a	4.62
		female silkworm reared on an artificial diet	full-fat	72.36	55.11	n/a	n/a	n/a	5.46
		female silkworm reared on leaf	full-fat	56.29	42.87	n/a	n/a	n/a	5.02
Rodriguez-Ortega et al., 2016 [27]	Dried pupae. Larvae reared on mulberry tree leaves.		full-fat	64.31 ± 0.00 (true protein = 46.87 ± 0.00)	48.98 ± 0.00	20.63 ± 0.105	4.89 ± 0.107	3.92	6.25 ± 0.017
Tomotake et al., 2010 [28]	SP were purchased from Nishiki Food Ltd. (Nagano, Japan). SP were lyophilized followed by grinding into a fine powder.		full-fat	55.60	42.34.00	32.20	n/a	n/a	n/a
Pereira et al., 2003 [29]	Silkworm pupae fed with younger leaves of mulberry tree, and dried at 130 °C, until 4–5% of moisture.		full-fat	51.1 ± 1.8	38.92 ± 1.37	34.4 ± 0.8	n/a	n/a	3.64 ± 0.09
Ghosh et al., 2020 [30]	Pupae of *B. mori* were ground in a mortar and pestle and centrifuged adding some distilled water. Then the liquid centrifugate was spray-dried and ground into fine powdered material.		full-fat	55.87 ± 1.16	42.55 ± 0.88	23.45 ± 0.28	3.77 ± 1.16	1.76 ± 1.19	5.38 ± 0.07
Mishyna et al., 2020 [31]	Insects were stored at −20 °C prior to grinding or drying. After drying, the insects were ground in a food blender and stored in a double plastic bag at −20 °C prior to use, but no longer than 1 month.	freeze-dried	full-fat	60.29 ± 0.38	45.91 ± 0.29	n/a	n/a	n/a	n/a
		oven-dried	full-fat	59.33 ± 0.12	45.19 ± 0.09	n/a	n/a	n/a	n/a
		microwave-dried	full-fat	61.57 ± 0.21	46.89 ± 0.16	n/a	n/a	n/a	n/a

Table 2. *Cont.*

Reference	Treatment, Strain, Diet		Separation	Protein (Kp = 6.25)	Protein (Kp = 4.76)	Fat	Crude Fiber	NFE (Often Indicates as Carbohydrates in the Original Paper)	Ash
Hirunyophat et al., 2021 [32]	Frozen and stored at −18 °C.	Nangnoi (NN) strain	full-fat	54.83 ± 0.43	41.76 ± 0.33	23.61 ± 0.56	2.67 ± 0.06	14.99 ± 1.05	3.99 ± 0.11
		Siwtui (ST) strain	full-fat	55.39 ± 0.19	42.19 ± 0.14	21.26 ± 0.92	2.91 ± 0.92	17.45 ± 0.83	3.12 ± 0.06
		Luang Saraburi (LS) strain	full-fat	50.36 ± 0.21	38.35 ± 0.16	27.59 ± 0.43	3.29 ± 0.07	13.42 ± 0.43	5.41 ± 0.09
		Ubon Ratchathani 60–35 (U 60–35) strain	full-fat	48.31 ± 0.55	36.79 ± 0.42	29.93 ± 0.85	2.81 ± 0.03	14.87 ± 0.08	4.23 ± 0.47
	Frozen and stored at −18 °C, thawed at 5 °C for 12 h, and dried at 60 °C for 16 h; ground to pass through 177-micron mesh sieve; stored at −18 °C; composition dry weight.	Nangnoi (NN) strain	full-fat	52.94 ± 0.11	40.32 ± 0.08	25.53 ± 0.01	4.78 ± 0.06	12.00 ± 0.09	4.69 ± 0.05
		Siwtui (ST) strain	full-fat	56.00 ± 0.4	42.65 ± 0.3	23.95 ± 0.15	5.58 ± 0.09	10.78 ± 0.7	3.94 ± 0.02
		Luang Saraburi (LS) strain	full-fat	49.19 ± 0.26	37.46 ± 0.2	30.47 ± 0.33	5.46 ± 0.14	9.93 ± 0.76	4.73 ± 0.06
		Ubon Ratchathani 60–35 (U 60–35) strain	full-fat	46.22 ± 0.15	35.2 ± 0.11	34.69 ± 0.27	4.92 ± 0.19	9.82 ± 0.03	4.25 ± 0.02

The principal method utilized to remove lipids from the samples was chemical extraction, utilizing ethanol or hexane as the organic solvent. Full-fat SP was the subject of 16 investigations, yielding a total of 31 data entries. Protein levels in full-fat SP ranged from 34.4 [5] to 72.4% [26]. The lowest value was reported by Mishra et al. [5] and related to SP that had been boiled, dissected, and had the inner content removed. Lamberti et al. [26] discovered female pupae raised on an artificial diet displayed the greatest protein value. Rodriguez-Ortega et al. [27] determined that the highest protein value was 64.3% when only pupae raised on mulberry leaves were used. Lamberti et al. [26] also investigated the protein differences between males and females, and the results highlighted the highest values for females. When microwave-dried SP was compared to oven-dried SP, freeze-dried SP, and other types of dried SP, Mishyna et al. [31] discovered that microwave-dried SP had a considerably higher protein value.

The total lipid content ranged from 19.2 [24] to 57.6% [5] and was available for 22 data entries pertaining to full-fat SP. The majority of the research that dealt with lipid extraction methods used standard laboratory extraction procedures using n-hexane, chloroform, methanol, and ethanol as the extraction solvents.

It should be noted that the study by Wei et al., was the only one to investigate supercritical CO_2 as the extraction solvent [33], while Hu et al. employed a microwave-assisted extraction using a mixture of ethanol and n-hexane as the extraction solvent [4]. The crude fiber content of full-fat pupae ranged from 0.54 [5] to 6.38% [24]. For this nutrient, only 14 data entries were available. Nitrogen-free extract content was provided for 17 data entries and ranged from 0.92% [25] to 28.2% [18]. In most of the studies, NFE was referred to as "carbohydrate" and calculated by difference. Ash content ranged from 0.9 [23] to 7.94% [22] for 22 data entries.

3.2. Amino Acid Profile

Table 3 presents a total of 11 data entries from six original investigations about the AA profile. The relative abundance of each AA was given as a percentage of the total amount, albeit the values were biased because only three studies quantified all the AA. A complete comparison was impossible since two studies failed to provide the tryptophan value and one failed to provide tryptophan, cysteine, and methionine values [21,30,31].

We further assumed that asparagine and glutamine values were included in aspartic and glutamic acid values, respectively, as a result of the widespread AA measurement methods [34]. However, research by Kwon et al. [35] emphasized how the value of certain AA varied over the course of pupae development; for instance, histidine values declined with pupal age while threonine and serine values increased. There are a few differences between the AA reported in this study for pupae that are 12–13 days old (D12–13) and for pupae that are 13 days old (D–13), but the comparison was not possible since the size and variability of the samples are unknown.

3.3. Fatty Acid Profile

Seventeen data points about the FA profiles were taken from nine papers and reported in Table 4. The single FA appears to be influenced by a number of variables, including SP strain, nutrition, pupal age, sex, drying and extraction procedure, and analytical methodologies. According to research by Kwon et al. [35] on the influence of pupal age on the composition of fatty acids, linoleic acid was the most variable FA between days 10 and 13, when its relative abundance varied from 5.7 to 8.7% total FAME.

Table 3. Amino acid composition. When available in the original data, values given as a percentage of the total AA along with their respective SD.

Reference		Khoeler et al., 2019 [17]		Akande et al., 2020 [18]	Anootthato et al., 2019 [21]	Tomotake et al., 2010 [28]	Kwon et al., 2012 [36]				Shi et al., 2018 [37]	
Treatment		SH (ready-to-eat, deep-fried in cooking oil. Samples were frozen and freeze-dried for 24 h, powdered and conserved at −80 °C)	SM (non-fried snack, cooked using steam and hot air. Samples were frozen and freeze-dried for 24 h, powdered and conserved at −80 °C)	SP boiled without seasonings, blended	Frozen SP were thawed at 4 °C overnight and washed three times before blanching at 95 °C for 10 min. Then, the sample was dried in a tray dryer at 60 °C for 18 h	SP were lyophilized, followed by grinding into a powder	Frozen SP were extracted and then the protein extract was hydrolyzed with hydrochloric acid at 110 °C for 24 h				SP protein hydrolysate, obtained through enzymatic hydrolysis	
							Pupal age: Day 12–13	Pupal age: Day D7	Pupal age: Day D10	Pupal age: Day D13	Spray dried	Freeze dried
Amino acid	Val	5.95	5.99	4.94 ± 0.04	6.18 ± 0.12	5.53	5.7	4.2	4	2.8	5.27	4.86
	Thr	4.82	4.72	4.62 ± 0.06	5.08 ± 0.07	4.59	3.3	4.6	8.5	9.7	3.89	5.19
	Ile	4.16	4.35	4.92 ± 0.09	4.84 ± 0.10	4	2.7	2.3	1.8	2.1	4.27	2.61
	Leu	7.44	7.39	8.16 ± 0.07	8.16 ± 0.07	7.29	3.5	2.4	2.1	2.5	7.07	6.04
	Lys	7.4	7.19	9.00 ± 0.05	9.09 ± 0.21	7.18	4.5	2.9	3.1	3.9	7.78	7.93
	Met	3.37	4.01	2.28 ± 0.22	-	4	5	6.5	6.4	5	2.42	2.3
	Phe	5.04	5.71	4.21 ± 0.07	5.25 ± 0.02	5.41	3	2.6	1.4	1.6	7.2	7.33
	His	2.98	3.04	2.63 ± 0.06	3.25 ± 0.02	3.18	14.6	25.1	16	13.8	3.16	3.66
	Trp	1.49	1.55	1.28 ± 0.06	-	1.76	-	-	-	-	-	-
	Glu	10.99	11.31	17.02 ± 0.46	13.58 ± 0.21	11.18	18.3	18.9	21.5	18.1	12.44	12.68
	Gln	-	-	-	-	-	-	-	-	-	-	-
	Gly	4.6	4.18	5.16 ± 0.04	4.51 ± 0.05	4.24	7.3	5.2	7	6.1	8.73	10.09
	Ala	5.47	4.85	7.85 ± 0.04	5.80 ± 0.07	4.59	10.2	6.4	3.8	5.1	5.14	4.77
	Ser	4.95	4.61	4.63 ± 0.09	5.30 ± 0.10	4.35	5.2	3.4	9	10.6	3.78	3.1
	Pro	4.55	4.09	3.61 ± 0.05	4.87 ± 0.10	8.24	4.1	6.1	4.4	3.4	9.58	9.76
	Asp	11.69	11.48	8.57 ± 0.04	-	10.71	1.5	1.5	2.2	6	11.23	11.04
	Asn	-	-	-	12.12 ± 0.31	-	-	-	-	-	-	-
	Arg	5.47	5.77	6.62 ± 0.06	6.61 ± 0.10	5.53	3.6	1.3	1.5	2.5	3.25	3.68
	Tyr	8.23	8.46	2.62 ± 0.41	5.35 ± 0.14	6.59	6.2	6.2	6.8	6.7	3.79	4.09
	Cys	1.4	1.31	1.90 ± 0.05	-	1.65	1.5	0.2	0.6	0.4	1	0.88

Table 4. Fatty acid profile (% total FAME) of selected studies and data.

Reference	Treatment		C12:0	C14:0	C15:0	C15:1	C16:0	C16:1, C16:1n-7	C16:2n-6	C16:3n-3	C17:0	C17:1	C18:0	C18:1, C18:1n-7, C18:1n-9	C18:2, C18:2(n-6)	C18:3, C18:3(n-3)	C20:0	C20:1, C20:1n-9	C20:3 (and C20:3n-6, C20:3n-3)	C20:4n-6	C20:5, C20:5n-3	C22:0	C24:0	others
Hu et al., 2017 [4]	BMP were dried in a vacuum drier at 70 °C for 24h to a moisture content of <5% and ground into a fine powder. The powder was sieved and stored at 4 °C.	Microwave assisted extraction, using a mixed solvent consisting of ethanol and n-hexane (1:1, v/v)	-	0.18 ± 0	-	-	23.18 ± 0.52	1.07 ± 0.09	-	-	0.15 ± 0	0.1 ± 0	4.69 ± 0.17	28.32 ± 0.63	3.88 ± 0.13	38.25 ± 0.75	0.16	-	-	-	-	-	-	-
		Soxhlet extraction using n-hexane, at 80 °C for 6 h. The oil was dried at 100 °C ± 5 °C for 15 min. in a drying oven	-	0.19 ± 0	-	-	23.04 ± 0.58	1.05 ± 0.07	-	-	0.17 ± 0	nd ± nd	4.68 ± 0.19	28.15 ± 0.54	3.85 ± 0.54	38.06 ± 0.68	0.16	-	-	-	-	-	-	-
Tomotake et al., 2010 [28]	Silkworm pupae were lyophilized, followed by grinding into a powder. Lipid content was determined according to the methods of AOAC.		-	0.1	-	-	24.2	1.7	-	-	-	-	4.5	26	7.3	36.3	-	-	-	-	-	-	-	-
Pereira et al., 2003 [29]	Chrysalis (worm) toast (originated from the silkworm pupae fed with younger mulberry tree leaves and dried at 130 °C until 4–5% of moisture). Lipids were chemically extracted using chloroform-methanol.		-	0.164 ± 0.02	-	-	24.6 ± 2.1	0.656 ± 0.04	-	-	0.192 ± 0.06	-	7.56 ± 1.54	34.8 ± 3.3	7.03 ± 1.08	24.4 ± 6.7	-	-	0.275 ± 0.1	0.334 ± 0.08	-	-	-	-
Wei et al., 2009 [33]	Silkworm pupae were vacuum dried at 60 °C to a moisture content of <5%, then finely ground to powder, and finally oil was extracted by supercritical CO_2 extraction.		-	-	-	-	21.77	-	-	-	-	-	7.02	33.26	7.12	27.99	-	-	-	-	-	-	-	2.84

Table 4. *Cont.*

Reference	Treatment		C12:0	C14:0	C15:0	C15:1	C16:0	C16:1, C16:1n-7	C16:2n-6	C16:3n-3	C17:0	C17:1	C18:0	C18:1, C18:1n-7, C18:1n-9	C18:2, C18:2(n-6)	C18:3, C18:3(n-3)	C20:0	C20:1, C20:1n-9	C20:3 (and C20:3n-6, C20:3n-3)	C20:4n-6	C20:5, C20:5n-3	C22:0	C24:0	others
Kwon et al., 2012 [35]	The frozen silkworm pupae were ground, blended, mixed with ethanol and stirred. The sample was filtered, and the particles were extracted with ethyl acetate. The extracts were vacuum-concentrated and dissolved in hexane. Activated carbon was added, and the mixture was heated for 1 h and then cooled and filtered before vacuum-concentration.	Pupal period days 12–13	-	-	-	-	19.7	2.5	-	-	-	-	8.6	19.9	7.4	41.6	-	-	-	-	0.3	-	-	-
		Pupae at day 7	-	-	-	-	22.6	2	-	-	-	-	9.6	21.1	8.2	36.4	-	-	-	-	-	-	-	-
		Pupae at day 10	-	-	-	-	21	1.7	-	-	-	-	8.3	24.3	5.7	38.9	-	-	-	-	0.1	-	-	-
		Pupae at day 13	-	-	-	-	20	1.8	-	-	-	-	8.8	21.8	8.7	38.6	-	-	-	-	0.3	-	-	-
Chieco et al., 2019 [37]	Cocoons were dried for 3 days at 60 °C. The pupae were extracted and stored at −80 °C. Frozen pupae were milled to a fine powder. Lipids were chemically extracted using a chloroform/methanol solution according to the Folch's method.	White polyhibrid on artificial diet	-	-	-	-	29.2 ± 0.4	1.5 ± 0	-	-	-	-	10.9 ± 0.1	35.1 ± 0.3	11.4 ± 0.5	11.7 ± 0.3	-	-	-	-	-	-	-	-
		Golden nistari on artificial diet	-	-	-	-	25.1 ± 0.3	1.5 ± 0	-	-	-	-	9.3 ± 0.2	39.8 ± 0.3	11.8 ± 0.1	12.3 ± 0.2	-	-	-	-	-	-	-	-
		White polyhibrid on mulberry leaf of the Florio cultivar	-	-	-	-	25.2 ± 0.7	0.8 ± 0.1	-	-	-	-	7.2 ± 0.2	35.2 ± 0.3	7.1 ± 0.4	24.3 ± 0.2	-	-	-	-	-	-	-	-
		Golden nistari on mulberry leaf of the Florio cultivar	-	-	-	-	21.6 ± 0.4	1.4 ± 0	-	-	-	-	4.8 ± 0.1	32.3 ± 0.1	10.4 ± 0.2	29.3 ± 0.2	-	-	-	-	-	-	-	-

Table 4. *Cont.*

Reference	Treatment	C12:0	C14:0	C15:0	C15:1	C16:0	C16:1, C16:1n-7	C16:2n-6	C16:3n-3	C17:0	C17:1	C18:0	C18:1, C18:1n-7, C18:1n-9	C18:2, C18:2(n-6)	C18:3, C18:3(n-3)	C20:0	C20:1, C20:1n-9	C20:3 (and C20:3n-6, C20:3n-3)	C20:4n-6	C20:5, C20:5n-3	C22:0	C24:0	others
Tong et al., 2011 [38]	Silkworms were directly frozen on the third day of the 5th instar and then lyophilized. Lipid content was measured using Soxhlet extraction.	-	-	0.15	-	16.03	0.39	-	-	-	-	9.45	28.12	12.24	31.91	0.7	0.15	-	-	-	0.7	0.15	-
Usub et al., 2008 [39]	Sun-dried pupae. Five grams were grounded and the lipid extracted with 50 mL of chloroform-methanol (2:1, *v/v*) containing 10 mg L^{-1} of butylated hydroxytoluene and 0.1 mg L^{-1} of tricosanoic acid.	0.1 ± 0	0.1 ± 0	0.1 ± 0	0.1 ± 0	17 ± 0.1	0.6 ± 0	-	-	0.2 ± 0.1	-	6.2 ± 0	20.4 ± 0.1	8.8 ± 0.1	45.6 ± 0.2	0.4 ± 0	0.1 ± 0	0.1 ± 0	0.1 ± 0	-	-	-	-
	Solar-tunnel dried pupae. Five grams were ground and the lipid extracted with 50 mL of chloroform-methanol (2:1, *v/v*) containing 10 mg L^{-1} of butylated hydroxytoluene and 0.1 mg L^{-1} of tricosanoic acid.	0.1 ± 0	0.1 ± 0	0.1 ± 0	0.1 ± 0	17.5 ± 0.6	0.6 ± 0	-	-	0.2 ± 0	-	6.2 ± 0.1	21 ± 0.2	8.8 ± 0.1	44.6 ± 0.5	0.4 ± 0	0.1 ± 0	0.1 ± 0	0.1 ± 0	-	-	-	-
Wang et al., 2020 [39]	Silkworm pupae were dried in a vacuum at 60 °C for 24 h to a moisture content of <5% and then slightly ground and sieved through a 60-mesh sieve. Lipid extraction was conducted using the Folch and Sloane method.	-	-	-	-	22.04 ± 0.49	0.92 ± 0.01	-	-	-	-	6.84 ± 0.23	33.91 ± 0.08	5.48 ± 0.14	30.81 ± 0.21	-	-	-	-	-	-	-	-

Table 4. *Cont.*

Reference	Treatment		C12:0	C14:0	C15:0	C15:1	C16:0	C16:1, C16:1n-7	C16:2n-6	C16:3n-3	C17:0	C17:1	C18:0	C18:1, C18:1n-7, C18:1n-9	C18:2, C18:2(n-6)	C18:3, C18:3(n-3)	C20:0	C20:1, C20:1n-9	C20:3 (and C20:3n-6, C20:3n-3)	C20:4n-6	C20:5, C20:5n-3	C22:0	C24:0	others
Yu et al., 2018 [40]	Pupae were frozen in liquid nitrogen and stored at −80 °C before analysis. Pupae were homogenized and oil was chemically extracted using hexane and boron trifluoride-methanol solution.	Three-day old silkworm pupae, strain Dazao, fed with an artificial diet (Silkmate 2S), Table 2 in the original paper	-	-	-	-	16.6 ± 2.38	0.51 ± 0.04	0.55 ± 0	0.38 ± 0	-	-	11.26 ± 0.7	28.53 ± 1.83	20.54 ± 1.68	21.13 ± 1.7	0.51 ± 0.12	-	-	-	-	-	-	-
		Three-day-old silkworm pupae, strain Dazao, fed with an artificial diet (Silkmate 2S), Table 3 in the original paper	-	-	-	-	15.61 ± 1.89	0.46 ± 0.09	0.46 ± 0.1	0.23 ± 0.17	-	-	12.87 ± 1.4	26.14 ± 1.41	22.24 ± 1.25	21.31 ± 1.21	0.67 ± 0.2	-	-	-	-	-	-	-
		Three-day-old silkworm pupae, strain Dazao, female	-	-	-	-	17.37 ± 0.53	0.53 ± 0.06	0.48 ± 0.06	0.33 ± 0.05	-	-	12.55 ± 1.32	26.63 ± 1.4	21.32 ± 0.53	20.26 ± 0.59	0.53 ± 0.19	-	-	-	-	-	-	-
		Three-day-old silkworm pupae, strain Dazao, male	-	-	-	-	14.34 ± 1.19	0.45 ± 0.08	0.55 ± 0.06	0.36 ± 0.05	-	-	12.54 ± 1.46	25.38 ± 1.72	23.11 ± 1.09	22.57 ± 0.92	0.71 ± 0.11	-	-	-	-	-	-	-
		Three-day-old silkworm pupae, strain 305, female	-	-	-	-	18.44 ± 1.59	1.19 ± 0.57	0.53 ± 0.03	0.29 ± 0.15	-	-	12.93 ± 2.14	28.23 ± 2.08	18.72 ± 1.05	19.14 ± 1.3	0.53 ± 0.18	-	-	-	-	-	-	-
		Three-day-old silkworm pupae, strain 305, male	-	-	-	-	15.23 ± 0.36	1.16 ± 0.58	0.57 ± 0.08	0.38 ± 0.06	-	-	13.12 ± 0.96	25.97 ± 1.57	21.2 ± 1.3	21.54 ± 0.64	0.83 ± 0.38	-	-	-	-	-	-	-

The two extraction techniques—microwave and Soxhlet—used in the study by Hu et al. [4] and the two drying techniques—sun drying and solar tunnel drying—tested by Usub et al. [39] appear to have minimal impact on the final FA profile. Additionally, the FA profile is significantly impacted by the use of an artificial diet during the larval stage. The use of an artificial diet, as well as the usage of various silkworm strains, were both examined in the study by Chieco et al. [37]. The former had a significant impact on the SP's FA composition by raising stearic acid and lowering alfa-linolenic acid. Regarding the silkworm strain, it was shown that the pupae of polyhybrid and Nistari (a tropical pure strain) show notable differences in FA profile, with the Nistari strain having the largest concentration of linoleic and linolenic acids.

Two distinct strains were raised on an artificial diet in the study by Yu et al. [40] and the FA composition was assessed in relation to strain and sex. The findings suggest that higher levels of unsaturated FA (UFA) characterize female pupae, and they probably account for the largest quantities of energy devoted to oviposition. According to this study, the investigated strains were similar in terms of FA profile.

When comparing the findings obtained by Yu et al. [40], to those from other research where silkworms were grown on mulberry leaves, it was found that the concentration of the palmitic (C16:0), linolenic (C18:3), and linoleic (C18:2) acids decreased while the content of linoleic (C18:2) increased. All the investigations came to the same conclusion about the five primary FAs that characterized the SP. In fact, the acids palmitic (C16:0), stearic (C18:0), oleic (C18:1), linoleic (C18:2), and alfa-linolenic (C18:3) make up between 97.1 percent [41] and 99.1 percent [39] of total FAME. Among the five FAs, linoleic and alfa-linolenic acids are essential FAs. Their amount fluctuates between 23.1% and 54.4% of total FAME, with pupae grown on an artificial diet having the lowest value [37].

Furthermore, palmitoleic acid (C16:1 n-7) contributes a small but consistent amount to total FA quantity (0.51%–2.00% total FAME). A few FAs were found in single studies. Among unsaturated FA, C16:2n-6 and C16:3n-3 were identified only by Yu et al. [40]. Considering saturated FA, C22:0, and C24:0 were identified only by Tong et al. [38], while C12:0 appears only in the study by Usub et al. [39]. The percentage of total saturated FA (SFA) in total FAME ranges from 24.1 to 40.1%. In particular, the highest values were reached by those silkworms reared on an artificial diet [37].

3.4. Mineral Content

From six studies, a total of seven data entries about mineral content were accumulated and presented in Table 5. Iron, zinc, manganese, magnesium, and calcium were among the minerals found and quantified in the majority of the publications. It is important to note that the magnitudes of the minerals identified in the study by Akande et al. [23] differed greatly from the other studies. This difference could be because pupae were boiled before analysis, and leaking mineral components might have affected the results.

Iron concentrations ranged from 2.83 to 4.95 mg/100 g, calcium from 92.0 to 181 mg/100 g, manganese from 1.08 to 2.30 mg/100 g, zinc from 1.39 to 24.4 mg/100 g, and magnesium from 89 to 280 mg/100 g, excluding the values found in the study by Akande et al. [23]. Other minerals were quantified in a few studies and included sodium, which ranged from 29.6 [29] to 363 mg/100 g [17].

Three investigations showed potassium content, which ranged from 477 to 672 mg/100 g. Table 5 lists additional minerals that were mentioned in the selected papers.

Table 5. The mineral content of selected studies and data. Akande et al., 2020, originally expressed it as ppm. Kim et al., 2016, originally expressed it as mg/kg. Rodriguez-Ortega et al. 2016, originally expressed it as a percentage of dry matter. Köhler et al., 2019, expressed it as mg/100g of edible portion (EP).

	Ref.	Köhler et al., 2019 [17]		Akande et al., 2020 [18]	David-Birman et al., 2019 [19]	Kim et al., 2016 [24]	Kuntadi et al., 2018 [25]	Rodriguez-Ortega et al., 2016 [27]	Pereira et al., 2003 [29]
		mg/100 g, (Edible Portion)		M ± SD; mg/100 g	M ± SD; mg/100 g (dry weight)	M ± SD; mg/100 g (dry weight)	mg/100 g (dry weight)	Mean ± SD; mg/100 g (dry weight)	M ± SD; mg/100 g (dry weight)
		Street Hawker	Supermarket						
Minerals	Cr	-	-	-	N/D	-	-	-	0.253 ± 0.023
	Hg	<0.005	<0.005	-	N/D	-	-	-	-
	Pb	0.0138	0.0044	-	N/D	-	-	-	-
	Cd	<0.005	<0.005	-	N/D	-	-	-	-
	As	0.0432	0.0165	-	-	-	-	-	-
	Se	0.0534	0.2285	-	-	-	-	-	-
	Mo	0.0151	0.0516	-	-	0.02 ± 0.00	-	-	-
	B	-	-	-	-	1.46 ± 0.42	-	-	-
	Cu	0.711	0.943	-	-	0.94 ± 0.01	-	-	1.52 ± 0.07
	Na	128.4	362.6	-	41.86 ± 3.45	-	-	-	29.60 ± 2.10
	K	492.9	672	-	513.8 ± 1.13	-	-	-	477 ± 16
	Mg	157.7	174.7	-	280.8 ± 1.13	252.29 ± 10.26	-	-	89 ± 0
	Fe	2.83	3.18	0.142 ± 0.063	N/D	4.95 ± 0.05	3.54	-	3.00 ± 0.88
	Zn	9.8	15.4	0.203 ± 0.004	1.39 ± 0.07	14.70 ± 1.45	-	-	24.40 ± 0.96
	Mn	1.26	1.86	0.021 ± 0.021	1.08 ± 0.05	1.68 ± 0.06	-	-	2.30 ± 0.17
	Ca	92.2	107.6	0.249 ± 0.013	137.01 ± 1.34	98.72 ± 0.00	29.17	760 ± 0	181 ± 9
	P	492.9	672	5.57 ± 0.144	-	870.92 ± 15.77	-	720 ± 0	-

M ± SD = mean ± standard deviation; N/D = not detected.

4. Discussion

Protein content was the nutritional value that was most frequently reported, which is indicative of the fact that silkworm pupae were primarily seen as a promising protein source. The removal of the inner material, an unusual method in insect processing that could have altered the nutrient content of SP, may have caused the high value of lipids and low value of protein discovered by Mishra et al. [5]. Most of the protein values for full-fat SP were between 45 and 60% (percentage of the dry matter), which was comparable to other raised insects such as *Tenebrio molitor* and *Grillus bimaculatus* (53.2 and 58.3%, respectively) [42] and chicken eggs (54.1%) [43].

For the majority (86%) of the samples, the full-fat SP's total lipid content ranges from 20 to 35%. These figures could be in line with those discovered for chicken (34.8%), swine (26.3%), tuna (24.9%), and *T. molitor* (34.5%) expressed on a dry matter basis [42,43]. It should be noted that when comparing nutritional data for insects to that of other traditional animal sources, the nutritional content of insects is expressed as referring to the full organism, as opposed to traditional foods of animal origin, where only a portion of the animal is consumed.

A nitrogen-to-protein conversion factor (Kp) is frequently used to determine the protein content from the total nitrogen determined using analytical techniques. Most of the studies used a Kp of 6.25, whereas only two reported an insect corrected Kp of 4.76. The 6.25 conversion factor is the most commonly used for protein estimation in meat. However, it is based on the assumption that all of the nitrogen found in a food matrix comes from protein and that nitrogen accounts for 16 percent of protein [44]. This could result in an overestimation of protein content, particularly in insects that are characterized by the presence of chitin. As a result, Janssen et al. [16] proposed using a Kp conversion factor of 4.76 based on the quantification of non-protein nitrogen (i.e., chitin and chitosan, nucleic acids, and urea) in three insect species: *Hermetia illucens, Tenebrio molitor,* and *Alphitobius diaperinus*. Moreover, because NFE is usually calculated by difference, an incorrect estimation of protein content could lead to an incorrect estimation of NFE content. For these reasons, and in the absence of a specific Kp for SP, we agree to use the 4.76 Kp value proposed by Janssen et al. [16]. The presence of all essential AA, characterized SP's amino acid content, but the variability of the retrieved data precludes further considerations. Kwon et al. [35] found variation in AA profile as a function of pupal age, which should be taken into account during the setup of rearing and further processing of SP in relation to the desired AA profile in the final product.

The main issue with the AA dataset was that only two of the six studies performed replications of their analyses, making data comparison even more difficult. Despite the variability of FA identified in the selected studies, five main FAs (palmitic, stearic, oleic, linoleic, and alfa-linolenic acid) characterized SP oil, making it a good source of UFA, whose content ranges between 59.7 and 75.8%, with the lowest values (from 59.7 to 65.4%) in SP reared on an artificial diet [37], which also have the highest SFA.

The majority of the UFA retrieved values were between 65 and 75%; however, the proportion of monounsaturated FA (MUFA) and polyunsaturated FA (PUFA) that contributes to UFA content varies across papers. The SP reared on an artificial diet had the highest MUFA values as well as the lowest PUFA values [37], implying that the use of an artificial diet could change the proportion of FA classes, reducing PUFA in favor of MUFA and SFA. Nonetheless, the SP reared on an artificial diet by Yu et al. [40] had PUFA and MUFA values comparable to all other SP reared on mulberry leaves. The findings of Chieco et al. [37] and Yu et al. [40] indicate that the use of an artificial diet has a significant impact on the SP FA profile.

Domesticated silkworms are strictly monophagous insects that feed primarily on fresh mulberry leaves (*Morus alba* L.) but can also be reared on artificial diets [44]. In artificial diets, mulberry leaf powder is typically mixed with plant-based meals (soybean, wheat, or corn meals), vitamins, antioxidants, and inorganic salts [44,45]. As previously demonstrated [46], the type and proportion of the different plant matrices included in the diet determine the FA profile of the diet and are likely to influence the SP FA profile.

When we looked at the proportion of n-3 and n-6 FA in SP fed mulberry leaves versus an artificial diet, we discovered that SP fed mulberry leaves have an n-6/n-3 FA ratio ranging from 2.80:1 to 9.88:1, whereas those fed an artificial diet have an n-6/n-3 FA ratio ranging from 0.96:1 to 1.05:1. This is primarily due to the comparable amounts of linoleic and alfa-linolenic acid and a much lower amount of n-3 FAs in the artificial diet fed to SP, indicating that the diet appears to be the main variable influencing the SP's FA profile.

Since they are not synthesizable by the human body, n-3 and n-6 FA play an important role in the human diet and are thus considered essential FA. Furthermore, their ratio influences body fat metabolism and systemic inflammation, with an increase in the n-6/n-3 ratio associated with an increased risk of obesity, atherosclerosis, and diabetes [47,48]. As a result, an n-6/n-3 FA ratio of 1:1 to 5:1 is thought to be optimal for human health [49]. The average Western diet has an n-6/n-3 ratio of 15.0:1 to 16.7:1, so it would be preferable to include food supplements or nutraceuticals with a lower n-6/n-3 ratio in the Western diet. Furthermore, lowering the n-6/n-3 ratio in broiler diets has been shown to improve growth performance and immune response in these birds [50]. In the case of other farmed animals' meat, n-6/n-3 values of 1.3, 2.1, 7.2, and 13.7 were found in the muscle tissues of sheep, cattle, pigs, and chickens, respectively [51,52]. The n-6/n-3 ratio is always lower in marine species such as herring, tuna, pollock, salmon, and rainbow trout because their diet is based on marine phytoplankton [53]. The values discovered for mulberry leaves fed SP are thus comparable to those found in sheep, cattle, and pigs.

The linoleic and alfa-linolenic acid content of mulberry leaves fed with SP varied greatly, but it is unclear which variable could explain these results. Consequently, in order to achieve a better optimal n-6/n-3 ratio, we must identify the most relevant variables and their cause-effect relationships.

In terms of the mineral content of SP, the available literature indicates that mulberry is important in determining the mineral content of silkworms, which bioaccumulate minerals through nutrition. The high variability in SP mineral content observed could thus be linked to mineral availability in the soil where the mulberry tree grows. Mineral uptake in the mulberry tree may also vary in response to heavy metal contamination in the soil [54]. Mulberry cultivar and harvest time are also important factors in mineral accumulation [55].

The high sodium value discovered by Kohler et al. [17] was found in a supermarket sample, and it cannot be excluded that it was due to the addition of additives. The *B. mori* strain used is another variable that influences the SP nutrient composition, as demonstrated by Hirunyophat et al. [32]. This information is frequently missing, except in papers in which the strain was one of the primary factors investigated. However, when not reported, the analysis would have most likely been performed on hybrid silkworms, which are the most common type in experimental and production settings. Although the impact of several intrinsic factors on the nutritional value of SP has been investigated, it is not always clear how the analytical methods were implemented, and especially if and how the analyzed pupae were preprocessed.

When silkworms were not reared in a laboratory or under controlled conditions, the process they went through to reach the pupal stage was sometimes unclear. This is especially true for SP recovered from silk reeling industries (by-product pupae), for which precise data on industrial processing, storage time, and temperature were not available [56]. Silkworm pupae from reeling industries have often been dried and boiled, which may have altered the nutritional content. Furthermore, once recovered from the reeling plants, they were most likely dried and stored under highly oxidative conditions. These factors may have a significant impact on the composition and must be considered if the pupae's final destination is the feed or food market.

Finally, as reported by Kwon et al. [35], pupal age may also play a role in determining the composition. It is worth noting that because only 64% of the data has a standard deviation or other measures of variation associated with mean values, a large portion of them are statistically weak and difficult to compare.

5. Conclusions

The literature on SP as food has increased recently, and the topic is expected to be investigated further, given the growing interest in insects as food. Our systematic review, which focuses on the nutrient composition of SP, highlights the wide range of data obtained. On the one hand, the variables involved in sample preparation and analysis are frequently well described in the studies; on the other hand, variables such as the mulberry cultivar used to feed the larvae, silkworm strain, and rearing environmental conditions are rarely reported. When SP was obtained from the silk reeling industry, pre-treatments, and storage conditions were also poorly described or unknown. Furthermore, studies reported data in various units of measurement or formats, making quick comparison difficult.

For all these reasons, new studies with more detailed and better described analytical methodologies, normalized data on a dry matter basis, and discussion about those values that significantly differ from previously reported literature are urgently needed. It is recommended that standard methods from other product sectors be established or adapted to analyze the nutritional values of edible insects for the food market (and feed as well) and that a reference benchmark for the most commonly used analytical procedures be established. In terms of nutritional composition, SP is primarily distinguished by its protein content, which is the most commonly reported variable in papers. The wide range of values found in the literature for protein content and AA profile requires further study, as do the variables that influence those values. When it comes to the FA profile, it is worth noting that the available literature agrees on identifying five major FAs that account for more than 97.1% of the total SP lipid content. The balance of these five main FAs has been shown to be influenced by diet and pupal age, but the sum of these five FAs is very consistent regardless of upstream conditions.

Author Contributions: Conceptualization, L.T., S.C., S.B., A.S. and A.D.Z.; methodology, S.B., F.M. and P.A.; software, S.B., F.M. and P.A.; validation, S.B., L.T., A.S. and S.C.; formal analysis, L.T. and A.S.; investigation, L.T. and A.S.; resources, S.B., F.M. and P.A.; data curation, L.T. and A.S.; writing—original draft preparation, L.T., A.S. and S.C.; writing—review and editing, L.T., A.S., S.C., A.D.Z., F.M., P.A. and S.B.; visualization, L.T. and A.S.; supervision, S.C. and A.D.Z.; project administration, S.C. and S.B; funding acquisition, S.C., L.T., S.B. and A.D.Z. All authors have read and agreed to the published version of the manuscript.

Funding: This research was funded by the funding programme "POR FESR 2014–2020 Regione del Veneto, Bando per il sostegno a progetti sviluppati da aggregazioni di imprese". Project name: "SILK-PLUS: Rivalutazione dei sottoprodotti della filiera serica in campo cosmetico e alimentare". Grant ID: 10223675.

Institutional Review Board Statement: Not applicable.

Informed Consent Statement: Not applicable.

Data Availability Statement: Not applicable.

Conflicts of Interest: The authors declare no conflict of interest. The funders had no role in the design of the study; in the collection, analyses, or interpretation of data; in the writing of the manuscript; or in the decision to publish the results.

References

1. Yang, S.-Y.; Han, M.-J.; Kang, L.-F.; Li, Z.-W.; Shen, Y.-H.; Zhang, Z. Demographic History and Gene Flow during Silkworm Domestication. *BMC Evol. Biol.* **2014**, *14*, 185. [CrossRef]
2. Meng, X.; Zhu, Y.; Chen, K. Silkworm: A Promising Model Organism in Life Science. *J. Insect Sci.* **2017**, *17*, 97. [CrossRef]
3. Xu, S.; Wang, F.; Wang, Y.; Wang, R.; Hou, K.; Tian, C.; Ji, Y.; Yang, Q.; Zhao, P.; Xia, Q. A Silkworm Based Silk Gland Bioreactor for High-Efficiency Production of Recombinant Human Lactoferrin with Antibacterial and Anti-Inflammatory Activities. *J. Biol. Eng.* **2019**, *13*, 61. [CrossRef] [PubMed]

4. Hu, B.; Li, C.; Zhang, Z.; Zhao, Q.; Zhu, Y.; Su, Z.; Chen, Y. Microwave-Assisted Extraction of Silkworm Pupal Oil and Evaluation of Its Fatty Acid Composition, Physicochemical Properties and Antioxidant Activities. *Food Chem.* **2017**, *231*, 348–355. [CrossRef]
5. Mishra, N.; Hazarika, N.C.; Narain, K.; Mahanta, J. Nutritive Value of Non-Mulberry and Mulberry Silkworm Pupae and Consumption Pattern in Assam, India. *Nutr. Res.* **2003**, *23*, 1303–1311. [CrossRef]
6. Zhang, F.; Zhang, Z. Study on Edible Insect Resources and Their Exploitation and Utilization. *Resour. Sci.* **2001**, *23*, 58–61.
7. Mitsuhashi, J. Insects as Traditional Foods in Japan. *Ecol. Food Nutr.* **1997**, *36*, 187–199. [CrossRef]
8. Han, R.; Shin, J.T.; Kim, J.; Choi, Y.S.; Kim, Y.W. An Overview of the South Korean Edible Insect Food Industry: Challenges and Future Pricing/Promotion Strategies: S. Korean Edible Insect Food Industry. *Entomol. Res.* **2017**, *47*, 141–151. [CrossRef]
9. Yhoung-Aree, J.; Puwastien, P.; Attig, G.A. Edible Insects in Thailand: An Unconventional Protein Source? *Ecol. Food Nutr.* **1997**, *36*, 133–149. [CrossRef]
10. Aznar-Cervantes, S.D.; Monteagudo Santesteban, B.; Cenis, J.L. Products of Sericulture and Their Hypoglycemic Action Evaluated by Using the Silkworm, *Bombyx mori* (Lepidoptera: Bombycidae), as a Model. *Insects* **2021**, *12*, 1059. [CrossRef]
11. Ratcliffe, N.A.; Mello, C.B.; Garcia, E.S.; Butt, T.M.; Azambuja, P. Insect Natural Products and Processes: New Treatments for Human Disease. *Insect Biochem. Mol. Biol.* **2011**, *41*, 747–769. [CrossRef] [PubMed]
12. Zimian, D.; Yonghua, Z.; Xiwu, G. Medicinal Insects in China. *Ecol. Food Nutr.* **1997**, *36*, 209–220. [CrossRef]
13. EFSA Scientific Committee. Risk Profile Related to Production and Consumption of Insects as Food and Feed. *EFSA J.* **2015**, *13*, 4257. [CrossRef]
14. Wu, X.; He, K.; Velickovic, T.C.; Liu, Z. Nutritional, Functional, and Allergenic Properties of Silkworm Pupae. *Food Sci. Nutr.* **2021**, *9*, 4655–4665. [CrossRef] [PubMed]
15. Thomas, J.; Brunton, J.; Graziosi, S. *EPPI-Reviewer 4.0: Software for Research Synthesis. EPPI-Centre Software*; Social Science Research Unit, Institute of Education, University of London: London, UK, 2010.
16. Janssen, R.H.; Vincken, J.-P.; van den Broek, L.A.M.; Fogliano, V.; Lakemond, C.M.M. Nitrogen-to-Protein Conversion Factors for Three Edible Insects: *Tenebrio Molitor*, *Alphitobius Diaperinus*, and *Hermetia Illucens*. *J. Agric. Food Chem.* **2017**, *65*, 2275–2278. [CrossRef]
17. Köhler, R.; Kariuki, L.; Lambert, C.; Biesalski, H.K. Protein, Amino Acid and Mineral Composition of Some Edible Insects from Thailand. *J. Asia-Pac. Entomol.* **2019**, *22*, 372–378. [CrossRef]
18. Akande, O.A.; Falade, O.O.; Badejo, A.A.; Adekoya, I. Assessment of Mulberry Silkworm Pupae and African Palm Weevil Larvae as Alternative Protein Sources in Snack Fillings. *Heliyon* **2020**, *6*, e03754. [CrossRef]
19. David-Birman, T.; Moshe, H.; Lesmes, U. Impact of Thermal Processing on Physicochemical Properties of Silk Moth Pupae (*Bombyx mori*) Flour and in-Vitro Gastrointestinal Proteolysis in Adults and Seniors. *Food Res. Int.* **2019**, *123*, 11–19. [CrossRef]
20. Kim, H.-W.; Setyabrata, D.; Lee, Y.J.; Jones, O.G.; Kim, Y.H.B. Pre-Treated Mealworm Larvae and Silkworm Pupae as a Novel Protein Ingredient in Emulsion Sausages. *Innov. Food Sci. Emerg. Technol.* **2016**, *38*, 116–123. [CrossRef]
21. Anootthato, S.; Therdthai, N.; Ritthiruangdej, P. Characterization of Protein Hydrolysate from Silkworm Pupae (*Bombyx mori*). *J. Food Process. Preserv.* **2019**, *43*. [CrossRef]
22. Felix, M.; Bascon, C.; Cermeño, M.; FitzGerald, R.J.; de la Fuente, J.; Carrera-Sánchez, C. Interfacial/Foaming Properties and Antioxidant Activity of a Silkworm (*Bombyx mori*) Pupae Protein Concentrate. *Food Hydrocoll.* **2020**, *103*, 105645. [CrossRef]
23. Akande, A.O.; Jolayemi, O.S.; Adelugba, V.A.; Akande, S.T. Silkworm Pupae (*Bombyx mori*) and Locusts as Alternative Protein Sources for High-Energy Biscuits. *J. Asia-Pac. Entomol.* **2020**, *23*, 234–241. [CrossRef]
24. Kim, S.-K.; Weaver, C.M.; Choi, M.-K. Proximate Composition and Mineral Content of Five Edible Insects Consumed in Korea. *CyTA–J. Food* **2016**, *15*, 143–146. [CrossRef]
25. Kuntadi, K.; Adalina, Y.; Maharani, K.E. Nutritional Composition of Six Edible Insects in Java. *Indones. J. For. Res.* **2018**, *5*, 57–68. [CrossRef]
26. Lamberti, C.; Gai, F.; Cirrincione, S.; Giribaldi, M.; Purrotti, M.; Manfredi, M.; Marengo, E.; Sicuro, B.; Saviane, A.; Cappellozza, S.; et al. Investigation of the Protein Profile of Silkworm (*Bombyx mori*) Pupae Reared on a Well-Calibrated Artificial Diet Compared to Mulberry Leaf Diet. *PeerJ* **2019**, *7*, e6723. [CrossRef] [PubMed]
27. Rodríguez-Ortega, A.; Pino-Moreno, J.M.; Ángeles-Campos, S.C.; García-Pérez, Á.; Barrón-Yánez, R.M.; Callejas-Hernández, J. Valor Nutritivo de Larvas y Pupas de Gusano de Seda (*Bombyx mori*) (Lepidoptera: Bombycidae). *Rev. Colomb. Entomol.* **2016**, *42*, 69. [CrossRef]
28. Tomotake, H.; Katagiri, M.; Yamato, M. Silkworm Pupae (*Bombyx mori*) Are New Sources of High Quality Protein and Lipid. *J. Nutr. Sci. Vitaminol.* **2010**, *56*, 446–448. [CrossRef]
29. Pereira, N.R.; Ferrarese-Filho, O.; Matsushita, M.; de Souza, N.E. Proximate Composition and Fatty Acid Profile of *Bombyx mori* L. Chrysalis Toast. *J. Food Compos. Anal.* **2003**, *16*, 451–457. [CrossRef]
30. Ghosh, A.; Ray, M.; Gangopadhyay, D. Evaluation of Proximate Composition and Antioxidant Properties in Silk-Industrial Byproduct. *LWT* **2020**, *132*, 109900. [CrossRef]
31. Mishyna, M.; Haber, M.; Benjamin, O.; Martinez, J.J.I.; Chen, J. Drying Methods Differentially Alter Volatile Profiles of Edible Locusts and Silkworms. *J. Insects Food Feed* **2020**, *6*, 405–415. [CrossRef]
32. Hirunyophat, P.; Chalermchaiwat, P.; On-nom, N.; Prinyawiwatkul, W. Selected Nutritional Quality and Physicochemical Properties of Silkworm Pupae (Frozen or Powdered) from Two Species. *Int. J. Food Sci. Technol.* **2021**, *56*, 3578–3587. [CrossRef]

33. Wei, Z.-J.; Liao, A.-M.; Zhang, H.-X.; Liu, J.; Jiang, S.-T. Optimization of Supercritical Carbon Dioxide Extraction of Silkworm Pupal Oil Applying the Response Surface Methodology. *Bioresour. Technol.* **2009**, *100*, 4214–4219. [CrossRef] [PubMed]
34. Walker, V.; Mills, G.A. Quantitative Methods for Amino Acid Analysis in Biological Fluids. *Ann. Clin. Biochem. Int. J. Lab. Med.* **1995**, *32*, 28–57. [CrossRef] [PubMed]
35. Kwon, M.-G.; Kim, D.-S.; Lee, J.-H.; Park, S.-W.; Choo, Y.-K.; Han, Y.-S.; Kim, J.-S.; Hwang, K.-A.; Ko, K.; Ko, K. Isolation and Analysis of Natural Compounds from Silkworm Pupae and Effect of Its Extracts on Alcohol Detoxification: Nutritional Value of Silkworm Component. *Entomol. Res.* **2012**, *42*, 55–62. [CrossRef]
36. Chieco, C.; Morrone, L.; Bertazza, G.; Cappellozza, S.; Saviane, A.; Gai, F.; Di Virgilio, N.; Rossi, F. The Effect of Strain and Rearing Medium on the Chemical Composition, Fatty Acid Profile and Carotenoid Content in Silkworm (*Bombyx mori*) Pupae. *Animals* **2019**, *9*, 103. [CrossRef]
37. Tong, L.; Yu, X.; Liu, H. Insect Food for Astronauts: Gas Exchange in Silkworms Fed on Mulberry and Lettuce and the Nutritional Value of These Insects for Human Consumption during Deep Space Flights. *Bull. Entomol. Res.* **2011**, *101*, 613–622. [CrossRef]
38. Usub, T.; Lertsatitthanakorn, C.; Poomsa-ad, N.; Wiset, L.; Yang, L.; Siriamornpun, S. Experimental Performance of a Solar Tunnel Dryer for Drying Silkworm Pupae. *Biosyst. Eng.* **2008**, *101*, 209–216. [CrossRef]
39. Wang, W.; Xu, L.; Zou, Y.; Pang, D.; Shi, W.; Mu, L.; Li, E.; Lan, D.; Wang, Y.; Liao, S. Comprehensive Identification of Principal Lipid Classes and Tocochromanols in Silkworm (*Antheraea pernyi* and *Bombyx mori*) Pupae Oils. *Eur. J. Lipid Sci. Technol.* **2020**, *122*, 1900280. [CrossRef]
40. Yu, X.; Shen, Y.; Cui, Q.; Chen, Y.; Sun, W.; Huang, X.; Zhu, Y. Silkworm (*Bombyx mori*) Has the Capability to Accumulate C_{20} and C_{22} Polyunsaturated Fatty Acids. *Eur. J. Lipid Sci. Technol.* **2018**, *120*, 1700268. [CrossRef]
41. Ghosh, S.; Lee, S.-M.; Jung, C.; Meyer-Rochow, V.B. Nutritional Composition of Five Commercial Edible Insects in South Korea. *J. Asia-Pac. Entomol.* **2017**, *20*, 686–694. [CrossRef]
42. CREA Centro Alimenti e Nutrizione Tabelle Di Composizione Degli Alimenti, Aggiornamento. 2019. Available online: https://www.alimentinutrizione.it/tabelle-nutrizionali (accessed on 15 April 2022).
43. World Health Organization; Food and Agriculture Organization of the United Nations. *Nitrogen and Protein Content Measurement and Nitrogen to Protein Conversion Factors for Dairy and Soy Protein-Based Foods: A Systematic Review and Modelling Analysis*; World Health Organization: Geneva, Switzerland, 2019; ISBN 978-92-4-151698-3.
44. Cappellozza, L.; Cappellozza, S.; Saviane, A.; Sbrenna, G. Artificial Diet Rearing System for the Silkworm *Bombyx mori* (Lepidoptera: Bombycidae): Effect of Vitamin C Deprivation on Larval Growth and Cocoon Production. *Appl. Entomol. Zool.* **2005**, *40*, 405–412. [CrossRef]
45. Li, J.; Chen, C.; Zha, X. Midgut and Head Transcriptomic Analysis of Silkworms Reveals the Physiological Effects of Artificial Diets. *Insects* **2022**, *13*, 291. [CrossRef] [PubMed]
46. Horie, Y.; Nakasone, S. Effects of the Levels of Fatty Acids and Carbohydrates in a Diet on the Biosynthesis of Fatty Acids in Larvae of the Silkworm, *Bombyx mori*. *J. Insect Physiol.* **1971**, *17*, 1441–1450. [CrossRef]
47. Patel, A.; Desai, S.S.; Mane, V.K.; Enman, J.; Rova, U.; Christakopoulos, P.; Matsakas, L. Futuristic Food Fortification with a Balanced Ratio of Dietary ω-3/ω-6 Omega Fatty Acids for the Prevention of Lifestyle Diseases. *Trends Food Sci. Technol.* **2022**, *120*, 140–153. [CrossRef]
48. Simopoulos, A. An Increase in the Omega-6/Omega-3 Fatty Acid Ratio Increases the Risk for Obesity. *Nutrients* **2016**, *8*, 128. [CrossRef]
49. Lupette, J.; Benning, C. Human Health Benefits of Very-Long-Chain Polyunsaturated Fatty Acids from Microalgae. *Biochimie* **2020**, *178*, 15–25. [CrossRef]
50. Ibrahim, D.; El-Sayed, R.; Khater, S.I.; Said, E.N.; El-Mandrawy, S.A.M. Changing Dietary N-6:N-3 Ratio Using Different Oil Sources Affects Performance, Behavior, Cytokines MRNA Expression and Meat Fatty Acid Profile of Broiler Chickens. *Anim. Nutr.* **2018**, *4*, 44–51. [CrossRef]
51. Wood, J.D.; Enser, M.; Fisher, A.V.; Nute, G.R.; Sheard, P.R.; Richardson, R.I.; Hughes, S.I.; Whittington, F.M. Fat Deposition, Fatty Acid Composition and Meat Quality: A Review. *Meat Sci.* **2008**, *78*, 343–358. [CrossRef]
52. Trembecka, L.; Hascik, P.; Cubon, J.; Bobko, M.; Pavelkova, A. Fatty Acids Profile of Breast and Thigh Muscles of Broiler Chickens Fed Diets with Propolis and Probiotics. *J. Cent. Eur. Agric.* **2016**, *17*, 1179–1193. [CrossRef]
53. Strobel, C.; Jahreis, G.; Kuhnt, K. Survey of N- 3 and n-6 Polyunsaturated Fatty Acids in Fish and Fish Products. *Lipids Health Dis.* **2012**, *11*, 144. [CrossRef]
54. Zeng, P.; Guo, Z.; Xiao, X.; Peng, C.; Liu, L.; Yan, D.; He, Y. Physiological Stress Responses, Mineral Element Uptake and Phytoremediation Potential of *Morus alba* L. in Cadmium-Contaminated Soil. *Ecotoxicol. Environ. Saf.* **2020**, *189*, 109973. [CrossRef] [PubMed]
55. Levickienė, D.; Vaitkevičienė, N.; Jarienė, E.; Mažeika, R. The Content of Macroelements in White Mulberry (*Morus alba* L.) Leaves. *Žemės Ūkio Moksl.* **2019**, *25*. [CrossRef]
56. Saviane, A.; Tassoni, L.; Naviglio, D.; Lupi, D.; Savoldelli, S.; Bianchi, G.; Cortellino, G.; Bondioli, P.; Folegatti, L.; Casartelli, M.; et al. Mechanical Processing of Hermetia Illucens Larvae and *Bombyx mori* Pupae Produces Oils with Antimicrobial Activity. *Animals* **2021**, *11*, 783. [CrossRef] [PubMed]

insects

MDPI

Article

A *Bombyx mori* Infection Model for Screening Antibiotics against *Staphylococcus epidermidis*

Aurora Montali [1,†], Francesca Berini [1,†], Alessio Saviane [2], Silvia Cappellozza [2], Flavia Marinelli [1] and Gianluca Tettamanti [1,3,*]

[1] Department of Biotechnology and Life Sciences, University of Insubria, 21100 Varese, Italy
[2] Council for Agricultural Research and Economics, Research Centre for Agriculture and Environment (CREA-AA), 35143 Padova, Italy
[3] Interuniversity Center for Studies on Bioinspired Agro-Environmental Technology (BAT Center), University of Napoli Federico II, 80055 Portici, Italy
* Correspondence: gianluca.tettamanti@uninsubria.it; Tel.: +39-0332-421312
† These authors contributed equally to the work.

Simple Summary: The use and misuse of antibiotics in the past decades have contributed to the wide spread of antibiotic resistance, which currently represents a major issue and threat to human health. Consequently, the discovery of new anti-infective molecules is of primary importance. In vivo studies are crucial for testing the efficacy of novel antibiotics. Invertebrate models look promising to reduce the large-scale use of mammalians, which is mainly limited by high costs and ethical concerns. In this scenario, the silkworm proved to be an interesting alternative among insects. Here, we developed a silkworm infection model by challenging the larvae with *Staphylococcus epidermidis*, one common cause of infections in hospitals, and assessing the curative effects of three life-saving glycopeptide antibiotics that are used to treat infections caused by multidrug-resistant Gram-positive pathogens.

Abstract: The increasing number of microorganisms that are resistant to antibiotics is prompting the development of new antimicrobial compounds and strategies to fight bacterial infections. The use of insects to screen and test new drugs is increasingly considered a promising tool to accelerate the discovery phase and limit the use of mammalians. In this study, we used for the first time the silkworm, *Bombyx mori*, as an in vivo infection model to test the efficacy of three glycopeptide antibiotics (GPAs), against the nosocomial pathogen *Staphylococcus epidermidis*. To reproduce the human physiological temperature, the bacterial infection was performed at 37 °C and it was monitored over time by evaluating the survival rate of the larvae, as well the response of immunological markers (i.e., activity of hemocytes, activation of the prophenoloxidase system, and lysozyme activity). All the three GPAs tested (vancomycin, teicoplanin, and dalbavancin) were effective in curing infected larvae, significantly reducing their mortality and blocking the activation of the immune system. These results corroborate the use of this silkworm infection model for the in vivo studies of antimicrobial molecules active against staphylococci.

Keywords: *Bombyx mori*; infection model; *Staphylococcus epidermidis*; insect immune response; antimicrobial compounds; glycopeptide antibiotics; vancomycin; teicoplanin; dalbavancin

Citation: Montali, A.; Berini, F.; Saviane, A.; Cappellozza, S.; Marinelli, F.; Tettamanti, G. A *Bombyx mori* Infection Model for Screening Antibiotics against *Staphylococcus epidermidis*. *Insects* **2022**, *13*, 748. https://doi.org/10.3390/insects13080748

Academic Editor: Toru Shimada

Received: 29 July 2022
Accepted: 16 August 2022
Published: 19 August 2022

1. Introduction

Antibiotics are considered one of the greatest discoveries of the 20th century and continue to be extremely important in modern medicine, which relies on them both to treat and to prevent infections in high-risk patients [1]. However, the massive use and/or misuse of antibiotics have contributed to the development and spread of resistant bacteria [2]. The need to find and create novel antibacterial compounds is urgent given the increasing problem of antibiotic resistance. In vivo screening of novel antibiotics might accelerate the identification of promising molecules favoring the prioritization of those endowed by an acceptable therapeutic index (i.e., good efficacy versus low toxicity) [3]. The large-scale use of mammalian animal models, primarily mice and rats, at such preclinical phase is not feasible due to the high costs of their handling and maintenance and the ethical considerations that limit their use, as recommended by the European directive on animal protection guided by the 3Rs rules (i.e., Replacement, Reduction, and Refinement) [4]. To develop suitable alternatives, different invertebrate models have been recently proposed, such as amoebas and nematodes [5,6]. Among them, holometabolous insects represent a promising option due to their minimal cost requirements and convenient larval handling [7]. Moreover, the use of insects is not subjected to ethical restrictions [8] and, although they lack acquired immunity, their innate response is evolutionally and functionally analogous to that of mammals [9].

Different insect species, as *Drosophila melanogaster*, *Galleria mellonella*, and *Bombyx mori*, have recently been used to investigate the action of novel antimicrobials, and to test their efficacy [8,10]. A particular advantage in developing silkworm as an infection model is represented by the access to germplasm banks, where *B. mori* is maintained in genetic stock collections; these centers adopt rearing on artificial diet, thus contributing to standardize the quality of this insect supplies [8]. In addition, genome sequencing of *B. mori* [11] has led to the development of molecular tools and gene editing systems, such as RNA interference-based gene silencing, transposon-mediated transgenesis, and CRISPR/Cas9 [12], which proved useful to set up tailored disease models for investigating drug mode of action. Silkworm infection models have been already employed to evaluate antimicrobial drugs against Gram-negative and Gram-positive bacteria of the ESKAPE group (such as *Escherichia coli*, *Pseudomonas aeruginosa*, *Staphylococcus aureus*, and *Klebsiella pneumoniae*) [13–16], and towards other bacterial pathogens, such as *Francisella tularensis*, *Listeria monocytogenes*, *Bacillus cereus*, and *Mycobacterium abscessus* [17–20], and fungi (*Candida* spp. and *Aspergillus fumigatus*) [21,22]. However, to date, no study in this setting has been performed by using one of the most common causes of nosocomial infections, the coagulase-negative *Staphylococcus epidermidis*. *S. epidermidis* is an abundant harmless colonizer of human skin and mucosa [23]. However, it becomes an opportunistic pathogen that can cause virulence once it invades the human body via medical and prosthetic devices, such as peripheral or central intravenous catheters or orthopedic implants. Bacteremia arises most commonly by indwelling medical device contamination by *S. epidermidis*, which in addition can produce biofilms resistant to the host defense and antimicrobial treatments [23,24].

In the present study, the effects of three life-saving glycopeptide antibiotics (GPAs) (i.e., vancomycin, teicoplanin, and dalbavancin), used in clinical settings to treat severe infections caused by Gram-positive bacteria [25], were investigated in *B. mori* larvae following infection with a clinical isolate of *S. epidermidis*. By monitoring the insect survival rate and the responses of three relevant immunological markers (i.e., hemocyte metabolic rate, lysozyme activity, and phenoloxidase (proPO) system activity), we demonstrated the curative efficacy of the three GPAs, confirming the robustness of this insect-based infection model for testing antimicrobial compounds.

2. Materials and Methods

2.1. Experimental Model

The larvae of *B. mori* (polyhybrid (126 × 57) (70 × 90)) used in this study were provided by CREA-AA Sericulture Laboratory (Padova, Italy). Insects were reared in glass Petri dishes at 25 ± 0.5 °C, with 70 ± 5% relative humidity, under a 12L:12D photoperiod. Larvae were fed on artificial diet [26] until the end of the 4th larval instar. After animals had ecdysed to the 5th larval instar, they were synchronized [27] and fed with a germ-free diet [28].

2.2. Bacterial Strain

S. epidermidis strain 4, a clinical isolate kindly provided by Laboratorio di Microbiologia Clinica-Ospedale di Circolo (Varese, Italy), was grown in 10 mL of Müller–Hinton Broth 2

(MHB2; VWR International Srl, Radnor, PA, USA), overnight at 37 °C under 200 rpm shaking. Cells harvested by centrifuging 1 mL of culture at 1900× *g* at 4 °C for 10 min were then resuspended at the desired concentration with saline solution (0.6% *w*/*v* NaCl). To determine the volume of saline solution to be used, the optical density of the culture was measured, considering that one unit of OD_{600nm} corresponded to 7.03×10^8 CFU (colony forming units)/mL.

2.3. Minimum Inhibitory Concentrations (MICs) and Minimum Bactericidal Concentrations (MBCs)

MICs of vancomycin (Sigma-Aldrich, St. Louis, MO, USA), teicoplanin (Sigma-Aldrich), and dalbavancin (kindly provided by Sanofi, Brindisi, Italy) towards *S. epidermidis* strain 4 were determined by the broth dilution method, following Clinical and Laboratory Standards Institute guidelines [29]. A total of 5×10^5 viable bacterial cells were inoculated into MHB2 medium (VWR International Srl) in 96-well plates and supplemented with increasing antibiotic concentrations (from 0 to 128 µg/mL) up to 100 µL final volume. Plates were then incubated for 16–20 h at 37 °C and 100 rpm. MIC was defined as the minimal concentration of antibiotic at which no turbidity could be detected. For MBC determination, bacterial cultures used for the MIC test were plated onto Müller–Hinton Agar (MHA; VWR International Srl) and incubated at 37 °C for 24 h. MBC was defined as the lowest concentration of antibiotic at which no visible growth was detected on plates. All experiments were performed at least in triplicate.

2.4. Injection of Larvae and Collection of Hemolymph

Insects were infected at the second day of the fifth larval instar by injecting bacteria in the second right proleg with autoclaved Hamilton 1702 LT 25 µL syringes (Hamilton Company, Reno, NV, USA), under a sterile hood. After the injection, larvae were reared at 37 °C. Hemolymph collection was performed by puncturing the larvae or cutting the second left proleg 6 or 24 h after the infection depending on the analysis.

2.5. Lethal Dose 50 (LD_{50}) for S. epidermidis

To determine the LD_{50}, larvae were infected with *S. epidermidis* at different concentrations (from 3×10 CFU to 3×10^5 CFU in a final volume of 10 µL). The mortality rate was monitored every 24 h for three days. Larvae were considered dead when no reaction to a stimulation with a plastic tip was observed. Uninjected larvae and larvae injected with 10 µl of saline solution (0.6% *w*/*v* NaCl) were used as controls. Forty larvae for each experimental condition were used. LD_{50}, calculated by Probit analysis [30], was defined as the concentration of bacteria at which 50% of animals died within 3 days after the infection.

2.6. Administration of Antibiotics

The effects of GPA administration were evaluated by injecting 10 µL of *S. epidermidis* at LD_{50} into the larvae, and 10 µL of vancomycin, teicoplanin, or dalbavancin (at a concentration equal to 8.75 µg/g body weight) after 2 h, as previously reported [15]. Control groups were represented by uninjected larvae, larvae injected once or twice with saline solution (10 µL of 0.6% *w*/*v* NaCl eventually repeated after two hours), and healthy larvae injected with 10 µL of antibiotic. Larval mortality was monitored for three days. Fifty larvae were used for each experimental condition.

2.7. Analysis of the Immunological Markers

Untreated larvae, larvae injected once or twice with 10 µL of 0.6% *w*/*v* NaCl, larvae injected with 10 µL of antibiotic (i.e., vancomycin, teicoplanin, or dalbavancin at 8.75 µg/g body weight), larvae injected with 3×10^3 CFU of *S. epidermidis* in 10 µL, and larvae injected with 3×10^3 CFU of *S. epidermidis* in 10 µL and two hours later with 10 µL of antibiotic (8.75 µg/g body weight) were used for the evaluation of the immunological markers (i.e., cells viability, lysozyme, and prophenoloxidase activity).

2.7.1. Hemocyte Viability

Hemolymph was collected from larvae 24 h after the first injection and diluted 1:50 with Saline Solution for Lepidoptera (sucrose 210 mM, KCl 45 mM, Tris-HCl 10 mM, pH 7.0) CellTiter-Glo Luminescent Cell Viability Assay (Promega, Madison, WI, USA) was used to analyze the hemocyte viability. A total of 100 µL of CellTiter-Glo were added to 100 µl of diluted hemolymph into a 96-well plate, and then incubated for 5 min at room temperature on an orbital shaker. The measurement of luminescence was performed by using an Infinite F200 96-well plate-reader (Tecan, Männedorf, Switzerland). Ten surviving larvae for each experimental group were analyzed.

2.7.2. Prophenoloxidase (proPO) System Activation

Hemolymph was collected from pools of three larvae and centrifuged at 250× *g* for 5 min at 4 °C. A total of 100 µL of hemolymph were loaded into 96-well plates and the absorbance was measured by reading OD_{450nm} every 10 min for 50 min using an Infinite F200 96-well plate-reader (Tecan) [31]. A total of 100 µL of hemolymph supplemented with 2.5 mM N-phenylthiourea (PTU, Sigma-Aldrich) dissolved in 100% *v*/*v* EtOH were used as negative controls. The activation of proPO system was evaluated by incubating hemolymph with β-glucans from *Saccharomyces cerevisiae* (Zymosan, Sigma-Aldrich) and 4 mM $CaCl_2$ [32], too. ΔOD_{450nm} was calculated subtracting the OD_{450nm} recorded at time zero from the values obtained at the different time points. To evaluate the melanization rate, linear regression was performed for each ΔOD_{450nm} measurement obtained versus time [31].

2.7.3. Lysozyme Activity

Hemolymph was extracted from a pool of three larvae 6 h after infection and added with a few crystals of PTU to avoid melanization. After two centrifugations at 250× *g* for 5 min and one at 1600× *g* for 10 min at 4 °C, the supernatant was collected and diluted 1:10 with sterile phosphate-buffered saline (PBS: 1370 mM NaCl, 27 mM KCl, 100 mM $Na_2HPO_4 \times 12H_2O$, 19.8 mM KH_2PO_4). The activity of lysozyme was determined according to Bruno et al. [33]. Briefly, 100 µL of diluted hemolymph were added to 150 µL of 0.45 mg/mL lyophilized *Micrococcus lysodeikticus* (Sigma-Aldrich) in 30 mM phosphate buffer (38 mM KH_2PO_4, 61.4 mM K_2HPO_4, pH 7.2) (OD_{600nm} of 0.6–0.7). *M. lysodeikticus* and hemolymph both added with PBS were used as controls. Finally, absorbance at 450 nm was recorded every 30 s for 10 min by using an Infinite F200 96-well plate-reader (Tecan, Switzerland).

2.8. Statistical Analysis

Statistical analysis was performed using ANOVA, followed by Tukey's Honestly Significant Difference (HSD) test (significance $p < 0.05$).

3. Results

3.1. Larval Survival and Calculation of LD_{50}

To determine the LD_{50} for infection experiments, *B. mori* larvae were injected with increasing concentrations of *S. epidermidis*. Daily monitoring revealed a normal development of control groups (i.e., untreated larvae and larvae injected with 0.6% *w*/*v* NaCl), with a survival rate of 100% after 72 h (Figure 1). A correlation between the mortality rate of larvae and the concentration of injected bacteria was instead observed after *S. epidermidis* injection. In detail, 70% of the larvae survived 72 h after the injection of 3×10^2 CFU of *S. epidermidis* (Figure 1), while only 27% and 11% of insects were still alive after the infection with 3×10^3 and 3×10^4 CFU of bacteria, respectively. Finally, no larvae infected with 3×10^5 CFU survived after 72 h (Figure 1). The lethal bacterial dose that killed 50% of the infected larvae (LD_{50}), calculated with Probit analysis, proved to be 1.05×10^3 CFU.

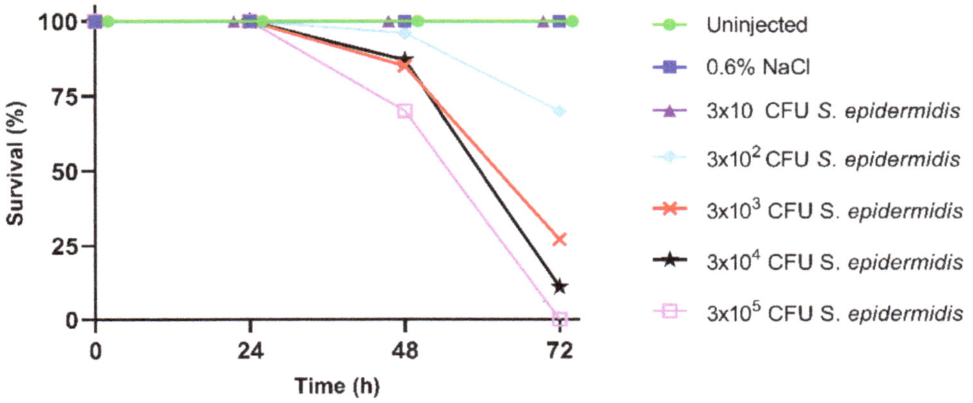

Figure 1. Survival rate of *B. mori* larvae injected with increasing CFU (colony forming unit) of *S. epidermidis*.

3.2. Effect of GPA Administration to the Larvae

The curative efficacy of the three selected GPAs (i.e., vancomycin, teicoplanin, and dalbavancin) was assessed by administering the antibiotics to the larvae infected with *S. epidermidis* at LD_{50}. Larvae injected with only GPAs were monitored to exclude any toxic effect. Moreover, insects were injected twice with saline solution to exclude any potential side effect due to the double injection (bacteria plus antibiotic). In accordance with previous observations [15], all the control groups showed 100% survival (Figure 2) confirming that GPAs were not toxic to larvae and that the experimental procedure of injection was reliable. All the GPAs tested improved the survival rate of infected silkworms. Indeed, while only 51% of the infected larvae survived 72 h after the infection, the administration of vancomycin, teicoplanin, and dalbavancin increased the survival rate to 90%, 93%, and 97%, respectively (Figure 2).

The minimum inhibitory concentration (MIC) and the minimum bactericidal concentration (MBC) of the three GPAs towards *S. epidermidis* (Table 1) were in agreement with the results reported above. In fact, the selected strain was sensitive to all three GPAs and dalbavancin proved to be, among the three of them, the most effective antibiotic also in vitro.

Table 1. Minimum inhibitory concentration (MIC) and minimum bactericidal concentration (MBC) of vancomycin, teicoplanin, and dalbavancin towards *S. epidermidis*. Values represent the average of data from three independent experiments.

Antibiotic	MIC (µg/mL)	MBC (µg/mL)
Vancomycin	2	16
Teicoplanin	1	32
Dalbavancin	0.25	8

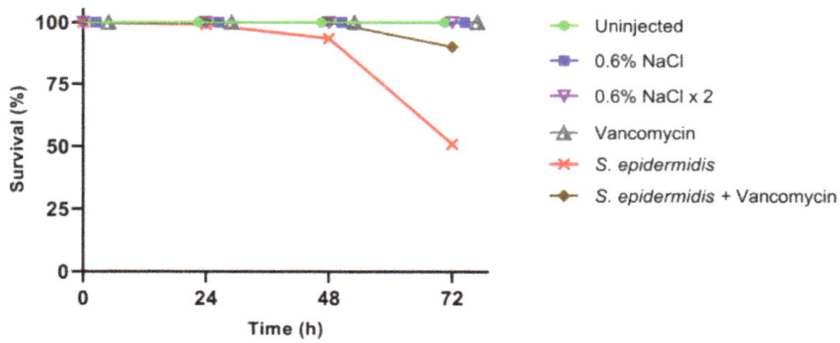

(**a**)

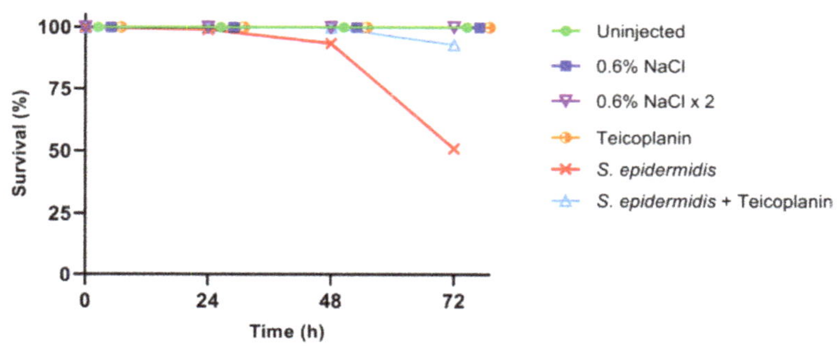

(**b**)

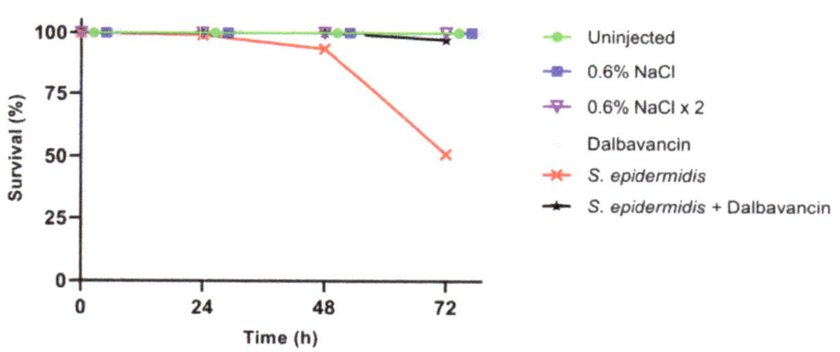

(**c**)

Figure 2. Survival rate. Curative efficacy of vancomycin (**a**), teicoplanin (**b**), and dalbavancin (**c**) (8.75 μg/g larval body weight) on larvae infected with *S. epidermidis* (1.05×10^3 CFU). Results are from the same experiment run in parallel, using the same control groups (uninjected larvae and healthy larvae injected once or twice with saline solution) and infected larvae.

3.3. Analysis of the Immunological Markers

The insect immune system consists of a fine and balanced crosstalk between hemocytes and humoral molecules, which aims at maintaining the hemolymph devoid of pathogens. Herein, we evaluated the cellular and humoral responses of *B. mori* larvae following *S. epidermidis* infection.

3.3.1. Hemocyte Viability

Hemocytes are the main mediators of the cellular immune response in insects [34]. To evaluate the recruitment/activation of these immune cells following pathogen invasion, their viability was evaluated by quantifying the ATP content through a luminescence assay. In insects infected by *S. epidermidis*, a significant increase in luminescence was observed, whereas the treatment with the three GPAs restored the ATP amount to levels comparable to those of the control groups (Figure 3). These results indicated that the antibiotics blocked the infection and, consequently, the involvement of hemocytes in the immune response.

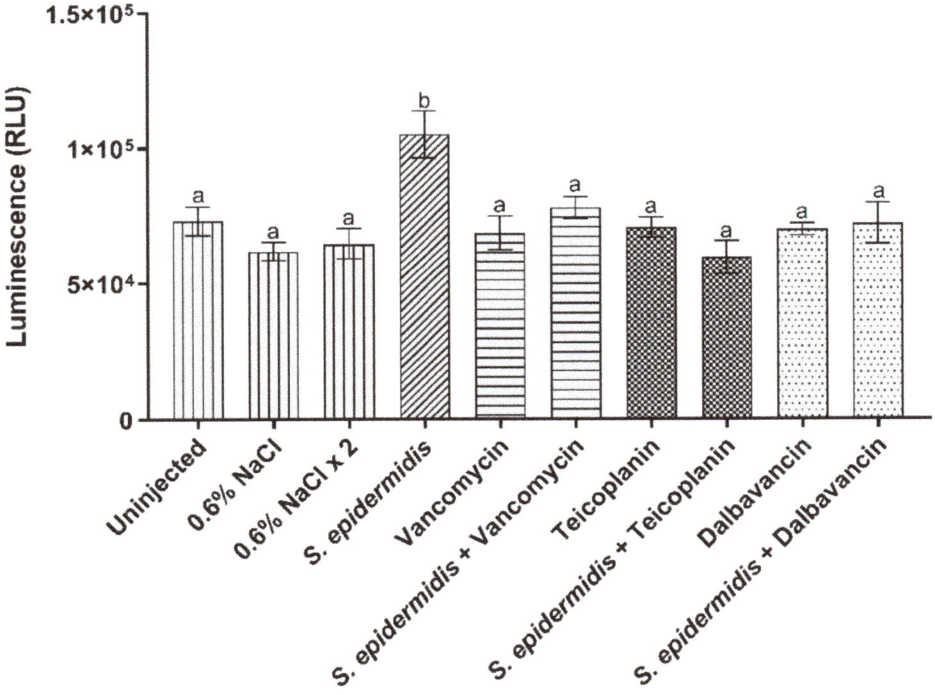

Figure 3. Hemocytes activity detected using a luminescence assay. Values represent mean ± s.e.m. Different letters indicate statistically significant differences among treatments ($p < 0.05$).

3.3.2. Activation of Prophenoloxidase System

In insects, the entrance of pathogens in the hemocoel can induce the activation of the proPO cascade, resulting in the production of melanin to isolate the foreign agents [35]. No significant variation in the activity of proPO system was instead observed after the infection of larvae with 3×10^5 CFU of *S. epidermidis*, compared to the control (Figure 4). Given these results, Zymosan, a specific activator of the proPO system, was used in combination with $CaCl_2$ to check if the enzyme activity responded to specific stimuli. A marked increase in the absorbance values confirmed a consistent activation of phenoloxidase under these last conditions (Figure 4), contrarily to what observed following bacterial infection.

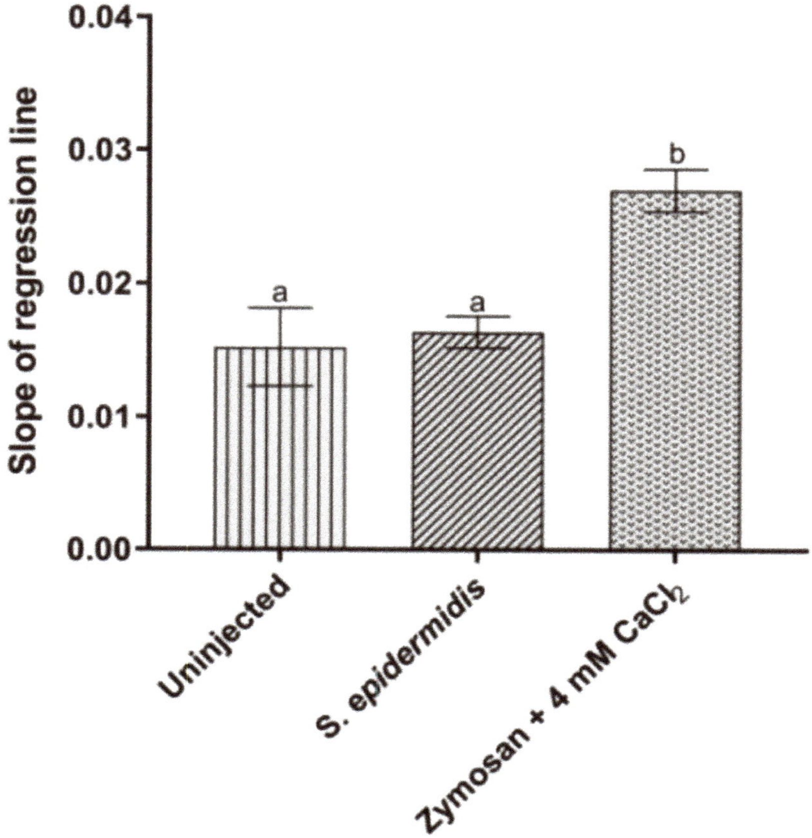

Figure 4. Analysis of proPO system activation. Values represent mean ± s.e.m. Different letters indicate statistically significant differences among treatments ($p < 0.05$).

3.3.3. Lysozyme Activity

Lysozyme cleaves the peptidoglycan component of the bacterial cell wall, leading to microbial cell death. The production of this enzyme generally increases after the pathogen attack [36]. In uninjected larvae, as well as in other control groups (i.e., healthy larvae injected once or twice with 0.6% w/v NaCl or with the antibiotics), the activity of lysozyme was comparable. The enzyme activity markedly increased after bacterial infection (Figure 5). Lysozyme activity was restored to the basal levels of the control groups following the administration of vancomycin, teicoplanin, or dalbavancin (Figure 5).

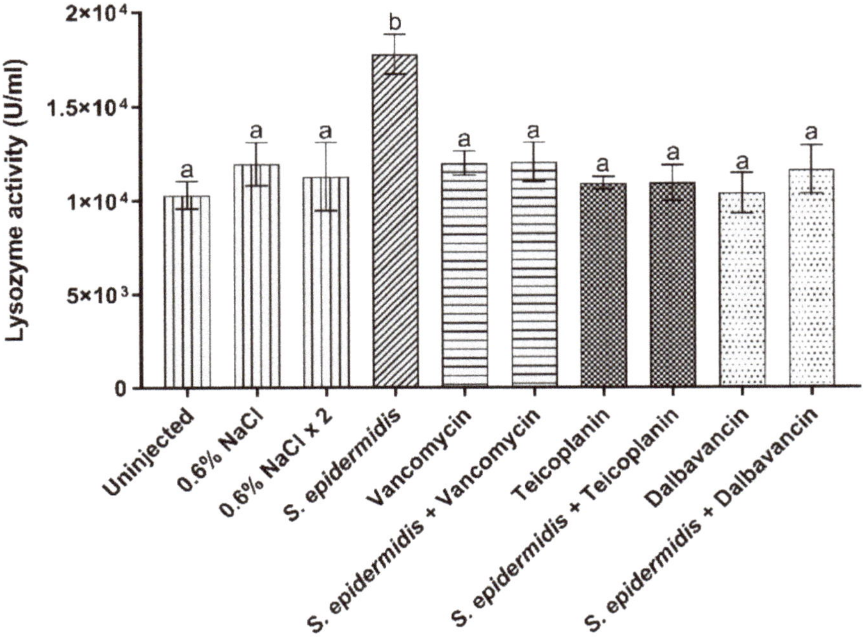

Figure 5. Lysozyme relative activity. Values represent mean ± s.e.m. Different letters indicate statistically significant differences among treatments ($p < 0.05$).

4. Discussion

Various invertebrate models have been proposed over the years to overcome the drawbacks associated with the use of mammals in drug discovery and development, which include not only the ethical and regulatory issues, but also the high costs for animal handling, and the need of ad hoc facilities for their maintenance [37,38]. In this scenario, the silkworm *B. mori* has emerged as a suitable model for mimicking infections by various human pathogens, including bacteria [13,15,39–41] as well as filamentous fungi [22] and yeasts [39]. Moreover, this insect has been recently used to test the efficacy of antibacterial agents, such as kanamycin, tetracycline, fluconazole, and vancomycin, in curing infections [7,13,15,39,42].

The optimal temperature for rearing silkworms is commonly 25 °C [26], but this condition significantly differs from the human body temperature. The resistance of *B. mori* larvae to higher temperatures (37 °C) for a limited period of time without any apparent side effect was recently proved [15]. This feature supported the development of a silkworm infection model, which can operate at the human body temperature, and better mimics in this way the trend of bacterial infections in humans [7,15,42]. Although a few papers reported on *B. mori* infected with the coagulase-positive *S. aureus* [13,15,39–41], to our knowledge, no publication has previously described infections determined by coagulase-negative staphylococci (CoNS), such as *S. epidermidis* and *S. haemolyticus*. Opportunistic CoNS, and *S. epidermidis* above all, are emerging as major sources of nosocomial infections, especially of foreign body-related infections (FBBIs) associated with indwelling or implanted devices and transmitted by medical and/or nursing procedures [43,44]. *S. epidermidis*-related infections include prosthetic valve endocarditis, keratitis associated with contact lens use, and intravascular catheter- and prosthesis-associated infections. Therapeutic options against CoNS are limited, due to the large number of methicillin-resistant strains or strains with reduced susceptibility to quinolones or to GPAs. In addition, these staphylococci form multilayered and difficult-to-eradicate biofilms, which are highly impenetrable to the majority of antibiotics [43,44].

In our work, we reported on the silkworm model infected at 37 °C with a clinical isolate of *S. epidermidis* and, to validate it, we investigated the curative effect of three molecules belonging to the GPA class. GPAs are defined as 'drugs of last resort' for their role in treating life-threating infections caused by staphylococci, enterococci, and *Clostridium difficile* [25]. Their use in fighting clinical infections by *S. epidermidis* is consolidated, too [45]. The selected molecules for this study included the first-generation GPAs vancomycin and teicoplanin, introduced in 1958 and 1988, respectively, and still largely used in clinics, and the second-generation dalbavancin, approved in 2014 and designated as Qualified Infection Disease Product by the Food and Drug Administration (FDA) for its potency and extended dosing interval [25]. We determined the susceptibility profile of our *S. epidermidis* clinical isolate to the three selected GPAs by measuring their MICs and MBCs following in vitro standard protocols. Dalbavancin showed the highest efficacy in inhibiting the growth of the target strain, with MIC and MBC values lower than those recorded for vancomycin and teicoplanin, as also demonstrated in previous publications [46–49]. Consistently with previous data [46], teicoplanin was confirmed to have a weaker bactericidal effect on *S. epidermidis* cells than the other two GPAs. As regards the in vivo infection model, the dosage of antibiotics administered to the larvae (i.e., 8.75 μg/g body weight) was determined according to Montali et al. [15] and was comparable to the amount of vancomycin used to treat severe staphylococcal infections in humans. It is noteworthy that this dosage, as well as higher concentrations of all three GPAs (up to 35 μg/g body weight), did not exert any significant toxicity in the larvae [15]. Under the tested conditions, all the three antibiotics proved to be effective in counteracting the infection caused by our *S. epidermidis* clinical isolate.

Previous studies explored the use of the silkworm as infection model mainly monitoring the survival rate of the larvae and, only in some cases, evaluating additional markers, such as the quantification of bacteria and/or their localization in the insect body [13,39,42]. However, detailed studies on the immune system of *B. mori* [50] might provide new tools to finely dissect the physiological mechanisms triggered by invading bacteria. Although additional markers (e.g., analysis of the bacterial load and expression of antimicrobial peptides) were monitored in our previous study [15], the complexity of the procedure and the limited reproducibility of the results obtained by following their temporal trend, prompted us to look for new ones. Since our final goal was to use the silkworm model in the early antibiotic discovery and development phase as a means to select molecules with a promising therapeutic index, we looked for robust and trustable markers that could be easy and rapid to monitor. Our results demonstrate that the administration of GPAs in infected silkworm stopped the activation of the insect's immune response and that the protective effect of these antibiotics could be appreciated at different levels. First, hemocytes, which play a primary role against the invasion of microorganisms in insects [51], confirmed to be a reliable marker, showing a reduced activity in larvae treated with GPAs, caused by a block in their recruitment/activation [15]. In relation to the humoral response that cooperates with the cellular processes to maintain the hemolymph devoid of pathogens [52], we examined the possible induction of two different enzyme activities, i.e., lysozyme and proPO. Lysozyme hydrolyzes peptidoglycan β-(1,4) glycosidic bonds in the cell wall of Gram-positive bacteria [52]. Our results confirm the activation of this enzyme following the injection of *S. epidermidis* in the larvae [53] and its return to control levels after GPA administration, demonstrating the antibiotic efficacy in reducing the effects of bacterial infection. To the best of our knowledge, this marker was used herein for the first time in an insect infection model. It is a rapid, simple, and reliable assay, which could be thus recommended for further studies on antibiotic response in insects. On the contrary, no difference in proPO activity between the control and infected larvae was observed, suggesting that this marker is not suitable for such type of investigations. In insects, proPO leads to melanin formation by generating quinones and other reactive intermediates, with the final aim of trophically isolating and killing pathogens and parasites [54]. The lack of activation of this enzymatic cascade might depend on the bacterial load in the hemolymph, since it is known that proPO system is induced only over a certain threshold concentration of bacteria [33] to avoid the unneeded production of melanin, which is toxic for the insect itself [55]. An alternative possible explanation might involve the ability of some specific pathogens to produce serpin-type inhibitors or non-proteinaceous factors (e.g., polyphenol derivatives), which specifically interfere with the proteolytic activation of proPO, increasing their survival in the hosts [54,56]. Further investigations in this direction would be useful to understand the role of proPO system following bacterial infections.

In conclusion, this study paves the way to use silkworm as a trustable infection model for testing antimicrobial compounds with therapeutic potential against staphylococci. In addition to monitoring the larval survival rate, two additional immunological markers were validated (i.e., hemocyte viability and lysozyme activation) and a third one proved useless (proPO system) for this purpose. The use of these markers could allow to better analyze positive or negative responses of the silkworm to the administration of different antimicrobial products, although a preliminary and accurate verification of their behavior at different temperatures is recommended before introducing them into the screening protocol. Moreover, the reduced number of used insects and the reduced time required for the analysis of these markers could lead to identify new tools useful to improve the procedures for screening antimicrobial compounds on large scale. Our final hope is that this insect infection model might be helpful to accelerate the discovery and development of novel compounds that are urgently needed to contrast the antibiotic resistance.

Author Contributions: Conceptualization, G.T. and F.M.; Data curation, A.M. and A.S.; Investigation, A.M., F.B. and A.S.; Methodology, F.B. and A.M.; Supervision, G.T. and F.M.; Validation, A.M. and F.B.; Writing—original draft, A.M., F.B. and G.T.; Writing—review and editing, A.M., F.B., S.C., F.M. and G.T. All authors have read and agreed to the published version of the manuscript.

Funding: The work was partially supported by FAR 2020–2021 (Fondi di Ateneo per la Ricerca, University of Insubria) to G.T., F.M. and F.B., and by the University of Insubria 'Starting Grant' 2020–2021 to F.B.

Institutional Review Board Statement: Not applicable.

Data Availability Statement: The datasets generated for this study are available on reasonable request to the corresponding author.

Acknowledgments: The authors are grateful to Silvia Caccia for assistance with Probit analysis.

Conflicts of Interest: The authors declare no conflict of interest.

References

1. Wencewicz, T.A. Crossroads of antibiotic resistance and biosynthesis. *J. Mol. Biol.* **2019**, *431*, 3370–3399. [CrossRef] [PubMed]
2. Berini, F.; Orlandi, V.; Gornati, R.; Bernardini, G.; Marinelli, F. Nanoantibiotics to fight multidrug resistant infections by Gram-positive bacteria: Hope or reality? *Biotechnol. Adv.* **2022**, *57*, 107948. [CrossRef] [PubMed]
3. Paudel, A.; Panthee, S.; Urai, M.; Hamamoto, H.; Ohwada, T.; Sekimizu, K. Pharmacokinetic parameters explain the therapeutic activity of antimicrobial agents in a silkworm infection model. *Sci. Rep.* **2018**, *8*, 1578. [CrossRef] [PubMed]
4. Richmond, J. The 3Rs–Past, present and future. *Scand. J. Lab. Anim. Sci.* **2000**, *27*, 84–92.
5. Mylonakis, E.; Casadevall, A.; Ausubel, F.M. Exploiting amoeboid and non-vertebrate animal model systems to study the virulence of human pathogenic fungi. *PLoS Pathog.* **2007**, *3*, e101. [CrossRef]
6. Kong, C.; Eng, S.A.; Lim, M.P.; Nathan, S. Beyond traditional antimicrobials: A *Caenorhabditis elegans* model for discovery of novel anti-infectives. *Front. Microbiol.* **2016**, *7*, 1956. [CrossRef]
7. Matsumoto, Y.; Miyazaki, S.; Fukunaga, D.H.; Shimizu, K.; Kawamoto, S.; Sekimizu, K. Quantitative evaluation of cryptococcal pathogenesis and antifungal drugs using a silkworm infection model with *Cryptococcus neoformans*. *J. Appl. Microbiol.* **2012**, *112*, 138–146. [CrossRef]
8. Panthee, S.; Paudel, A.; Hamamoto, H.; Sekimizu, K. Advantages of the silkworm as an animal model for developing novel antimicrobial agents. *Front. Microbiol.* **2017**, *8*, 373. [CrossRef]
9. Buchmann, K. Evolution of innate immunity: Clues from invertebrates via fish to mammals. *Front. Immunol.* **2014**, *5*, 459. [CrossRef]

10. Ménard, G.; Rouillon, A.; Ghukasyan, G.; Emily, M.; Felden, B.; Donnio, P.Y. *Galleria mellonella* larvae as an infection model to investigate srna-mediated pathogenesis in *Staphylococcus aureus*. *Front. Cell. Infect. Microbiol.* **2021**, *11*, 631710. [CrossRef]
11. The International Silkworm Genome Consortium. The genome of a lepidopteran model insect, the silkworm *Bombyx mori*. *Insect Biochem. Mol. Biol.* **2008**, *38*, 1036–1045. [CrossRef] [PubMed]
12. Xu, H.; O'Brochta, D.A. Advanced technologies for genetically manipulating the silkworm *Bombyx mori*, a model Lepidopteran insect. *Proc. Biol. Sci.* **2015**, *282*, 20150487.
13. Kaito, C.; Akimitsu, N.; Watanabe, H.; Sekimizu, K. Silkworm larvae as an animal model of bacterial infection pathogenic to humans. *Microb. Pathog.* **2002**, *32*, 183–190. [CrossRef] [PubMed]
14. Uchida, R.; Hanaki, H.; Matsui, H.; Hamamoto, H.; Sekimizu, K.; Iwatsuki, M.; Kim, Y.P.; Tomoda, H. In vitro and in vivo anti-MRSA activities of nosokomycins. *Drug Discov. Ther.* **2014**, *8*, 249–254. [CrossRef] [PubMed]
15. Montali, A.; Berini, F.; Brivio, M.F.; Mastore, M.; Saviane. A.; Cappellozza, S.; Marinelli, F.; Tettamanti, G. A silkworm infection model for in vivo study of glycopeptide antibiotics. *Antibiotics* **2020**, *9*, 300. [CrossRef]
16. Tuba, T.; Chowdhury, F.R.; Hossain, T.; Farzana, M.; Ahad, I.; Hossain, M.M.; Hossain, M.I.; Saleh, N.; Nawaar, N.; Uddin, M.A.; et al. *Klebsiella pneumoniae* pathogenicity in silk moth larvae infection model. *FEMS Microbial. Lett.* **2022**, *368*, 21–24. [CrossRef]
17. Usui, K.; Miyazaki, S.; Kaito, C.; Sekimizu, K. Purification of a soil bacteria exotoxin using silkworm toxicity to measure specificactivity. *Microb. Pathog.* **2009**, *46*, 59–62. [CrossRef]
18. Suzuki, J.; Uda, A.; Watanabe, K.; Shimizu, T.; Watarai, M. Sym-biosis with *Francisella tularensisprovides* resistance to patho-gens in the silkworm. *Sci. Rep.* **2016**, *6*, 31476. [CrossRef]
19. Castillo, Y.; Suzuki, J.; Watanabe, K.; Shimizu, T.; Watarai, M. Effect of vitamin A on *Listeria monocytogenes* infection in a silkwormmodel. *PLoS ONE* **2016**, *11*, e0163747. [CrossRef]
20. Hosoda, K.; Koyama, N.; Hamamoto, H.; Yagi, A.; Uchida, R.; Kanamoto, A.; Tomoda, H. Evaluation of anti-mycobacterial compounds in a silkworm infection model with *Mycobacteroides abscessus*. *Molecules* **2020**, *25*, 4971. [CrossRef]
21. Uchida, R.; Namiguchi, S.; Ishijima, H.; Tomoda, H. Therapeutic effects of three trichothecenes in the silkworm infection assay with *Candida albicans*. *Drug Discov. Ther.* **2016**, *10*, 44–48. [CrossRef] [PubMed]
22. Nakamura, I.; Kanasaki, R.; Yoshikawa, K.; Furukawa, S.; Fujie, A.; Hamamoto, H.; Sekimizu, K. Discovery of a new antifungal agent ASP2397 using a silkworm model of *Aspergillus fumigatus* infection. *J. Antibiot.* **2017**, *70*, 41–44. [CrossRef] [PubMed]
23. Oliveira, W.F.; Silva, P.M.S.; Silva, R.C.S.; Silva, G.M.M.; Machado, G.; Coelho, L.C.B.B.; Correia, M.T.S. *Staphylococcus aureus* and *Staphylococcus epidermidis* infections on implants. *J. Hosp. Infect.* **2018**, *98*, 111–117. [CrossRef] [PubMed]
24. Otto, M. Molecular basis of *Staphylococcus epidermidis* infections. *Semin. Immunopathol.* **2012**, *34*, 201–214. [CrossRef] [PubMed]
25. Marcone, G.L.; Binda, E.; Berini, F.; Marinelli, F. Old and new glycopeptide antibiotics: From product to gene and back in the post-genomic era. *Biotechnol. Adv.* **2018**, *36*, 534–554. [CrossRef] [PubMed]
26. Cappellozza, L.; Cappellozza, S.; Saviane, A.; Sbrenna, G. Artificial diet rearing system for the silkworm *Bombyx mori* (Lepidoptera: Bombycidae): Effect of vitamin C deprivation on larval growth and cocoon production. *Appl. Entomol. Zool.* **2005**, *40*, 405–412. [CrossRef]
27. Franzetti, E.; Romanelli, D.; Caccia, S.; Cappellozza, S.; Congiu, T.; Rajagopalan, M.; Grimaldi, A.; de Eguileor, M.; Casartelli, M.; Tettamanti, G. The midgut of the silkmoth *Bombyx mori* is able to recycle molecules derived from degeneration of the larval midgut epithelium. *Cell Tissue Res.* **2015**, *361*, 509–528. [CrossRef]
28. Casati, B.; Terova, G.; Cattaneo, A.G.; Rimoldi, S.; Franzetti, E.; de Eguileor, M.; Tettamanti, G. Molecular cloning, characterization and expression analysis of *ATG1* in the silkworm, *Bombyx mori*. *Gene* **2012**, *511*, 326–337. [CrossRef]
29. CLSI. *Performance Standards for Antimicrobial Susceptibility Testing*, 28th ed.; Clinical and Laboratory Standards Institute: Wayne, PA, USA, 2018.
30. Finney, D.J. *Probit Analysis*, 3rd ed.; Cambridge University Press: Cambridge, UK, 1971.
31. Brady, D.; Saviane, A.; Romoli, O.; Tettamanti, G.; Sandrelli, F.; Cappellozza, S. Oral infection in a germ-free *Bombyx mori* model. In *Immunity in Insects*; Sandrelli, F., Tettamanti, G., Eds.; Humana: New York, NY, USA, 2020; pp. 217–231.
32. Ashida, M.; Ishizaki, Y.; Iwahana, H. Activation of pro-phenoloxidase by bacterial cell walls or β-1,3-glucans in plasma of the silkworm, *Bombyx mori*. *Biochem. Biophys. Res. Commun.* **1983**, *113*, 562–568. [CrossRef]
33. Bruno, D.; Montali, A.; Mastore, M.; Brivio, M.F.; Mohamed, A.; Tian, L.; Grimaldi, A.; Casartelli, M.; Tettamanti, G. Insights into the immune response of the black soldier fly larvae to bacteria. *Front. Immunol.* **2021**, *12*, 745160. [CrossRef]
34. Eleftherianos, I.; Heryanto, C.; Bassal, T.; Zhang, W.; Tettamanti, G.; Mohamed, A. Haemocyte-mediated immunity in insects: Cells, processes, and associated components in the fight against pathogens and parasites. *Immunology* **2021**, *164*, 401–432. [CrossRef] [PubMed]
35. González-Santoyo, I.; Córdoba-Aguilar, A. Phenoloxidase: A key component of the insect immune system. *Entomol. Exp. Appl.* **2012**, *142*, 1–16. [CrossRef]
36. Hultmark, D. Insect lysozymes. *EXS* **1996**, *75*, 87–102. [PubMed]
37. Doke, S.K.; Dhawale, S.C. Alternatives to animal testing: A review. *Saudi Pharm. J.* **2015**, *23*, 223–229. [CrossRef]
38. Kaito, C.; Murakami, K.; Imai, L.; Furuta, K. Animal infection models using non-mammals. *Microbiol. Immunol.* **2020**, *64*, 585–592. [CrossRef]

39. Hamamoto, H.; Kurokawa, K.; Kaito, C.; Kamura, K.; Manitra Razanajatovo, I.; Kasuhara, H.; Santa, T.; Sekimizu, K. Quantitative evaluation of the therapeutic effects of antibiotics using silkworms infected with human pathogenic microorganisms. *Antimicrob. Agents Chemother.* **2004**, *48*, 774–779. [CrossRef]
40. Kaito, C.; Kurokawa, K.; Matsumoto, Y.; Terao, Y.; Kawabata, S.; Hamada, S.; Sekimizu, K. Silkworm pathogenic bacteria infection model for identification of novel virulence genes. *Mol. Microbiol.* **2005**, *56*, 934–944. [CrossRef]
41. Barman, T.K.; Arora, P.; Rao, M.; Bhadauriya, T.; Upadhyay, D.J. Utilization of *Bombyx mori* larvae as a surrogate animal model for evaluation of the anti-infective potential of oxazolidinones. *J. Infect. Chemother.* **2008**, *14*, 166–169. [CrossRef]
42. Tabuchi, F.; Matsumoto, Y.; Ishii, M.; Tatsuno, K.; Okazaki, M.; Sato, T.; Moriya, K.; Sekimizu, K. Synergistic effects of vancomycin and β-lactams against vancomycin highly resistant *Staphylococcus aureus*. *J. Antibiot.* **2017**, *70*, 771–774. [CrossRef]
43. Becker, K.; Heilmann, C.; Peters, G. Coagulase-negative staphylococci. *Clin. Microbiol. Rev.* **2014**, *27*, 870–926. [CrossRef]
44. Sabaté Brescó, M.; Harris, L.G.; Thompson, K.; Stanic, B.; Morgenstern, M.; O'Mahony, L.; Richards, R.G.; Moriarty, T.F. Pathogenic mechanisms and host interactions in *Staphylococcus epidermidis* device-related infection. *Front. Microbiol.* **2017**, *8*, 1401. [CrossRef] [PubMed]
45. Knafl, D.; Tobudic, S.; Cheng, S.C.; Bellamy, D.R.; Thalhammer, F. Dalbavancin reduces biofilms of methicillin-resistant *Staphylococcus aureus* (MRSA) and methicillin-resistant *Staphylococcus epidermidis* (MRSE). *Eur. J. Clin. Microbiol. Infect. Dis.* **2017**, *36*, 677–680. [CrossRef] [PubMed]
46. Candiani, G.; Abbondi, M.; Borgonovi, M.; Romanò, G.; Parenti, F. In-vitro and in-vivo antibacterial activity of BI 397, a new semi-synthetic glycopeptide antibiotic. *J. Antimicrob. Chemother.* **1999**, *44*, 179–192. [CrossRef] [PubMed]
47. Chen, A.Y.; Zervos, M.J.; Vazquez, J.A. Dalbavancin: A novel antimicrobial. *Int. J. Clin. Pract.* **2007**, *61*, 853–863. [CrossRef] [PubMed]
48. Di Pilato, V.; Ceccherini, F.; Sennati, S.; D'Agostino, F.; Arena, F.; D'Atanasio, N.; Di Giorgio, F.P.; Tongiani, S.; Pallecchi, L.; Rossolini, G.M. In vitro time-kill kinetics of dalbavancin against *Staphylococcus* spp. biofilms over prolonged exposure times. *Diagn. Microbiol. Infect. Dis.* **2020**, *96*, 114901. [CrossRef] [PubMed]
49. Goldstein, B.P.; Draghi, D.C.; Sheehan, D.J.; Hogan, P.; Sahm, D.F. Bactericidal activity and resistance development profiling of dalbavancin. *Antimicrob. Agents. Chemother.* **2007**, *51*, 1150–1154. [CrossRef]
50. Chen, K.; Lu, Z. Immune responses to bacterial and fungal infections in the silkworm, *Bombyx mori*. *Dev. Comp. Immunol.* **2018**, *83*, 3–11. [CrossRef]
51. Li, T.; Yan, D.; Wang, X.; Zhang, L.; Chen, P. Hemocyte changes during immune melanization in *Bombyx mori* infected with *Escherichia coli*. *Insects* **2019**, *10*, 301. [CrossRef]
52. Tsakas, S.; Marmaras, V.J. Insect immunity and its signalling: An overview. *Invertebr. Surviv. J.* **2010**, *7*, 228–238.
53. Isao, M.; Tomoki, H.; Yoshiaki, Y. Lysozyme activity in immunized and non-immunized hemolymph during the development of the silkworm, *Bombyx mori*. *Comp. Biochem. Physiol.* **1994**, *108*, 311–314. [CrossRef]
54. Cerenius, L.; Lee, B.L.; Söderhäll, K. The proPO-system: Pros and cons for its role in invertebrate immunity. *Trends Immunol.* **2008**, *29*, 263–271. [CrossRef] [PubMed]
55. An, C.; Budd, A.; Kanost, M.R.; Michel, K. Characterization of a regulatory unit that controls melanization and affects longevity of mosquitoes. *Cell. Mol. Life Sci.* **2011**, *68*, 1929–1939. [CrossRef] [PubMed]
56. Ma, Z.; Li, C.; Pan, G.; Li, Z.; Han, B.; Xu, J.; Lan, X.; Chen, J.; Yang, D.; Chen, Q.; et al. Genome-wide transcriptional response of silkworm (*Bombyx mori*) to infection by the microsporidian *Nosema bombycis*. *PLoS ONE* **2013**, *8*, e84137. [CrossRef] [PubMed]

insects

Article

The Effect of Feeding with Central European Local Mulberry Genotypes on the Development and Health Status of Silkworms and Quality Parameters of Raw Silk

Andreja Urbanek Krajnc [1], Tamas Bakonyi [2], Istvan Ando [3], Eva Kurucz [3], Norbert Solymosi [4], Paula Pongrac [5,6] and Rebeka Lucijana Berčič [2,7,*]

1 Faculty of Agriculture and Life Sciences, University of Maribor, Pivola 10, 2311 Hoče, Slovenia
2 Department of Microbiology and Infectious Diseases, University of Veterinary Medicine Budapest, István u. 2, H-1078 Budapest, Hungary
3 Institute of Genetics, Biological Research Center of the Hungarian Academy of Sciences, P.O. Box 521, H-6701 Szeged, Hungary
4 Centre for Bioinformatics, University of Veterinary Medicine Budapest, István u. 2, H-1078 Budapest, Hungary
5 Department of Biology, Chair of Botany and Plant Physiology, Biotechnical Faculty, University of Ljubljana, Jamnikarjeva 101, 1000 Ljubljana, Slovenia
6 Jožef Stefan Institute, Jamova 39, 1000 Ljubljana, Slovenia
7 Institute for Sericulture, Rebecca Luciana Bercic, Koroška c. 65, 2000 Maribor, Slovenia
* Correspondence: rebekalucijana@yahoo.com; Tel.: +386-41-776-711

Citation: Urbanek Krajnc, A.; Bakonyi, T.; Ando, I.; Kurucz, E.; Solymosi, N.; Pongrac, P.; Berčič, R.L. The Effect of Feeding with Central European Local Mulberry Genotypes on the Development and Health Status of Silkworms and Quality Parameters of Raw Silk. *Insects* **2022**, *13*, 836. https://doi.org/10.3390/insects13090836

Academic Editors: Silvia Cappellozza, Morena Casartelli, Federica Sandrelli, Alessio Saviane and Gianluca Tettamanti

Received: 30 August 2022
Accepted: 8 September 2022
Published: 14 September 2022

Publisher's Note: MDPI stays neutral with regard to jurisdictional claims in published maps and institutional affiliations.

Simple Summary: Several regions of Slovenia and Hungary retained numerous centuries-old white mulberry trees, evidence of past sericultural activities, being traditionally used to feed the silkworm larvae. Attempts for the reintroduction of sericulture in these countries are ongoing. The current study assessed the suitability of the locally adapted mulberry trees for contemporary sericultural needs. Silkworm hybrids were fed with leaves of the selected local mulberry genotypes and the larvae performance parameters (bodyweight, spinning success, cocoon quantity, and quality) were compared to those fed with reference mulberry varieties. The chemical contents and nutritive parameters of the mulberry leaves were determined, and connections were predicted between selected leaf compounds and silkworm performance parameters. The local mulberries had higher total protein contents, and lower total phenolic contents and differed in some individual phenolics, macro- and microelements compared to the reference sericultural and fruit varieties. A combined positive influence of proteins, specific phenolics, and microelements on larval growth and silk thread parameters was predicted. The health status and gut microbiome compositions of larvae were also analyzed. The results of the study indicate that selected local Slovenian and Hungarian mulberry varieties are suitable for high-quality silk cocoon and raw silk production.

Abstract: Silkworm rearing activities ceased in the 1970′s in several European countries. Attempts on the re-establishment of ecological and sustainable sericulture in Slovenia and Hungary are ongoing. The aim of the study was to assess the usability of locally adapted mulberry genotypes for sericulture and to estimate connections between leaf compound and silkworm performance parameters. A controlled feeding experiment of silkworms was performed to test the influence of leaves from selected trees on the growth of larvae, the health and microbiological status of larvae (e.g., gut bacterial microbiome, *Bombyx mori nucleopolyhedrovirus* infection), weight of cocoons and raw silk parameters. The Slovenian and Hungarian mulberry genotypes had significantly higher total protein contents, and lower total phenolic contents and differed significantly in some individual phenolics compared to the reference sericultural and fruit varieties. Significant differences were found in the contents of the macro- and microelements, namely S, Mn, Fe, and Sr. Based on correlative statistics and multivariate analysis, a combined positive influence of proteins, specific phenolics, and microelements on larval growth and silk thread parameters was predicted. The results of the study indicate that selected local Slovenian and Hungarian mulberry varieties are suitable for high-quality silk cocoon and raw silk production.

Keywords: silkworm; mulberry leaves; chemical composition; feeding experiment; bacterial microbiome; BmNPV

1. Introduction

Although China and India are the current monopolists in silk production, representing more than 97% of the global sourcing of this natural fiber, there are initiatives for the re-establishment of silk production in European countries which were major silk producers in the past centuries [1]. In the time of the Austro-Hungarian Empire, sericulture was highly developed in the territories of today's Slovenia and Hungary. Due to the competition in the global silk market, highly developed silkworm rearing in these two countries ceased approximately 60 years ago [2–4]. Nowadays, all around these countries, we can still see up to 600 years old white mulberry (*Morus alba* L.) trees—living monuments of sericultural history, as the leaves of these trees were used for feeding the larvae of the domestic silk moth (*Bombyx mori* L.) over centuries. These mulberry genotypes are adapted to the local climatic and environmental conditions and were appropriate for feeding silkworm larvae in the past. Sericulture subsided in the previous decades in Italy, and nowadays production is increasing [5]. In Italy currently, high-performance hybrids of *B. mori* are used for silk production, which are typically fed by selected mulberry genotypes. One of the most important practical questions of the sericulture re-establishment attempt is, whether the locally adapted genotypes in other European countries are appropriate for feeding the modern hybrids of silkworm, or the reference varieties are necessary to achieve appropriate performance (i.e., quality of silk cocoons according to the standards and market's expectations, and quantities comparable with rearing farms in Italy).

Recently, a joint Slovenian-Hungarian basic research project was implemented to define the conditions for the re-establishment of ecologically and economically sustainable sericulture. The main research question was whether the development and health status of silkworm larvae, production of silk cocoons, and quality of raw silk are affected by the diversity and nutritive characteristics of locally adapted mulberry trees native in Hungary and Slovenia compared to reference varieties currently used for silkworm rearing. As a multidisciplinary research approach, local mulberry genetic resources were catalogued and their biochemical characteristics were analysed [6,7]. The aim of the presented research was to test the selected genotypes in a silkworm feeding experiment and to compare their effect on silkworm development, cocoon production, the health status of silkworm larvae, and raw silk parameters, compared to reference mulberry varieties.

2. Materials and Methods

2.1. Mulberry Material for Silkworm Feeding Experiment

Mulberry genotypes from the mulberry collection were chosen for the feeding experiment based on previous screening of metabolites in leaves [6,7], and were organized into three sections: the first part is represented by Italian sericultural *M. alba* varieties ('Giazzola', 'Florio', 'Morettiana') and the Japanese variety 'Kokusou-20' obtained from the gene bank at the Centro di Ricerca per l'Agricoltura e Ambiente (CREA-AA), Laboratorio di Gelsibachicoltura di Padova, Italy. The second part comprises vegetatively propagated trees derived from the local historical Slovenian and Hungarian mulberries, which were obtained during the classification of the mulberry gene pool. The third part of the collection is intended for growing recent varieties of *M. alba*, *M. nigra* and *M. australis* as well as hybrids of *M. alba* × *rubra* suitable for fruit production.

The choice of mulberry trees for reference plantation feeding experiments was conducted according to: (1) geographical distribution of the original old local mulberry genotypes in Hungary and Slovenia; (2) multivariate analyses of previous biochemical values of original trees that allowed us to define seven chemotypes in more details in terms of the composition of individual amino acids and phenolics [6,7], out of which genotypes rich in total proteins, total phenolics, chlorogenic acid, and particular flavonoids were chosen. For the experiment, three to five trees of the same genotype were harvested.

Mulberry trees are grouped into four main sections:

(A) Slovenian mulberry genotypes: selected, locally adapted genotypes, obtained by cuttings from old (trunk diameter > 180 cm) Slovenian *M. alba* trees (n = 10);

(B) Hungarian mulberry genotypes: selected, locally adapted genotypes, obtained by cuttings from old (trunk diameter > 180 cm) Hungarian *M. alba* trees (n = 16);

(C) reference sericultural *M. alba* varieties, obtained from the mulberry gene bank of the CREA-AA ('Kokusou-20', 'Morettiana', 'Florio', 'Giazzola');

(D) varieties of *M. alba*, *M. alba* × *rubra*, *M. nigra* and *M. australis* grown for fruit production.

The list of mulberry genotypes along with their location and geographic coordinates of historical mulberry trees are presented in Supplementary Table S1.

2.2. Determination of the Chemical Composition of Mulberry Leaves

For biochemical analyses, five to seven fully developed sun-exposed leaves (5th to 7th leaf below the apex) from a one-year branch were collected randomly per each sampled tree (3–5 trees per selected genotype) and used as one sample. The trees were sampled twice during the experiment on 16th and 23rd of June 2021. Samples were immediately stored on dry ice, later transferred to a freezer at −80 °C, and subsequently freeze-dried and grounded. The prepared samples were then stored in airtight vials at −20 °C prior to biochemical analyses. The concentrations of the analyzed nutrients are calculated on a dry weight (DW) basis.

Total proteins were determined spectrophotometrically at 595 nm, following the Bradford procedure [8]. The total protein content was calculated based on a standard curve, which was prepared using bovine serum albumin (BSA, 0.05–0.25 mg mL^{-1}) and expressed as mg BSA equivalent per 100 g of dried mulberry leaves (mg BSA/100 g DW).

Total phenolics of the methanolic extracts (3% formic acid in 95% methanol) were determined using the Folin-Ciocalteu method following the procedure of Ainsworth and Gillespie [9]. The absorbance was measured at 765 nm against a reagent blank (3% formic acid in 95% methanol). Gallic acid was used as the reference standard (GA, 0.025–0.25 mg mL^{-1}). The total PH was expressed as mg gallic acid equivalents per 100 g of dried leaf sample (mg GAE/100 g DW).

Furthermore, the methanolic extracts were subjected to the gradient HPLC analysis as described in detail in a previous study [6]. All considered phenolic compounds were previously identified by a mass spectrometer (Thermo Finigan, San Jose, CA, USA) with an electrospray interface (ESI) operating in negative ion mode as previously described in Šelih et al. [7]. The identification of compounds was confirmed by comparing their spectra, retention times, and fragmentation as well as by adding the standard solution to the sample. Quantification was achieved by comparing with corresponding external standards (chlorogenic acid, kaempferol, p-coumaric acid, rutin, and quercetin; all obtained from Sigma Aldrich) of known concentrations. For the compounds for which the standards were not available, related compounds were used as standards. Therefore, quercetin-3-O-glucoside (isoquercetin), quercetin dirhamnosylhexoside and quercetin malonylhexoside were quantified in quercetin equivalents, kaempferol acetylhexoside in kaempferol equivalents, caffeoylquinic acid derivatives in equivalent of chlorogenic acid and p-coumaric and p-coumaroylquinic acid derivatives in equivalent of p-coumaric acid. The contents of individual compounds were expressed in mg/g DW.

Macro- and microelement analyses (phosphorus (P), sulphur (S), chlorine (Cl), potassium (K), calcium (Ca), manganese (Mn), iron (Fe), nickel (Ni), zinc (Zn), rubidium (Rb) and strontium (Sr)) were performed with a tabletop X-ray fluorescence (XRF) spectrometer PEDUZO T02 (Jožef Stefan Institute, Slovenia) with rhodium-anode X-ray tube. X-ray fluorescence was detected by a silicon drift diode detector (Amptek Inc., Bedford, MA, USA). The energy resolution of the spectrometer at count rates below 1000 cps was 140 eV at 5.9 keV. Measurements were performed in air and the samples were irradiated for

1000–5000 s to ensure sufficiently low statistical error [10,11]. The spectra were analysed by software operating in LabVIEW [12]. Element analysis was validated using standard reference material NIST SRM 1573a (tomato leaves).

2.3. Silkworm Rearing

2.3.1. Silkworm Genetic Material

Certified, traceable silkworm genetic material (eggs of a polyhybrid strain) was obtained from CREA-AA, Laboratory of Sericulture in Padua, Italy. The polyhybrid (4-way-hybrid) strain was generated by crossing a parental Chinese strain (SC2 × SC3) × a parental Japanese strain (SG1 × SG3). Eggs proved to be negative to the microscopic pebrine analysis, which is routinely carried out to detect *Nosema bombycis* infection.

2.3.2. Silkworm Rearing Technology

Estimation of the optimal timing for the start of silkworm rearing was determined by observation of the vegetative status of mulberry trees according to the reference guidelines of Brion [13].

The eggs of silkworm hybrids were transported from CREA-AA to the Faculty of Agriculture and Life Sciences, University of Maribor, Slovenia in controlled temperature conditions (25 °C). The eggs were weighed to have similar quantities for each lot, i.e., about 5000 larvae per thesis. In all five instars of silkworm development silkworms were reared in faculties facility near reference plantation in cardboard boxes, changed after each moult. Photoperiod was 12 h day-night. Rearing temperature and relative humidity all the time varied between 23–26 °C and 45–60%, respectively. Rearing place was constantly aerated.

Until the beginning of the 4th developmental instar larvae were reared together in a common box and fed with the mixture of leaves of Slovenian (Slo), Hungarian (Hu) mulberry genotypes and reference sericultural varieties (ref) of *Morus alba* trees from the mulberry collection (groups 1–4). After the 3rd moult larvae were randomly distributed to 38 cardboard boxes (25 × 55 × 5 cm) in groups of 30 larvae to test individual mulberry genotypes. For studies on gut microbiome analysis and haemocyte parameters, groups of larvae were fed with mixtures of leaves according to the four main groups (i.e., Slo, Hu, ref, and fruit varieties).

Larvae were fed ad libitum with leaves of selected genotypes/species of mulberry trees. When in the rearing box about 10% of leaves from the previous feeding were left, a new portion of food was added every 3–4 h during the day and every 6 h, during the night. Leaves were offered until the last larva started to spin or for 12 days after the last moult to those larvae which did not start to spin. After the start of each moult, food was not offered for two days, in order to obtain complete moulting from all the individuals of each lot and to start the following instar homogeneously.

The cumulative body weight of the larvae within one group was measured every 24 h. Larvae were observed for activities (e.g., eating, moulting, spinning) and health status (signs of any disease).

Cocoons were collected eight days after beginning of spinning. The number and cumulative weight of the cocoons were measured in each experimental group. Cocoons were dried to 42–45% of the original fresh weight at 60 °C.

2.3.3. Monitoring the Health and Microbiological Status of Silkworm Larvae

The activity of larvae was checked at least six times per day, during feeding and bodyweight measuring. Larvae showing signs of illness (e.g., inactivity, cessation of eating, sluggishness, or flaccidity, colour changes, swollen body, fragile intersegmental membranes, diarrhoea) or dead larvae were immediately removed from the group and were stored at −20 °C until processing.

Molecular Detection of Bombyx Mori Nucleopolyhedrovirus

Bombyx mori nucleopolyhedrovirus (BmNPV, *Baculoviridae*) (the causative agent of Grasserie disease) has previously been detected in larvae at CREA-AA [14]. Vertical (transovarial) transmission occurs with BmNPV, and the virus can persist in some larvae as sublethal infection [15,16]. Therefore, diseased larvae were tested for infection by in-house developed, qualitative real-time PCR (rt-PCR) assay using the TaqMan technology. Larval specimens were homogenised in sterile ceramic mortars, and 10× volume sterile phosphate-buffered saline (PBS) was added. Homogenates were centrifuged at 1500× *g* for five minutes and total DNA was extracted from 200 μL supernatants, using the Qiagen Viral DNA extraction kit (Qiagen, Hilden, Germany). The rt-PCR assay was targeting the polymerase gene region (genomic primer: 5′-GCCACCGTAATCACRCGTCTTT-3′, complementary primer: 5′-CGATAACCCGGGCAAAAA-3′ and TaqMan probe: 5′-FAM-ACCTTCATTATTATCGTCAGCCGATTTGCG-TAMRA-3′. Samples were tested in duplicates. The specificity of the assay was confirmed by direct sequencing of the amplification products and identification through BLAST search in the gene bank databases. Relative quantification of BmNPV DNA loads in the samples was based on cT values.

Identification of Gut Microbiome and Potential Bacterial Pathogens by Metagenomic Studies

Three randomly selected, healthy 5th instar larvae were collected from groups fed with Slovenian and Hungarian mulberry genotypes, reference varieties, *M. australis*, and *M. nigra*. The guts of larvae from each pool were removed, homogenized, and total DNA was extracted. The DNA samples were submitted to a metagenomic investigation targeting the 16S rRNA gene of bacteria to reveal the bacterial gut microbiomes of silkworms, fed with different genotypes and species of mulberry leaves.

After merging the paired-end reads by PEAR quality-based filtering, trimming was performed by Trimmomatic, using 20 as the quality threshold, and reads longer than 50 bp were retained only [17,18]. The remaining reads after deduplication by VSEARCH [19] were taxonomically classified using Kraken2 (k = 35) with the NCBI non-redundant nucleotide database for the shotgun sequenced samples [20,21]. In the case of 16S rRNA sequenced samples for the classification the Greengenes database was used following chimera filtering by VSEARCH [19,22]. The taxon classification data was managed in R (R Core Team) using functions of package phyloseq [23,24].

Comparison of Selected Qualitative and Quantitative Hemogram Parameters in the Main Feeding Groups of Larvae

For the description of the immune cells, 10 fifth instar larvae were collected from different experimental groups. Haemolymph was collected by cutting a proleg into Schneider's insect medium supplemented with 5% foetal bovine serum and phenylthiourea. Tenfold dilution of the haemolymph in the same medium was used to determine total cell count in a Bürker counting chamber and 30 μL aliquots of the diluted cell suspensions were placed on the spots of Hendley Essex—Diagnostic Microscope Slides for phenotype analysis. The haemocytes were allowed to settle and adhere to the slide for 60 min in a humidity chamber. After the one-hour incubation, the haemocytes were fixed with acetone for six minutes and were classified according to morphological criteria [25,26]. The proportions of three main haemocyte classes, the granular cells, the plasmacytes, and the prohemocytes were determined.

2.4. Methods for Evaluation of Quality of Silk Cocoons

The number and cumulative weight of the cocoons were measured in each experimental group. Cocoons were dried until they reached approximately 45% of their original weight at 60 °C and sent to CREA-AA, Laboratory of Sericulture, for the analysis.

The whole cocoon weight, the silk shell weight, and the silk percentage were calculated after cutting cocoons and extracting pupae on around 20 cocoons per sample. The silk percentage was calculated according to Lee [27].

Ten cocoons for each experimental sample were used for experimental reeling on a testing reeling machine apt to reel individual cocoons, by recording the silk thread length, the title in deniers, the number of breakages, and the wastes.

2.5. Statistical Data Analysis

The results of biochemical analyses of mulberry leaves are shown as mean (average) values (±standard deviation, SD) of the analyses on two sampling dates (16th and 23rd of June 2021) during the feeding experiments, in which three to five trees were sampled and used as one sample. Measurements were performed at least four times for each sample and in duplicate. Assumptions of normality for all chemical traits were checked with Kolmogorov–Smirnov test.

The chemical traits of mulberry leaves, silkworm weight (5–7th day/5th instar), cocoon weight (fresh), silk thread parameters, and reeling wastes were presented by means and SD and were statistically evaluated by one-way analysis of variance (ANOVA), followed by post-hoc comparison according to Duncan. Letters describe significant differences among genotypes and origin-dependent genotype groups.

The Pearson correlation coefficient (Sig. 2-tailed) was calculated between evaluated chemical parameters, silkworm weight, cocoon weight, and silk thread parameters, to analyze the correlative relationship between the measured parameters and to find out the most effective differentiating traits.

Principal component analysis (PCA) enabled us to perform a comprehensive assessment of chemical traits of individual trees, silkworm weight, cocoon weight, and silk thread parameters by discriminating geographical distribution with respect to reference varieties.

IBM SPSS Statistics 25 (Armonk, New York, NY, USA; 2017), StatSoft, Inc. Statistica 8.0 (Victoria, Australia; 2007) and Past 3.17 (Zürich, Switzerland; 2020) software were used for statistical analysis [28].

3. Results

3.1. Chemical Composition of the Leaves of Selected Slovenian and Hungarian Mulberry Genotypes, Reference Sericultural and Fruit Varieties

The results of biochemical analyses of the total proteins, total phenolics and individual phenolic components in leaves are presented in Figure 1A,B and Supplementary Table S2. The highest total protein content was determined in genotype SM 101.1 (239.42 mg/g DW), the lowest in the Bulgarian fruit genotype. Phenolics were the highest in *M. alba* × *rubra* (21.51 mg/g DW) and the lowest in Slovenian genotype SM 6 (11.26 mg/g DW) (Figure 1A, Supplementary Table S2).

By evaluating the individual phenolics in mulberry leaves, eight different hydroxycinnamic acids and eleven flavonols were identified (Figure 1B, Supplementary Table S2). The main phenolic acids, from the hydroxycinnamic group were caffeoylquinic derivatives (with chlorogenic acid predominating) and p-coumaroylquinic acid derivatives. The predominant flavonoids were quercetin and kaempferol glycosides. The main quercetin glycosides were rutin, quercetin malonyl-hexoside, and quercetin-3-glucoside (isoquercetin), whereas the predominant kaempferol glycoside was kaempferol acetyl-hexoside (Figure 1B, Supplementary Table S2).

The maximum concentration of chlorogenic acid was determined in *M. nigra* (18.05 mg/g DW), the minimum concentration in Hungarian genotype SO 1042 (4.00 mg/g DW). Within the Slovenian mulberry genotypes, the highest concentration was determined in SE 5 (11.73 mg/g DW), whereas within the Hungarian genotypes in SO 1035 (13.52 mg/g DW). Among the reference varieties 'Kokusou-20' yielded the highest chlorogenic acid content (17.67 mg/g DW), whereas the concentration of 4-caffeoylquinic acid (4.36 mg/g DW) was the highest in *M. alba* × *rubra*. The above-mentioned genotypes superior in chlorogenic acid were also characterized by the highest total caffeoylquinic acid derivatives (Figure 1B, Supplementary Table S2).

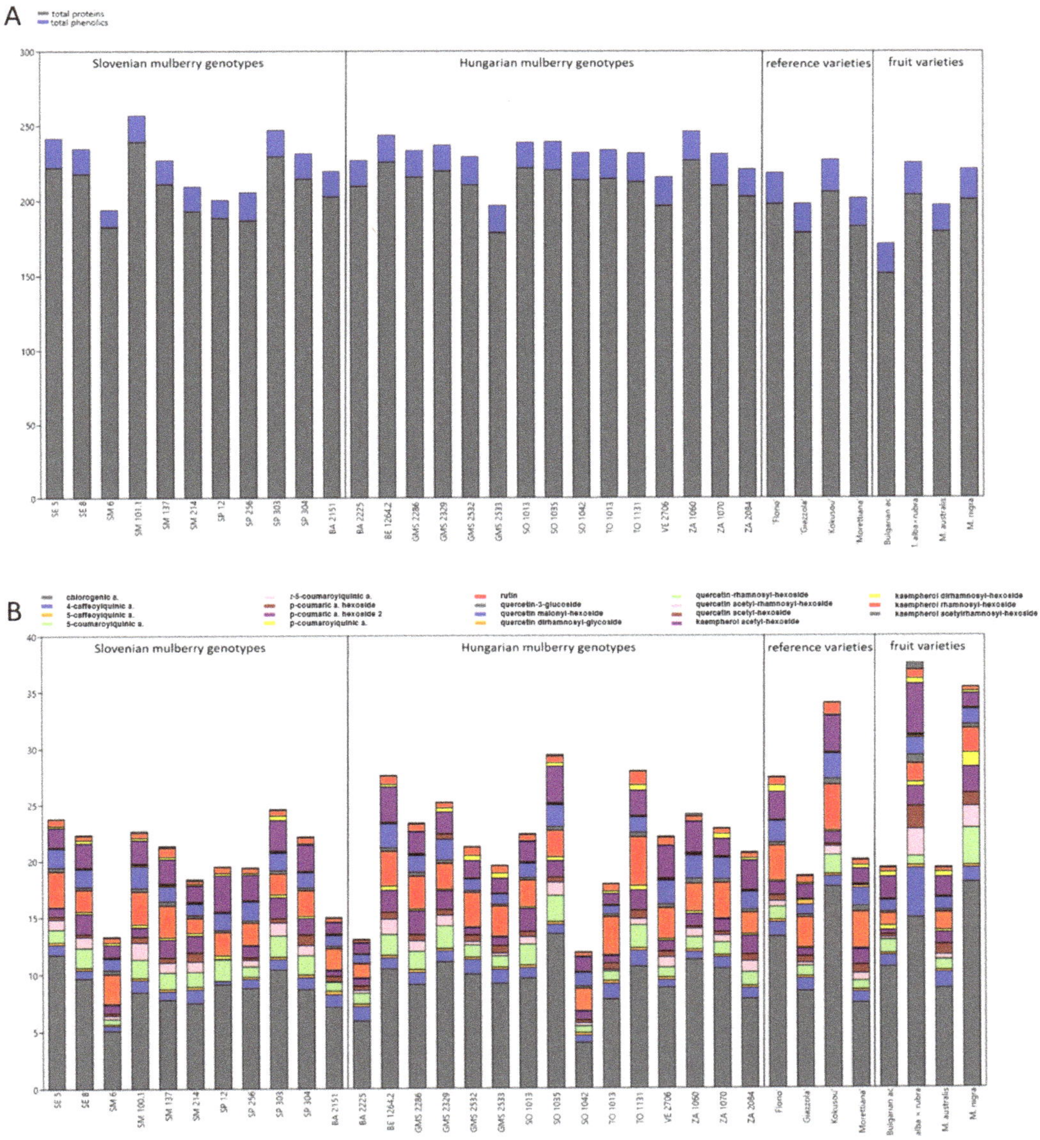

Figure 1. The mean concentrations of the (**A**) total proteins and total phenolics (mg/g DW), and (**B**) the concentrations of individual phenolics (mg/g DW) in leaves of Slovenian, Hungarian old mulberry genotypes, reference sericultural, and fruit varieties. For detailed data and statistics see Table 1 and Supplementary Table S2.

Table 1. Mean concentrations (±SD) of the total proteins, total phenolics, and predominant phenolics (mg/g DW) in Slovenian, Hungarian mulberry varieties, reference sericultural varieties, and fruit varieties. Different letters (a–c) indicate significant differences ($p < 0.05$) in the concentrations of specific compounds between the analysed groups, as determined by the post hoc Duncan test. *n*, number of repetitions.

	n	Total Proteins	Total PH	Chlorogenic a.	4-CQA	c-CQA	Rutin	Q-3-glu	QMH	KAH *
Slovenian genotypes	19	211.48 ± 18.51 a	17.03 ± 2.07 b	9.07 ± 1.66 c	0.9 2± 0.30 b	1.50 ± 0.32 a	2.32 ± 0.63 b	0.2 6 ± 0.06 b	1.51 ± 0.30 ab	2.15 ± 0.48 a
Hungarian genotypes	25	211.89 ± 13.77 a	18.19 ± 2.02 b	9.97 ± 1.61 bc	0.91 ± 0.25 b	1.52 ± 0.48 a	2.81 ± 0.64 a	0.31 ± 0.08 ab	1.63 ± 0.43 a	2.29 ± 0.65 a
reference varieties	4	191.38 ± 12.72 ab	19.95 ± 1.21 a	11.72 ± 4.36 ab	1.03 ± 0.14 b	1.12 ± 0.37 a	3.27 ± 0.57 a	0.40 ± 0.15 a	1.63 ± 0.53 a	2.09 ± 0.87 a
fruit varieties	4	183.95 ± 24.20 b	20.07 ± 2.03 a	13.43 ± 3.61 a	2.47 ± 1.64 a	1.53 ± 1.07 a	1.63 ± 0.40 c	0.41 ± 0.26 a	1.26 ± 0.24 b	2.35 ± 1.31 a

* t PH, total phenolics; 4-CQA, 4-caffeoylquinnic acid; c-CQA, c-5-coumaroyilquinnic acid; Q-3-glu, quercetin-3-glucoside; QMH, quercetin-malonyl-hexoside; KAH, kaempherol-acetyl-hexoside.

The predominant coumaroylquinic acid derivative was 5-coumaroylquinic acid, whose concentration was the highest in *M. nigra*, yielding a total coumaroylquinic acid amount of 10 mg/g DW, followed by *M. alba* × *rubra* with 7.39 mg/g DW.

Among quercetin glycosides, rutin concentration was the highest in Hungarian genotype TO 1131. Maximum quercetin malonylhexoside was determined in 'Kokusou-20' (2.25 mg/g DW). Quercetin-3-glucoside and kaempherol acetyl-hexoside were the highest *M. alba* × *rubra*. This hybrid was also characterized by the highest total kaempherol glycoside derivatives (6.27 mg/g DW) (Figure 1B, Supplementary Table S2).

The results of the element analysis of mulberry leaves are shown in Figure 2. P concentrations ranged between 1.07 and 2.04 g/kg DW and were the highest in genotype SE 5. Some mulberries contained more than 1 g/kg DW of S with the highest concentration in genotype GMS 2532 (1.26 g/kg DW). K reached maximum concentration in genotype BA 2225 (22.1 g/kg DW) and Ca in genotype GMS 2532 (20 g/kg DW) (Figure 2A, Supplementary Table S3).

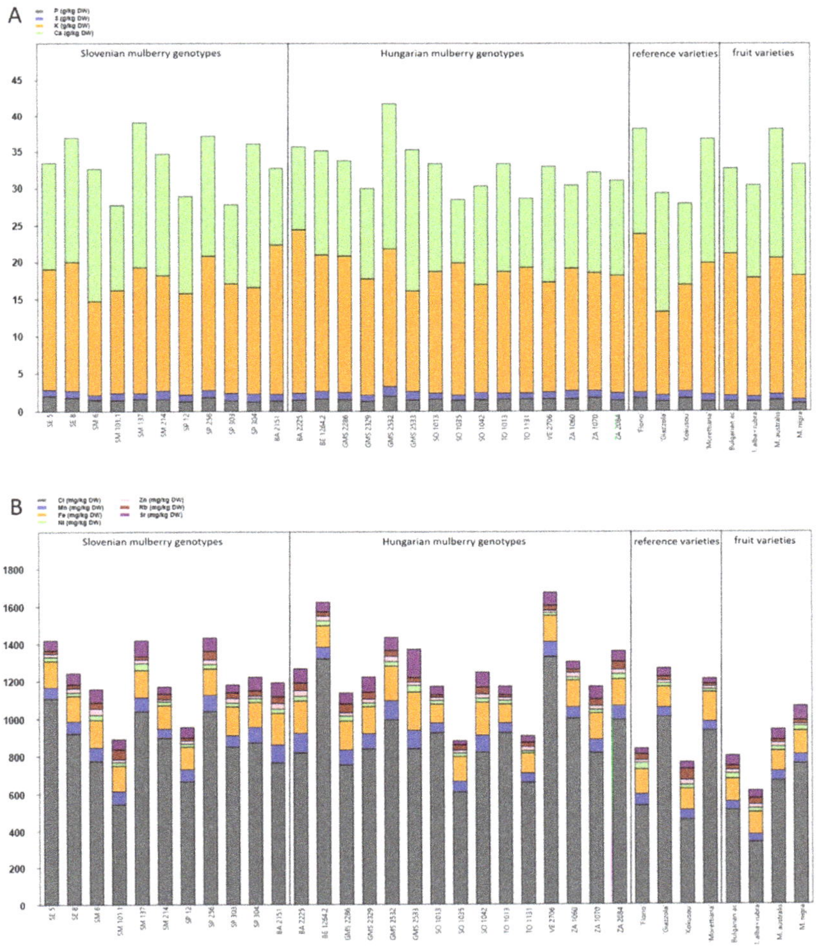

Figure 2. The mean concentrations of the (**A**) macroelements (mg/kg DW) and (**B**) microelements (mg/kg DW) in leaves of Slovenian, Hungarian old mulberry genotypes, reference sericultural and fruit varieties. For detailed data and statistics see Table 2 and Supplementary Table S3.

Table 2. Mean concentrations (±SD) of the macroelements (g/kg DW) and the predominant microelements (mg/kg DW) in mulberry leaves in Slovenian, Hungarian mulberry varieties, reference sericultural varieties, and fruit varieties. Different letters (a, b) indicate significant differences ($p < 0.05$), which were determined using the post hoc Duncan test. *n*, number of repetitions.

	n	P	S	K	Ca	Cl	
Slovenian genotypes	19	1.75 ± 0.75	0.88 ± 0.23 a	15.92 ± 3.59	17.14 ± 4.90	788.16 ± 36.26	
Hungarian genotypes	25	1.42 ± 0.32	0.83 ± 0.19 ab	16.47 ± 3.16	14.03 ± 3.47	953.20 ± 34.69	
reference varieties	4	1.52 ± 0.23	0.84 ± 0.80 ab	16.15 ± 4.39	14.63 ± 2.57	735.25 ± 27.79	
fruit varieties	4	1.30 ± 0.18	0.62 ± 0.12 b	17.58 ± 1.48	14.20 ± 2.72	569.25 ± 18.45	
		Mn	**Fe**	**Ni**	**Zn**	**Rb**	**Sr**
Slovenian genotypes	19	66.04 ± 14.58 a	147.42 ± 26.27 ab	28.35 ± 11.88	23.91 ± 11.34	36.23 ± 28.72	54.02 ± 25.16 ab
Hungarian genotypes	25	67.32 ± 21.29 a	154.36 ± 29.89 a	24.19 ± 9.40	19.98 ± 5.66	26.75 ± 9.73	72.01 ± 31.40 a
reference varieties	4	52.03 ± 6.16 ab	128.25 ± 21.42 ab	21.68 ± 8.06	18.20 ± 5.32	31.20 ± 20.27	35.10 ± 7.16 b
fruit varieties	4	45.38 ± 4.22 b	119.50 ± 7.05 b	23.65 ± 4.20	18.40 ± 4.59	22.40 ± 7.73	57.53 ± 15.07 ab

Among microelements, maximum Cl concentrations were found in genotype BE 1264.2 (1320 mg/kg DW). Fe and Ni concentrations were the highest in genotype GMS 2533. The highest Zn concentration was found in Slovenian genotype SM 6. 'Kokusou-20' was found to be rich in Rb (59.4 mg/kg DW), whereas Sr concentrations ranged from 22.8 to 152 mg/kg DW with the maximum value in GMS 2533 (Figure 2B, Supplementary Table S3)

3.2. Origin Dependent Differences in Chemical Composition of Mulberry Leaves

The mulberry genotypes were categorized according to their origin to Hungarian and Slovenian old mulberry genotypes and compared to the reference and fruit varieties. Based on the statistical evaluation of the chemical compounds we determined the significant highest amounts of proteins in both Slovenian and Hungarian genotypes of *M. alba*, when compared to fruit varieties, which had the lowest protein contents. The sericultural reference varieties were intermediate in proteins (Table 1, Supplementary Table S2).

The significant highest amount of total phenolics was observed in classical contemporary sericultural genotypes and fruit varieties. In Table 1 predominant individual phenolics are summarised as mean values of Slovenian, Hungarian local genotypes, reference sericultural, and fruit varieties, whereas the ranges of all analysed phenolics of individually selected mulberry genotypes are presented in Supplementary Table S2. The least mean content of chlorogenic acid was found in Slovenian genotypes, the most in fruit varieties. Similarly, the highest amount of 4-caffeoylquinic acid was found in fruit varieties. The highest amount of rutin was analysed in reference and Hungarian genotypes of *M. alba*, the lowest in fruit varieties. Reference sericultural genotypes and fruit varieties have the highest amount of quercetin-3-glucoside, the least concentration was determined in Slovenian genotypes of mulberry trees. The highest amount of quercetin malonyl-hexoside was found in the leaves of Hungarian and reference mulberry trees, the lowest in fruit varieties (Table 1, Supplementary Table S2).

Significant differences between the mulberry groups in the contents of the main elements were determined for S, Mn, Fe, and Sr. Slovenian mulberry genotypes were characterized by the highest S content, whereas the lowest was determined in fruit varieties. Both, Slovenian and Hungarian genotypes were rich in Mn content. Fe was significantly highest in Hungarian genotypes and lowest in fruit varieties. Sr was significantly enhanced in Hungarian genotypes, whereas the lowest concentrations were found in reference varieties (Table 2, Supplementary Table S3).

3.3. Description of the Influence of Feeding Silkworms with Leaves of Various Genotypes of Mulberry Trees on Larval Development

The highest silkworm weight was obtained when silkworms were fed with the Hungarian genotype BA 2151 and the lowest with ZA 1060 (Supplementary Table S4). Out of Slovenian varieties, the maximum weight was obtained when larvae were fed with leaves of SP12 mulberry genotype, the lowest weight was determined for larvae that were fed with Submediterrean varieties (SM6, SM101.1, SM137). Among the reference varieties, significantly higher weight was obtained by feeding them with 'Florio'. Among the fruit varieties, larvae fed with *M. nigra* gave the highest weight (Figure 3, Supplementary Table S4).

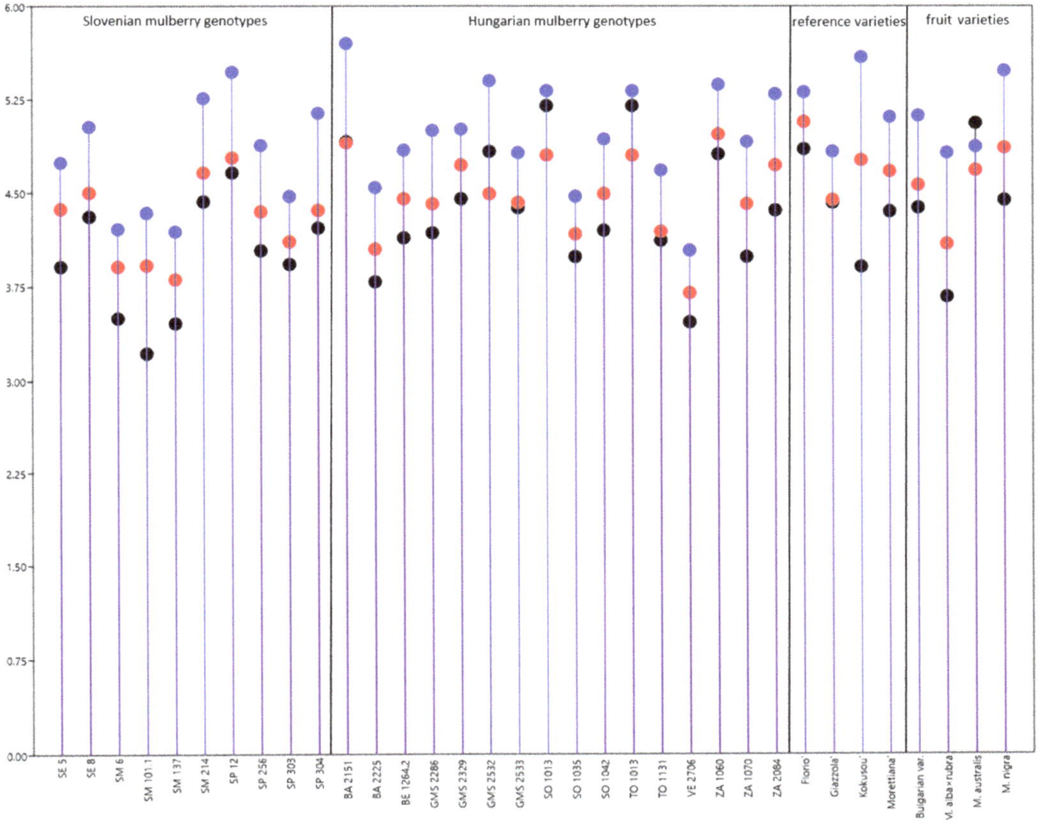

Figure 3. Mean weight of a larvae (±SD) on day 5 (black dots), 6 (red dots), 7 (blue dots) of the fifth instar when fed with Slovenian, Hungarian mulberry varieties, reference sericultural varieties, and fruit varieties. For detailed data and statistics see Table 3 and Supplementary Table S4.

Table 3. Mean weight (±SD) of a larvae and cocoon fresh weight (FW, g) when fed with Slovenian, Hungarian mulberry varieties, reference sericultural varieties, and fruit varieties. Different letters (a, b) indicate significant differences ($p < 0.05$), which were determined using the post hoc Duncan test. FW, fresh weight. *n*, number of repetitions.

	n	5th Instar/5th D	5th Instar/6th D	5th Instar/7th D	Cocoon FW
Slovenian mulberry genotypes	19	3.97 ± 0.46 b	4.28 ± 0.33 b	4.77 ± 0.46 b	2.14 ± 0.12 a
Hungarian mulberry genotypes	25	4.32 ± 0.46 ab	4.46 ± 0.34 b	4.96 ± 0.43 b	2.26 ± 0.09 a
reference sericultural varieties	4	4.54 ± 0.38 a	4.85 ± 0.23 a	5.36 ± 0.36 a	2.25 ± 0.14 a
fruit varieties	4	4.39 ± 0.57 ab	4.55 ± 0.33 ab	5.07 ± 0.30 ab	2.20 ± 0.14 a

3.4. Origin Dependent Differences in the Growth of the Silkworm Larvae and Cocoon Weight

Based on statistical evaluation of the four main groups (Slovenian, Hungarian, reference sericultural and fruit varieties) reference sericultural varieties were characterized by the highest larval growth, whereas no statistical differences were found between mean cocoon fresh weight (Table 3).

3.5. Description of the Influence of Feeding Silkworms with Leaves of Various Mulberry Genotypes on the Quality of Raw Silk

Among Slovenian varieties, the maximum length was obtained when larvae were fed with Slovenian variety SP 256. Superior length was also obtained with GMS 2286, whereas the lowest was obtained with the reference varieties 'Giazzola', Bulgarian fruit genotype, and *M. alba* × *rubra*, which also gave the lowest raw silk weight. Significant highest weight was obtained with Hungarian genotype SO 1013. The Hungarian genotype BE 1264.2 was characterized by the highest thickness and the Slovenian genotype SM 137 with the lowest. When analysing the reeling wastes, the highest struse was found in most of the Hungarian varieties and the significant highest telette was determined in Hungarian variety GMS 2533 (Supplementary Table S4).

3.6. Origin Dependent Differences in the Quality of Raw Silk

There was a trend for longer silk thread production when larvae were fed with Slovenian and Hungarian varieties in comparison to reference sericultural varieties, although the trend was insignificant. Fruit varieties were characterized by a shorter length and weight of raw silk. Struse was significantly largest when larvae were fed with Hungarian varieties, thickness and telette did not significantly differ among the groups. When analysing reeling wastes struse was highest in Hungarian genotypes and lowest in Slovenian genotypes and fruit varieties. There were no significant differences in telette (Table 4).

Table 4. Mean (±SD) raw silk and reeling waste parameters (g) of Slovenian, Hungarian mulberry varieties, reference sericultural varieties, and fruit varieties. Different letters (a, b) indicate significant differences ($p < 0.05$), which were determined using the post hoc Duncan test. *n*, number of repetitions.

	n	Silk Thread Parameters			Reeling Wastes	
		Length	Weight	Thickness	Struse	Telette
Slovenian genotypes	95	1419.09 ± 64.14 a	0.43 ± 0.03 a	2.72 ± 0.13 a	0.036 ± 0.013 b	0.017 ± 0.009 a
Hungarian genotypes	211	1429.18 ± 94.82 a	0.44 ± 0.03 a	2.82 ± 0.15 a	0.047 ± 0.022 a	0.021 ± 0.019 a
reference varieties	70	1363.20 ± 135.94 a	0.43 ± 0.02 a	2.84 ± 0.19 a	0.043 ± 0.016 ab	0.017 ± 0.016 a
fruit varieties	30	1244.80 ± 44.03 b	0.37 ± 0.03 b	2.73 ± 0.13 a	0.040 ± 0.020 b	0.022 ± 0.016 a

3.7. Correlation between Mulberry Metabolites, Silkworm, and Raw Silk Parameters

A high correlation was found between silkworm weight and raw silk thickness, as well as raw silk length, and weight; whereas a negative between silk thread parameters (length, weight) and reeling waste (telette, struse). The results of the chemical analysis of mulberry leaves indicate a strong correlation between total proteins and raw silk length and weight. Among phenolic acids, t-5-coumaroylquinic acid correlated strongly with the weight of silkworms. There was a medium negative correlation between p-coumaric acid hexoside and raw silk length. Among flavonols, rutin correlated positively with struse. There was a negative correlation between kaempherol-dirhamnosyl-hexoside and raw silk length and weight, whereas a positive correlation between kaempherol-rhamnosyl-hexoside and struse (Table 5).

Table 5. Pearson's correlation coefficient between silkworm, cocoon, and raw silk parameters in relation to the main phenolics (n = 32) in mulberry leaves listed according to their retention times on HPLC.

	Silkworm Weight (Larvae/g)				Cocoon Weight		Silk Thread		Reeling Waste		
Correlations	**5th C/5th D**	**6th D**	**7th D**	**Mean 3d Weight**	**Fresh**	**Dry**	**Length**	**Weight**	**Thickness**	**Struse**	**Telette**
5th C/5th D	1	00.856**	0.764 **	0.934 **	0.777 **	0.766 **	−0.11	0.191	0.438 *	0.234	0.121
5th C/6th D	0.856 **	1	0.907 **	0.967 **	0.713 **	0.715 **	−0.088	0.21	0.452 **	0.18	0.066
5th C/7th D	0.764 **	0.907 **	1	0.935 **	0.607 **	0.618 **	−0.032	0.166	0.298	0.226	0.133
mean 3d weight	0.934 **	0.967 **	0.935 **	1	0.744 **	0.745 **	−0.082	0.199	0.418 *	0.229	0.116
cocoon FW	0.777 **	0.713 **	0.607 **	0.744 **	1	0.992 **	0.024	0.376 *	0.562 **	0.193	0.081
cocoon DW	0.766 **	0.715 **	0.618 **	0.745 **	0.992 **	1	0.016	0.378 *	0.583 **	0.189	0.076
length	−0.11	−0.088	−0.032	−0.082	0.024	0.016	1	0.785 **	−0.075	−00.368 *	−0.183
weight	0.191	0.21	0.166	0.199	0.376 *	0.378 *	0.785 **	1	0.555 **	−00.455 **	−0.07
thickness	0.438 *	0.452 **	0.298	0.418 *	0.562 **	0.583 **	−0.075	0.555 **	1	−0.269	0.099
struse	0.234	0.18	0.226	0.229	0.193	0.189	−0.368 *	−0.455 **	−0.269	1	0.176
telette	0.121	0.066	0.133	0.116	0.081	0.076	−0.183	−0.07	0.099	0.176	1
total proteins	−0.198	−0.139	−0.116	−0.164	0.083	0.073	0.448 **	0.495 **	0.175	−0.024	−0.262
total phenolics	−0.001	0.131	0.137	0.087	0.096	0.108	−0.197	−0.12	0.069	0.191	0.13
chlorogenic a.	0.02	0.202	0.228	0.148	0.205	0.205	−0.198	−0.068	0.165	0.211	0.15
4-caffeoyl-QA	−0.092	−0.12	0.006	−0.071	−0.125	−0.136	−0.329	−0.312	−0.098	0.211	0.246
5-caffeoyl-QA	−0.031	−0.068	−0.002	−0.034	0.004	0.019	0.109	0.079	−0.023	0.046	−0.15
total caffeoyl-QA	−0.003	0.154	0.206	0.117	0.157	0.155	−0.248	−0.127	0.126	0.236	0.186
c-5-CQA	0.023	0.034	0.037	0.033	0.231	0.225	0.21	0.272	0.176	0.034	−0.167
t-5-CQA	−0.451 **	−0.328	−0.281	−0.382	−0.265	−0.28	−0.01	0.002	0.003	−0.1	−0.092
p-CAH	−0.048	−0.15	−0.059	−0.084	−0.123	−0.13	−0.378 *	−0.322	−0.045	0.158	0.196
p-CAH2	−0.044	−0.057	−0.1	−0.07	0.028	0.008	0.15	0.168	0.066	0.05	−0.281
p-CQA	−0.147	−0.181	−0.154	−0.168	−0.094	−0.09	−0.283	−0.313	−0.097	0.121	−0.034
total coumaroyl-QA	−0.183	−0.171	−0.146	−0.177	−0.04	−0.055	0.007	0.05	0.065	0.044	−0.145
rutin	−0.179	−0.054	−0.05	−0.108	0.081	0.101	−0.02	−0.041	−0.033	0.364 *	−0.018
Q-3-glu	−0.326	−0.105	−0.056	−0.185	−0.163	−0.153	−0.247	−0.342	−0.225	0.298	0.033
QMH	−0.268	0.02	−0.047	−0.122	0.098	0.11	0.085	0.177	0.193	−0.029	−0.161
Q-diR-gly	−0.082	−0.098	−0.132	−0.109	−0.155	−0.128	−0.209	−0.228	−0.088	0.086	0.011
QRH	−0.299	−0.301	−0.235	−0.295	−0.25	−0.27	−0.16	−0.304	−0.303	0.127	0.155
Q-acetyl-RH	−0.207	−0.185	−0.084	−0.168	−0.044	−0.065	−0.082	−0.125	−0.1	0.046	−0.047
QAH	0.009	0.135	0.006	0.045	0.284	0.275	−0.132	0.054	0.264	0.189	−0.12
total quercetin-gly	−0.27	−0.044	−0.071	−0.15	0.077	0.097	−0.037	−0.018	0.033	0.298	−0.077
KAH	−0.197	−0.071	−0.007	−0.105	−0.043	−0.044	−0.077	−0.028	0.066	0.086	0.054
K-diRH	0.126	0.054	−0.044	0.052	0.05	0.01	−0.525 **	−0.407 *	0.042	0.284	0.231
KRH	−0.199	−0.175	−0.075	−0.16	0.003	0.012	−0.152	−0.204	−0.125	0.482 **	0.106
K-acetyl-RH	−0.192	−0.175	−0.091	−0.163	−0.216	−0.23	−0.247	−0.324	−0.22	0.163	0.167
total K-gly. deriv. *	−0.194	−0.101	−0.038	−0.123	−0.046	−0.053	−0.202	−0.162	0.011	0.231	0.118

** Correlation is significant at the 0.01 level (2-tailed); * Correlation is significant at the 0.05 level (2-tailed). FW, fresh weight; DW, dry weight. * 4-caffeoyl-QA, 4-caffeoylquinic acid; 5-caffeoyl-QA, 5-caffeoylquinic acid; 5-total CQA, total caffeoylquinic acid derivatives; c-5-CQA, c-5-coumaroylquinic acid; t-5-CQA, t-5-coumaroylquinic acid; p-CAH, p-coumaric acid hexoside; p-CAH2, p-coumaric acid hexoside 2; p-CQA, p-coumaroylquinic acid; total CQA, total coumaroylquinic acid derivatives; Q-3-glu, quercetin-3-glucoside; QMH, quercetin malonyl-hexoside; Q-diR-gly, quercetin dirhamnosyl-glycoside; QRH, quercetin rhamnosyl-hexoside; Q-acetyl-RH, quercetin acetyl-rhamnosyl hexoside; QAH.

There was no significant correlation between the macroelements and silkworm weight, silk thread parameters, and reeling waste. Among microelements, Cl correlated positively with silk weight, while Fe and Sr correlated with telette (Table 6).

Table 6. Pearson's correlation coefficient between silkworm, cocoon, and raw silk parameters in relation to the macro- and microelements (n = 32) in **mulberry leaves**.

	Silkworm Weight (Larvae/g)				Cocoon Weight		Silk Thread Parameters			Reeling Waste	
Correlations	5th C/5th D	5th C/6th D	5th C/7th D	3d Weight	Fresh	Dry	Length	Weight	Thickness	Struse	Telette
5th C/5th D	1	0.856 **	0.764 **	0.934 **	0.777 **	0.766 **	−0.11	0.191	0.438 *	0.234	0.121
5th C/6th D	0.856 **	1	0.907 **	0.967 **	0.713 **	0.715 **	−0.088	0.21	0.452 **	0.18	0.066
5th C/7th D	0.764 **	0.907 **	1	0.935 **	0.607 **	0.618 **	−0.032	0.166	0.298	0.226	0.133
mean 3d weight	0.934 **	0.967 **	0.935 **	1	0.744 **	0.745 **	−0.082	0.199	0.418 *	0.229	0.116
cocoon FW	0.777 **	0.713 **	0.607 **	0.744 **	1	0.992 **	0.024	0.376 *	0.562 **	0.193	0.081
cocoon DW	0.766 **	0.715 **	0.618 **	0.745 **	0.992 **	1	0.016	0.378 *	0.583 **	0.189	0.076
length	−0.11	−0.088	−0.032	−0.082	0.024	0.016	1	0.785 **	−0.075	−0.368 *	−0.183
weight	0.191	0.21	0.166	0.199	0.376 *	0.378 *	0.785 **	1	0.555 **	−0.455 **	−0.07

Table 6. *Cont.*

Correlations	Silkworm Weight (larvae/g)				Cocoon Weight		Silk Thread Parameters			Reeling Waste	
	5th C/5th D	5th C/6th D	5th C/7th D	3d Weight	Fresh	Dry	Length	Weight	Thickness	Struse	Telette
thickness	0.438 *	0.452 **	0.298	0.418 *	0.562 **	0.583 **	−0.075	0.555 **	1	−0.269	0.099
struse	0.234	0.18	0.226	0.229	0.193	0.189	−0.368 *	−0.455 **	−0.269	1	0.176
telette	0.121	0.066	0.133	0.116	0.081	0.076	−0.183	−0.07	0.099	0.176	1
P	−0.016	0.001	−0.001	−0.006	0.144	0.129	0.269	0.261	0.052	−0.068	−0.097
S	0.207	0.212	0.347	0.271	0.075	0.08	0.228	0.287	0.135	−0.017	0.331
K	0.288	0.236	0.208	0.262	0.282	0.296	0.178	0.234	0.176	−0.14	−0.155
Ca	0.074	−0.108	−0.086	−0.032	−0.103	−0.141	−0.16	−0.1	0.021	−0.108	0.166
Cl	−0.003	−0.136	−0.194	−0.11	0.042	0.063	0.237	0.352 *	0.249	−0.211	−0.022
Mn	−0.058	−0.152	−0.05	−0.086	−0.035	−0.041	0.334	0.324	0.058	−0.106	0.226
Fe	−0.074	−0.115	−0.062	−0.086	−0.184	−0.202	0.167	0.113	−0.062	−0.019	0.434 *
Ni	−0.035	−0.027	−0.049	−0.04	−0.161	−0.211	0.027	0.012	−0.02	−0.151	0.185
Zn	−0.052	−0.048	0.025	−0.027	−0.139	−0.126	0.081	0.037	−0.032	−0.105	−0.016
Rb	−0.367	−0.094	0.056	−0.159	−0.276	−0.267	0.26	0.097	−0.185	−0.119	−0.16
Sr	−0.083	−0.205	−0.168	−0.154	−0.083	−0.123	−0.003	0.005	−0.024	−0.042	0.533 **

** Correlation is significant at the 0.01 level (2-tailed); * Correlation is significant at the 0.05 level (2-tailed); FW, fresh weight; DW, dry weight.

3.8. Principal Component Analysis of Mulberry Metabolites, Silkworm, and Raw Silk Parameters

To enable a comprehensive assessment of the chemical composition of mulberry leaves, with respect to their genotype, larval growth and silk parameters, a principal component analysis (PCA) has been conducted. The discriminant function 1, which accounts for 78.14% of the variance, as explained by the model, was weighted most strongly by total phenolics, followed by rutin, quercetin glucoside, larval weight, and chlorogenic acid. It is further negatively associated with silk length, total proteins, Mn and Sr.

Thus, function 1 clearly (over 89%) separates reference varieties, which are characterized by higher larval weight, which coincides with high values of chlorogenic acid, rutin, and other quercetin glycosides, from Hungarian and Slovenian varieties, being higher in proteins, length, and weight of silk thread. Furthermore, it must be pointed out that some old Hungarian and Slovenian varieties coincide with high concentrations of Cl, Mn, Zn, and Sr.

The second discriminant function accounts for another 21.86% of the variance. It is positively associated with cocoon fresh and dry weight, length, total phenolics, Mn, and K and negatively with Ca and quercetin glucoside. Function 2 clearly separates Hungarian mulberry (by over 87%), which is characterized by higher cocoon fresh and dry weight and the above-mentioned elements from Slovenian genotypes being high in Ca (Figure 4, Supplementary Table S5).

Genotypes with the highest average cocoon fresh and dry weight are GMS 2532 and GMS 2329. Genotype TO1131 contained high total phenolic concentration, in particular quercetin-3-rutinoside which coincided with high 5th instar/5th day weight. BA 2225 is highest in Mn and K concentrations. Slovenian genotype A6 was high in Ca and certain quercetin glycosides as well as NG 214 in Ca content and caffeoylquinic acid glycosides (Figure 4, Table 1).

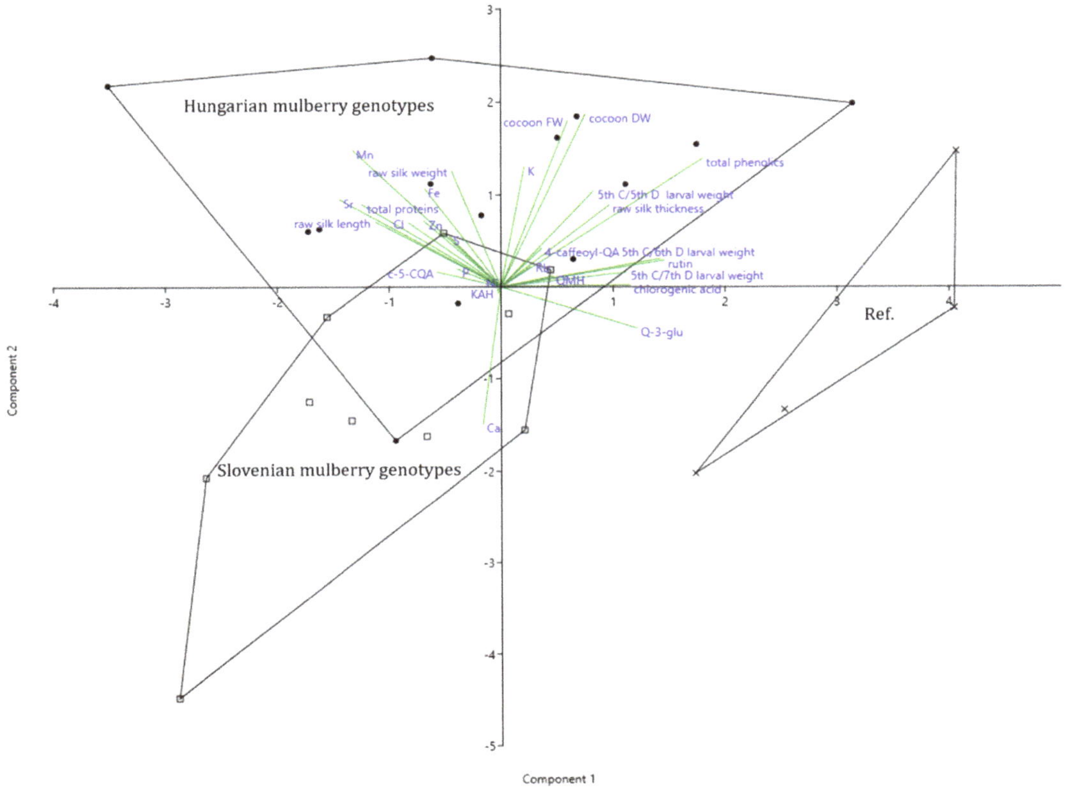

Figure 4. Principal component analysis of silkworm, cocoon, and raw silk parameters along with the chemical composition of leaves of different origins. The convex hulls delimit the space that includes the samples of mulberry trees from different Slovenian regions (squares), Hungarian regions (dots) compared to traditional high yielded sericultural varieties (Ref, crosses). * 4-caffeoyl-QA, 4-caffeoylquinnic acid; c-5-CQA, c-5-coumaroylquinnic a.; Q-3-glu, quercetin-3-glucoside; QMH, quercetin malonyl hex; KAH, kaemph acetyl-hexoside.

3.9. Health and Microbiological Status of Silkworm Larvae

All larvae were moulting synchronously, no prolonged instar duration was observed. Obvious unequal growth and development were observed by three larvae of various groups. During rearing healthy, but cachectic larvae were found only in groups fed with Hungarian (Hu) mulberry genotypes of trees GMS 2532 (fourth instar), SO 1042 (fourth instar), and TO 1131 (fifth instar).

Six ill or dead larvae (0.47% of all larvae) were found in the following developmental stages: 1st day of third instar (one dead larva); 2nd day of third instar (one ill cachectic larva); 1st day of fourth instar (one dead larva); 1st day of fifth instar (one dead larva in group Hu-BE 1264/2); end of fifth instar (one dead larva from group Hu-ZA 2084; one dead larva from group Hu-mix).

In the groups of larvae fed with Slovenian mulberry genotypes (n = 10) all larvae successfully spun in three groups. In five groups 5%, in one group 15% and in one group 20% of the larvae failed to spin. All larvae successfully spinned in the group fed with the mixture of Slovenian mulberry genotypes.

In the groups of larvae fed with Hungarian *M. alba* genotypes (n = 16) all larvae successfully spinned in seven groups. In another seven groups 5% and in two groups

15% of the larvae failed to spin. In the group fed with the mixture of Hungarian mulberry genotypes 20% of the larvae failed to spin.

The groups fed with reference sericultural *M. alba* varieties 'Florio' and 'Morettiana' had 10% losses, while the larvae fed with leaves of variety 'Kokusou' successfully spinned.

All larvae in groups fed with *M. alba* Bulgarian fruit genotype and *M. australis* leaves have successfully spinned. 5% losses were observed in the group fed with the *M. alba* × *rubra*. None of the larvae fed with *M. nigra* leaves spinned cocoons, however, food was offered until 12 days after the last moult.

3.9.1. Molecular Detection of BmNPV by Real-Time PCR

The rt-PCR assays resulted in amplification signals (cT values at 17–21) in DNA extracts of diseased larvae. The nucleotide sequences of the amplification products have shown >99% similarity to an 81 bp long region of the polymerase gene region (between nt positions 47,965 and 48,045) of the BmNPV reference sequence (GenBank accession number: NC_001962). No amplification was detected in the negative control samples (healthy larvae).

Clinically healthy larvae were collected randomly from each experimental group before each moulting and spinning and were tested for BmNPV infection by real-time PCR: all real time-PCR assays in regularly sampled healthy larvae from each group were negative.

BmNPV DNA was detected in the dead or cachectic body of third and fourth instar larvae from the common box:

- 1st day, third instar: 1 dead (liquid rotting) larva; cT mean 40.63, standard deviation (SD) 0.84;
- 2nd day, third instar: 1 cachectic larva; cT mean 40.70, SD 1.49;
- 1st day, fourth instar: 1 larva died during moulting, showed signs of Grasserie; cT mean 18.32, SD 0.17.
- In the fifth instar, BmNPV DNA was detected in three out of the 1340 larvae (0.22%). Positive larvae were ill or were found dead:
- 1st day: Hu-BE 1264/2 (signs of Grasserie); cT mean 37.38, cT SD 0.69;
- end of 5th instar: Hu-ZA 2084.2 (moribund larva with black stripes), cT mean 20-09, SD 0.14;
- end of 5th instar: Hu-mix (dead larva, signs of Grasserie), CT mean 37.43, SD 0.92.

BmNPV DNA was not detected in any of the 39 larvae extracted from cocoons showing signs of failed spinning.

3.9.2. Identification of Gut Microbiome and Potential Bacterial Pathogens by Metagenomic Studies

In the group of Slovenian genotypes mixture, approximately 102,489 different bacterial species were detected. In the group of Hungarian genotypes mixture, the number of detected species exceeded 106,000, while in the group reference genotypes mixture it exceeded 123,958. In group *M. australis* approximately 143,366, and in group *M. nigra* 144,551 different bacterial species were detected. Core bacterial microbiomes are shown in Figure 5.

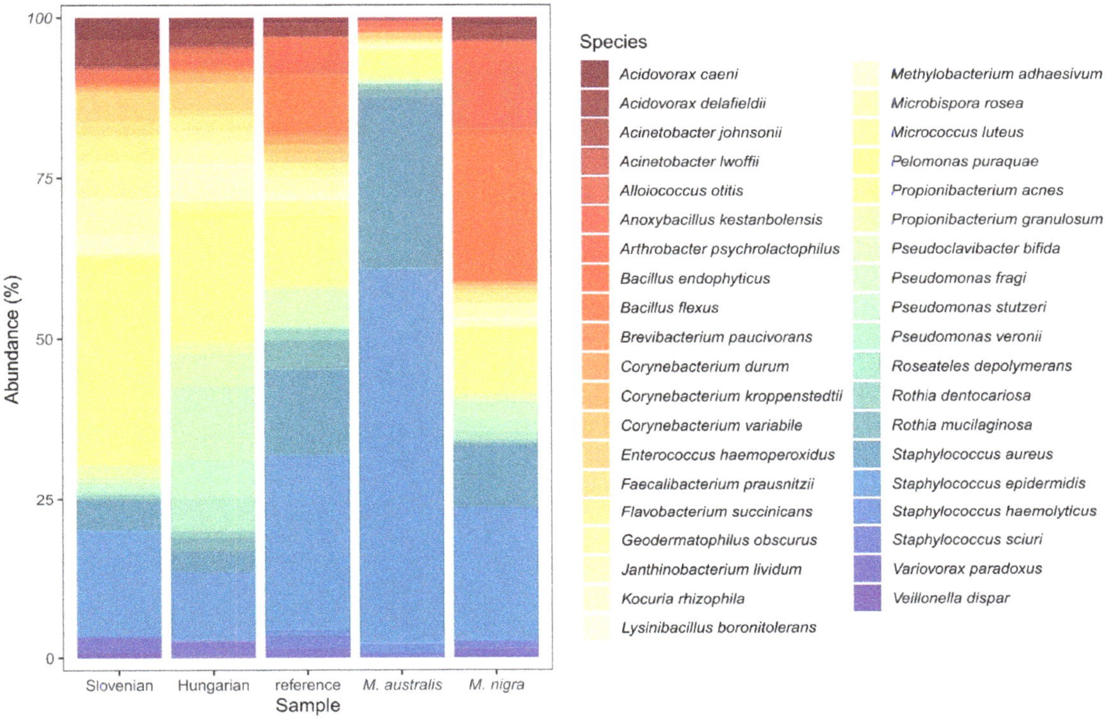

Figure 5. Species compositions of core gut bacterial microbiomes in the different experimental groups.

While on the 5th day of fifth instar all larvae were still eating, the gut microbiomes differed considerably in the samples of larvae fed with mixtures of Slovenian, Hungarian, and reference genotypes of *Morus alba*, as well as the fruit varieties *M. australis* and *M. nigra*.

Major differences in silkworm gut bacterial microbiome in different groups on the level of bacterial classes and Chloroplast are the followings:

(1) 57% of representatives of Chloroplast (99% Streptophyta) were found by Slovenian genotypes and 47% by Hungarian genotypes, almost half less (22%) by reference genotypes, while *M. australis* and *M. nigra* had only 16% and 12%.

(2) The range was quite opposite by class Betaproteobacteria (mainly representatives from order Burkholderiales): 44% by *M. nigra*, 39% by reference genotypes, 24% by Slovenian genotypes, 22% by Hungarian genotypes, and 18% by *M. australis*.

(3) The representatives from Alphaproteobacteria (99% of representatives coming from order Rickettsiales) were in the following ranges: *M. nigra* and Hungarian genotypes 8%, Slovenian genotypes 6%, reference genotypes 3%, and *M. australis* 2% (representatives mainly from orders: Rhodobacterales 40%, Rhizobiales 24%, Sphingomonadales 12%, Rhodospirillales 12%, 9% Rickettsiales).

(4) The representatives from class Gammaproteobacteria were in following ranges: Hungarian mulberry genotypes 4% (Pseudomonadales 3%, Enterobacteriaceae 1%, Xanthomonadales 0.1%), *M. nigra* 3% (Pseudomonadales 2%, Enterobacteriaceae 0.07%, Xanthomonadales 0.1%, Oceanospirillales 0.1%, Alteromonadales 0.1%), Slovenian genotypes 1% (Pseudomonadales 1%, Enterobacteriaceae 0.06%), reference genotypes 1% (Pseudomonadales 0.7%, Enterobacteriaceae 0.4%), *M. australis* 1% (Pseudomonadales 0.8%, Enterobacteriaceae 0.3%).

(5) The representatives from class Actinobacteria were in the following ranges: reference genotypes 14%, Hungarian genotypes 9%, *M. australis* 9%, *M. nigra* 5%, and Slovenian genotypes 4%.
(6) The representatives from class Bacilli were in the following ranges: Slovenian genotypes 4%, Hungarian genotypes 9%, reference genotypes 14%, *M. australis* 9%, and *M. nigra* 5%.
(7) The representatives from class Clostridia were in the following ranges: *M. nigra* 3%, *M. australis* 2%, Slovenian genotypes 1%, Hungarian genotypes 1%, reference genotypes 0.8%.
(8) The representatives from class Flavobacteriales were in the following ranges: *M. nigra* 3%, *M. australis* 2%, Slovenian genotypes 1%, Hungarian genotypes 1%, reference genotypes 0.8%.
(9) The representatives from other classes were in the following ranges: Slovenian genotypes 4%, Hungarian genotypes 3%, reference genotypes 2.2%, *M. nigra* 2%, *M. australis* 1.2 %.

In comparison with other groups, Slovenian genotypes had the highest percentage of Chloroplast and lowest percentage of Gammaproteobacteria, Actinobacteria, Bacilli, Clostridia, and Flavobacteriale.

Hungarian genotypes had the highest percentages of Chloroplast, Alphaproteobacteria, and Gammaproteobacteria and the lowest of Bacilli, Clostridia, and Flavobacteriales.

In reference genotypes, the highest percentage of Actinobacteria and low percentage of Gammaproteobacteria, as well as Clostridia and Flavobacteriales, were described.

The group of *M. australis* had the lowest percentage of Chloroplast, Betaproteobacteria, Alphaproteobacteria, Gammaproteobacteria, and Flavobacteriales while the highest percentage of Bacilli of all groups.

The group of *M. nigra* had the highest percentage of Alphaproteobacteria and Clostridia second highest of Betaproteobacteria, and second lowest percentage of Chloroplast.

Rikettsiales were the most representative order of class Alphaproteobacteria by all groups. The highest percentage of order Enterobacteriaceae was found by the Hungarian genotypes group, the lowest by *M. nigra*. The highest percentage of Lactobacillales was described by the group fed with *M. australis*.

The lowest percentage of family Microbacteriaceae was found by Slovenian genotypes and *M. nigra*, and the highest by reference genotypes and Hungarian genotypes. The highest percentage of family Micrococcaceae was found by reference genotypes and *M. australis*; the highest percentage of family Bifidobacteriaceae was found by *M. australis* and the lowest by Hungarian genotypes and reference genotypes groups. The lowest percentage of family Staphylococcaceae was found by Slovenian genotypes and Hungarian genotypes, highest by *M. australis*.

3.9.3. Comparison of Selected Qualitative and Quantitative Hemogram Parameters in the Main Feeding Groups of Larvae

The highest total number of haemocytes was found in larvae fed with Hungarian mulberry genotypes (74.3×10^4/mL; SE 5.8×10^4/mL), followed by Slovenian genotypes (62.8×10^4/mL; SE 7×10^4/mL), reference genotypes (61.2×10^4/mL; SE 2.6×10^4/mL), *M. nigra* (58.3×10^4/mL; SE 11.2×10^4/mL) and *M. australis* (45.9×10^4/mL; SE 6.6×10^4/mL).

The qualitative analysis of the different haemocyte subsets revealed the dominance of granular cells, followed by plasmatocytes and small spheric cells (Figure 6).

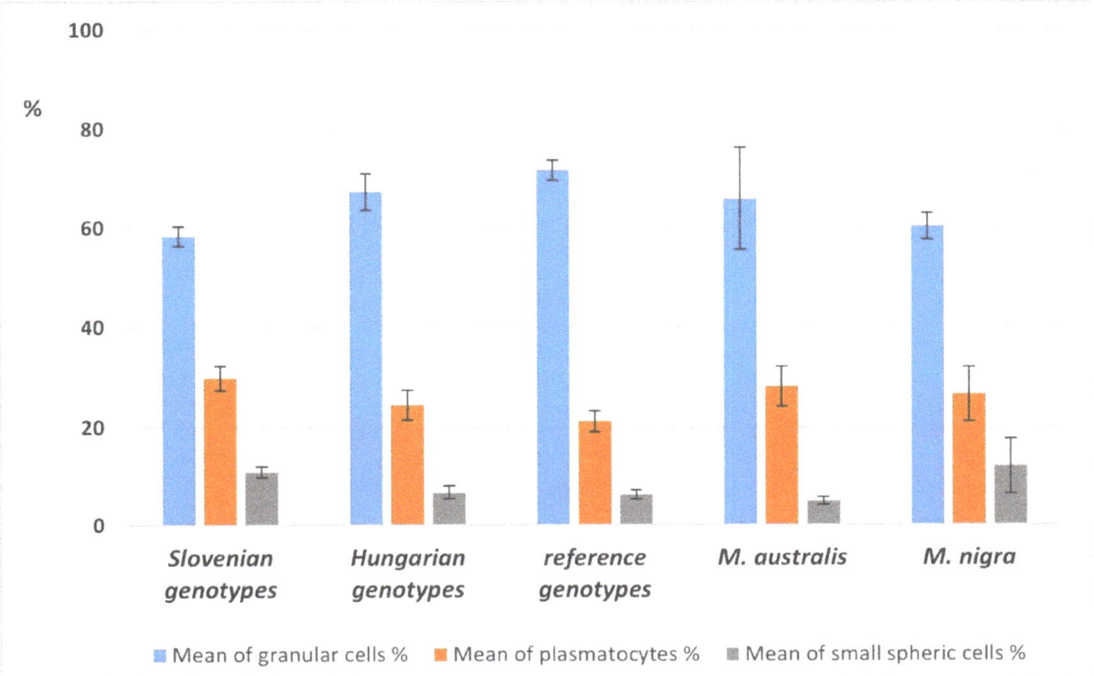

Figure 6. Differences in average haemocyte counts of larvae, fed with Slovenian, Hungarian, and reference sericultural genotypes and fruit varieties of mulberry trees.

4. Discussion

The main factors that contribute to the success of cocoon harvest are mulberry feed, environmental conditions, silkworm cultivation techniques, and silkworm strains [29]. Mulberry leaves are known as a rich source of proteins with unique amino acid composition and are highly palatable and digestible for herbivorous animals [6,30–32]. It has been previously confirmed that high protein content in leaves has a direct impact on the growth of larvae and cocoon production [28,33–36].

Mulberry leaves are further known by their unique composition of phenolic compounds, which were shown to have biological properties [6,37–44]. The identification of phenolics in mulberry leaves with UPLC-MS has been the subject of only a few recent studies so far [44,45].

The main phenolic compound in mulberry leaves was identified as chlorogenic acid followed by other hydroxycinnamic acids, including caffeoylquinic acid, p-coumaric acid derivatives, and p-coumaroylquinic acid derivatives [6,7,44–47]. The flavonol's fraction mostly contains quercetin-3-O-rutinoside (rutin) as well as other quercetin and kaempferol glycosides [6,7,46,48–52].

It has been previously confirmed that chlorogenic acid enhances the rate of silkworm development. Furthermore, quercetin and kaempferol glycosides have been found to be transferred from the larval diet into the hemolymph and cocoons, where they act as UV shields and antimicrobial barriers and might therefore increase the survival rate of developing insects inside the cocoons [53–58]. However, when treated artificially, flavonoids may act as antinutrients by binding to amino acids and proteins as well as digestive gut enzymes they may reduce the nutrient value of mulberry leaves [50].

Based on mulberry inventory in Slovenia and Hungary, we established a collection of historical mulberry trees and screened their biochemical patterns regarding important primary (proteins and amino acids) metabolites and phenolics; we aimed for a definition of high-yielding and nutritive richer mulberry genotypes from the local Slovenian and Hungarian gene pool, which were included in the presented feeding experiment and compared with reference sericultural and fruit varieties. The analysis of study result data indicates possible influences of mulberry chemical parameters on silkworm development and production.

4.1. The Influence of Proteins and Phenolics on Larval, Cocoon and Raw Silk Parameters

The silkworm's growth, cocoon, and raw silk quality can be affected by their feed sources to a great extent. The results of the screening of the metabolites in the leaves of Slovenian and Hungarian genotypes from their local origin were previously published by Urbanek Krajnc et al., and Šelih et al. [6,7]. These data served as the basis for the current research into the impact of feeding silkworms with selected mulberry genotypes out of the local gene pool. Based on biochemical analyses, we selected those genotypes of old local mulberries that proved to be the most favorable for silkworm rearing in terms of nutritional value and leaf yield.

The protein content of mulberry leaves ranges between 13 to 31% [29,51,59,60]. In the current experiment, the highest total protein content was determined in the Slovenian genotype SM 101.1 (239.42 mg/g DW), the lowest in the fruit variety Bulgarian accession (151.50 mg/g DW). The results further coincide with our previous study comprising local trees of the Gorizian region alone and Slovenian old mulberry genotypes sampled at the place of their origin with respect to pruning management and eco-geographical origin [6,7]. We confirmed a strong positive correlation between total proteins and raw silk length and weight. It has been previously confirmed that high protein content in leaves has a direct impact on the growth of larvae and cocoon production [29,61].

Mulberry leaves are also known to have high contents of phenolics, that in our varieties ranged between 11.26 and 21.51 mg/g DW. They were highest in the *M. alba* × *rubra* variety (21.51 mg/g DW) and lowest in SM 6 (11.26 mg/g DW). The analyzed concentrations were in accordance with those of Sánchez-Salcedo et al., who found phenolics in mulberry leaves in the range between 12.8 and 15.5 g GAE/g DW [62].

As previously reported, the predominant caffeoylquinic acid derivative was chlorogenic acid followed by 4-caffeoylquinic acid. In our feeding experiment, the average concentration of chlorogenic acid ranged between 4.00 and 18.05 mg/g DW, with the maximum content analyzed in *M. nigra* [6,7]. Amongst the Slovenian genotypes, the maximum content was found in genotype SE 5. Screening of phenolics on Slovenian genotypes sampled at the place of their origin revealed chlorogenic acid concentration in the range between 1.80 and 6.89 mg/g DW, whereas the maximum content was determined in genotype from the SM region. The concentrations are in accordance with other authors reporting the concentrations of chlorogenic acid in the range between 3 and 10 mg/g DW [43,49,62].

Generally, the predominant phenolic acids are known to have a stimulative effect on feeding, growth, and development [63]. Chlorogenic acid and other dihydroxybenzoid compounds isolated from mulberry leaves were found to be beneficial for silkworms' growth and development. Chlorogenic acid is sensed by chemosensory organs in the mouthparts and stimulates feeding; hence, high concentrations of chlorogenic acid in mulberry leaves significantly promote feeding and correlate with growth parameters of silkworm larvae [64–68]. Furthermore, Yamagishi et al., identified an additional role of chlorogenic acid in the mid-gut lumen as a cue inducing the tachykinin-related peptide secretion from enteroendocrine cells [63]. These peptide hormones are known to modulate physiological processes such as the release of other hormones, secretion of digestive enzymes, gut motility, feeding behaviour, and energy homeostasis. Thus, the silkworm might use chlorogenic acid in differentially directed functions as a food marker in both the mouthparts and mid-gut.

The beneficial effect also correlates with increased silk production and quality. Naik et al., reported that supplementation of chlorogenic acid increased silk productivity and had a positive influence on different silk parameters, such as silk filament length and weight as well as silk protein fibration [69]. Our results did not show a correlative relationship of silkworm parameters with chlorogenic acid, but out of phenolic acids, there was a negative correlation between 5-coumaroylquinic acid and silkworm weight and p coumaroylquinic acid hexoside and the silk thread length.

In our study, the predominant flavonoids were quercetin and kaempferol glycosides. The main quercetin glycosides were rutin, quercetin malonyl-hexoside, and quercetin-3-glucoside (isoquercetin), whereas the predominant kaempferol glycoside was kaempferol acetyl-hexoside, that were generally recognized as the main flavonols in mulberry leaves [6,7,44,45,70,71]. The nutritional effect of flavonols on the growth and development of silkworm larvae and cocoon formation has been intensively studied [53–58,71–74]. In low concentrations, they have a beneficial effect on growth and development, whereas high doses might have an antinutrient effect [75].

Quercetin malonyl-hexoside has been recognized as the main quercetin glycoside with antioxidant activities [43,45,49,71,76–79]. In our sampled genotypes the mean concentrations ranged between 0.74 and 2.25 mg/g DW with the highest value in sericultural variety 'Kokusou-20'. A high correlation was found between the length and weight of raw silk.

The second predominant flavonol was rutin, the maximum concentration was 4.31 mg/g DW determined in one Hungarian genotype (TO 1131). We found a medium correlation only with the struse of the raw silk. By reviewing the literature studying the effect of rutin on silkworm larvae, it was reported that rutin has no stimulative effect on the behaviour of silkworm larvae, although it stimulates feeding on many insects. Furthermore, a significant effect of rutin on the growth of silkworm larvae was not confirmed [74]. However, the authors were able to confirm that larvae can differentiate among quercetin glycosides of mulberry leaves. Quercetin-3-glucoside was recognized as a feeding stimulant but the rhamnose conjugate may deter feeding [74]. Furthermore, a positive effect of quercetin-3-glucoside on the growth and development of silkworm larvae was determined [66,74]. In the presented study, the trend towards a weak negative correlation was determined for silkworm weight and silk thread parameters.

The main kaempferol glycoside that was analyzed in leaves of local mulberries was identified as kaempferol acetyl-hexoside (0.94–4.42 mg/g DW), which reached the highest amount in *M. alba* × *rubra*. Other authors determined kaempferolhexoside in concentrations up to 0.75 mg/g DW, whereas kaempferol malonyl-hexoside was found in traces [71].

Interestingly, we found a negative correlation between kaempherol dirhamnosyl-hexoside and raw silk length and weight, whereas kaempherol rhamnosyl-hexoside correlated positively with struse. The studies on biologically active kaempferol derivatives in mulberry leaves are scarce, and it remains to be seen whether these derivatives may deter feeding.

4.2. The Influence of Macro- and Microelements on Larval, Cocoon and Raw Silk Parameters

Besides phenolics as bioactive compounds, minerals are among the important biochemical components of mulberry leaves, and they may have a high influence on silkworm, cocoon, and raw silk parameters. Previously, positive correlations of nitrogen, phosphorus, potassium, calcium, magnesium, and sulphur were obtained with larval, cocoon, and egg production parameters of mulberry silkworms [80,81]. Shifa et al. considered these macroelements as basic parameters for the evaluation of mulberry varieties for mulberry silkworms rearing in the future [81]. Based on this, we hypothesized that a high amount of these macroelements in mulberry leaves of selected mulberry genotypes will significantly contribute to silkworm and silk thread parameters.

Phosphorus (P) is an important major nutrient in the mulberry plant. It is a component of the complex nucleic acid structure of plants, which regulates protein synthesis. Therefore, it is very important in cell division and the development of new tissue. Phosphorus is also associated with complex energy transformations such as ATP [80,82]. An inadequate amount of P level affects the uptake of other nutritive elements in mulberry leaves for various other physiological activities, in turn, it hampers the growth and economic characteristics of silkworms [83,84]. The highest concentration of P was found in Slovenian genotype SE 5 (2.04 g/kg DW), which was superior in silk thread weight and thickness. When compared to other authors, Shifa et al., determined a minimum value of 1.11 g/kg in 'Jimma cell' and a maximum record of 3.22 g/kg in M-4 accession [81]. Similar to Shifa et al., we found no significant correlation between P and silk parameters or with silkworm weight [81].

Sulphur (S) is known to have an important role in the synthesis of proteins, oils, and vitamins [85]. It plays a vital role in the N metabolism and thus proper development of mulberry [86]. It is a constituent of S-containing amino acids, cysteine (contains 27% of S), and methionine (contains 21% of S). Methionine forms one of the ten essential amino acids for silk formation in silkworms. Cystine and cysteine are among the non-essential amino acids, the quantitative presence of which influences the formation of fibroin over sericin [87]. Deficiency of S level leads to low levels of S-containing amino acids, thus reducing protein synthesis. As a result, amino acids without S and amides of nitrate ions accumulate in the plant tissue and lead to a decrease in sugar as well as insoluble N (protein) in plants [86]. Similar to Ca, the highest content of S was found in the Hungarian genotype GMS 2532 (1260 mg/kg DW). The S concentrations of mulberry varieties analysed by Shifa et al., ranged from 0.15 g/kg in K-2 to 0.34 g/kg in M-4 accession [81]. In the presented experiment S ranged between a minimum value of 0.46 g/kg and a maximum value of 1.26 g/kg, which is up to four-fold higher than what was reported by Shifa et al. [81]. These could be due to the different soil conditions and because the mulberry gene bank is on the silicate geological basis of the southern slopes of Pohorje mountain (central alpine region) which might positively affect the uptake of several minerals (Zn, S, Fe), but negatively Ca, Mg, K, P [82]. Shifa et al., found S to have a significant positive correlation with larval weight, cocoon weight, and shell weight. In contrast to these authors, we did not find any correlation with measured parameters [81].

Potassium (K) plays important regulatory roles mostly in cell ion homeostasis, and stomatal conductance, and thus in maintaining water potential on the cell and whole plant level. Furthermore, it is known that the starch synthetase is activated by K. Thus, with inadequate K, the level of starch declines while soluble carbohydrates and N compounds accumulate. Therefore, it also plays a significant role in the high yield and quality of leaves [88]. It is also involved in the translocation of carbohydrates, protein metabolism, and pathogen tolerance in mulberry [80]. In the silkworm body, the strong alkalinity of the gastric juice originates from potassium and sodium compounds present in the haemolymph. The high alkaline condition of digestive fluid has strong germicidal power against pathogens. K is a unique element that contributes to the growth of silkworms to the maximum extent. In addition, K has a stimulating effect on protein synthesis including silk protein in the silk glands [89]. In the presented experiment, K was increased in fruit varieties, although insignificant. The highest concentration of K was present in Hungarian genotype BA 2225 (22.10 g/kg DW), which was also characterized by superior raw silk length, weight, and thickness as well as silk waste parameters. Shifa et al., reported the K contents of mulberry varieties with ranges from 11.35 g/kg in local varieties to 18.61 g/kg in M-4 accession [81].

Calcium (Ca), in the form of calcium pectate, is important for the cell wall structure in plant. Its deficiency causes incomplete cell division or mitosis, without the formation of a new cell wall resulting in multi-nuclear cells. Calcium is also important in activating certain enzymes and to acts as second messengers in cell signalling that coordinates certain cellular activities. Calcium acts as a detoxifying agent by neutralizing organic acids such as oxalic acid which helps in membrane stability and maintenance of chromosome structure, the activity of enzymes, and translocation of carbohydrates. It is also involved in the differential permeability of membranes [82]. Superior Ca contents were found in genotype GMS 2532 (20.00 g/kg DW). Shifa et al., reported calcium concentrations in the range from 13.45 mg/kg (local check) to 20.52 mg/kg (M-4), which is similar to our findings [81].

Micronutrients are needed in small quantities and they play a pivotal role in the enzymatic reactions and thus govern the growth, development, and yield of mulberries. Chloride (Cl) is involved in the hydrolysis of water in photosynthesis, the synthesis of starch, cellulose, and lignin. It influences cell homeostasis (water holding capacity) of plant tissues. It stimulates the activities of some enzymes [82]. Cl correlated with the weight of raw silk. The highest content of Cl was in the Hungarian genotype BE 1264.2.

Manganese (Mn) is essential for the synthesis of chlorophyll and the hydrolysis of water in photosynthesis, and its principal function is to activate some of the enzyme systems in plant physiology and regulation of Fe metabolism. In addition, it has a close relation with N metabolism, assimilation of carbohydrates, and formation of ascorbate. It is involved in redox processes and electron transport systems [82]. Similar to Fe, Mn has the potential to enhance larval development, filament length, cocoon weight, and yield [90]. Mn was the highest in Hungarian genotype BA 2225 (105 mg/kg DW).

Iron (Fe) is present in the chloroplast proteins and several enzymes. It plays a dominant role in protein metabolism and N fixation [82]. Fe has the potential to enhance larval (silkworm) development, filament length of a single cocoon, cocoon weight, and yield [90]. The altered Fe content in mulberry foliage resulted in reduced larval weight, cocoon weight, and silk filament length [89]. In the presented experiment, Fe correlated strongly with telette. The highest content of Fe was found in genotype GMS 2533.

Zinc (Zn) correlates strongly with silk filament length and pupal weight, whereas the excess Zn content in mulberry leaves leads to a reduction in cocoon yield [84,90]. In our experiment, the highest Zn content was found in Slovenian genotype SM6. However, no correlative relationship with Zn was detected in our analysis. We further found a negative correlation between Rb and silkworm weight and a positive correlation between Sr and telette.

In the presented experiment, significant differences between the mulberry groups in the contents of the main elements were determined for S, Mn, Fe, and Sr, which were based on the PCA analysis considered as important markers in the selection of mulberry feed source.

4.3. Correlations between Test Parameters in Multivariate Analysis

The foliar protein, phenolics, and mineral composition of mulberry varieties resulted in significant inter-relationship with larval, cocoon, and silk thread parameters when their leaves served as feeds. This relationship between leaf composition values and important silkworm and raw silk traits has been worked out through correlation and multivariate (PCA) analysis.

Function 1 clearly separated reference varieties, which were characterized by higher larval weight, which coincided with high values of chlorogenic acid, rutin, and other quercetin glycosides, from Slovenian and Hungarian varieties with higher in proteins, length, and weight of raw silk. The second discriminant function was positively associated with cocoon weight, length of raw silk, total phenolics, Mn, and K, and negatively with Ca and quercetin glucoside. Function 2 clearly separated Hungarian mulberries (by over 87%) from Slovenian genotypes.

Positive correlation of coumaroylquinic acid derivatives, certain flavonols, phosphorus, sulphur, Cl, Ca, Mn, Fe, Ni and Rb were obtained with larval, cocoon, and silk thread parameters, whereas caffeoylquinic acid derivatives affected only the length of raw silk. Therefore, it is likely that the levels of these bioactive compounds and elements in mulberry leaves are important feed markers (basis parameters) to the gains on important mulberry silkworm parameters when these leaves served as feed sources.

4.4. Development and Health Status of Larvae

The measurement of body weight gain was used within the experiment to monitor larval development in the different groups. Bodyweight was most successfully increasing in groups of larvae fed with reference varieties of mulberry trees, followed by fruit varieties closely together with the Hungarian genotypes. The slowest gain of bodyweight and lowest weight on day seven of the fifth instar were observed by larvae fed with Slovenian genotypes of mulberry trees. Nonetheless, larvae with lower body weights at this instar started spinning approximately one day later. The additional one-day eating mainly compensated for their backlogs and they have started spinning with approximately the same weight as the other ones.

A correlation was observed between larval body weight in the last days of fifth instar and the weight of the fresh cocoons; however, differences between the averages of sub-groups of cocoons within Slovenian, Hungarian, reference, and fruit variety groups were only 5% (2.14–2.25 g of average fresh cocoon weights). The highest proportion of heavier cocoons was produced in the Hungarian groups. Additionally, the best silk thread parameters (length, weight, and thickness) were also described by cocoons from Hungarian groups, while most cocoon values from reference sericultural varieties and Slovenian groups were positioned within intermediate values. The lowest values of cocoon weight and silk thread parameters were described by fruit groups. Larvae fed with leaves of *M. nigra* showed the worst possible performance, as none of them started spinning (despite rapid larval bodyweight gains).

The general health status of silkworm groups was sufficient, as more than 99% of larvae reached the spinning stage healthy and started to spin (i.e., only three larvae out of 1270 died). By some (about 30) of larvae in the last day or two before spinning some common, general, mild signs of disease were observed: inactivity, cessation of eating, laying on the side of the rearing box, transparent skin, yellow/ivory colour. However, differentiation between signs of disease and the physiological changes connected to preparation for spinning was not obvious.

Within larvae that failed spinning BmNPV DNA was not detected, and all other randomly sampled larvae were also negative. However, BmNPV DNA was detected already in a dead and in a cachectic/ill larva in the third instar, though with very low viral DNA content (cT values > 40). BmNPV DNA was also detected in a dead larva after the third moult and in three ill/dead larvae in the fifth instar. Two of these larvae contained high amounts of BmNPV (cT 18–20). These molecular data indicate that BmNPV infection was present in (some of the) the larvae from the beginning of their life. Silkworm rearing has never been performed in the place of the experiment, and there is no known alternative, wild insect hosts of BmNPV, so the infection could not come from the environment, fomites, or contaminated leaves. Nevertheless, the virus amount in the vast majority of the larvae remained under detectable levels, and the larvae stayed healthy.

Whole genome sequencing of silkworm body tissue and 16S rRNA gene sequencing of bacteria in guts did not reveal DNA of BmDNV or relevant amounts of facultatively pathogenic bacteria (e.g., *Enterococcus (Streptococcus) faecalis*, *E. (S.) faecium*, Staphylococcal species, *Serratia marcescens*); however, there were considerable differences in proportions of detected sequences on the level of bacterial classes and orders between the groups (detailed analysis will be described elsewhere). Attempts on the identification of probiotic components of *B. mori* gut microbiota were reviewed by Barretto et al.: studies indicated the impact of Lactobacillus, Enterococcus, and Bacillus spp. as major gut microbiota components [91]. Besides competitive and antimicrobial effects on enteric pathogens, the probiotic effects of Actinobacteria (e.g., Actinomycetales, Bifidobacteriaceae) and Betaproteobacteria (e.g., Burkholderiales) contribute to the digestion with enzymes (e.g., protease, amylase, and lipase production) [91]. In the experimental groups, the highest abundance of DNA sequences from Burkholderiales was detected, followed by Actinomycetales and Bacillales. Lactobacillales (including Streptococcaceae, Lactobacillaceae, and Enterococcaceae) were found in the lowest amounts. The relative amounts of these detected sequences were higher in the reference, *M. australis* and *M. nigra* groups, however, it was mainly attributed to the high amounts of chloroplast-related sequences in the groups of Slovenian and Hungarian varieties. When chloroplast DNA was excluded from the analysis, the relative abundance of bacterial groups with suspected probiotic effects was similar in the different groups.

Although the trees were cultured in the same collection (i.e., in a common microbial environment), the lower relative amounts of Burkholderiales, and higher relative amounts of Bacillales, Lactobacillales, and Bifidobacteriaceae were detected in the gut microbiome of larvae fed with *M. australis*, compared to the other four groups. No clear connection was identified between the gut microbiome compositions, larval bodyweight gain, cocoon production, and quality in this study.

The mean total haemocyte counts (THC) in the different groups ranged between 45.9 and 74.3 × 104 cells/mL; which is considerably higher than THCs reported by Nematollahian et al. [92]. However, the haemocyte subsets (based on cell morphology) were found similar. No significant differences were found in the haemocyte subset rations between the different feeding groups. Neither was any correlation identified between THCs, haemocyte subset ratios and larval bodyweight gain, cocoon production, and quality in this study.

5. Conclusions

The results of the present investigation showed that mulberry varieties of local genetic origin as compared to reference sericultural and fruit varieties showed wide qualitative and quantitative variation in chemical traits with respect to proteins, phenolics, and minerals.

As a result, positive correlations of total proteins and kaempferol derivatives were obtained with silk thread parameters (i.e., length, weight). Coumaric acid correlated negatively with raw silk length, whereas 5-coumaroylquinic acid with larval weight. In addition, a positive correlation was found between Cl and raw silk weight, whereas Rb had a negative correlation with larval weight. Fe and Sr correlated positively with reeling waste.

Hence, the present study reveals that the selection of mulberry varieties out of the local gene pool for rearing silkworms based up on foliar protein, specific phenolics and mineral constituents of the mulberry varieties is very important to optimise larval development, cocoon production, and raw silk parameters. However, more research should be carried out to support the current findings in consideration of varying periods of leaf picking and nutrient analysis, pruning management, and field performances of mulberry varieties in different regions when using these varieties as feed sources. Although larvae fed with reference varieties were quickest reaching the final body weight, individual Hungarian genotypes (BA2151, SO 1013, TO1131) showed promising results as the mean larval weight was the highest. Furthermore, we were able to recognize Slovenian and Hungarian varieties which gave superior raw silk parameters. The lowest was obtained when larvae were fed with the reference varieties 'Giazzola' and some fruit genotypes. When analysing the reeling wastes, the highest was obtained when larvae were fed with Hungarian varieties.

Besides the above results, it is important to consider that some of the reference sericultural and fruit varieties of the mulberry trees are starting to develop earlier in the season than the Slovenian and Hungarian genotypes, which is a risk factor for increased losses due to early spring freezes, a relatively frequent climatic condition in these countries. So later spring development of local genotypes of Slovenian and Hungarian trees and the resistance to freezing can compensate for the bigger leaf yields of reference sericultural varieties in field conditions.

While Saxena et al., in sericulture worldwide reported annual losses of almost 20% of potential cocoon production, our experiment demonstrates that the sustainable production of quality silk cocoons is possible in Slovenia and Hungary providing the selection of superior local genotypes and suitable, locally adapted rearing technology is applied [93].

Supplementary Materials: The following supporting information can be downloaded at: https://www.mdpi.com/article/10.3390/insects13090836/s1, Table S1: List of local mulberry genotypes form Slovenia and Hungary planted in the mulberry collection of the Faculty of Agriculture University Maribor with location, geographic coordinates and detailed data along with the list and specification of reference sericultural and fruit varieties; Table S2: The mean concentrations of the total proteins, total phenolics and individual phenolics (mg/g DW) in leaves of Slovenian, Hungarian old mulberry genotypes, reference sericultural and fruit varieties. Among predominant phenolics different letters (a–n) indicate significant differences ($p < 0.05$), which were determined using the post hoc Duncan test; Table S3: The mean concentrations of the macro- and micronutrients in leaves of Slovenian Hungarian old mulberry genotypes, reference sericultural and fruit varieties; Table S4: Differences in mean silkworm weight, fresh cocoon weight and silk thread parameters among Slovenian, Hungarian mulberry varieties, reference sericultural varieties and fruit varieties. *n*, number of repetition; Table S5 PCA scores of the main PC1 and PC2 axes.

Author Contributions: Conceptualization, R.L.B., A.U.K. and T.B.; methodology, A.U.K., R.L.B and T.B.; software, A.U.K., N.S., T.B.; validation, R.L.B., T.B., A.U.K., I.A., E.K., N.S., P.P.; formal analysis, A.U.K., T.B., I.A., N.S., P.P., R.L.B.; investigation, R.L.B., T.B., A.U.K., I.A., E.K., P.P.; data curation, R.L.B., T.B., I.A., E.K., N.S., P.P.; writing—original draft preparation, R.L.B., A.U.K. and T.B. writing—review and editing, T.B., A.U.K., R.L.B., I.A., E.K.; visualization, A.U.K., T.B., N.S., I.A., E.K. funding acquisition, R.L.B., A.U.K. All authors have read and agreed to the published version of the manuscript.

Funding: This research was funded by Hungarian Scientific Research Fund (OTKA SNN 116993) and Slovenian Research Agency (ARRS, N1-0041, P-0164, P1-0212, J4-3091).

Institutional Review Board Statement: Not applicable.

Data Availability Statement: Nucleotide sequence data were deposited in the NCBI SRA database under the BioProject accession number PRJNA879405. Raw data of chemical and microbiological analyses will be provided by the authors upon request.

Acknowledgments: The support of Silvia Cappellozza and Alessio Saviane (CREA-AA) through the provision of silkworm hybrids, analysis of cocoon and raw silk parameters and scientific advice is highly appreciated and acknowledged. Mateja Šelih and Staš Miljuš are acknowledged for assistance in mulberry leaf sampling and feeding experiment. The authors would further like to thank Špela Jelen and Maja Mikulič Petkovšek for assistance in phenolic analyses. Johannes Rabensteiner and Anton Ivančič are acknowledged for assistance in mulberry collection maintaining and valuable suggestions during the research. The authors also thank the reviewers for valuable comments and suggestions.

Conflicts of Interest: The authors declare no conflict of interest. The funders had no role in the design of the study; in the collection, analyses, or interpretation of data; in the writing of the manuscript, or in the decision to publish the results.

References

1. Inserco 2022, International Sericultural Commission Web Page. Available online: https://www.inserco.org/en/ (accessed on 13 September 2022).
2. Žontar, J. *Svilogojstvo in Svilarstvo na Slovenskem od 16. do 20. Stoletja*; SAZU, Razred za Zgodovinske in Družbene Vede: Ljubljana, Slovenia, 1957; pp. 1–160.
3. Žontar, J. Gojitev sviloprejk. In *Gospodarska in Družbena Zgodovina Slovencev: Zgodovina Agrarnih Panog; I. Zvezek Agrarno Gospodarstvo*; Blaznik, P., Grafenauer, B., Vilfan, S., Eds.; DZS: Ljubljana, Slovenia, 1970; pp. 409–415.
4. Papp, Z. A selyemhernyó-tenyésztés történeti áttekintése és újrahonosításának időszerűsége. *Erdészeti Lapok* **2000**, *85*, 1–12.
5. Tzenov, P.; Cappellozza, S.; Saviane, A. Black, Caspian Seas and Central Asia Silk Association (BACSA) for the Future of Sericulture in Europe and Central Asia. *Insects* **2022**, *13*, 44. [CrossRef] [PubMed]
6. Urbanek Krajnc, A.; Ugulin, T.; Paušič, A.; Rabensteiner, J.; Bukovac, V.; Mikulič Petkovšek, M.; Janžekovič, F.; Bakonyi, T.; Berčič, R.L.; Felicijan, M. Morphometric and biochemical screening of old mulberry trees (*Morus alba* L.) in the former sericulture region of Slovenia. *Acta Soc. Bot. Pol.* **2019**, *88*, 3614. [CrossRef]
7. Šelih, M.; Mikulič Petkovšek, M.; Krajnc, D.; Berčič, R.L.; Urbanek Krajnc, A. Screening of leaf metabolites in historical mulberry trees (*Morus alba* L.) from different eco-geographical regions of Slovenia 2020. *Trees* **2020**, *34*, 971–986. [CrossRef]
8. Bradford, M.M. A rapid and sensitive method for quantitation of icrogram quantities of protein utilizing the principle of proteindye binding. *Anal. Biochem.* **1976**, *72*, 248–254. [CrossRef]
9. Ainsworth, E.A.; Gillespie, K.M. Estimation of total phenolic content and other oxidation substrates in plant tissues using Folin–Ciocalteu reagent. *Nat. Protoc.* **2007**, *2*, 875–877. [CrossRef]
10. Nečemer, M.; Kump, P.; Ščančar, J.; Jaćimović, R.; Simčič, J.; Pelicon, P.; Budnar, M.; Jeran, Z.; Pongrac, P.; Regvar, M.; et al. Application of X-ray fluorescence analytical techniques in phytoremediation and plant biology studies. *Spectrochim. Acta Part B* **2008**, *63*, 1240–1247. [CrossRef]
11. Nečemer, M.; Kump, P.; Vogel-Mikuš, K. Energy Dispersive X-ray Fluorescence Analysis of Biological Materials. In *X-ray Fluorescence in Biological Sciences: Principles, Instrumentation, and Applications*; Singh, V.K., Kawai, J., Tripathi, D.K., Eds.; John Wiley & Sons: Newark, NJ, USA, 2022; pp. 311–325. [CrossRef]

12. Kump, P.; Nečemer, M.; Rupnik, Z.; Pelicon, P.; Ponikvar, D.; Vogel-Mikuš, K. Improvement of the XRF quantification and enhancement of the combined applications by EDXRF and micro-PIXE. In *Integration of Nuclear Spectrometry Methods as a New Approach to Material Research*; International Atomic Energy Agency: Vienna, Austria, 2012; pp. 101–109. Available online: https://www.iaea.org/publications/8423/integration-of-nuclear-spectrometry-methods-as-a-new-approach-to-material-research (accessed on 13 September 2022).
13. Brion, H. *Praxishandbuch Seidenraupen-Aufzucht*, 2nd ed.; Vereinigung Schweizer Seidenproduzenten: Wohlen bei Bern, Switzerland, 2016; p. 19.
14. Cappellozza, S. (Council for Agricultural Research and Economics, Research Center for Agriculture and Environment, Padua, Italy). Personal communication, 2015.
15. Gani, M.; Chouhan, S.; Lal, B.; Gupta, R.K.; Khan, G.; Kumar, N.B.; Saini, P.; Ghosh, M.K. *Bombyx mori* nucleopolyhedrovirus (BmBPV): Its impact on silkworm rearing and management strategies. *J. Biol. Control* **2017**, *31*, 189–193. [CrossRef]
16. Khurad, A.M.; Mahulikara, A.; Rathod, M.K.; Rai, M.M.; Kanginakudru, S.; Nagaraju, J. Vertical transmission of nucleopolyhedrovirus in the silkworm, *Bombyx mori* L. *J. Invertebr. Pathol.* **2004**, *87*, 8–15. [CrossRef]
17. Zhang, J.; Kobert, K.; Flouri, T.; Stamatakis, A. PEAR: A fast and accurate Illumina Paired-End reAd mergeR. *Bioinformatics* **2013**, *30*, 614–620. [CrossRef] [PubMed]
18. Bolger, A.M.; Lohse, M.; Usadel, B. Trimmomatic: A flexible trimmer for Illumina sequence data. *Bioinformatics* **2014**, *30*, 2114–2120. [CrossRef]
19. Rognes, T.; Flouri, T.; Nichols, B.; Quince, C.; Mahé, F. VSEARCH: A versatile open source tool for metagenomics. *PeerJ* **2016**, *4*, e2584. [CrossRef] [PubMed]
20. Wood, D.E.; Lu, J.; Langmead, B. Improved metagenomic analysis with Kraken 2. *Genome Biol.* **2019**, *20*, 257. [CrossRef] [PubMed]
21. Pruitt, K.D.; Tatusova, T.; Maglott, D.R. NCBI Reference Sequence (RefSeq): A curated non-redundant sequence database of genomes, transcripts and proteins. *Nucleic Acids Res.* **2005**, *33*, D501–D504. [CrossRef] [PubMed]
22. McDonald, D.; Price, M.N.; Goodrich, J.; Nawrocki, E.P.; DeSantis, T.Z.; Probst, A.; Andersen, G.L.; Knight, R.; Hugenholtz, P. An improved Greengenes taxonomy with explicit ranks for ecological and evolutionary analyses of bacteria and archaea. *ISME J.* **2012**, *6*, 610–618. [CrossRef] [PubMed]
23. R Core Team. *R: A Language and Environment for Statistical Computing*; R Foundation for Statistical Computing: Vienna, Australia, 2021; Available online: http://www.R-project.org/ (accessed on 13 September 2022).
24. McMurdie, P.J.; Holmes, S. phyloseq: An R package for reproducible interactive analysis and graphics of microbiome census data. *PLoS ONE* **2013**, *8*, e61217. [CrossRef]
25. Nittono, Y.; Tomabechi, S.; Onodera, N. Formation of haemocytes near the imaginal wingdisc in the silkworm, *Bombyx mori* L. *J. Sericult. Sci. Jpn.* **1964**, *33*, 43–45. [CrossRef]
26. Yamashita, M.; Iwabuchi, K. *Bombyx mori* prohemocyte division and differentiation in individual microcultures. *J. Insect Physiol.* **2001**, *47*, 325–331. [CrossRef]
27. Lee, Y.W. *Silk Reeling and Testing Manual*; FAO Agricultural Services Bulletin, No. 136; Food and Agriculture Organization of the United Nations: Rome, Italy, 1999; p. 139.
28. Hammer, Ø.; Harper, D.A.T.; Ryan, P.D. PAST: Paleontological Statistics Software Package for Education and Data Analysis. *Palaeontol. Electron.* **2001**, *4*, 9. Available online: https://palaeo-electronica.org/2001_1/past/issue1_01.htm (accessed on 13 September 2022).
29. Andadari, L.; Minarningsih, M.; Suwandi. The effect of feeding various species of mulberry (*Morus* spp.) on the growth of silkworm and quality of cocoon hybrid BS 09. *IOP Conf. Ser. Earth Environ. Sci.* **2021**, *914*, 012017. [CrossRef]
30. Kumar, R.V.; Chauhan, S.; Kumar, D.; More, N. Nutritional composition in leaves of some mulberry varieties: A comparative study. In Proceedings of the International Conference of Bioinformatics and Biomedical Technology, Chengdu, China, 16–18 April 2010; pp. 438–442. [CrossRef]
31. Radjabi, R. Effect of mulberry leaves enrichment with amino acid supplementary nutrients on silkworm, *Bombyx mori* L. at north of Iran. *Acad. J. Entomol.* **2010**, *3*, 45–51.
32. Wang, C.; Yang, F.; Wang, Q.; Zhou, X.; Xie, M.; Kang, P.; Wang, Y.; Peng, X. Nutritive Value of Mulberry Leaf Meal and its Effect on the Performance of 35-70-Day-Old Geese. *Int. J. Poult. Sci.* **2016**, *54*, 41–46. [CrossRef] [PubMed]
33. Rajabi, K.; Ebadi, R.; Mirhosseini, R.; Seidavi, S.Z.; Zolfaghari, M.; Etebari, K. A review on nutritive effect of mulberry leaves enrichment with vitamins on economic traits and biological parameters of silkworm *Bombyx mori* L. *Invert. Surviv. J.* **2007**, *4*, 86–91.
34. Mendonça, G.A.; de Marchini, L.C.; Macedo, L.P.M. Cocoon production of the silkworm, *Bombyx mori* L. (Lepidoptera: Bombycidae), fed on leaves of mulberry hybrids. *Caatinga* **2010**, *23*, 118–122.
35. El-Banna, A.A.; Moustafa, M.N.; Mahmoud, S.M.; El-Shafei, A.M.; Moustafa, A.A. Effect of feeding different mulberry varieties on some biological characteristics of the silkworm, *Bombyx mori* L. *Egypt J. Pure Appl. Sci.* **2013**, *51*, 55–60. [CrossRef]
36. Adeduntan, S. Influence of different varieties of mulberry leaves (*Morus alba*) on growth and cocoon performance of biovoltine strain of silkworm (*Bombyx mori*). *Int. J. Biol. Chem. Sci.* **2015**, *9*, 751–757. [CrossRef]
37. Radojković, M.M.; Zeković, Z.P.; Vidović, S.S.; Kočar, D.D.; Mašković, P.Z. Free radical scavenging activity and total phenolic and flavonoid contents of mulberry (*Morus* spp. L., Moraceae) extracts. *Hem. Ind.* **2012**, *66*, 552. [CrossRef]

38. Chung, H.I.; Kim, J.; Kim, J.Y.; Kwon, O. Acute intake of mulberry leaf aqueous extract affects postprandial glucose response after maltose loading: Randomized double-blind placebo-controlled pilot study. *J. Funct. Foods* **2013**, *5*, 1502–1506. [CrossRef]
39. Hunyadi, A.; Liktor-Busa, E.; Márki, A.; Martins, A.; Jedlinszki, N.; Hsieh, T.J.; Báthori, M.; Hohmann, J.; Zupkó, I. Metabolic Effects of Mulberry Leaves: Exploring Potential Benefits in Type 2 Diabetes and Hyperuricemia. *Evid.-Based Complement. Altern. Med.* **2013**, *2013*, 948627. [CrossRef]
40. Jeszka-Skowron, M.; Flaczyk, E.; Jeszka, J.; Krejpcio, Z.; Król, E.; Buchowski, M.S. Mulberry leaf extract intake reduces hyperglycaemia in streptozotocin (STZ)-induced diabetic rats fed high-fat diet. *J. Funct. Foods* **2014**, *8*, 9–17. [CrossRef]
41. Chen, H.; He, X.; Liu, Y.; Li, J.; He, Q.; Zhang, C.; Wei, B.; Zhang, Y.; Wang, J. Extraction, purification and anti-fatigue activity of γ-aminobutyric acid from mulberry (*Morus alba* L.) leaves. *Sci. Rep.* **2016**, *6*, 18933. [CrossRef] [PubMed]
42. Gryn-Rynko, A.; Bazylaka, G.; Olszewska-Slonina, D. New potential phytotherapeutics obtained from white mulberry (*Morus alba* L.) leaves. *Biomed. Pharmacother.* **2016**, *84*, 628–636. [CrossRef]
43. Sugiyama, M.; Katsube, T.; Koyama, A.; Itamura, H. Seasonal Changes in Functional Component Contents in Mulberry (*Morus alba* L.) Leaves. *Hortic. J.* **2017**, *86*, 534–542. [CrossRef]
44. Ju, W.-T.; Kwon, O.-C.; Kim, H.-B.; Sung, G.-B.; Kim, H.-W.; Kim, Y.-S. Qualitative and quantitative analysis of flavonoids from 12 species of Korean mulberry leaves. *J. Food Sci. Technol.* **2018**, *55*, 1789–1796. [CrossRef] [PubMed]
45. Ju, W.-T.; Kwon, O.-C.; Lee, M.-K.; Kim, H.-B.; Sung, G.-B.; Kim, Y.-S. Quali-Quantitative Analysis of Flavonoids for Mulberry Leaf and Fruit of 'Suhyang'. *Korean J. Environ. Agric.* **2017**, *36*, 249–255. [CrossRef]
46. Memon, A.A.; Memon, N.; Luthria, D.L.; Bhanger, M.I.; Pitafi, A.A. Phenolic acids profiling and antioxidant potential of mulberry (*Morus laevigata* W., *Morus nigra* L., *Morus alba* L.) leaves and fruits grown in Pakistan. *Pol. J. Food Nutr. Sci.* **2010**, *60*, 25–32.
47. Zou, Y.; Liao, S.; Shen, W.; Liu, F.; Tang, C.; Chen, C.-Y.O.; Sun, Y. Phenolics and Antioxidant Activity of Mulberry Leaves Depend on Cultivar and Harvest Month in Southern China. *Int. J. Mol. Sci.* **2012**, *13*, 16544–16553. [CrossRef]
48. Iqbal, S.; Younas, U.; Sirajuddin; Chan, K.W.; Sarfraz, R.A.; Uddin, K. Proximate Composition and Antioxidant Potential of Leaves from Three Varieties of Mulberry (*Morus* sp.): A Comparative Study. *Int. J. Mol. Sci.* **2012**, *13*, 6651–6664. [CrossRef]
49. Lee, W.J.; Choi, S.W. Quantitative Changes of Polyphenolic Compounds in Mulberry (*Morus alba* L.) Leaves in Relation to Varieties, Harvest Period and Heat Processing. *Prev. Nutr. Food Sci.* **2012**, *17*, 280–285. [CrossRef]
50. Thabti, I.; Elfalleh, W.; Hannachi, H.; Ferchichi, A.; Campos, M.D.G. Identification and quantification of phenolic acids and flavonol glycosides in Tunisian Morus species by HPLC-DAD and HPLC–MS. *J. Funct. Foods* **2012**, *4*, 367–374. [CrossRef]
51. Flaczyk, E.; Kobus-Cisowska, J.; Przeor, M.; Korczak, J.; Remiszewski, M.; Korbas, E.; Buchowski, M. Chemical characterization and antioxidative properties of Polish variety of *Morus alba* L. leaf aqueous extracts from the laboratory and pilot-scale processes. *J. Agric. Sci.* **2013**, *4*, 141–147. [CrossRef]
52. Choi, S.W.; Lee, Y.J.; Ha, S.B.; Jeon, Y.H.; Lee, D.H. Evaluation of Biological Activity and Analysis of Functional Constituents from Different Parts of Mulberry (*Morus alba* L.) Tree. *J. Korean Soc. Food Sci. Nutr.* **2015**, *6*, 823–831. [CrossRef]
53. Germanò, M.P.; D'Angelo, V.; Catania, S.; Miano, T.C.; Perna, V.; Farago, S.; Cappellozza, L.; Cappellozza, S. Phenolic content of leaf of different mulberry cultivars affects growth in the silkworm. In Proceedings of the 21st International Sericultural Congress of ISC, Athens, Greece, 3–6 November 2008; pp. 25–29.
54. Hirayama, C.; Ono, H.; Tamura, Y.; Konno, K.; Nakamura, M. Regioselective formation of quercetin 5-O-glucoside from orally administered quercetin in the silkworm, *Bombyx mori*. *Phytochemistry* **2008**, *69*, 1141–1149. [CrossRef] [PubMed]
55. Daimon, T.; Hirayama, C.; Kanai, M.; Ruike, Y.; Meng, Y.; Kosegawa, E.; Nakamura, M.; Tsujimoto, G.; Katsuma, S.; Shimada, T. The silkworm *Green b* locus encodes a quercetin 5-*O*-glucosyltransferase that produces green cocoons with UV-shielding properties. *Proc. Natl. Acad. Sci. USA* **2010**, *107*, 11471–11476. [CrossRef]
56. Vihakas, M. Flavonoid and Other Phenolic Compounds: Characterization and Interactions with Lepidopteran and Sawfly Larvae. Doctoral Thesis, University of Turku, Turku, Finland, 2014.
57. Kusurkar, T.S.; Tandon, I.; Sethy, N.K.; Bhargava, K.; Sarkar, S.; Singh, S.K.; Das, M. Fluorescent silk cocoon creating fluorescent diatom using a "Water glass-fluorophore ferry". *Sci. Rep.* **2013**, *3*, 3290. [CrossRef] [PubMed]
58. Shimada, T. Basic and Applied Genetics of Silkworms. In *Silkworm Biofactory; Silk to, Biology*; Maenaka, K., Park, E.Y., Eds.; CRC Press: Boca Raton, FL, USA, 2018; p. 15. [CrossRef]
59. Srivastava, S.; Kapoor, R.; Thathola, A.; Srivastava, R.P. Nutritional quality of leaves of some genotypes of mulberry (*Morus alba*). *Int. J. Food Sci. Nutr.* **2006**, *57*, 305–313. [CrossRef]
60. Al-kirshi, R.A.; Alimon, A.R.; Zulkifli, I.; Zahari, M.W.; Sazili, A.Q. The chemical composition and nutritive value of mulberry leaf as a protein source in poultry diets. In *Feed and Nutrition: The 1st International Seminar on Animal Industry, Bogor, Indonesia*; Bogor Agricultural University: Bogor, Indonesia, 2009; pp. 98–102.
61. Kumar, K.; Mohan, M.; Tiwari, N.; Kumar, S. Production potential and leaf quality evaluation of selected mulberry (*Morus alba*) clones. *J. Pharmacogn. Phytochem.* **2018**, *7*, 482–486.
62. Sánchez-Salcedo, E.M.; Mena, P.; García-Viguera, C.; Martínez, J.J.; Hernández, F. Phytochemical evaluation of white (*Morus alba* L.) and black (*Morus nigra* L.) mulberry fruits, a starting point for the assessment of their beneficial properties. *J. Funct. Foods* **2015**, *12*, 399–408. [CrossRef]
63. Yamagishi, T.; Endo, H.; Fukumura, K.; Nagata, S.; Hayakawa, T.; Adegawa, S.; Kasubuchi, M.; Sato, R. Glucose, some amino acids and a plant secondary metabolite, chlorogenic acid induce the secretion of a regulatory hormone, tachykinin-related peptide, from the silkworm midgut. *Peptides* **2018**, *106*, 21–27. [CrossRef]

54. Kato, M.; Yamada, H. Chlorogenic acid as a growth factor of silkworm. *J. Silkworm* **1964**, *16*, 85. [CrossRef]
55. Naito, K.; Hayashiya. Studies on the micro constituents in mulberry leaves. IV. Isolation of chlorogenic acid. *J. Agric. Chem. Soc. Jpn.* **1965**, *39*, 237–238.
56. Hamamura, Y.; Kuwata, K.; Masuda, H. Effect of Gallic Acid on the Growth of the Silkworm Larvae *Bombyx mori* L. *Nature* **1966**, *212*, 1386–1387. [CrossRef] [PubMed]
57. Yamada, H.; Kato, M. Chlorogenic Acid as an Indispensable Component of the Synthetic Diet for the Silkworm. *Proc. Jpn. Acad.* **1966**, *42*, 399–403. [CrossRef]
58. Koike, S.; Iizuka, T.; Mizutani, J. Determination of Caffeic Acid in the Digestive Juice of Silkworm Larvae and Its Antibacterial Activity against the Pathogenic Streptococcus faecalis AD-4. *Agric. Biol. Chem.* **1979**, *43*, 1727–1731. [CrossRef]
59. Naik, D.; Patil, G.M.; Biradar, R. Impact of fortification of chlorogenic acid rich botanical on silkworm growth and cocoon parameters of silkworm *Bombyx mori* L. *Plant Arch.* **2017**, *17*, 661–665.
70. Jarienė, E.; Levickienė, D.; Danilčenko, H.; Vaitkevičienė, N.; Kulaitienė, J.; Jakštas, V.; Ivanauskas, L.; Gajewski, M. Effects of biodynamic preparations on concentration of phenolic compounds in the leaves of two white mulberry cultivars. *Biol. Agric. Hortic.* **2018**, *35*, 132–142. [CrossRef]
71. Pothinuch, P.; Tongchitpakdee, S. Phenolic Analysis for Classification of Mulberry (*Morus* spp.) Leaves according to Cultivar and Leaf Age. *J. Food Qual.* **2019**, *2019*, 2807690. [CrossRef]
72. Kurioka, A.; Yamazaki, M. Purification and Identification of Flavonoids from the Yellow Green Cocoon Shell (Sasamayu) of the Silkworm, *Bombyx mori*. *Biosci. Biotechnol. Biochem.* **2002**, *66*, 1396–1399. [CrossRef]
73. Kurioka, A.; Yamazaki, M. Antioxidant in the cocoon of the silkworm, *Bombyx mori*. *J. Insect Biotechnol. Sericol.* **2002**, *71*, 177–180. [CrossRef]
74. Preiß, S.; Degenhardt, J.; Gershenzon, J. Plant animal dialogues. In *Ecological Biochemistry: Environmental and Interspecies Interactions*; Krauss, G.J., Nies, D.H., Eds.; Wiley-Blackwell: Oxford, UK, 2015; pp. 313–330. ISBN 978-3-527-31650-2.
75. Zhang, Y.-E.; Ma, H.-J.; Feng, D.-D.; Lai, X.-F.; Chen, Z.-M.; Xu, M.-Y.; Yu, Q.-Y.; Zhang, Z. Induction of Detoxification Enzymes by Quercetin in the Silkworm. *J. Econ. Entomol.* **2012**, *105*, 1034–1042. [CrossRef] [PubMed]
76. Enkhmaa, B.; Shiwaku, K.; Katsube, T.; Kitajima, K.; Anuurad, E.; Yamasaki, M.; Yamane, Y. Mulberry (*Morus alba* L.) Leaves and Their Major Flavonol Quercetin- 3-(6-Malonylglucoside) Attenuate Atherosclerotic Lesion Development in LDL Receptor-Deficient Mice. *J. Nutr.* **2005**, *135*, 729–734. [CrossRef] [PubMed]
77. Katsube, T.; Imawaka, N.; Kawano, Y.; Yamazaki, Y.; Shiwaku, K.; Yamane, Y. Antioxidant flavonol glycosides in mulberry (*Morus alba* L.) leaves isolated based on LDL antioxidant activity. *Food Chem.* **2006**, *97*, 25–31. [CrossRef]
78. Katsube, T.; Tsurunaga, Y.; Sugiyama, M.; Furuno, T.; Yamasaki, Y. Effect of air-drying temperature on antioxidant capacity and stability of polyphenolic compounds in mulberry (*Morus alba* L.) leaves. *Food Chem.* **2009**, *113*, 964–969. [CrossRef]
79. Katsube, T.; Yamasaki, M.; Shiwaku, K.; Ishijima, T.; Matsumoto, I.; Abe, K.; Yamasaki, Y. Effect of flavonol glycoside in mulberry (*Morus alba* L.) leaf on glucose metabolism and oxidative stress in liver in diet-induced obese mice. *J. Sci. Food Agric.* **2010**, *90*, 2386–2392. [CrossRef] [PubMed]
80. Mahadeva, A. Nutritive Elemental Status in Mulberry (*Morus* sp.) Foliage under Jassids (*Empoasca flavescenes* F.) Infestation. *Indian J. Nat. Sci.* **2016**, *7*, 11537.
81. Shifa, K.; Terefe, M.; Tilahun, A.; Ibrahim, A.; Nigusu, Y. Leaf macronutrient composition of mulberry (*Morus indica*) varieties and its relationship with productivity of mulberry silkworm (*Bombyx mori* L.). *J. Agric. Sci. Res.* **2020**, *8*, 390–398.
82. Biskin, D.P.; Bloom, A. Mineral Nutrition. In *Plant Physiology*, 5th ed.; Taiz, L., Zeiger, P.E.E., Eds.; Sinauer Associatiates Publishers: Munderland, MA, USA, 2010; pp. 107–125.
83. Ito, T.; Nimura, M. Nutrition of the silkworm *Bombyx mori* L., its specific nutrient requirements and its nutrition in relation to the mineral nutrition of its host plant mulberry, *Morus indica* L. *Indian J. Exptl. Biol.* **1966**, *4*, 31–36.
84. Chakrabarti, S.; Subramanyam, M.R.; Singhal, B.K.; Datta, R.K. Nutrient deficiency management in mulberry. In *Handbook of Mulberry Nutrition*; Shankar, M.A., Shetty, G.P., Eds.; Multiplex, Karnataka Agro Chemicals: Bangalore, India, 1997; pp. 19–75.
85. Miller, A.J. Plant Mineral Nutrition. In *eLS*; John Wiley & Sons, Ltd.: Chichester, UK, 2014. [CrossRef]
86. Munirathnam Reddy, M.; Subbaswamy, M.R.; Sinha, A.K. Sulphur nutrition in mulberry plants. *Indian Silk* **1990**, *29*, 45–46.
87. Mahadevappa, L.; Magadi, S.P.; Prabhuraj, D.K.; Thimmareddy, H.; Bongale, U.D. Available sulphur in some mulberry growing soils of Karnataka. In Nutritional Management and Quality Improvement in Sericulture, Proceedings of the National Seminar on Mulberry Sericulture Research in India, Karnataka, India, 26–28 November 2001; State Sericulture Research and Development Institute: Bangalore, India, 2001; pp. 161–165.
88. Shree, M.P.; Anuradha, R.; Nagaveni, V. Impact of rust disease on the mineral nutrition of mulberry plants. *Sericologia* **2005**, *45*, 115–121.
89. Shankar, M.A. *Handbook of Mulberry Nutrition*; Shetty, G.P., Ed.; Multiplex, Karnataka Agro Chemicals: Bangalore, India, 1997.
90. Lokanath, R.; Shivashankar, K.; Kasiviswanathan, K. Effect of foliar application of magnesium and micronutrients to mulberry on the quality and production of cocoons. *Indian J. Seric* **1986**, *24*, 40–45.
91. Barretto, D.A.; Gadwala, M.; Vootla, S.K. Chapter 1—The silkworm gut microbiota: A potential source for biotechnological applications. *Methods Microbiol.* **2021**, *49*, 1–26. [CrossRef]

92. Nematollahian, S.; Bagheri, M.; Yousefi, Z.; Khezrian, A.; Zahmatkesh, Z. Haematological Changes in *Bombyx mori* (Lepidoptera: Bombycidae) Infected by Nosema bombycis. *J. Agric. Urban Entomol.* **2021**, *37*, 38–47. [CrossRef]
93. Saxena, S.; Chandra, R.T.; Kallare, P.S.; Arunkumar, K.P. MicroRNAs in the silkworm-pathogen interactions. *Methods Microbiol.* **2021**, *49*, 97–113. [CrossRef]

insects

Article

Global Profiling of Genes Expressed in the Silk Glands of Philippine-Reared Mulberry Silkworms (*Bombyx mori*)

Pauline Nicole O. de la Peña, Adria Gabrielle D. Lao and Ma. Anita M. Bautista *

National Institute of Molecular Biology and Biotechnology, University of the Philippines Diliman, Quezon City 1101, Philippines; podelapena1@up.edu.ph (P.N.O.d.l.P.); adlao@up.edu.ph (A.G.D.L.)
* Correspondence: mmbautista20@up.edu.ph

Simple Summary: The sericulture industry in the Philippines remains to be enhanced through efforts to improve local strains of the silkworm *Bombyx mori*, such as selective hybridization. Genetic trait markers could help improve breeding; however, there is a scarcity of genetic data on local strains. This study aims to bridge this gap by analyzing the gene expression profiles of Philippine-reared silkworm strains through next-generation sequencing. Transcriptome assemblies were generated, and gene expressions were compared between silkworm strains reared in different temperatures, revealing genes that may be important for development in a tropical environment. This study is the first to provide transcriptome datasets for the Philippine-reared *B. mori* strains, which may serve as a resource for improving local strains and increasing silk production.

Abstract: RNA sequencing was used to assemble transcriptome data for Philippine-reared silkworm and compare gene expression profiles of strains reared in high- and low-temperature environments. RNA was isolated from the silk glands of fifth instar larvae and mRNA-enriched libraries were sequenced using Illumina NextSeq 500. Transcriptome reads were assembled using reference-based and de novo assemblers, and assemblies were evaluated using different metrics for transcriptome quality, including the read mapping rate, E90N50, RSEM-eval, and the presence of single-copy orthologs. All transcriptome assemblies were able to reconstruct >40,000 transcripts. Differential expression analysis found 476 differentially expressed genes (DEGs; 222 upregulated, 254 downregulated) in strains reared in different temperatures. Among the top DEGs were myrosinase, heat shock proteins, serine protease inhibitors, dehydrogenases, and regulators of the juvenile hormone. Validation of some of the top DEGs using qPCR supported the findings of the in silico analysis. GO term enrichment analysis reveals an overrepresentation of GO terms related to nucleotide metabolism and biosynthesis, lipid and carbohydrate metabolic processes, regulation of transcription, nucleotide binding, protein binding, metal binding, catalytic activity, oxidoreductase activity, and hydrolase activity. The data provided here will serve as a resource for improving local strains and increasing silk production of Philippine-reared *B. mori* strains.

Keywords: silkworm; *Bombyx mori*; RNA-seq; silk gland; transcriptome analysis

Citation: de la Peña, P.N.O.; Lao, A.G.D.; Bautista, M.A.M. Global Profiling of Genes Expressed in the Silk Glands of Philippine-Reared Mulberry Silkworms (*Bombyx mori*). *Insects* **2022**, *13*, 669. https://doi.org/10.3390/insects13080669

Academic Editor: Silvia Cappellozza

Received: 24 June 2022
Accepted: 20 July 2022
Published: 24 July 2022

1. Introduction

The mulberry silkworm *Bombyx mori* is an economically important insect for the commercial production of silk, which is highly valued for its texture and luster, and for being one of the strongest natural fibers [1]. Silk has many uses such as in garments, upholstery, parachute linings, insulation, and as suturing material in surgeries. It is also considered a promising biomaterial in tissue engineering due to its mechanical strength and biocompatibility [2].

Since sericulture was introduced to the Philippines in the 1970s, local silkworm stocks have originated from Chinese, Japanese, and European strains [3]. Previous studies determined that temperate strains produce finer and stronger silk fibers, while tropical strains are more robust and disease-resistant but produce coarser and weaker silk fibers [4]. There have been efforts to breed hybrids that are acclimatized to the Philippine climate with desired phenotypic traits such as a high hatch rate, cocoon yield, cocoon weight, and silk filament quality [5–7]. Local silk production, however, has been reported to continue lagging behind demand, with 10 metric tons of demand annually and only 1 metric ton of raw silk production [8]. Low production has been attributed to the quality of local silkworm strains and poor rearing management [9]. The low silk yield may be improved by understanding how it is produced. The molecular mechanisms behind the different silk yields in silkworm strains, for example, need to be investigated as they are not yet well-understood [10,11]. Knowledge to be generated from molecular studies may therefore help in breeding local silkworm strains with a capacity to produce high silk yield. Markers associated with high silk yield and other economically important traits may also be developed for marker-assisted selection and breeding. Thus, this endeavor would need genetic resources, which can be sourced from genome and transcriptome sequences of foreign strains of *B. mori* in public databases such as the Silkworm Genomic Database (SilkDB) [12], KAIKObase [13], SilkTransDB (transcriptome) [14], SilkSatDb (microsatellites) [15], and BmTEdb (transposable elements) [16].

While there are already several genomic resources for the silkworm, transcriptome analysis can be used to gain insight into the functional elements of the genome and its biological pathways. One such technique is RNA sequencing (RNA-seq), which utilizes next-generation sequencing to provide high-throughput and high-resolution data to analyze expressed genes in a cost-efficient manner. RNA-seq can be used to identify novel transcripts, splice variants, and single nucleotide polymorphisms [17]. It also allows quantification over a dynamic range since it detects both highly and lowly expressed genes, making it ideal for gene expression analysis [18]. Recently, RNA-seq has been used by Zhang et al. to reveal gene expression changes in silkworms exposed to hydrogen sulfide [19]. Yokoi et al. also used RNA-seq to generate reference transcriptome data for major tissues of the silkworm p50T strain [20]. However, available transcriptome resources currently lack representative sequences from Philippine-reared *B. mori* strains; hence, the need to generate such resources is deemed important.

This study aimed to generate transcriptome resources for Philippine-reared mulberry silkworms, which can be utilized in future studies for its potential industrial valorization and to address the issue of low silk yield in the Philippines. This investigation focused on genes found in the silk glands of different strains of Philippine-reared silkworm.

2. Materials and Methods

2.1. Insect and Tissue Collection

Silkworms reared at the Philippine Textile Research Institute Technology Center in Misamis Oriental (PTRI-TCMO, Villanueva, Misamis Oriental) and the Technology Center of the Department of Science and Technology Cordillera Administrative Region (DOST-CAR, Benguet) were collected in 2018. The ambient temperature in TCMO and CAR at the time of collection was 31.7 °C and 23.9 °C, respectively. For each site, priority strains (MO204, LAT51, ITA) and a poorly performing strain (DMMMSU119) were used as representative strains. The strains were classified by the rearing facilities as either priority strains or poorly performing strains based on their rearing performance (hatching ratio, larval mortality, and pupation ratio). Economic characteristics such as silk yield and quality were not considered in the classification of priority strains. Three individuals per strain were used as bioreplicates. Silkworms were collected on the 3rd day of the 5th instar larval stage and were characterized morphologically. Larvae were preserved by flash-freezing in liquid nitrogen and stored at −80 °C until ready for dissection. Larvae were dissected in a petri dish on ice to collect whole silk glands. Silk glands were chopped using sterile surgical blades and homogenized mechanically using a mortar and pestle.

2.2. RNA Extraction and Sequencing

Total RNA was extracted from 100 μg of the tissue with the TRIzol reagent (Invitrogen, Life Technologies, Carlsbad, CA, USA). Total RNA was treated with TURBO DNAse (Ambion, Life Technologies, Carlsbad, CA, USA) and purified with the RNA Clean and Concentrator kit (Zymo Research, Irvine, CA, USA). The quality of RNA extracts was visually estimated by running them in 1% agarose gel at 90 V for 40 min. RNA purity was estimated by measuring the 260/280 and 260/230 absorbance ratios in a NanoDrop 2000/2000c Spectrophotometer (Thermo Fisher Scientific, Wilmington, DE, USA), and RNA was quantified by fluorometry using a Qubit RNA BR Assay kit (Thermo Fisher Scientific) according to the manufacturer's instructions. Finally, extracts were analyzed in the Agilent 2200 TapeStation (Agilent Technologies, Santa Clara, CA, USA) to evaluate the RNA Integrity Number (RIN). Samples with RIN ≥ 6 were used for library preparation and sequencing.

The RNA samples were diluted and used as input (0.1–4 μg) for mRNA-enrichment library preparation using the TruSeq Stranded RNA Library Prep Kit (Illumina, San Diego, CA, USA). Twelve libraries were prepared (4 strains $\times$ 3 bioreplicates) with fragment sizes ranging 131–279 bp. Libraries were denatured, diluted to 1.8 pM, and combined with a spike-in of 1.0% PhiX sequencing control. The libraries were sequenced in the Philippine Genome Center using a NextSeq 500 (Illumina, San Diego, CA, USA) platform and a NextSeq v2 high output kit (Illumina, San Diego, CA, USA) with 300 cycles (2 $\times$ 150 paired-end sequencing).

2.3. Preprocessing, Alignment, and Transcriptome Assembly

Illumina bcl2fastq (v. 2.2.0) was used to convert sequencing base call files to FASTQ format. FastQC was used to evaluate reads and visualize read quality. Raw reads were processed using Trimmomatic (0.39) and fastp (0.20.0) to trim adapter sequences and filter low-quality reads.

The splice-aware aligner STAR (2.7.0f) was used to align reads to the *B. mori* reference genome from NCBI (RefSeq GCF_000151625.1). The BAM file was processed with RNA-SeQC (1.1.9) to evaluate RNA-Seq data bias based on the total read count, coverage, and expression correlation (RPKM-based estimation of expression levels). Another aligner, HISAT (2.1.0), was used for alignment in preparation for generating the StringTie assembly. HISAT alignment files were used for transcriptome assembly evaluation and expression quantification.

Reference-based assembly was performed using Cufflinks (2.2.1) and StringTie (1.3.6). In addition to reference-based assembly, de novo assembly was also performed using Trinity (2.8.6). The percentage of aligned reads was determined using Bowtie. Contig statistics (number of transcripts, N50, mean, and median contig length) were generated using Trinitystats.pl. Assemblies were evaluated using DETONATE (1.11) to estimate their RSEM-EVAL scores. The completeness of the assemblies was evaluated using BUSCO (v. 3.0.2).

2.4. Expression Quantification and Differential Expression Analysis

FeatureCounts (1.4.6-p5) was used for gene-level quantification and DESeq2 (1.30.1) was used for gene-level differential expression analysis. Differentially expressed genes (DEGs) were filtered based on the false discovery rate (FDR)/adjusted p-value < 0.1 and $|\log_2 \text{fold change}| > 1$. Figures to visualize DEGs were generated from the plotMA function of DESeq2 and R packages EnhancedVolcano (1.8.0) and pheatmap (1.0.12).

The annotation of DEGs was performed against the NCBI nr protein database with an e-value cut-off of 1×10^{-5}. GO terms were mapped to DEGs by matching the protein IDs to the GO annotations in the newest release of the *B. mori* genome annotations from KAIKObase 4.1.0 (https://kaikobase.dna.affrc.go.jp/KAIKObase_download.html (accessed on 4 April 2021). The R package goseq (1.42.0) was used to determine GO term enrichment among DEGs. Significantly enriched GO terms were summarized and visualized using REVIGO (http://revigo.irb.hr/ (accessed on 8 April 2021).

2.5. Quantitative Real-Time PCR (qPCR)

Quantitative real-time PCR assays were carried out using the Bio-Rad CFX96™ Touch system (Bio-Rad Laboratories, Inc., Hercules, CA, USA). Each 10-μL reaction contained 5 μL of 2X iTaq Universal SYBR Green Supermix (Bio-Rad Laboratories, Inc., Hercules, CA, USA), 5 pmol each of forward and reverse primers, and 25 ng of cDNA, following a thermal profile comprising initial denaturation at 95 °C for 5 min, 40 cycles of denaturation at 95 °C for 10 s, annealing and extension at 60 °C for 30 s, and a melt curve assay from 95 °C to 65 °C in 0.5 °C stepwise increments. Primers for select genes differentially expressed between CAR and TCMO were designed using Primer-BLAST and are shown in Table 1.

Table 1. Primers used for qPCR validation of differentially expressed genes in *Bombyx mori* strains.

Primers	Sequences (5′ to 3′) Forward Reverse	Product Length (bp)	Protein ID
Downregulated genes (CAR vs. TCMO)			
Hsp 20	GCCAACGATGTCCAGAGATT CTGCCTCTCCTCGTGCTTAC	196	heat shock protein hsp20.1
odcdl	AAATGTCTGGGTGGAAGCAG GGGTGGAATCATCAGGTTGT	235	oxygen-dependent choline dehydrogenase-like
myr	GGAGACCGAGTGAAGACCTG GCTGTGGCATGAGCTAACAA	152	myrosinase 1
Upregulated genes (CAR vs. TCMO)			
pl	TGTTGTCGCTTTTCATCTGC AGTTTTTGAGCGCCGTATTG	207	peroxidase-like
LOC105842393	AAATGGCGCACAAATAGAGG CCAAGCTCTGCGTAAGGTTC	177	uncharacterized protein
ts	AAACCGAGCGTCCACTTATG TATTGTGATGGCAGCGGTAA	186	uncharacterized protein (tetraspanin family)
Housekeeping gene for normalization			
rp49	CAGGCGGTTCAAGGGTCAATAC TGCTGGGCTCTTTCCACGA	213	ribosomal protein 49

Total RNA was extracted from silkworms reared at the PTRI-TCMO and PTRI-CAR, followed by TURBO DNase treatment as previously described. RNA from two individual silkworms from each location was pooled and used to generate cDNA using the ProtoScript II First Strand cDNA Synthesis Kit (New England Biolabs, Ipswich, MA, USA) according to the manufacturer's recommendations. The cDNA was quantified by fluorometry using the Qubit ssDNA Assay kit (Thermo Fisher Scientific). Serial dilutions (10×) of pooled cDNA samples were prepared to generate a standard curve for each primer set used for the qPCR assays, starting at 100 ng of cDNA. The standard curves were used in conjunction with cycle threshold values to calculate starting quantities of the transcripts, and the expression levels and fold changes of each gene were normalized to *B. mori* ribosomal protein 49 [20]. All reactions were performed in triplicate. Calculations were performed using the Bio-Rad CFX Manager (v. 3.1). Statistical significance was determined using Student's t-test with GraphPad Prism 6.

3. Results

3.1. Selection of Representative Silkworm Strains

Representative strains from the two collection sites were selected based on their phenotypic and economic characteristics (Table 2). Each strain can be differentiated by its larval markings. MO204 and DMMMSU119 are milky white, while LAT51 and ITA are beige.

Table 2. Characteristics of silkworm strains used in the study.

Strain	Origin	Strain Classification	Cocoon	Cocoon Shell Percentage	Raw Silk Percentage
MO204	TCMO	Priority	Oval, white	22.92%	22.42%
DMMMSU119	TCMO	Poor performing	Peanut, white	21.81%	21.12%
LAT51	CAR	Priority	Peanut, white	21.57%	17.44%
ITA	CAR	Priority	Oval, yellow gold	20.27%	no data

Cocoon shell percentage: Ratio of the weight of silk shell to the weight of the whole cocoon; Raw silk percentage: Ratio of the weight of raw silk to the weight of the whole cocoon.

The rearing sites had different conditions, particularly temperature and relative humidity (RH), the most important environmental factors in silkworm rearing. During sample collection, the temperature in TCMO was higher by 7.8 ± 0.1 °C and RH was also higher by 12%.

3.2. RNA Sequencing and Quality Control

High-quality RNA was extracted (RIN $\geq$ 6), and the sequencing libraries had fragment sizes ranging from 131 to 279 bp, as determined by the Agilent TapeStation. Sequencing yielded 50.1 Gb of 169.2 M reads passing filter, with 96.1% of reads having >Q30.

The number of raw reads from RNA sequencing is shown in Table 3. All ITA bioreplicates yielded lower than the expected number of reads. Because the number of reads was deemed insufficient, ITA was no longer included in downstream analyses.

Table 3. Statistics of raw RNA-seq read sequences of *Bombyx mori* strains.

Sample	Read Count	% $\geq$ Q30	Mean Q Score	GC Content (%)
MO204-1	17,531,032	94.11	34.60	41
MO204-3	8,590,562	93.49	34.47	43
MO204-4	13,085,562	94.10	34.59	44
DMMMSU-0	11,406,238	92.68	34.29	46
DMMMSU-2	21,789,720	92.38	34.22	46
DMMMSU-4	15,464,914	91.51	34.04	46
LAT51-1	12,321,992	93.70	34.51	42
LAT51-2	22,767,422	93.91	34.56	41
LAT51-4	8,793,834	94.74	34.73	44
ITA-1	3,621,664	86.80	33.05	48
ITA-2	4,633,324	88.96	33.51	48
ITA-3	3106	65.36	28.55	75

Trimmomatic was used for trimming Illumina adapters that were ligated to tag each sample during sequencing, and for quality trimming on leading and trailing sequences. Sequences that were under 36 bp were also dropped. A sliding window of 4:20 was used, where the sequences will be trimmed if the quality drops below Q20 in a 4-base window. After processing with Trimmomatic, most sequences were still retained, indicating that only a small number of sequences were dropped due to having low-quality bases (Figure 1). After Trimmomatic processing, poly G sequences still appeared in overrepresented sequences in FastQC, particularly in forward reads. Poly G sequences are artifacts in NextSeq sequencing due to its two-color filter that calls for G bases when there is no emission detected. Poly G sequences were filtered out using fastp, after which 77.9–83.1% of the raw reads were retained.

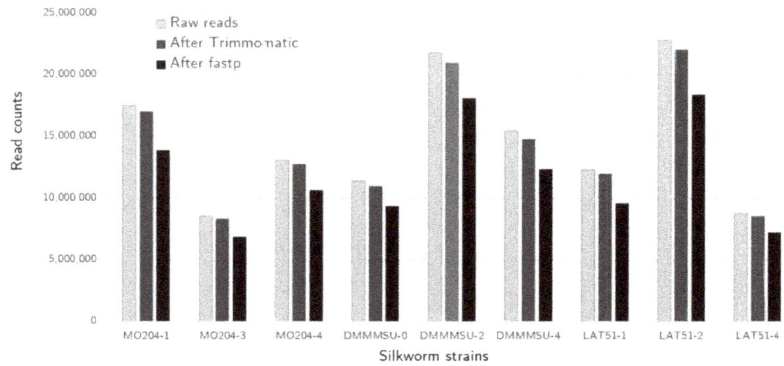

Figure 1. *Bombyx mori* raw read counts and proportion of reads after processing.

3.3. RNA-seq Read Alignment to the Reference Genome

Two samples per strain with the most read counts were aligned to the *B. mori* reference genome RefSeq GCF_000151625.1 [21] using the splice-aware aligner STAR. Most single and paired reads were uniquely mapped to the reference with only a small percentage (0.11–0.53%) of unmapped reads (Figure 2). The DMMMSU replicates had a higher percentage of reads that were mapped to multiple loci (22.73% and 14.70% for DMMMSU-2 and DMMMSU-4, respectively), while for LAT51 and MO204 replicates, more than 90% of reads were uniquely mapped.

Moreover, most of the splice sites detected by STAR were in the GTF annotation of the reference genome. It has been reported that STAR generates a large number of putative novel splice sites [22], but in this case, the splice sites were validated by the reference genome annotation.

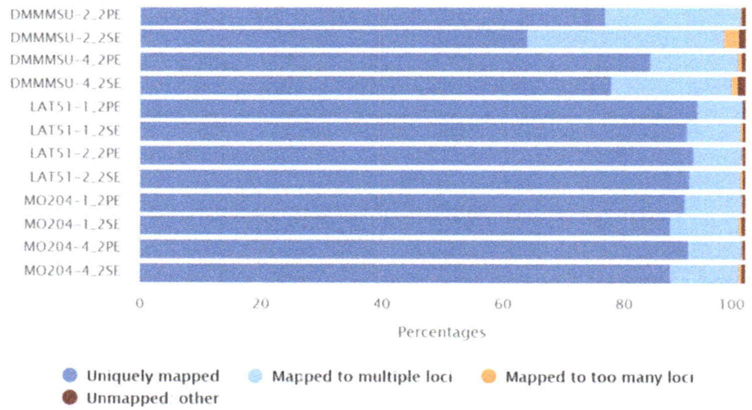

Figure 2. Summary of STAR alignment statistics of *Bombyx mori* RNA-seq reads to the reference genome. (PE: Paired-end reads, SE: Single-end reads). Generated using STAR.log files and MultiQC [23].

Samtools flagstat was also run to analyze data from the FLAG field of the generated BAM files. According to flagstat, most of the mapped reads were properly paired and mapped with their mate. No duplicates were detected, but secondary alignments were produced (ranging from 9.6 to 31%). This coincides with the STAR statistics indicating that there were several reads mapped to multiple loci (ranging from 7.43 to 22.73%). In the

BAM files of the paired-end reads, less than 1% were singletons, and 0 reads had mates mapped to a different chromosome.

3.4. Transcriptome Assembly and Evaluation Metrics

After assembling the reads using different assemblers, the assemblies were mapped back to the input reads to evaluate their read support. Figure 3 shows that the reads had the highest alignment rates with the Trinity assembly, while the Cufflinks and StringTie assemblies had comparable alignment rates. Since the reads from the DMMMSU samples had a lower percentage of uniquely mapped reads, it could be expected that the reference-based assemblies would have less support from these samples.

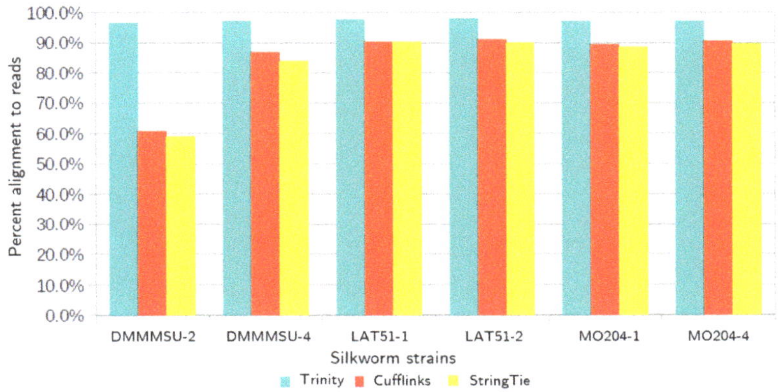

Figure 3. Overall alignment rate of RNA-seq reads to the transcriptome assemblies.

Transcriptome assemblies are also evaluated based on the number and length of unigenes (unique sequences reconstructed in the assembly; includes not only genes but all assembled contigs). According to the contig statistics in Table 4, the Trinity assembly had the lowest N50 (1313) and lowest median contig length (468). Cufflinks had the highest contiguity with the highest N50 (3414), highest median contig length (1738), and average contig length (2305.35). However, in a transcriptome assembly, long contig lengths and high N50 are not as important as in genome assemblies. The most highly expressed transcripts are not necessarily the longest ones, and the majority of transcripts are expected to have low expression levels. Transcript lengths are also not indicative of a good transcriptome assembly since transcripts may be present in a wide range of sizes.

Table 4. Contig statistics of the different transcriptome assemblies from *Bombyx mori* strains. Statistics generated from TrinityStats.pl.

	Trinity	**Cufflinks**	**StringTie**
Total Trinity "genes"	47,562	40,593	41,851
Total Trinity transcripts	69,948	40,593	41,851
Percent GC	38.12%	42.08%	41.66%
N10	3257	7944	7333
N20	2487	6108	5595
N30	1992	4987	4491
N40	1635	4126	3640
N50	1313	3414	2981
Median contig length	468	1738	1226
Average contig length	800.21	2305.35	1789.3

Trinity assemblies had the highest overall alignment rates (96.60–98.00%), while the alignment rate for the Cufflinks and StringTie assemblies were comparable (60.5–91.10% and 58.80–90.2%, respectively).

A metric more appropriate for transcriptome assemblies is the ExN50 statistic, which represents N50 based on the most highly expressed transcripts that represent x% of the normalized expression data (e.g., E90N50 is the N50 value for 90% of the normalized expressed transcripts, excluding the lowly expressed transcripts) [24]. To obtain the ExN50 scores, the expression data were first normalized using kallisto, then count matrices were generated. Finally, the Trinity scripts contig_ExN50_statistic.pl and plot_ExN50_statistic.Rscript were run to compute and visualize ExN50 contig statistics.

Comparing the N50 of the assemblies with their E90N50, a reversal is observed. When all contigs were considered (E100), Cufflinks and StringTie had a higher N50 than Trinity, indicating longer contig lengths. However, when N50 is confined to 90% of expression data, the Trinity assembly resulted in a higher value for E90N50. This suggests that lowly expressed genes make up longer contigs in reference-based assemblers but not in the de novo assembler. The ExN50 profiles in Figure 4 show that the peak N50 occurs at higher Ex values, indicating that the sequencing depth was enough to assemble the lowly expressed transcripts.

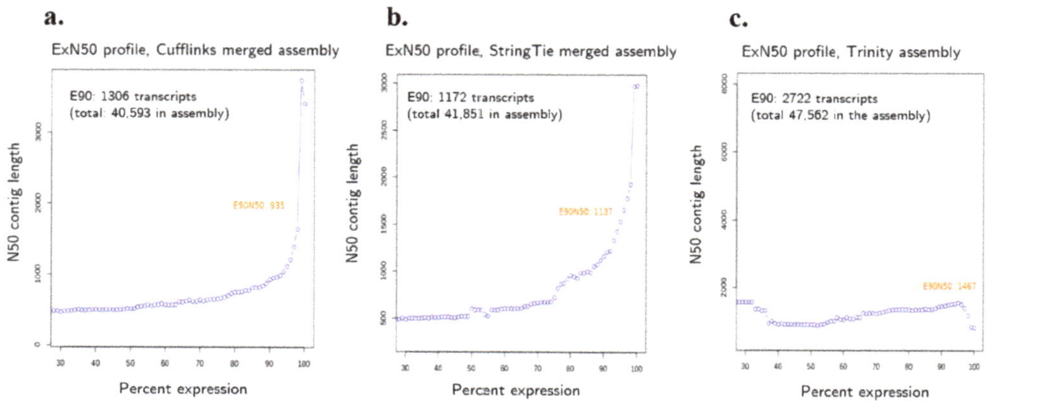

Figure 4. ExN50 profiles of the (**a**) Cufflinks, (**b**) StringTie, and (**c**) Trinity merged *Bombyx mori* transcriptome assemblies.

A reference-free evaluation metric for transcriptome assemblies was developed by Li et al. [25] in the package DETONATE (De novo Transcriptome RNA-seq Assembly with or without the Truth Evaluation). DETONATE has two component packages, RSEM-EVAL and REF-EVAL. RSEM-EVAL is based on the RSEM (RNA-Seq by Expectation Maximization) algorithm. RSEM-EVAL is based on a probabilistic model that combines different factors including the compactness of the assembly and read support [25]. The RSEM-EVAL score is a log joint probability based on three components: Likelihood of the assembly, assembly prior, and a Bayesian information criterion penalty. Penalties are imposed when assemblies have too many bases or contigs or have an unusual distribution of contig lengths relative to the expected read coverage. RSEM-EVAL scores are always reported to be negative, but higher scores are indicative of better assemblies. Out of the three transcriptome assemblies, Trinity obtained the highest RSEM-EVAL score (Table 5). Aside from the RSEM-EVAL scores, REF-EVAL scores are also reported in Table 5. REF-EVAL reports the recall and precision of the assemblies at the contig and nucleotide levels. The recall is the fraction of reference elements (contigs, scaffolds, nucleotides) that are recovered in the assembly, while precision is the fraction of assembly elements that recover a reference element. The F_1 score represents the harmonic mean of the recall and precision scores:

$$F_1 = \frac{2 \times recall \times precision}{recall + precision}. \quad (1)$$

Table 5. DETONATE scores of the different *Bombyx mori* transcriptome assemblies.

Assembly	RSEM-Eval	Nucleotide F1 (Unweighted)	Contig F1 (Unweighted)	Weighted k-Mer Recall	k-Mer Compression Score
Trinity	-8.35×10^9	0.38019	0.00239295	0.678414	−7.32371
Cufflinks	-1.34×10^{10}	0.682955	0.690038	0.545637	−22.5079
StringTie	-1.55×10^{10}	0.775391	0.677147	0.545727	−17.3472

The k-mer compression (*KC*) score is computed by the formula:

$$KC = WKR\ (weighted\ kmer\ recall) - ICR\ (inverse\ compression\ rate), \quad (2)$$

where *WKR* measures an assembly's recall of the k-mer content in the reference, with each k-mer weighted by relative frequency, and *ICR* measures the compression of the RNA-seq data.

To evaluate the completeness of assemblies, Benchmarking Universal Single-Copy Orthologs (BUSCO) was employed using the dataset insecta_odb9 (2016). BUSCOs are core genes from OrthoDB that are expected to be present as single copies in orthologous groups. Recovered genes are classified as 'complete' when their lengths are within two standard deviations of the group mean length; 'fragmented' when they are only partially recovered; and 'missing' if they are unrecovered [26]. Both reference-guided assemblies had a high percentage (>95%) of completely recovered BUSCOs, while the Trinity assembly had the lowest number of BUSCOs at 80.10% (Figure 5). The StringTie assembly had more single-copy and fewer duplicate orthologs. Duplicated BUSCOs should be rare and could indicate erroneous assembly of haplotypes [26]. Running BUSCO for unmerged transcriptome assemblies (per sample) resulted in a small percentage of duplicated BUSCOs, so the duplication may be due to heterozygous alleles that were not collapsed when the assemblies were merged.

To summarize the results of the different transcriptome assembly evaluation metrics, the Trinity de novo assembly performed best in terms of alignment rate, assembled the greatest number of transcripts with high contig length represented by E90N50, and had the best RSEM-eval score. However, when it comes to the completeness of assembly based on BUSCO, Cufflinks outperformed the other assemblers.

3.5. Expression Quantification, Normalization, and Differential Expression Analysis

The reads were counted according to gene-level features using FeatureCounts (1.4.6-p5), which is part of the R Subread package [27]. Quantification was run at the meta-feature level (gene level), where reads that overlap multiple exons of the same gene are counted exactly once, provided there is no overlap with another gene (i.e., exon spanning reads will be counted once). Read count normalization was performed with DE analysis using DESeq2's size factor normalization. After running DESeq2, 1647 DEGs were identified within the threshold $p < 0.1$, but after FDR correction, only 476 DEGs were within the adjusted threshold $p_{adj} < 0.1$. Factor levels were set by location, where TCMO is the reference condition and CAR is set as the contrast condition. Among the 476 DEGs, there were 222 upregulated genes and 254 downregulated genes in CAR strains.

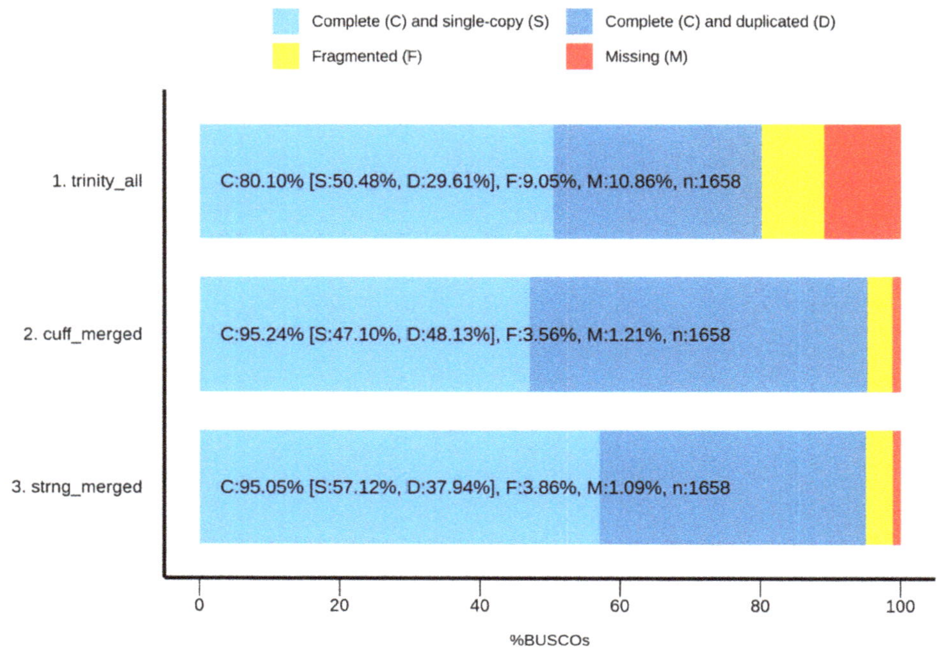

Figure 5. Measure of completeness of transcriptome assemblies from *Bombyx mori* strains using BUSCO (dataset: insecta_odb9).

A heat map of the 476 DEGs that passed the significance threshold is shown in Figure 6. Clustering of the DEGs by location is apparent in the heat map. The replicates for each strain show similar DEGs, except for a cluster of DEGs that are upregulated in bioreplicate MO204-4 but not in other MO204 replicates. This DEG cluster is composed of loci with uncharacterized proteins (LOC101741578, LOC101739373, LOC101736556, LOC101741319, LOC101738771, LOC101740400, LOC101735788). LOC101741578 has a conserved domain (COG5240 SEC21; vesicle coat complex COPI, gamma subunit) that is found in proteins involved in intracellular traffic and secretion [12]. Since these genes are highly upregulated in MO204-4 but not in other TCMO strains, this may be due to individual differences that were not taken into account (e.g., sex differences).

DEGs were filtered by the criteria $p_{adj} < 0.1$ and $|\log_2 FC| > 1$ to select statistically and biologically significant genes; genes passing this significance threshold are represented by red points in the volcano plot shown in Figure 7. The filtered DEGs with their corresponding gene and protein IDs are listed in Table S1.

3.6. Functional Annotation of Select DEGs

The top DEGs were selected based on statistical significance and high fold changes and are summarized in Table 6. This list includes fold changes, p_{adj} values, as well as the protein names (if available) and the functions derived from Pfam/InterPro annotations available in UniProt [29].

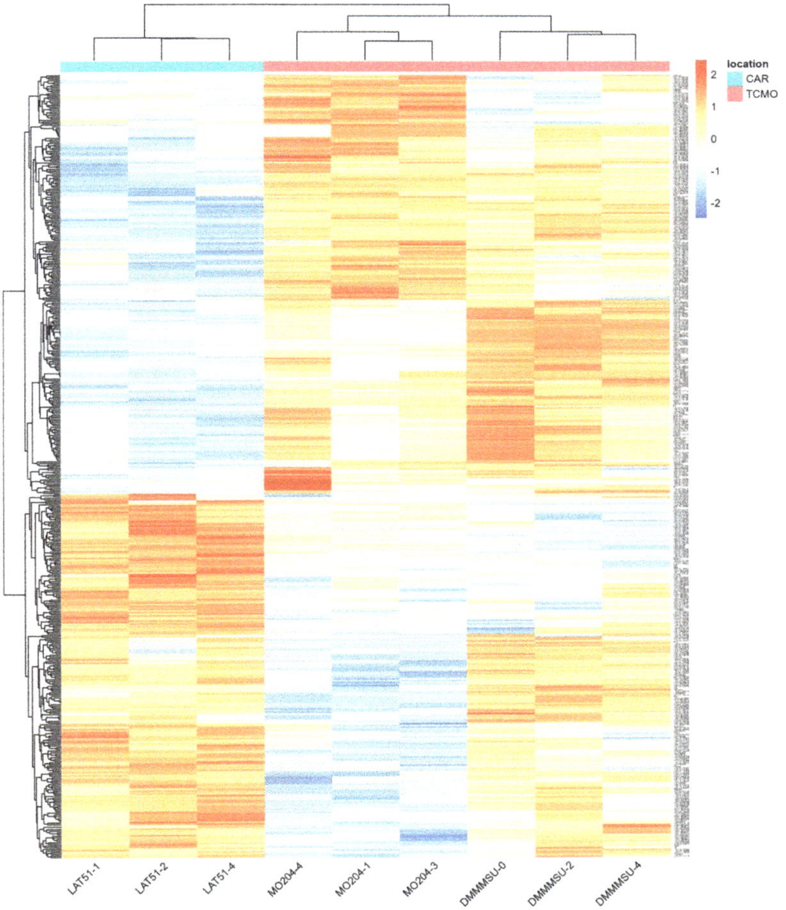

Figure 6. Heat map of the 112 DEGs from *Bombyx mori* with $p_{adj} < 0.1$. The values shown are scaled to the distances from the average of each row (gene).

Gene Ontology (GO) terms were mapped to the gene sets by matching the protein IDs to the GO annotations in the newest release of *B. mori* genome annotations from KAIKObase (ver. 4.1.0, July 2020). Their annotation contains the newest predicted gene models, descriptions including InterPro IDs, GO terms, and best hit in the NCBI nr database. Out of 14,124 assayed genes, 11,792 were annotated with protein IDs from nr, and 3948 were annotated with GO terms.

GO term enrichment was determined using the R package goseq. Significantly overrepresented GO terms are summarized in Tables S1 and S3 for upregulated and downregulated genes, respectively. Among the upregulated genes, the biological processes that were enriched are related to nucleotide metabolism and biosynthesis, lipid and carbohydrate metabolic processes, and regulation of transcription. Among down-regulated genes, similar biological processes were found to be enriched, but the ontology for carbohydrate metabolic process stands out with a low p-value. For cellular components, there is no overlap of GO terms for up- and downregulated genes. For molecular functions, nucleotide, protein, and metal binding are enriched in upregulated genes, while DNA binding, ATP binding, and zinc binding are enriched in downregulated genes. Catalytic activity, oxidoreductase, and hydrolase activities are also enriched molecular functions found in both sets.

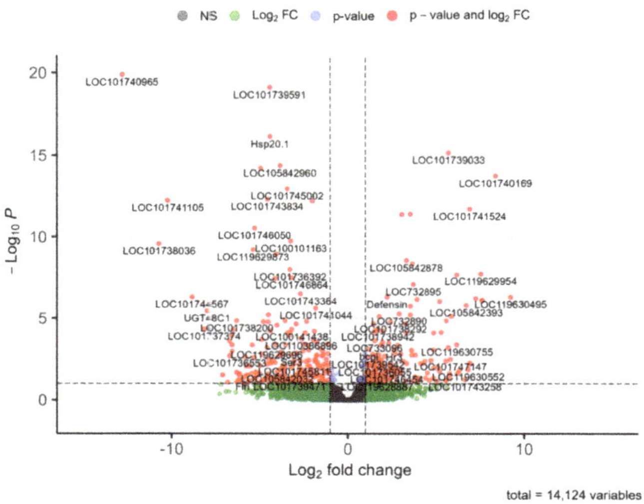

Figure 7. Volcano plot of DEGs from *Bombyx mori*, log fold change vs. $-\log P$. Gray: Nonsignificant genes; green: Genes with significant fold changes but insignificant adjusted *p*-values; blue: Genes that passed the significance threshold according to adjusted *p*-value but did not have significant fold changes; red: Genes that have adjusted *p*-value < 0.1 and significant fold change ($|\log_2 FC| > 1$). Graph generated in R package EnhancedVolcano [28].

Table 6. Summary of select top DEGs between different *Bombyx mori* strains (CAR vs. TCMO).

Downregulated Genes (CAR vs. TCMO)				
Gene ID	**$\log_2$FC**	**p_{adj}**	**Protein ID**	**Function**
LOC101740965	−12.8030	1.31×10^{-20}	myrosinase 1 isoform X2	hydrolysis of glucosinolates, defense enzyme
Hsp20.1	−4.4182	7.45×10^{-17}	heat shock protein hsp20.1	chaperone protein
LOC101739591	−4.4249	7.92×10^{-20}	transient receptor potential channel pyrexia isoform X1	ion transport, contains ankyrin repeat
LOC101741105	−10.2153	5.85×10^{-13}	myrosinase 1	hydrolysis of glucosinolates, defense enzyme
LOC101738036	−10.7102	2.66×10^{-10}	serine protease inhibitor swm-1	serine protease inhibitor, trypsin inhibitor
LOC105842960	−3.8422	4.35×10^{-15}	oxygen-dependent choline dehydrogenase-like	oxidoreductase activity, FAD/NAD(P)-binding
Upregulated genes (CAR vs. TCMO)				
Gene ID	**$\log_2$FC**	**p_{adj}**	**Protein ID**	**Function**
LOC101739033	5.7239	7.09×10^{-16}	zinc finger protein 664	dimer formation, metal binding
LOC101740169	8.3899	1.81×10^{-14}	uncharacterized protein LOC101740169	tetraspanin family, scaffolding, cell adhesion, proliferation
LOC101741524	6.9289	1.99×10^{-12}	synaptic vesicle glycoprotein 2C	sugar transporter
LOC105842878	3.3250	2.85×10^{-9}	peroxidase-like	heme peroxidase, redox activity, oxidoreductase activity, defense enzyme
LOC119629954	7.5484	1.91×10^{-8}	zinc finger BED domain-containing protein 4-like isoform X2	contains hAT dimerization domain, ribonuclease H-like superfamily: nucleotide metabolism and transport
LOC119630495	9.2422	5.08×10^{-7}	uncharacterized protein LOC119630495	contains motifs: peptidase, Hsp70, TetR
LOC105842393	6.7307	1.61×10^{-6}	uncharacterized protein LOC105842393	contains motifs: DUF5641 (pfam18701), unknown function, found in retrotransposons
Defensin	2.2251	5.08×10^{-7}	defensin like protein 2 precursor	contains signal peptide

3.7. qPCR Validation of Select DEGs

Among the selected top DEGs in Table 6, six were selected for validation using qPCR: Heat shock protein 20 (Hsp20), oxygen-dependent choline dehydrogenase-like (odcdl), myrosinase 1 (myr), peroxidase-like (pl), uncharacterized protein (LOC105842393), and uncharacterized protein tetraspanin family (ts). The first three were observed in silico to be downregulated in CAR vs. TCMO silkworms, while the last three were calculated to be upregulated. The results for the qPCR found Hsp20 to be significantly downregulated in CAR vs. TCMO samples ($p < 0.0001$), as were odcdl ($p = 0.0005$) and myr ($p = 0.0008$). LOC105842393 was found to be significantly upregulated in CAR vs. TCMO ($p = 0.028$), as were pl ($p = 0.0001$) and ts ($p = 0.0093$). These data support the results of in silico analysis (Figure 8). Additionally, $\log_2$FC values were calculated and compared with the values produced using DESeq2, as presented in Figure 8. While differences were found between the $\log_2$FC values between qPCR and DESeq2, these may be due to the individual gene expression differences between samples used for sequencing and samples used for qPCR. Despite these differences, the trends of downregulation or upregulation in CAR vs. TCMO samples observed during qPCR appear to support those found during in silico analysis.

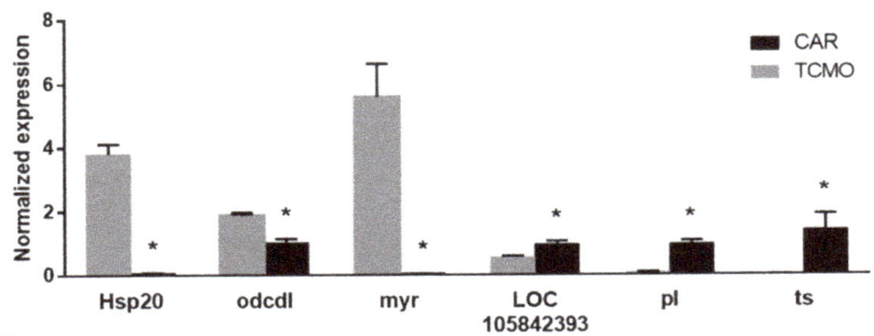

log2FC

qRT-PCR	-5.61	-0.88	-8.82	0.77	3.58	7.17
DESeq2	-4.42	-3.84	-12.8	6.73	3.33	8.39

Figure 8. Results of qPCR validation of differentially expressed genes. Heat shock protein 20 (Hsp20), oxygen-dependent choline dehydrogenase-like (odcdl), and myrosinase 1 (myr) were downregulated in CAR vs. TCMO, while uncharacterized protein (LOC105842393), peroxidase-like (pl), and uncharacterized protein tetraspanin family (ts) were upregulated in CAR vs. TCMO. $\log_2$FC values calculated from qPCR results are compared with values obtained during DESeq2 analysis. Error bars indicate standard deviation; asterisks denote significant change in CAR vs. TCMO (Student's *t*-test), $p < 0.05$.

4. Discussion

In this study, we compared gene expression in silk glands of *B. mori* strains reared in two different local facilities with different rearing temperatures. According to previous studies, the optimal temperature for silkworm growth is 20–28 °C and 23–28 °C for silk productivity [30]. The ambient temperature in the rearing sites during collection was 31.7 °C and 23.9 °C for TCMO and CAR, respectively. These data are based on bivoltine strains, although polyvoltine strains acclimatized in tropical countries are known to tolerate higher temperatures. Nevertheless, it has been shown that lower temperatures are better for silkworm productivity and larval duration [31].

Closely related to temperature is humidity, thus, an important factor to consider. It influences the withering of leaves in rearing beds, which affects larval feeding. However, there is no limiting range for humidity, and most insects can develop as long as they can control their water balance [30].

Among the downregulated genes in CAR were myrosinase, heat shock proteins, and serine protease inhibitors, which are related to defense and ensuring proper protein folding.

These were downregulated in CAR strains likely because chaperone and protease inhibitors are not needed as much in lower temperatures. Aside from Hsp20.1, other heat shock proteins that are significantly downregulated are Hsp 68, Hsp 70, Hsp 23.7 precursor, and Hsp 25.4 precursor.

On the other hand, there were upregulated genes found in CAR related to defense such as peroxidase-like protein, defensin, and other proteins related to oxidoreductase activity. Many of these protein products are yet to be characterized, but sequence motifs provide some hints about their function based on information from protein families and conserved domains.

There were also several dehydrogenases found among the DEGs, such as oxygen-dependent choline dehydrogenase-like protein, L-sorbose 1-dehydrogenase, glucose dehydrogenase [FAD, quinone], 15-hydroxyprostaglandin dehydrogenase, farnesol dehydrogenase, dihydropyrimidine dehydrogenase [NADP(+)], mitochondrial aldehyde dehydrogenase, and alcohol dehydrogenases. Dehydrogenases are a type of oxidoreductases that oxidize substrates by reducing an electron acceptor and are important for their role in detoxification of metabolites. Malate dehydrogenase (BmMDH1) and 6-phosphogluconate dehydrogenase (Bm6PGD) were previously found to be differentially expressed in Dazao silkworms exposed to different temperatures, which implies enhancement in energy metabolism and ATPase expression at low temperatures [32].

Another DEG of note is SOSS complex subunit B homolog ($\log_2$FC: −2.0042; p_{adj}: -2.004×10^{-13}). The SOSS complex binds to single-stranded DNA and is involved in DNA damage response, cell cycle checkpoint activation, homologous repair of double-stranded breaks, and ATM-dependent signaling pathways [33]. Downregulation of SOSS in CAR strains may indicate that there is less DNA damage response in silkworms reared in low-temperature environments.

Microvitellogenin ($\log_2$FC: 6.2283; p_{adj}: 0.023) was one of the genes found upregulated in CAR strains, which encodes a low molecular weight (30 kDa) lipoprotein (30 K proteins or 30 KPs). These lipoproteins are synthesized in the *B. mori* fat body and secreted in the hemolymph during the last instar larva stage. They are involved in several physiological processes, such as energy storage, embryonic development, and immune response [34] Upregulation of microvitellogenin may indicate higher energy storage and enhanced immunoprotection in low-temperature strains [35,36].

Another upregulated DEG was the protein eyes shut ($\log_2$FC: 5.7860; p_{adj}: 0.024), also known as EGF-like protein 10, an essential protein for the formation of photoreceptor cells in *Drosophila* [33]. Most EGF-like domains are found in extracellular domains of membrane-bound proteins or secreted proteins. Aside from its EGF-like domain, *B. mori* protein eyes shut also has a laminin G domain, which is associated with different functions such as cell adhesion, signaling, migration, assembly, and differentiation [37].

Three regulators of the juvenile hormone (JH) were found among the DEGs in this analysis: Juvenile hormone epoxide hydrolase-like protein 2 (Jheh-lp2), juvenile hormone epoxide hydrolase precursor (Jheh2), and cytosolic juvenile hormone binding protein 36 kDa subunit (Cjhbp). JHEH inactivates juvenile hormones by hydrolyzing the epoxide groups in JH. Differential expression of JH regulators could explain the observed differences in larval development between CAR and TCMO strains, wherein CAR strains have a longer larval duration (23 days for TCMO, 27 days for CAR). JH has long been known to affect silk production, and the application of JH analogs has been used to prolong the 5th instar stage to increase larval weight and silk secretion [38]. More recently, the JH regulatory pathway has been demonstrated to influence Fib-H expression through the action of transcription factors *Bmdimm* and *Bmsage* [39]. Zhou et al. [40] showed that JH biosynthesis and sexual maturation are delayed in cotton ballworm (*Helicoverpa armigera*) reared at 19 °C compared to those reared in higher temperatures. Liu et al. [41] showed that JH titers in Formosan termite (*Coptotermes formonasus*) were positively correlated with temperature. In contrast, Geister et al. [42] found no correlation between temperature and JH titers in the tropical butterfly *Bicyclus anynana*.

Caytaxin was downregulated in CAR strains (log_2FC: −2.8609; p_{adj}: 0.032). Caytaxin is a member of the BNIP2 (Bcl2-/adenovirus E1B nineteen kDa-interacting protein 2) protein family (pfam12496). Caytaxins interact with pro- and anti-apoptotic molecules in the cell and are involved in Rho GTPase regulation [43].

Another DEG found was dymeclin (log_2FC: 1.7453; p_{adj}: 0.030), which belongs to the Dymeclin (Dyggve-Melchior-Clausen syndrome protein) protein family (pfam09742) in plants and animals. Dymeclin proteins have a length of approximately 700 residues and contain many leucine and isoleucine in their conserved domain [12]. Mutations in human dymeclin cause Dyggve–Melchior–Clausen syndrome (DMC, MIM 223800), an autosomal recessive disorder. Dymeclin is a peripheral membrane protein dynamically associated with the Golgi apparatus. In *Caenorhabditis elegans*, a member of the Dymeclin protein family is hid1 (high-temperature-induced dauer-formation protein 1), which encodes a highly conserved transmembrane protein that could be involved in vesicle secretion or intercellular signaling [44].

Previous comparative studies found similar enrichment in GO terms as those in this study. In one study comparing wild and domestic silkworms, oxidoreductase activity (GO: 0016491) was the only enriched GO term in four pairwise comparisons [11]. In another study comparing Chinese silkworm strains JingSong and Lan10, which have different rates of silk production, the dominant GO term was membrane-enclosed lumen under the cellular component [10].

5. Conclusions

The present study provides, for the first time, valuable information on the transcriptome of *B. mori* strains found in the Philippines. The differential expression analysis performed on silk glands of *B. mori* strains grown in different sites gives us insight into the processes and functions linked to the down- and up-regulated genes. This molecular information could be used as a source of trait markers for the improvement of local silkworm strains to enhance the sericulture industry in the Philippines. For future directions, aside from protein annotations and GO terms, KEGG pathway analysis may also be performed to further elucidate the functions of DEGs in silkworms and in silk production. It is also recommended to perform further quantitative PCR assays to validate the other bioinformatically obtained DEGs, in addition to the DEGs validated here. Furthermore, this study could be extended to transcriptome analysis of other *B. mori* strains, tissues, and developmental stages.

Supplementary Materials: The following are available online at https://www.mdpi.com/article/10.3390/insects13080669/s1, Table S1: DEGs between *B. mori* CAR and TCMO strains within the *p*-adj threshold, Table S2: GO terms enriched among upregulated genes in the silk glands of selected *B. mori* strains from CAR, Table S3: GO terms enriched among downregulated genes in the silk glands of selected *B. mori* strains from CAR.

Author Contributions: P.N.O.d.l.P.: Methodology, investigation, data curation, formal analysis, visualization, writing—original draft; A.G.D.L.: Investigation, formal analysis, visualization, writing; M.A.M.B.: Conceptualization, funding acquisition, methodology, project administration, supervision, writing—review and editing. All authors have read and agreed to the published version of the manuscript.

Funding: This study was supported by in-house funding for 2020 provided by the National Institute of Molecular Biology and Biotechnology, University of the Philippines—Diliman for P.N.O.d.L.P.'s master's thesis.

Institutional Review Board Statement: Not applicable.

Informed Consent Statement: Not applicable.

Data Availability Statement: The RNA-seq dataset generated in this study has been deposited in the National Center for Biotechnology Information (NCBI) Gene Expression Omnibus [45] accessible through GEO Series Accession Number GSE184152.

Acknowledgments: The authors would like to thank the Department of Science and Technology—Philippine Textile Research Institute (DOST-PTRI) for providing administrative and material support for sample collection, and the PTRI Technology Center in Misamis Oriental and DOST Cordillera Administrative Region Technology Center for providing silkworm samples.

Conflicts of Interest: The authors declare no conflict of interest.

References

1. Trevisan, A.; Hart, M.; Yuste, V.; Hjoellund, A.; Galle, D.; Bergman, O. Cocoon Silk: A Natural Architecture. Available online: https://web.archive.org/web/20120507085636/http://www.senature.com/research/publications/cocoon-silk-a-natural-architecture (accessed on 19 October 2021).
2. Kunz, R.I.; Brancalhão, R.M.C.; de Ribeiro, L.F.C.; Natali, M.R.M. Silkworm Sericin: Properties and Biomedical Applications. *BioMed Res. Int.* **2016**, *2016*, 8175701. [CrossRef] [PubMed]
3. PhilFIDA. The Negros Silk Project. Available online: http://www.philfida.da.gov.ph/index.php/recent-news-articles/114-the-negros-silk-project (accessed on 4 November 2021).
4. Zanatta, D.B.; Bravo, J.P.; Barbosa, J.F.; Munhoz, R.E.F.; Fernandez, M.A. Evaluation of Economically Important Traits from Sixteen Parental Strains of the Silkworm Bombyx Mori L (Lepidoptera: Bombycidae). *Neotrop. Entomol.* **2009**, *38*, 327–331. [CrossRef] [PubMed]
5. Reyes, R.C. Silkworm Genetics Research: Developing Silkworm Strains Adapted to Lowland Philippine Conditions. Presented at the Development Fund Professorial Chair Lecture, University of the Philippines Los Baños, College, Laguna. 2003. Available online: https://agris.fao.org/agris-search/search.do?recordID=PH2004000177 (accessed on 4 November 2021).
6. Supsup, G.E.; Abuan, J.P.; Viduya, M.M.; Sanchez, M.L.; Ulat, M.E. Selection of Superior Bivoltine Silkworm Hybrids through Combining Ability Test and Multiple Traits Evaluation Index. *IAMURE Int. J. Ecol. Conserv.* **2018**, *23*, 1.
7. Velasco, A.B.; Bacuso, P.M.; Cabrito, F.P.; Fernandez, R.W.; Alvarez, V.B.; de Guzman, Z.I.; Villanueva, E.P. New Breeds of Filipino Silkworm Varieties Commercialized. *Philipp. Technol. J.* **1994**, *19*. Available online: https://agris.fao.org/agris-search/search.do?recordID=PH9710645 (accessed on 4 November 2021).
8. Aro, S.C. Silk Production on the Decline. *Sunstar*. 2015. Available online: https://web.archive.org/web/20220722062934/https://www.sunstar.com.ph//article/40526 (accessed on 22 July 2022).
9. Villanueva, R.C. Problems and Issues Affecting the Pace of Sericulture Development in the Philippines. *Don Marian. Marcos Meml. State Univ. Res. Ext. J. Bacnotan Campus*. 1999. Available online: https://agris.fao.org/agris-search/search.do?recordID=PH2007M00034 (accessed on 20 June 2022).
10. Li, J.; Qin, S.; Yu, H.; Zhang, J.; Liu, N.; Yu, Y.; Hou, C.; Li, M. Comparative Transcriptome Analysis Reveals Different Silk Yields of Two Silkworm Strains. *PLoS ONE* **2016**, *11*, e0155329. [CrossRef]
11. Fang, S.-M.; Hu, B.-L.; Zhou, Q.-Z.; Yu, Q.-Y.; Zhang, Z. Comparative Analysis of the Silk Gland Transcriptomes between the Domestic and Wild Silkworms. *BMC Genom.* **2015**, *16*, 60. [CrossRef]
12. Lu, F.; Wei, Z.; Luo, Y.; Guo, H.; Zhang, G.; Xia, Q.; Wang, Y. SilkDB 3.0: Visualizing and Exploring Multiple Levels of Data for Silkworm. *Nucleic Acids Res.* **2020**, *48*, D749–D755. [CrossRef]
13. Shimomura, M.; Minami, H.; Suetsugu, Y.; Ohyanagi, H.; Satoh, C.; Antonio, B.; Nagamura, Y.; Kadono-Okuda, K.; Kajiwara, H.; Sezutsu, H.; et al. KAIKObase: An Integrated Silkworm Genome Database and Data Mining Tool. *BMC Genom.* **2009**, *10*, 486. [CrossRef]
14. Li, Y.; Wang, G.; Tian, J.; Liu, H.; Yang, H.; Yi, Y.; Wang, J.; Shi, X.; Jiang, F.; Yao, B.; et al. Transcriptome Analysis of the Silkworm (Bombyx Mori) by High-Throughput RNA Sequencing. *PLoS ONE* **2012**, *7*, e43713. [CrossRef]
15. Prasad, M.D.; Muthulakshmi, M.; Arunkumar, K.P.; Madhu, M.; Sreenu, V.B.; Pavithra, V.; Bose, B.; Nagarajaram, H.A.; Mita, K.; Shimada, T.; et al. SilkSatDb: A Microsatellite Database of the Silkworm, Bombyx Mori. *Nucleic Acids Res.* **2005**, *33*, D403–D406. [CrossRef]
16. Xu, H.-E.; Zhang, H.-H.; Xia, T.; Han, M.-J.; Shen, Y.-H.; Zhang, Z. BmTEdb: A Collective Database of Transposable Elements in the Silkworm Genome. *Database* **2013**, *2013*. [CrossRef] [PubMed]
17. Xia, Q.; Li, S.; Feng, Q. Advances in Silkworm Studies Accelerated by the Genome Sequencing of Bombyx Mori. *Annu. Rev. Entomol.* **2014**, *59*, 513–536. [CrossRef] [PubMed]
18. Wang, Z.; Gerstein, M.; Snyder, M. RNA-Seq: A Revolutionary Tool for Transcriptomics. *Nat. Rev. Genet.* **2009**, *10*, 57–63. [CrossRef]
19. Zhang, R.; Cao, Y.-Y.; Du, J.; Thakur, K.; Tang, S.-M.; Hu, F.; Wei, Z.-J. Transcriptome Analysis Reveals the Gene Expression Changes in the Silkworm (Bombyx Mori) in Response to Hydrogen Sulfide Exposure. *Insects* **2021**, *12*, 1110. [CrossRef] [PubMed]
20. Yokoi, K.; Tsubota, T.; Jouraku, A.; Sezutsu, H.; Bono, H. Reference Transcriptome Data in Silkworm Bombyx Mori. *Insects* **2021**, *12*, 519. [CrossRef]
21. The International Silkworm Genome Consortium. The Genome of a Lepidopteran Model Insect, the Silkworm Bombyx Mori. *Insect Biochem. Mol. Biol.* **2008**, *38*, 1036–1045. [CrossRef]
22. Dundar, W.F.; Skrabanek, L.; Zumbo, P. Introduction to Differential Gene Expression Analysis Using RNA-Seq 2019. Available online: https://chagall.med.cornell.edu/RNASEQcourse/Intro2RNAseq.pdf (accessed on 2 December 2021).

23. Ewels, P.; Magnusson, M.; Lundin, S.; Käller, M. MultiQC: Summarize Analysis Results for Multiple Tools and Samples in a Single Report. *Bioinformatics* **2016**, *32*, 3047–3048. [CrossRef]
24. Haas, B. Transcriptome Contig Nx and ExN50 Stats Trinityrnaseq/Trinityrnaseq Wiki. Available online: https://github.com/trinityrnaseq/trinityrnaseq (accessed on 2 December 2021).
25. Li, B.; Fillmore, N.; Bai, Y.; Collins, M.; Thomson, J.A.; Stewart, R.; Dewey, C.N. Evaluation of de Novo Transcriptome Assemblies from RNA-Seq Data. *Genome Biol.* **2014**, *15*, 553. [CrossRef]
26. Simão, F.A.; Waterhouse, R.M.; Ioannidis, P.; Kriventseva, E.V.; Zdobnov, E.M. BUSCO: Assessing Genome Assembly and Annotation Completeness with Single-Copy Orthologs. *Bioinformatics* **2015**, *31*, 3210–3212. [CrossRef]
27. Liao, Y.; Smyth, G.K.; Shi, W. FeatureCounts: An Efficient General Purpose Program for Assigning Sequence Reads to Genomic Features. *Bioinformatics* **2014**, *30*, 923–930. [CrossRef]
28. Blighe, K.; Rana, S.; Lewis, M. EnhancedVolcano: Publication-Ready Volcano Plots with Enhanced Colouring and Labeling. 2018. Available online: https://github.com/kevinblighe/EnhancedVolcano (accessed on 2 December 2021).
29. The UniProt Consortium. UniProt: The Universal Protein Knowledgebase in 2021. *Nucleic Acids Res.* **2021**, *49*, D480–D489. [CrossRef] [PubMed]
30. Rahmathulla, V.K. Management of Climatic Factors for Successful Silkworm (Bombyx Mori L.) Crop and Higher Silk Production: A Review. *Psyche* **2012**, *2012*, e121234. [CrossRef]
31. Kumar, N.S.; Basavaraja, H.K.; Kumar, C.M.K.; Reddy, N.M.; Datta, R.K. On the Breeding of "CSR18xCSR19"- A Robust Bivoltine Hybrid of Silkworm, Bombyx Mori L. for the Tropics. *Int. J. Ind. Entomol.* **2002**, *5*, 153–162.
32. Guo, H.; Huang, C.; Jiang, L.; Cheng, T.; Feng, T.; Xia, Q. Transcriptome Analysis of the Response of Silkworm to Drastic Changes in Ambient Temperature. *Appl. Microbiol. Biotechnol.* **2018**, *102*, 10161–10170. [CrossRef]
33. Zahn-Zabal, M.; Michel, P.-A.; Gateau, A.; Nikitin, F.; Schaeffer, M.; Audot, E.; Gaudet, P.; Duek, P.D.; Teixeira, D.; Rech de Laval, V.; et al. The NeXtProt Knowledgebase in 2020: Data, Tools and Usability Improvements. *Nucleic Acids Res.* **2020**, *48*, D328–D334. [CrossRef]
34. Shi, X.-F.; Li, Y.-N.; Yi, Y.-Z.; Xiao, X.-G.; Zhang, Z.-F. Identification and Characterization of 30 K Protein Genes Found in Bombyx Mori (Lepidoptera: Bombycidae) Transcriptome. *J. Insect Sci.* **2015**, *15*, 71. [CrossRef]
35. Singh, N.K.; Pakkianathan, B.C.; Kumar, M.; Prasad, T.; Kannan, M.; König, S.; Krishnan, M. Vitellogenin from the Silkworm, Bombyx Mori: An Effective Anti-Bacterial Agent. *PLoS ONE* **2013**, *8*, e73005. [CrossRef]
36. Liu, Y.; Chen, M.; Su, J.; Ma, H.; Zheng, X.; Li, Q.; Shi, S.; Qin, L. Identification and Characterization of a Novel Microvitellogenin from the Chinese Oak Silkworm Antheraea Pernyi. *PLoS ONE* **2015**, *10*, e0131751. [CrossRef]
37. Sasaki, T.; Fässler, R.; Hohenester, E. Laminin: The Crux of Basement Membrane Assembly. *J. Cell Biol.* **2004**, *164*, 959–963. [CrossRef]
38. Daillie, J. Juvenile Hormone Modifies Larvae and Silk Gland Development in Bombyx Mori. *Biochimie* **1979**, *61*, 275–281. [CrossRef]
39. Zhao, X.-M.; Liu, C.; Jiang, L.-J.; Li, Q.-Y.; Zhou, M.-T.; Cheng, T.-C.; Mita, K.; Xia, Q.-Y. A Juvenile Hormone Transcription Factor Bmdimm-Fibroin H Chain Pathway Is Involved in the Synthesis of Silk Protein in Silkworm, Bombyx Mori *. *J. Biol. Chem.* **2015**, *290*, 972–986. [CrossRef] [PubMed]
40. Zhou, X.; Coll, M.; Applebaum, S.W. Effect of Temperature and Photoperiod on Juvenile Hormone Biosynthesis and Sexual Maturation in the Cotton Bollworm, Helicoverpa Armigera: Implications for Life History Traits. *Insect Biochem. Mol. Biol.* **2000**, *30*, 863–868. [CrossRef]
41. Liu, Y.; Henderson, G.; Mao, L.; Laine, R.A. Effects of Temperature and Nutrition on Juvenile Hormone Titers of Coptotermes Formosanus (Isoptera: Rhinotermitidae). *Ann. Entomol. Soc. Am.* **2005**, *98*, 732–737. [CrossRef]
42. Geister, T.L.; Lorenz, M.W.; Meyering-Vos, M.; Klaus, H.H.; Fischer, K. Effects of Temperature on Reproductive Output, Egg Provisioning, Juvenile Hormone and Vitellogenin Titres in the Butterfly Bicyclus Anynana. *J. Insect Physiol.* **2008**, *54*, 1253–1260. [CrossRef]
43. Gupta, A.B.; Wee, L.E.; Zhou, Y.T.; Hortsch, M.; Low, B.C. Cross-Species Analyses Identify the BNIP-2 and Cdc42GAP Homology (BCH) Domain as a Distinct Functional Subclass of the CRAL_TRIO/Sec14 Superfamily. *PLoS ONE* **2012**, *7*, e33863. [CrossRef]
44. Ailion, M.; Thomas, J.H. Isolation and Characterization of High-Temperature-Induced Dauer Formation Mutants in Caenorhabditis Elegans. *Genetics* **2003**, *165*, 127–144. [CrossRef]
45. Edgar, R. Gene Expression Omnibus: NCBI Gene Expression and Hybridization Array Data Repository. *Nucleic Acids Res.* **2002**, *30*, 207–210. [CrossRef]

insects

MDPI

Article

Bacterial Survey in the Guts of Domestic Silkworms, *Bombyx mori* L.

Ivan Y. Dee Tan [1,2] and Ma. Anita M. Bautista [2,*]

1 Research and Development Division, Philippine Textile Research Institute, Department of Science and Technology, Bicutan, Taguig City 1631, Philippines; iydeetan@upd.edu.ph

2 National Institute of Molecular Biology and Biotechnology, University of the Philippines—Diliman, Quezon City 1101, Philippines

* Correspondence: mmbautista20@up.edu.ph

Simple Summary: To enhance the sustainability of commercial production of high-quality silk, factors that affect the economic characteristics of the silkworm and the silk it produces have been widely studied. Among these are the gut microbiota, which have been linked to absorption and utilization of nutrients, and immunity to diseases in silkworms. Because the Philippines has yet to improve the silkworm strains it currently uses for silk production, sufficient biological data, including that of microbiota, are warranted. Profiling the bacterial communities in local silkworm strains through the use of high-throughput 16S rRNA gene amplicon sequencing would be a source of useful information. Results of the 16S rRNA gene amplicon sequencing in this study showed that four of the silkworm strains reared in the Philippines are abundant in five bacterial genera, which have also been found in other silkworm strains. Results also showed that bacterial diversity and evenness increase as larvae mature, which can be correlated to larval development and to the shift in the amount and age of mulberry leaves the larvae consume.

Citation: Dee Tan, I.Y.; Bautista, M.A.M. Bacterial Survey in the Guts of Domestic Silkworms, *Bombyx mori* L. *Insects* **2022**, *13*, 100. https://doi.org/10.3390/insects13010100

Academic Editors: Silvia Cappellozza, Morena Casartelli, Federica Sandrelli, Alessio Saviane and Gianluca Tettamanti

Received: 26 November 2021
Accepted: 14 January 2022
Published: 17 January 2022

Abstract: Silkworm, *Bombyx mori* L., research involves studies on improving strains for enhanced sustainability of high-quality silk production. Several of these have investigated the factors affecting growth and development of silkworm larvae and cocoon characteristics that subsequently affect the yield and quality of silk. The gut microbiota has been reported to impact growth and development of silkworms and has been linked, in particular, with absorption and utilization of nutrients and immunity to diseases. The silkworm strains maintained in the Philippines lack sufficient biological data for use in strain improvement. This prompted efforts to augment the data by profiling bacterial communities through high-throughput 16S rRNA gene amplicon sequencing and analysis in four of the local silkworm strains that are bred and maintained in the country. Results of the study showed that the four silkworm strains are abundant in bacteria that belong to the genera *Pseudomonas*, *Sphingomonas*, *Delftia*, *Methylobacterium* and *Acinetobacter*. Results also showed that bacterial diversity and evenness increase as larvae mature, which can be correlated to larval development and shifts in the amount and age of mulberry leaves the larvae consume.

Keywords: 16S rRNA gene; amplicon sequencing; *Bombyx mori*; gut microbiota

1. Introduction

Bombyx mori L., commonly referred to as the domestic silkworm [1] or mulberry silkworm [2], feeds exclusively on the leaves of the mulberry plant, *Morus alba* [3]. Belonging to the family Bombycidae [4], the insect is known for its ability to produce silk fibers through the formation of its cocoon [5]. *Bombyx mori* has been reared for as long as 5000 years [4] and has become central to the silk production industry [6]. *Bombyx mori* is also the model organism representative of the order Lepidoptera [7]. It is next to *Drosophila melanogaster* in terms of the number of genetic studies conducted on insects [4].

Countries in Asia remain the top producers of natural silk [8], contributing to 98% of the world's supply [2]. High demand for silk in the textile and fashion industries globally prompts sericulturists to improve the production of silk and enhance its quality [9]. Recent efforts to increase the capacity for silk production include breeding silkworms with improved economically important traits. In the Philippines, key qualitative traits used to assess strains for the production of silk include size, weight, and shape of cocoons, hatchability of fertilized eggs, larval development time, larval and pupal weight, moth emergence, reproductive capacity of moths, and mortality rate [10]. Of these, strains exhibiting faster larval development, accompanied by increased cocoon weight and yield are the most preferred. The Philippine silk industry, however, has been facing issues with the quality of silkworms, and these contribute to low cocoon production [11]. The local silk industry, thus, needs silkworms with improved quality. Genetic characterization of the 17 strains that are reared, bred, and maintained in germplasm centers in the country has already been initiated by Alcudia-Catalma et al. 2020 [9], but additional biological information is deemed necessary.

The growth and development of silkworm larvae and the economic characteristics of cocoons are influenced largely by nutritional and climatic factors, rearing techniques, silkworm races, and other factors [12]. Like most insects, silkworms also harbor a consortium of microorganisms that play crucial roles in their survival [13]. The presence of symbiotic bacteria in their gut, for example, allows for the absorption and utilization of nutrients and provides immunity to diseases [14]. These affect silkworm growth, development and survival, which in turn affect the yield and quality of cocoons. Microbial community investigation can be performed through approaches that utilize high-throughput next generation sequencing techniques. 16S rRNA gene amplicon sequencing is one approach that enables the identification of species of microorganisms present in environmental samples and provides information on the function of these microorganisms through the analysis of genomic data. This approach offers a way to bypass the challenge posed by bacterial species that cannot be cultured because it employs analyses to investigate bacterial diversity, abundance, processes, and possible interactions of an entire bacterial community based on genetic content of samples collected [15]. The process by which 16S rRNA gene amplicon sequencing allows "direct access" to the genetic information of various microorganisms from the environment [16] relies on the 16S rRNA gene conserved in prokaryotic organisms [17]. Bacterial species may be identified based on the sequences of this gene [17]. More recently, next-generation sequencing has dramatically sped up this process through different platforms that allow for sequencing of target regions in multiple samples, assembly of multiple reads into contiguous sequences, matching sequences to either a single target organism or to related species, and ultimately assigning the organisms to the selected taxonomic level [16].

The main goal of the present study was to profile and compare the bacterial communities found in the guts of Philippine silkworm strains and their larval instars using 16S rRNA gene amplicon sequencing. Results from this study provide sources of information that are associated with the target local strains that have been reared, bred, and maintained in the germplasm centers in the Philippines for decades [9], but biological data remain scarce to enhance commercial sustainability. The data generated here can be used in strain improvement through conventional hybridization and selective breeding, which is widely practiced in other silk-producing countries for enhanced sustainability of silk production.

2. Materials and Methods

2.1. Sample Collection, Preservation, and Selection

Four *B. mori* strains were considered for inclusion in this study. These silkworm strains (K203, MO202. MO204, and DMMMSU 119) were reared at the Philippine Textile Research Institute Technology Center in Misamis Oriental (PTRI-TCMO) and were chosen for analysis based on the available data describing each strain's larval development, as well as their cocoon and silk characteristics (Table S1). Based on the facility's characterization of strains and on the report of Basaen et al. (1982) [10], high hatch percentage, short larval duration, high cocoon weight, and high silk yield are considered as the most economically important and preferred silkworm traits.

All test strains were reared within the same facility under similar conditions and fed with mulberry leaves collected from the same field. However, early instars (first to third) were fed with young leaves, whereas later instars (fourth and fifth) were fed with fully developed leaves according to the practices of the facility. The second, third, fourth, and fifth instars of each of the silkworm strains were preserved in 95% ethanol and stored at 4 °C prior to gut dissection and genomic DNA extraction. Ethanol preservation was adopted based on Hammer et al. (2015) [18].

2.2. Silkworm Gut Dissection and DNA Extraction

One fifth instar larva, one fourth instar larva, three third instar larvae, and five second instar larvae from each strain were first washed with distilled water to remove excess ethanol used for preservation. These were then dissected on the dorsal side to extract the guts, which were washed with distilled water to remove excess plant material within. For the second and third instars, the guts of individuals belonging to the same strain were pooled to obtain 10–30 mg of starting material for DNA extraction. Two biological replicates were made in total using guts of different larvae from the same batch of eggs. Dissection and pooling of silkworm guts was done inside a laminar flow hood. Sterile scalpel blades and forceps used for dissection were also changed in between the dissection of each strain and instar.

DNA was extracted from the guts using the Quick-DNA Tissue/Insect Miniprep Kit (Zymo Research, Irvine, CA, USA) following the manufacturer's instructions. The extracted DNA was quantified using Qubit dsDNA BR assay (Invitrogen, Waltham, MA, USA). The quality and purity of the extracts was assessed using 1% agarose gel electrophoresis and Nanodrop spectrophotometer (Scientific Industries, Bohemia, NY, USA), respectively. DNA extracts were stored at −40 °C prior to library preparation.

2.3. Library Preparation and Sequencing on MiSeq FGx

Libraries were prepared and pooled from a total of 32 samples (4 silkworm strains × 4 instar stages × 2 biological replicates) based on the *16S* Metagenomic Sequencing Library Preparation guide by Illumina with a few modifications. DNA extracts were diluted to 10 ng/µL using nuclease-free water. Primers 341F and 805R containing 5′ overhang sequences (Table S2) used for *16S* V3-V4 PCR were from Herlemann et al. (2011) [19]. The amplicon PCR mix contained the following: 2.5 µL DNA template (15 ng/µL), 5 µL primer 341F (6 µM), 5 µL primer 805R (6 µM), and 12.5 µL MyFi PCR mix (Bioline Reagents Ltd., United Kingdom) for a total reaction volume of 25 µL. PCR was then performed under the following conditions: initial denaturation at 95 °C for 3 min; 35 cycles of denaturation at 95 °C for 30 s, annealing at 55 °C for 30 s; extension at 72 °C for 30 s; final extension at 72 °C for 5 min; and held at 4 °C.

Pairs of different indices (index 1 and index 2) from the Nextera XT Index kit (Illumina, San Diego, CA, USA) were selected such that each sample would have a unique pair of indices, as presented in Table S3. The index PCR mix contained the following: 5 µL amplicon PCR product, 5 µL Nextera XT index primer 1, 5 µL Nextera XT index primer 2, 25 µL MyFi PCR Mix, and 10 µL nuclease-free water, for a total reaction volume of 50 µL. PCR was done under the following conditions: initial denaturation at 95 °C for 3 min; 8 cycles of denaturation at 95 °C for 30 s; annealing at 55 °C for 30 s; extension at 72 °C for 30 s; final extension at 72 °C for 5 min; and held at 4 °C.

Individual library concentrations in nM were calculated based on Qubit concentration (ng/µL) and amplicon size based on TapeStation (Agilent, Santa Clara, CA, USA).

Calculated library concentrations were then normalized, pooled, and denatured. Pre-chilled HT1 hybridization buffer was then added to dilute the denatured library to 20 pM. The PhiX control (Illumina, San Diego, CA, USA) library was prepared following the above procedure. A final concentration of 8 pM was prepared with 15% PhiX spike-in. The combined library was loaded into the MiSeq v3 (600 cycles, 2 × 300 bp reads) reagent cartridge (Illumina, San Diego, CA, USA), and the sequencing run using the MiSeq FGx (Illumina, San Diego, CA, USA) was initiated.

2.4. 16S rRNA Amplicon Sequence Analysis

The summary of the sequencing run is detailed in Table S4. The percent Q30 score of the run was 93.0%, which is above the minimum of 70% recommended by Illumina. Raw DNA sequence data generated were deposited in the NCBI Sequence Read Archive (SRA) under the accession numbers listed in Table S5.

Sequence data were then converted to FASTQ format and used for downstream analyses using QIIME 2 version 2021.11 (Available online: https://qiime2.org/ (accessed on 2 January 2022)). The QIIME 2 workflow is summarized in Figure S1. Lengths of 300 bp were obtained for the paired-end reads, and raw sequence counts per sample ranged from 32,620 to 80,412 sequences. Quality control was performed using DADA2 (Available online: https://qiime2.org/ (accessed on 2 January 2022)) included in the QIIME 2 pipeline. The final sequence counts per sample, which ranged from 13,700 to 41,059 after quality filtering, denoising, merging forward and reverse reads, and removing chimeric sequences using DADA2 prior to QIIME 2 analysis, are detailed in Table S6. This allows for replicate sequences to be collapsed into representative sequences (rep-seqs) used for classification. The range of the final sequence counts of this study fell within the range reported by Suenami et al. (2019) [20] (i.e., 12,627 to 56,357 for QIIME 2 analysis of the gut microbiome of hornets).

The representative sequences generated were then used for naïve Bayes taxonomic classification using the Silva 138 classifier, which calculates the probability that a sequence belongs to a certain species instead of sequence alignment [21]. Classification using QIIME 2 in this study, therefore, did not require a percent identity threshold value. Sequences of mitochondria and chloroplasts were then filtered out prior to diversity analyses. A sampling depth of 2420 was used for diversity analyses. This value was used such that all samples and the most sequences possible would be included in the study. This sampling depth would enable the analysis of the majority of the silkworm gut microbiome as the generated alpha-rarefaction plot showed that most sequence counts begin to taper off at this point and sample count declines past this point (Figure S2). Kruskal–Wallis and PERMANOVA tests (999 permutations) were conducted to determine the significance of alpha-diversity and beta-diversity analyses, respectively. A heatmap showing the abundance of bacterial genera across all samples was also generated (Figure S3).

The feature table that was subsequently generated after rep-seqs generation by QIIME 2 was then used to determine the most frequently observed representative sequences across samples. These rep-seqs were used for BLAST queries of bacteria that were associated with each of the genera. NCBI BLAST queries were performed using the nt database on 10 January 2022. The top BLAST hits for the sequences corresponding to the five most abundant bacterial taxa according to QIIME 2 classification were determined.

3. Results

3.1. Analysis of Bacterial Abundance

Bacterial genera were identified based on abundance by QIIME 2 using the Silva classifier. Among the top five most abundant genera, *Pseudomonas* appeared to be the most dominant genus across all silkworm strains and instars (Figure 1). It displayed lower abundance percentages in later instars of DMMMSU 119 (fifth instar) and K201 (fourth and fifth instars), but was still ranked the most abundant. *Sphingomonas* and *Delftia* were typically ranked the second to fourth most abundant in most of the strains and instars. These are followed by *Methylobacterium* which was generally ranked third to fifth most abundant across all instars and strains. *Acinetobacter* was ranked the fourth to fifth most abundant bacterial genus overall.

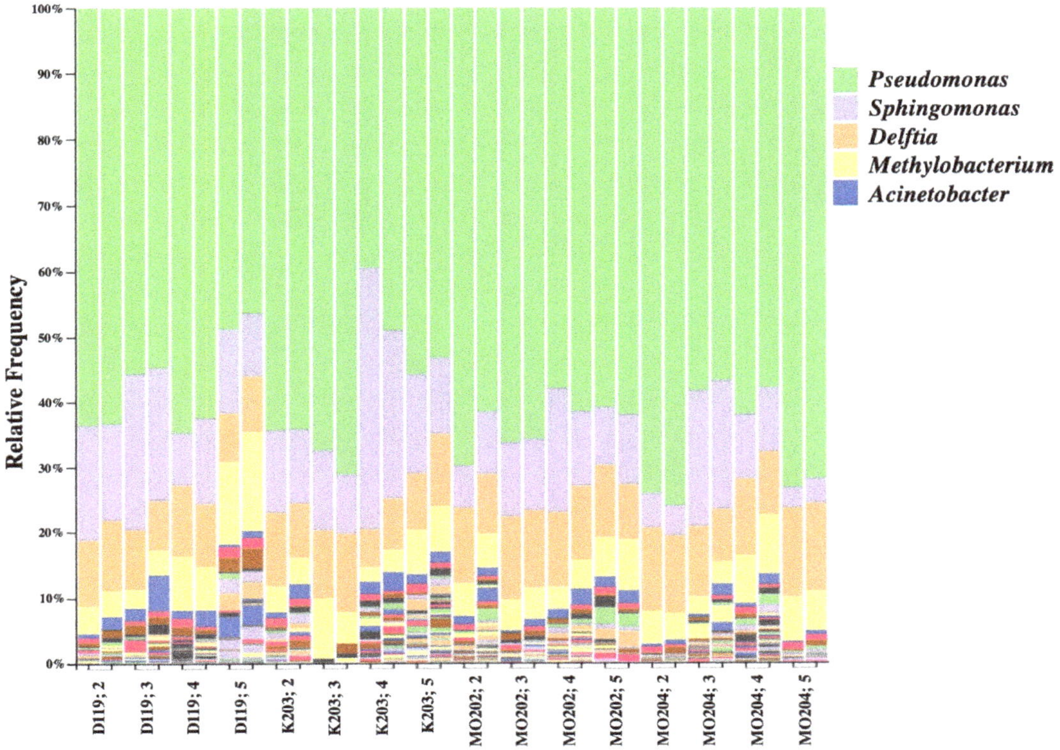

Figure 1. QIIME 2 bar plot of the top five most abundant bacterial genera per instar of each *Bombyx mori* L. strain based on two biological replicates using Silva classifier (top five genera: *Pseudomonas*, *Sphingomonas*, *Delftia*, *Methylobacterium*, *Acinetobacter*).

NCBI BLAST queries (Table S7) revealed that the most frequent representative sequences under each of the genera were species that have been isolated from various sources, including soil, plant, and insect gut [22]. BLAST queries also returned hits for contaminating uncultured and cultured bacteria that might have come from the reagents or the laboratory used to prepare the sequencing templates and libraries.

3.2. Alpha-Diversity Analysis

Bar plots detailing Shannon diversity index values per silkworm strain and instar are presented in Figure 2A,B, respectively. Individual Shannon diversity values for each strain and instar are listed in Table S8. Silkworm strains do not seem to differ significantly in terms of diversity based on *p*-values from Kruskal–Wallis tests for all groups (Table S9). Strain MO202 possesses the closest levels of bacterial diversity among its instars (Figure 2A). This strain, however, also appears to be the second least diverse on average compared to the other strains. DMMMSU 119 appears the most diverse on average compared to other strains, followed by K203 as the second most diverse strain. The individual instars of DMMMSU 119, K203, and MO204 appear to have different diversity values, suggesting that some instars of each strain are rich in the number of bacterial taxa within their guts and some instars have low bacterial diversity. MO204 has the lowest mean Shannon diversity index value, indicating that this strain is the least diverse on average.

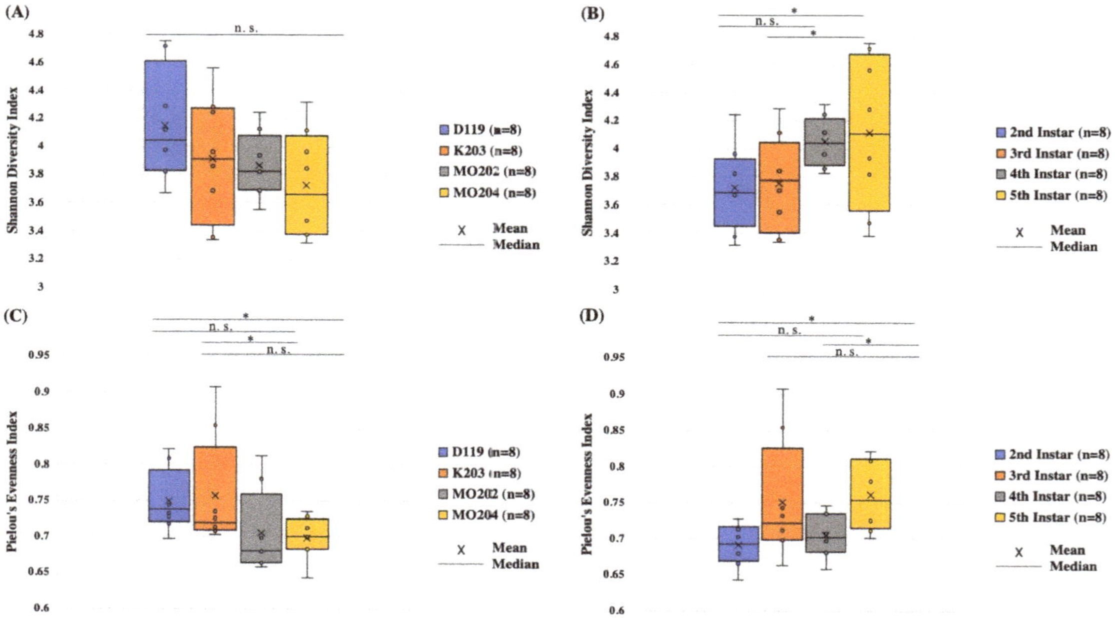

Figure 2. Alpha-diversity plots for *Bombyx mori* L. strains and instars reared at the Philippine Textile Research Institute Technological Center in Misamis Oriental (PTRI-TCMO). (**A**) Shannon diversity index boxplot for *Bombyx mori* L. strains; (**B**) Shannon diversity index boxplot for *Bombyx mori* L. instars; (**C**) Pielou's evenness index boxplot for *Bombyx mori* L. strains; (**D**) Pielou's evenness index boxplot for *Bombyx mori* L. instars. Significance ($p < 0.05$) is indicated with a (*). Comparisons that are not significant ($p > 0.05$) are indicated with "n.s."

Silkworm instars also do not appear significantly different in terms of diversity based on *p*-values from Kruskal–Wallis tests for all groups (Table S9). Among the silkworm instars across the four silkworm strains, the fifth instars seem to be the richest in terms of bacterial taxa, followed by the third instars (Figure 2B). Fourth instars, and subsequently the second instars, appear less diverse on average. Only fourth instars appear to differ significantly from second instars ($p = 0.0117$) and third instars ($p = 0.0460$), however. These significant differences in terms of bacterial diversity in the guts of some silkworm instars may correlate to the increase in mulberry leaf consumption and the potential changes in the gut bacterial community as a result. Chen et al. (2018) [3] also reported a shift in abundance of particular bacteria in the guts of silkworms from early to late instar stages.

Bar plots detailing Pielou's evenness index values per silkworm strain and instar are presented in Figure 2C,D, respectively. Individual Pielou's evenness values for each strain and instar are listed in Table S10. Silkworm strains appear to differ significantly in terms of evenness based on *p*-values from Kruskal–Wallis tests for all groups (Table S11). Strain K203 appears to be the most even in terms of the distribution of bacterial taxa on average (Figure 2C). DMMMSU 119 and MO204 appear as the second and third most even strains in terms of bacterial taxa distribution. DMMMSU 119 was also significantly different from MO204 in terms of evenness ($p = 0.0209$). The ranking of mean evenness values for DMMMSU 119 and K203 differs from those of their mean diversity values wherein DMMMSU 119 was more diverse than K203 (Figure 2A). MO202 and MO204 appear less even on average (Figure 2C), similar to their lower mean Shannon diversity values (Figure 2A).

Silkworm instars did not seem to differ significantly in terms of evenness based on p-values from Kruskal–Wallis tests for all groups (Table S11). Similar to the trend in diversity, fifth instars appear to be the most even in terms of bacterial taxa distribution (Figure 2D). Fifth instars only appeared significantly different from second instars in terms of evenness ($p = 0.0117$), however. Unlike the trend in diversity, third instars appeared more diverse than fourth instars (Figure 2D). Second instars were still the least even on average, however (Figure 2D), which is similar to their ranking as the least diverse instar on average (Figure 2B). Second and fourth instars were also more similar in terms of evenness within their respective groups compared to third and fifth instars. The trend of evenness values displayed by the instars matches that of their Shannon diversity index values (Figure 2B), suggesting that evenness possibly contributed to their diversity. This may also suggest that when fewer dominant bacterial taxa are present, there would be more taxa overall due to the absence of certain taxa that monopolize the environment and resources, thus, preventing others from thriving.

3.3. Beta-Diversity Analysis

Bray–Curtis dissimilarity 3D plots sorted by silkworm strain and instars are shown in Figure 3A,B. Silkworm strains appeared to differ significantly in terms of Bray–Curtis dissimilarity based on p-values from PERMANOVA tests (999 permutations) for all groups (Table S12). Figure 3A shows similar bacterial abundance within instars of strain MO202. Instars of strains DMMMSU 119, K203, and MO204 appear more dissimilar to other instars of their respective strain. Strain DMMMSU 119 also appears the most dissimilar compared to the other three strains, but only significantly different from strains MO202 ($p = 0.004$) and MO204 ($p = 0.021$). Silkworm instars also appeared significantly different in terms of Bray–Curtis dissimilarity based on p-values from PERMANOVA tests (999 permutations) for all groups (Table S12). Based on the Bray–Curtis dissimilarity values of the instars, diversity in second instars appears to be the most similar group and only varies significantly from the fourth instars ($p = 0.029$) (Figure 3B). Although there appears to be dissimilarity among the different instars, these are not significant when compared in a pairwise manner.

The Jaccard similarity index 3D plots sorted by silkworm strain and instar stages are shown in Figure 3C,D. Silkworm strains did not appear significantly different from each other in terms of Jaccard similarity based on p-values from PERMANOVA tests (999 permutations) for all groups (Table S13). Figure 3C shows no clustering within strains; however, the values are relatively close to one another. This further suggests that silkworm strains are similar in terms of the bacterial taxa that are present and absent (not considering the abundance of the taxa that are present). Despite the similarities observed, strain DMMMSU 119 appeared significantly different from strains MO202 ($p = 0.018$) and MO204 ($p = 0.002$) once again. Silkworm instars, in contrast, appeared significantly different from one another in terms of Jaccard similarity based on p-values from PERMANOVA tests (999 permutations) for all groups (Table S13). Once again, no clustering was observed within any of the instars. The values all appeared relatively similar to each other. Only the third and fifth instars appeared significantly different from one another ($p = 0.042$). Combined with the generally low Bray–Curtis dissimilarity values for most strains and instars, this suggests that the common bacterial taxa found among silkworm strains and instars possess close abundance values.

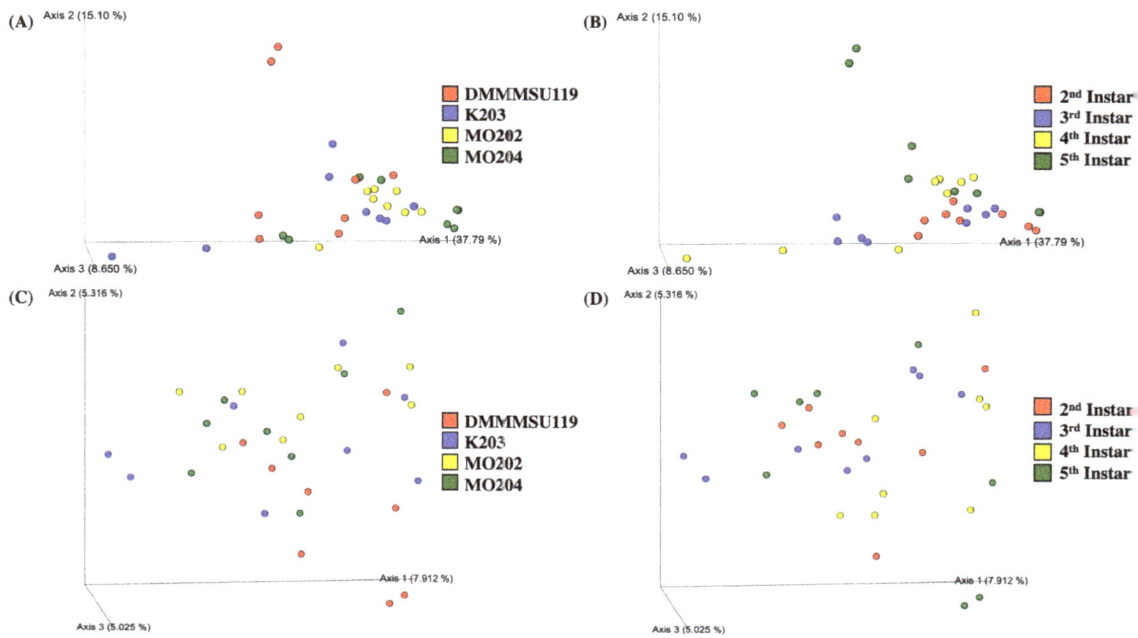

Figure 3. Beta-diversity plots for *Bombyx mori* L. strains and instars reared at the Philippine Textile Research Institute Technological Center in Misamis Oriental (PTRI-TCMO). (**A**) Bray–Curtis dissimilarity sorted by *Bombyx mori* L. strains (p = 0.03); (**B**) Bray–Curtis dissimilarity sorted by *Bombyx mori* L. instars (p = 0.048); (**C**) Jaccard similarity index sorted by *Bombyx mori* L. strains (p = 0.079) (**D**) Jaccard similarity index sorted by *Bombyx mori* L. instars (p = 0.024).

4. Discussion

4.1. Bacterial Abundance

The dominant bacterial communities associated with Philippine *B. mori* belong to the genera *Pseudomonas*, *Sphingomonas*, *Delftia*, *Methylobacterium*, and *Acinetobacter*. These genera have been associated with mulberries [23–25] and were also found prevalent in the guts of other strains of silkworms [3,17]. This indicates shared similarities between silkworms reared in the Philippines and silkworms reared elsewhere, in terms of the bacterial genera present in their guts. However, these bacterial communities still differ in terms of overall genera that are present as well as their relative abundance.

For the species identity of bacteria under each genus, NCBI BLAST queries revealed that the most frequent representative sequences under *Pseudomonas* were mostly identical to *Pseudomonas fluorescens* strains, which were isolated either from sediments or from gut of adult fruit-tree pinhole borer, *Xyleborinus saxesenii*, an ambrosia beetle. Like several bacteria of the genus *Pseudomonas*, the *P. fluorescens* group is found in diverse environments and well-known for its plant-beneficial properties, including pathogen suppression [26]. Thus, the *Pseudomonas* species found in silkworms in this study likely originated from mulberries and from the field where these host plants were cultivated. However, there were recent observations that some strains of this group can also colonize insects and cause severe infections leading to death [27]. *Pseudomonas* strains from other studies also showed amylolytic, cellulolytic, xylanolytic, lipolytic, and esterase activity in silkworms and beetles [28–30]. Hence, the sequence similarity of *Pseudomonas* rep-seqs in this study was notable in consideration of the ability to hydrolyze polymers of mulberries. BLAST queries also returned hits for uncultured bacterium isolated from a human body part, implicating the presence of contaminants. This could have been prevented to a certain extent if a negative control was included in the sequencing and analysis. BLAST also returned hits of most frequent rep-seqs for strains of *Sphingomonas* that were isolated from nectarine tree and paddy soil. Species of this genus have no reported biological function in the other strains of silkworm where they were detected, but some species have been observed to possess the ability to degrade organic matter [31] and also the ability to produce beta-carotene and gellan gum [32,33]. *Sphingomonas* has also been associated with diapause-destined female of the cabbage beetle, *Colaphellus bowringi* [34] and was found to be one of the genera of cellulolytic bacteria in the gut of the Chinese white pine beetle, *Dendroctonus armandi*, larvae [35]. Species of *Sphingomonas* have also been reported to be part of the core members of mulberry endophytes regardless of the season or cultivar [36]. Contaminating *Sphingomonas* species from sterile water was also detected, however, highlighting further the need for controls in this type of study.

Delftia species isolated from the gut of another moth and from soil were also among the BLAST hits for rep-seqs in this study. Besides finding *Delftia* strain in silkworms in other studies, a recent report indicated its presence in diapause-destined *C. bowringi* [34]. A species of *Delftia* (*D. acidovorans*) isolated from the American cockroach, *Periplaneta americana* L., which was found to produce bacteriocins against bacterial and fungal pathogens of humans [37] was also recently reported.

Functions linked to the presence of bacteria under the genus *Methylobacterium* include nitrogen-fixing in silkworm guts [3] and cellulolytic activity in the gut of the Chinese white pine beetle, *D. armandi*, larvae [35]. BLAST hits for *Methylobacterium* in the present study points to strains that might have been involved in nitrogen-fixation and -removal because the strains with hits were isolated from the rhizosphere and sewage sludge undergoing nitrogen removal. The strain similar to that isolated from a fish gut has no associated function, however.

Acinetobacter strains from other studies have shown amylolytic, cellulolytic, xylanolytic, lipolytic, and esterase activity in silkworms and beetles [28–30], but the strains most frequent in the silkworms used in this study were highly similar to those isolated from goose stools and soil.

Of interest in the data presented in this study is the relevance of these dominant bacterial communities in Philippine silkworm biology, which remains to be explored further.

4.2. Alpha-Diversity and Beta-Diversity

The increase in average bacterial diversity as the larvae mature suggests a shift in gut bacterial diversity. This occurs either in response to host development or an increase in leaf consumption and the age of mulberry leaves that were used. Strains that possessed high Shannon diversity values also possessed high Pielou's evenness values. This may suggest a potential link between bacterial diversity and evenness among the silkworm strains. This also indicates that bacterial diversity and evenness may be related given that the presence of fewer dominant bacterial taxa may allow more taxa to grow, as it is less likely that a certain taxon would monopolize resources or inhibit the growth of others. It should be noted, however, that neither silkworm strains nor instars appeared significantly different in terms of bacterial diversity overall, and only silkworm strains were significantly different in terms of bacterial evenness. This again relates to the similarity of the strains in terms of phenotypic traits and the most abundant bacterial taxa in their guts. These are also expected as the strains are genetically similar [9].

Quantitative measures of similarity among strains (Bray–Curtis dissimilarity) seem to set strain DMMMSU 119 apart from the other three silkworm strains. The other strains cluster together supporting the observed similar phenotypic traits. Whereas silkworm instars appeared significantly different from one another overall, only second and fourth instars appeared significantly different from each other. Both strains and instars appear similar in terms of qualitative measures of similarity that do not take bacterial abundance into account (Jaccard similarity). DMMMSU 119 was also significantly different from strains

MO202 and MO204, and third and fifth instars also appeared significantly different from each other. This indicates that strains and instars are similar in terms of the bacterial taxa present but are less similar in terms of the actual abundance of these bacterial taxa.

5. Summary and Conclusions

In this study, the bacterial communities of the four silkworm strains reared in the Philippines were investigated using high-throughput 16S rRNA gene amplicon sequencing and analysis. The results showed the abundance of five bacterial genera that were also found in other silkworm strains. Although the likely association of some of the bacterial species to biological activity in the gut as well as their involvement in the growth, development, and survival of silkworms have been reported, the exact roles of the bacteria within the silkworm strains in the present study and how these roles correlate to the production economically important silk characteristics. remain to be elucidated. Evidently, the results expand the biological data available for the silkworm strains present in the country and will be a useful source of information for future strain improvement to enhance the sustainability of quality silk production.

Supplementary Materials: The following supporting information can be downloaded at: https://www.mdpi.com/article/10.3390/insects13010100/s1: Table S1: Traits of *Bombyx mori* L. strains for selection, Table S2: Primer and Illumina adapter overhang sequences for *16S* V3-V4 PCR, Table S3: Nextera XT index pair assignments for each *16S* library prepared, Table S4: Sequencing run summary, Table S5: SRA accession numbers for *Bombyx mori* L. samples, Table S6: Sequence count summary for each *Bombyx mori* L. sample, Table S7: Representative sequences corresponding to top bacterial genera, their frequencies, top BLAST hits, and isolation sources, Table S8: Shannon's diversity boxplot values for each *Bombyx mori* L. strain and instar, Table S9: Comparison of Shannon's diversity index *p*-values, Table S10: Pielou's evenness boxplot values for each *Bombyx mori* L. strain and instar, Table S11: Comparison of Pielou's evenness index *p*-values, Table S12: Comparison of Bray–Curtis dissimilarity *p*-values, Table S13: Comparison of Jaccard similarity *p*-values, Figure S1: Diagram of QIIME 2 workflow, Figure S2: Alpha-rarefaction curves of each *Bombyx mori* L. strain, Figure S3: QIIME 2 heatmap of bacterial genera across the *Bombyx mori* L. samples.

Author Contributions: Conceptualization, M.A.M.B.; methodology and sampling, M.A.M.B. and I.Y.D.T.; investigation, M.A.M.B. and I.Y.D.T.; writing—original draft preparation, M.A.M.B. and I.Y.D.T.; writing—review and editing, M.A.M.B. and I.Y.D.T.; visualization, I.Y.D.T.; project administration, M.A.M.B. and I.Y.D.T. All authors have read and agreed to the published version of the manuscript.

Funding: This research was funded through the DOST—Philippine Textile Research Institute's Philippine Silkroad through Science and Technology Program's "Genome based Approach Silkworm Germplasm Maintenance and Hybridization in PTRI Technology Center Misamis Oriental" funded through Congressional Initiatives, and the National Institute of Molecular Biology and Biotechnology, University of the Philippines Diliman.

Institutional Review Board Statement: Not applicable.

Informed Consent Statement: Not applicable.

Data Availability Statement: The raw DNA sequence data generated and used in this study are available on NCBI's Sequence Read Archive (SRA) with accession numbers SRR17009362–SRR1700939[illegible].

Acknowledgments: The authors would like to thank the following: Philippine Textile Research Institute, Department of Science and Technology (DOST-PTRI) for the support and funding through their collaborative research project; PTRI Technology Center-Misamis Oriental (PTRI-TCMO) for providing the silkworm strains as well as the historical phenotypic data for the strains used in this study; National Institute of Molecular Biology and Biotechnology, University of the Philippines Diliman for the use of its facility during the conduct of the study and for publication support to M.A.M.B.; and John Michael Egana for his assistance in bioinformatics.

Conflicts of Interest: The authors declare no conflict of interest.

References

1. Mita, K.; Kasahara, M.; Sasaki, S.; Nagayasu, Y.; Yamada, T.; Kanamori, H.; Namiki, N.; Kitagawa, M.; Yamashita, H.; Yasukochi, Y.; et al. The genome sequence of silkworm, *Bombyx mori*. *DNA Res.* **2004**, *11*, 27–35. [CrossRef]
2. Mohanta, M.K.; Saha, A.K.; Saleh, D.K.M.A.; Islam, M.S.; Mannan, K.S.B.; Fakruddin, M. Characterization of *Klebsiella granulomatis* pathogenic to silkworm, *Bombyx mori* L. *Biotech* **2010**, *5*, 577–583. [CrossRef]
3. Chen, B.; Du, K.; Sun, C.; Vimalanathan, A.; Liang, X.; Li, Y.; Wang, B.; Lu, X.; Li, L.; Shao, Y. Gut bacterial and fungal communities of the domesticated silkworm (*Bombyx mori*) and wild mulberry-feeding relatives. *ISME J.* **2018**, *12*, 2252–2262. [CrossRef]
4. Nagaraju, J.; Goldsmith, M.R. Silkworm genomics—Progress and prospects. *Curr. Sci.* **2002**, *83*, 415–425.
5. Xia, Q.Y.; Zhou, Z.Y.; Lu, C.; Cheng, D.; Dai, F.-Y.; Liu, B.; Zhao, P.; Zha, X.; Cheng, T.; Biology Analysis Group; et al. A draft sequence for the genome of the domesticated silkworm (*Bombyx mori*). *Science* **2004**, *306*, 1937–1940. [PubMed]
6. Pereira, N.C.; Munhoz, R.E.F.; Bignotto, T.S.; Bespalhuk, R.; Garay, L.B.; Saez, C.R.N.; Fernandez, M.A. Biological and molecular characterization of silkworm strains from the Brazilian germplasm bank of *Bombyx mori*. *Genet. Mol. Res.* **2013**, *12*, 2138–2147. [CrossRef] [PubMed]
7. Zanatta, D.B.; Bravo, J.P.; Barbosa, J.F.; Munhoz, R.E.F.; Fernandez, M.A. Evaluation of economically important traits from sixteen parental strains of the silkworm *Bombyx mori* L. (Lepidoptera: Bombycidae). *Neotrop. Entomol.* **2009**, *38*, 327–331. [CrossRef]
8. Matienzo, L.H.; Gatica, R.A. Sericulture: Silk Production in the Philippines. Available online: https://aboutphilippines.org/files/Silk-Production-in-the-Philippines.pdf (accessed on 6 December 2018).
9. Alcudia-Catalma, M.N.; Conde, M.Y.E.D.; Dee Tan, I.Y.; Bautista, M.A.M. First report on the characterization of genetic diversity of Philippine-reared *Bombyx mori* strains based on *COI* and ITS2. *Philipp. J. Sci.* **2020**, *150* (Suppl. S1), 503–517.
10. Basaen, A.M.; Placido, A.; Bacuso, P.; Ladines, A.; Cabrito, F. Silkworm breeding for the development of Philippine purelines. *NSTA Technol. J.* **1982**, *7*, 72–85.
11. Villanueva, R.C. Problems and Issues Affecting the Pace of Sericulture Development in the Philippines. 1999. Available online: https://agris.fao.org/agris-search/search.do?recordID=PH2007M00034 (accessed on 11 January 2022).
12. Sarkar, K. Management of nutritional and climatic factors for silkworm rearing in West Bengal: A review. *Int. J. Agric. Environ. Biotechnol.* **2018**, *11*, 769–780. [CrossRef]
13. Barretto, D.A.; Gadwala, M.; Vootla, S.K. Chapter 1—The silkworm gut microbiota: A potential source for biotechnological applications. *Methods Microbiol.* **2021**, *49*, 1–26.
14. Sun, Z.; Kumar, D.; Cao, G.; Zhu, L.; Liu, B.; Zhu, M.; Liang, Z.; Kuang, S.; Chen, F.; Feng, Y.; et al. Effects of transient high temperature treatment on the intestinal flora of the silkworm *Bombyx mori*. *Sci. Rep.* **2017**, *7*, 3349. [CrossRef]
15. National Research Council (US). Committee on Metagenomics: Challenges and Functional Applications. In *The New Science of Metagenomics: Revealing Secrets of Our Microbial Planet*; National Academies Press (US): Washington, DC, USA, 2007. Available online: https://www.ncbi.nlm.nih.gov/books/NBK54011/ (accessed on 7 May 2019).
16. Thomas, T.; Gilbert, J.; Meyer, F. Metagenomics—A guide from sampling to data analysis. *Microb. Inform. Exp.* **2021**, *2*, 3. [CrossRef]
17. Sun, Z.; Lu, Y.; Zhang, H.; Kumar, D.; Liu, B.; Gong, Y.; Zhu, M.; Zhu, L.; Liang, Z.; Kuang, S.; et al. Effects of BmCPV infection on silkworm *Bombyx mori* intestinal bacteria. *PLoS ONE* **2016**, *11*, e0146313. [CrossRef]
18. Hammer, T.; Dickerson, J.; Fierer, N. Evidence-based recommendations on storing and handling specimens for analyses of insect microbiota. *PeerJ* **2015**, *3*, e1190. [CrossRef]
19. Herlemann, D.P.R.; Labrenz, M.; Jürgens, K.; Bertilsson, S.; Waniek, J.J.; Andersson, A.F. Transitions in bacterial communities along the 2000 km salinity gradient of the Baltic Sea. *ISME J.* **2011**, *5*, 1571–1579. [CrossRef]
20. Suenami, S.; Nobu, M.K.; Miyazaki, R. Community analysis of gut microbiota in hornets, the largest eusocial wasps, *Vespa mandarinia* and *V. simillima*. *Sci. Rep.* **2019**, *9*, 9830. [CrossRef]
21. Wang, Q.; Garrity, G.; Tiedje, J.; Cole, J. Naïve Bayesian classifier for rapid assignment of rRNA sequences into the new bacterial taxonomy. *Appl. Environ. Microbiol.* **2007**, *73*, 5261–5267. [CrossRef]
22. Schoch, C.L.; Ciufo, S.; Domrachev, M.; Hotton, C.L.; Kannan, S.; Khovanskaya, R.; Leipe, D.; McVeigh, R.; O'Neill, K.; Robbertse, B.; et al. NCBI Taxonomy: A comprehensive update on curation, resources and tools. *Database* **2020**, *2020*, 1–21. [CrossRef]
23. Ou, T.; Xu, W.-F.; Wang, F.; Strobel, G.; Zhou, Z.-Y.; Xiang, Z.-H.; Liu, J.; Xie, J. A microbiome study reveals seasonal variation in endophytic bacteria among different mulberry cultivars. *Comput. Struct. Biotechnol. J.* **2019**, *17*, 1091–1100. [CrossRef]
24. Lukša, J.; Servienė, E. White mulberry (*Morus alba* L.) fruit-associated bacterial and fungal microbiota. *J. Environ. Eng. Landsc. Manag.* **2020**, *28*, 183–191. [CrossRef]
25. Dong, H.L.; Zhang, S.X.; Chen, Z.H.; Tao, H.; Li, X.; Qiu, J.F.; Xu, S.Q. Differences in gut microbiota between silkworms (*Bombyx mori*) reared on fresh mulberry (*Morus alba* var. *multicaulis) leaves or an artificial diet. RSC Adv.* **2018**, *8*, 26188–26200.
26. Flury, P.; Aellen, N.; Ruffner, B.; Péchy-Tarr, M.; Fataar, S.; Metla, Z.; Maurhofer, M. Insect pathogenicity in plant-beneficial pseudomonads: Phylogenetic distribution and comparative genomics. *ISME J.* **2016**, *10*, 2527–2542. [CrossRef] [PubMed]
27. Flury, P.; Vesga, P.; Péchy-Tarr, M.; Aellen, N.; Dennert, F.; Hofer, N.; Kupferschmied, K.P.; Kupferschmied, P.; Metla, Z.; Ma, Z.; et al. Antimicrobial and insecticidal: Cyclic lipopeptides and hydrogen cyanide produced by plant-beneficial *Pseudomonas* strains CHA0, CMR12a, and PCL1391 contribute to insect killing. *Front. Microbiol.* **2017**, *8*, 100. [CrossRef]

28. Prem Anand, A.A.; Vennison, S.J.; Sankar, S.G.; Gilwax Prabhu, D.I.; Vasan, P.T.; Raghuraman, T.; Vendan, S.E. Isolation and characterization of bacteria from the gut of *Bombyx mori* that degrade cellulose, xylan, pectin and starch and their impact on digestion. *J. Insect. Sci.* **2010**, *10*, 107.
29. Briones-Roblero, C.I.; Rodríguez-Díaz, R.; Santiago-Cruz, J.A.; Zúñiga, G.; Rivera-Orduña, F.N. Degradation capacities of bacteria and yeasts isolated from the gut of *Dendroctonus rhizophagus* (Curculionidae: Scolytinae). *Folia Microbiol. (Praha)* **2017**, *62*, 1–9. [CrossRef]
30. Delalibera, I., Jr.; Handelsman, J.; Raffa, K.F. Contrasts in cellulolytic activities of gut microorganisms between the wood borer, *Saperda vestita* (Coleoptera: Cerambycidae), and the bark beetles, *Ips pini* and *Dendroctonus frontalis* (Coleoptera: Curculionidae). *Environ. Entomol.* **2005**, *34*, 541–547. [CrossRef]
31. Gong, B.; Wu, P.; Huang, Z.; Li, Y.; Dang, Z.; Ruan, B.; Kang, C.; Zhu, N. Enhanced degradation of phenol by *Sphingomonas* sp. GY2B with resistance towards suboptimal environment through adsorption on kaolinite. *Chemosphere* **2016**, *148*, 388–394. [CrossRef]
32. Silva, C.; Cabral, J.M.S.; van Keulen, F. Isolation of a beta-carotene over-producing soil bacterium, *Sphingomonas* sp. *Biotechnol. Lett.* **2004**, *26*, 257–262. [CrossRef]
33. Wang, X.; Xu, P.; Yuan, Y.; Liu, C.; Zhang, D.; Yang, Z.; Yang, C.; Ma, C. Modeling for gellan gum production by *Sphingomonas paucimobilis* ATCC 31461 in a simplified medium. *Appl. Environ. Microbiol.* **2006**, *72*, 3367–3374. [CrossRef]
34. Liu, W.; Li, Y.; Guo, S.; Yin, H.; Lei, C.; Wang, X. Association between gut microbiota and diapause preparation in the cabbage beetle: A new perspective for studying insect diapause. *Sci. Rep.* **2016**, *6*, 38900. [CrossRef]
35. Hu, X.; Yu, J.; Wang, C.; Chen, H. Cellulolytic bacteria associated with the gut of *Dendroctonus armandi* larvae (Coleoptera: Curculionidae: Scolytinae). *Forests* **2014**, *5*, 455–465. [CrossRef]
36. Xu, W.; Wang, F.; Zhang, M.; Ou, T.; Wang, R.; Strobel, G.; Xiang, Z.; Zhou, Z.; Xie, J. Diversity of cultivable endophytic bacteria in mulberry and their potential for antimicrobial and plant growth-promoting activities. *Microbiol. Res.* **2019**, *229*, 126328. [CrossRef] [PubMed]
37. Amer, A.; Hamdy, B.; Mahmoud, D.; Elanany, M.; Rady, M.; Alahmadi, T.; Alharbi, S.; AlAshaal, S. Antagonistic activity of bacteria isolated from the *Periplaneta americana* L. gut against some multidrug-resistant human pathogens. *Antibiotics* **2021**, *10*, 294. [CrossRef] [PubMed]

insects

MDPI

Article

Enterococci as Dominant Xylose Utilizing Lactic Acid Bacteria in Eri Silkworm Midgut and the Potential Use of *Enterococcus hirae* as Probiotic for Eri Culture

Kridsada Unban [1], Augchararat Klongklaew [2], Pratthana Kodchasee [2], Punnita Pamueangmun [2], Kalidas Shetty [3] and Chartchai Khanongnuch [1,2,4,*]

1 Division of Biotechnology, School of Agro-Industry, Faculty of Agro-Industry, Chiang Mai University, Mueang, Chiang Mai 50100, Thailand; kridsada_u@cmu.ac.th
2 Interdisciplinary Program in Biotechnology, The Graduate School, Chiang Mai University, Mueang, Chiang Mai 50200, Thailand; augchararat.taey@gmail.com (A.K.); pratthana_kod@cmu.ac.th (P.K.); pia_pannita39@hotmail.com (P.P.)
3 Department of Plant Sciences, Global Institute of Food Security and International Agriculture (GIFSIA), North Dakota State University, Fargo, ND 58108, USA; kalidas.shetty@ndsu.edu
4 Research Center for Multidisciplinary Approaches to Miang, Science and Technology Research Institute Chiang Mai University, Mueang, Chiang Mai 50200, Thailand
* Correspondence: chartchai.k@cmu.ac.th; Tel.: +66-53-948-261

Citation: Unban, K.; Klongklaew, A.; Kodchasee, P.; Pamueangmun, P.; Shetty, K.; Khanongnuch, C. Enterococci as Dominant Xylose Utilizing Lactic Acid Bacteria in Eri Silkworm Midgut and the Potential Use of *Enterococcus hirae* as Probiotic for Eri Culture. *Insects* **2022**, *13*, 136. https://doi.org/10.3390/insects13020136

Academic Editors: Silvia Cappellozza, Morena Casartelli, Federica Sandrelli, Alessio Saviane and Gianluca Tettamanti

Received: 12 November 2021
Accepted: 24 January 2022
Published: 27 January 2022

Publisher's Note: MDPI stays neutral with regard to jurisdictional claims in published maps and institutional affiliations.

Simple Summary: This study focused on isolation and identification of xylose utilizing lactic acid bacteria from the midgut of Eri silkworm to understand their characteristics such as tannin tolerance, production of cellulolytic enzymes, and antimicrobial activity against insect pathogenic bacteria. The *Enterococcus* was found as the dominant genus among xylose utilizing lactic acid bacteria. Within this genus, *Enterococcus hirae* SX2 showed the potential to be used as a probiotic in Eri silkworm culture due to its tannin tolerance and antimicrobial activity against insect pathogenic bacteria. The trial experiment for applying live *E. hirae* SX2 supplemented to castor leaves in Eri silkworm rearing showed a positive effect for improving larval weight and survival. These findings led to the development of a new probiotic for Eri culture and also could be the experimental model for screening of the potential probiotic from mulberry silkworm (*Bombyx mori*).

Abstract: A total of 51 pentose utilizing lactic acid bacteria (LAB) were isolated from acid-forming bacteria in the midgut of healthy mature Eri silkworm using de Man, Rogosa and Sharpe (MRS) agar containing 10 g/L xylose (MRS-xylose) as the carbon source supplemented with 0.04% (w/v) bromocresol purple. Further analysis of 16S rRNA gene sequences revealed the highest prevalence of up to 35 enterococci isolates, which included 20 isolates of *Enterococcus mundtii*, followed by *Entercoccus faecalis* (eight isolates), *Weissella cibaria* (four isolates), *Enterococcus hirae* (two isolates), *Enterococcus lactis* (one isolate), and *Enterococcus faecium* (one isolate). All 51 LAB isolates showed positive growth on MRS containing a range of polysaccharides as the sole carbon source. All isolates were able to grow and form clear zones on MRS supplemented with 1 g/L xylose, while *E. faecalis* SC1, *E. faecalis* SCT2, and *E. hirae* SX2 showed tannin tolerance ability up to 5 g/L. Moreover, five isolates showed antimicrobial activity against Eri silkworm pathogens, including *Bacillus cereus*, *Staphylococcus aureus*, and *Proteus vulgaris*, with *E. hirae* SX2 having the highest inhibitory effect. Supplementation of live *E. hirae* SX2 on castor leaves significantly improved the weight and reduced the silkworm mortality when compared with the control group ($p < 0.05$). This cocci LAB can be considered as the new probiotic for Eri culture. Additionally, this finding presented the perspective of non-mulberry silkworm that could also be used as the model for further applying to new trends of the sericulture industry.

Keywords: Eri silkworm; pentose-utilizing microbe; lactic acid bacteria; probiotic; Eri culture

1. Introduction

Eri silk is non-mulberry silk that is becoming more popular due to improvements in relevant silkworm domestication and multivoltine nature of silk. Eri silk is made from Eri silkworm (*Samia ricini*), which is believed to be originally found in northeast India, primarily in Assam and Meghalaya, and then distributed to other Asian countries [1]. The cultivation of Eri silkworm has benefits and relevance in rural-based on-farm and off-farm activity, which has been recognized as a beneficial approach for socio-economic development, particularly in some Asian developing countries [2]. Furthermore, the textile products from Eri silk are promoted as green products due to their natural process and eco-friendly properties [3]. Since the normal feed of Eri silkworm is a variety of plant leaves that are structurally composed of cellulose, hemicelluloses such as xylan and mannan, pectin, lignin, and the small quantities of protein and fat [4], the gut of this Lepidopteran has a major role in the digestion of leaf components for nutrition. The microbial community in the insect gut has been extensively studied and discussed regarding their symbiont characteristics and role as beneficial gut microbiota where their metabolically relevant enzymes are required for digestion of the substrates and release of nutritionally important compounds [5]. The bacterial community in the midgut of this Lepidopteran insect has also been investigated for the production of some digestive enzymes required for the digestion of leaf components [6]. Several reports concluded that gut microbiota influences the insects in several ways, such as supporting nutrient digestion and detoxification [7] improving innate immunity [8], providing nutrients and growth-promoting metabolites [9] and protecting against infectious pathogens and parasites [10]. Though several studies reported on the gut microbiota linked to insect metabolism and growth development, the relationship between the gut microbiota community and their possible role for advancing applications in Eri silk production are not fully understood.

Microbial biodiversity is now increasingly considered essential for the metabolism of higher eukaryotic systems, and therefore understanding relevant microbial community in Eri silkworm gut could advance these microbial resources in Eri silkworm production as a source of beneficial probiotic lactic acid bacteria. This has been specifically targeted as a source of optically pure L- or D-form producing microbes resulting in biological process for the production of optically pure L- or D-lactic acid, particularly from pentose substrates composed in lignocellulose that is required for the manufacturing of bioplastics such as polylactic acid (PLA) [11,12]. Besides the utilization as the biological catalyst for the production of lactic acid, some lactic acid bacteria (LAB) such as *Lactobacillus* sp. and *Bifidobacterium* sp. have been investigated and applied as probiotics for providing health benefits both for humans and animals. The effect of probiotic bacteria on the improvement of growth and quality of silkworm has also been reported [13–16]. However, most previous studies have investigated the mulberry silkworm (*Bombyx mori*), and none has focused on Eri silkworm. Since the important characteristic of LAB probiotics is the generation of antimicrobial metabolites for countering pathogenic or undesirable microbes in the host gastrointestinal tract [17], the antimicrobial activity against insect pathogens in response to LAB is also considered as an important characteristic of probiotics for silkworm cultivation. In addition, the application of live probiotics is more eco-friendly for the management of silkworm diseases rather than the use of antibiotics, which is widely used in the sericulture industry [18].

Therefore, based on the above background and rationale in this study, the isolation of xylose utilizing LAB and their molecular identification aligned with their antimicrobial activities against insect pathogenic bacteria were investigated. Further investigation of applying selected xylose utilizing LAB as probiotic in Eri culture was also advanced.

2. Materials and Methods

2.1. Isolation and Screening of Xylose Utilizing LAB

Ten mature Eri silkworm larvae (5th instar day 5) were collected from a local farm in Sankampang district, Chiang Mai province, Thailand. The larvae were rinsed twice with sterile water and surface sterilized in ethanol (70%, v/v) for 60 s, followed by a final rinsing in sterile distilled water before dissection. Then, the worm's gut tissue was aseptically dissected and immediately transferred to a sterile microcentrifuge tube containing sterile 0.85% (w/v) NaCl solution in a ratio of 1:10. A 10-fold serial dilution was carried out by pipetting 0.5 mL of homogenized sample into 4.5 mL of 0.85% (w/v) NaCl solution. The mixture was then mixed well, and 100 μL of each dilution was transferred to nutrient agar (NA) containing 10 g/L glucose and 0.04% (w/v) bromocresol purple using the spread plate technique. After 24 h incubation at 37 °C, the acid-forming bacterial isolates observed from the yellow clear zone around their colonies were then transferred into NA containing 10 g/L xylose and 0.04% (w/v) bromocresol purple. The viable bacterial cells were enumerated as colony-forming units (CFU) per gram of sample.

All acid-forming bacterial colonies capable of growth on NA xylose were transferred into modified de Man, Rogosa and Sharpe (MRS) agar containing 10 g/L beef extract, 10 g/L peptone, 5 g/L yeast extract, 5 g/L sodium acetate, 2 g/L K_2HPO_4, 2 g/L tri-ammonium citrate, 0.2 g/L $MgSO_4$, 0.2 g/L $MnSO_4$, 0.1% (v/v) Tween80, and 10 g/L xylose supplemented with 0.04% (w/v) bromocresol purple and then incubated at 37 °C for 24 h. The colonies that exhibited a yellow zone around their respective colony were presumptively considered to be LAB and were individually streaked on MRS-xylose agar plate to obtain single colonies. The cultures of pure isolates were stored at −80 °C in MRS broth mixed with 25% (v/v) glycerol.

2.2. Identification of LAB by 16S rDNA Sequence Analysis

The molecular identification of LAB isolates was carried out using the method described by Unban et al. [19] with some modification. Briefly, the bacterial cells were collected from 10 mL of an overnight bacterial culture by centrifugation at 10,000× g for 10 min, and the genomic DNA was extracted by using the Wizard Genomic DNA purification kit (Promega Corp., Madison, WI, USA), according to the manufacturer's instructions. The 16S rRNA gene was amplified by using genomic DNA as a template with a standard polymerase chain reaction (PCR). Forward primers 27F, 5′-AGAGTTTGATCMTGGCTCAG-3′, and reverse primers 1525R, 5′-AAGGAGGTGWTCCARCC-3′, were used for 16S rRNA gene amplification. The amplification cycles were initially performed at 94 °C for 5 min of denaturation, followed by 30 cycles of denaturation at 94 °C for 20 s, annealing at 55 °C for 20 s, extension at 72 °C for 1 min 30 s, and additional final extension at 72 °C for 5 min. All PCR reactions were performed using Phusion High-Fidelity PCR Master Mix (New England Biolabs, MA, USA). The PCR products were purified and sequenced by a sequencing service provider (1st BASE Laboratory Company, Singapore). The comparison of 16S rRNA gene sequences was performed with the genetic database from National Center for Biotechnology Information (NCBI) GenBank and EzBioCloud databases. The multiple sequence alignment was performed using BioEdit 7.0 software tool and MEGA 4.0 [20]. The phylogenetic tree was constructed by the neighbor-joining methods with 1000 replications bootstrap analysis. All 16S rRNA gene sequences generated in this study have been deposited in the NCBI GenBank database under accession number MZ127632–MZ127643 and OM090176–OM090214.

2.3. Growth and Acid Formation on Polysaccharides

The selected strains of xylose utilizing LAB from Eri silkworm midgut were investigated for their abilities to utilize the selected polysaccharides as the sole carbon source. Briefly, individual LAB isolate was spiked on MRS agar containing 0.5% (w/v) of single carbon source of individual polysaccharides including starch (Fisher Scientific, Loughborough, Leicestershire, UK), pectin (Fisher Scientific, Loughborough, Leicestershire, UK), beechwood xylan (Megazyme International, Bray, Co., Wicklow, Ireland), locust bean gum (Sigma Aldrich, St. Louis, MO, USA) or carboxymethyl cellulose (Sigma Aldrich, St. Louis, MO, USA) supplemented with bromocresol purple (0.04%, w/v). After incubation at 37 °C for 24 h, the bacterial growth and the yellow halo-formed surrounding colonies were observed. Further duplicated experiment was also carried out, but the bromocresol purple was replaced with trypan blue (0.01%, w/v) for detection of extracellular activities of amylase, pectinase, xylanase, β-mannanase, and cellulase, respectively. The clear zone formed surrounding colonies were observed after incubation at 37 °C for 24 h.

2.4. Tannin-Tolerance Test

The ability of LAB isolates to tolerate tannins was evaluated on MRS-xylose agar supplemented with tannin. To prepare tannin solution, 10 g of tannic acid (LOBA Chemie, Mumbai, India) was dissolved in 100 mL of 0.1 M sodium phosphate buffer (pH 7.0) and sterilized by filtering through a 0.2 μm filter cartridge (Millipore, Billerica, MA, USA). The tannin solution was aseptically added into the sterilized MRS-xylose agar to achieve the final concentrations of 1 and 5 g/L. A single colony of each LAB isolate was picked up and spiked on the supplemented MRS-xylose agar. Growth of bacterial colony and the yellow clear zone formation of the isolates were observed after incubation at 37 °C for 24 h.

2.5. Antimicrobial Activity against Insect Pathogens

The antimicrobial activity against insect pathogenic bacteria was investigated by independent cultivation of all selected LAB isolates in MRS-xylose broth at 37 °C for 24 h. Then, cell-free culture supernatants (CFCS) were collected by centrifugation at 10,000× g for 15 min. The CFCS were then neutralized to pH 6.5 with 5 N NaOH and filtered through a 0.22 μm filter cartridge (Millipore, Billerica, MA, USA). The antimicrobial activity was investigated by the disc-agar diffusion method following the method of Fehlberg et al. [21]. Briefly, 100 μL of the overnight grown pathogen including *B. cereus* TISTR 747, *S. aureus* TISTR 746, and *P. vulgaris* TISTR 100 cultures (10^6–10^7 CFU/mL) was spread onto NA agar plate, and paper discs (diameter 6 mm, Macherey-Nagel, Duren, Germany) were placed on each plate. Then, 50 μL of CFCS from LAB isolates were transferred into paper discs. The diameter of the inhibition zone against pathogenic bacteria was measured after incubation at 37 °C for 24 h.

2.6. Effect of *E. hirae* SX2 Supplementation on Eri Silkworm Growth

A pure culture of *E. hirae* SX2 was inoculated in MRS-xylose broth and incubated at 37 °C under static conditions for 24 h. The bacterial cells were collected by centrifugation at 8000× g, 4 °C for 10 min and cell pellets was washed twice with sterile 0.85% (w/v) NaCl solution, and the cell pellets were resuspended in NaCl solution and adjusted to an optical density (OD600) of around 0.5 corresponding to approximately 10^8 CFU/mL. The freshly picked castor leaves were cleaned by excess volume of tap water and finally washed with sterile water. The cleaned leaves targeted for feeding were then sprayed with a cell suspension of *E. hirae* SX2 on both sides of leaves and shade dried before feeding. The 1st instar larvae of Eri silkworm were divided into two groups for the treatment, with each group consisting of 50 larvae; one group was reared on castor leaves served as control, while the other was fed on probiotics treated leaves. Three replications were maintained for each treatment. The treatment was given the first feed on the first day of 1st instar larvae up to 5th instar stage (3 weeks) under laboratory conditions at 25 ± 1 °C, the humidity of about 75 ± 5%, and photoperiod of 16 h of light and 8 h of dark. The treatment was fed 2–3 times a day, and the unfed leaves were removed from the trays daily. The growth of larva was monitored (10 larvae per treatment were selected randomly) by recording the larva weight once a week. At the end of rearing (3 weeks), 10 larvae per treatment were randomly selected as the representative of each treatment for dissection, and the midgut content were collected and mixed well for use as the representative of each group for determination of total bacterial count on NA agar supplemented with 10 g/L glucose and 0.04% (w/v) bromocresol purple as an indicator. The viable cell number of acid-forming bacteria was also enumerated and observed from the yellow clear zone formed surrounding the colony. The viable cell number of *E. hirae* SX2 was enumerated by spread plating on MRS-xylose agar supplemented with 0.04% bromocresol purple.

2.7. Statistical Analysis

The collected data were analyzed using the statistical program SPSS/PC version 17.0 (SPSS Inc. Chicago, IL, USA). The data were analyzed for achieving both normality and homoscedasticity, and the statistical significance of the differences among the treatments was evaluated by one-way ANOVA followed by Tukey's multiple range test. Values not sharing a common letter are significantly different from each other at $p < 0.05$.

3. Results and Discussion

3.1. Xylose Utilizing LAB Isolation

The healthy mature fifth instar Eri silkworm larvae cultivated at the local farm in Sankampang district, Chiang Mai, Thailand (Figure 1a) was used as the source of midgut content. The numbers of acid-forming bacteria that grew on different kinds of media are presented in Figure 1b. The total viable bacterial population of the Eri silkworm midgut observed on NA glucose were found to be 3.95×10^7 CFU/g of larval gut, while acid-forming bacteria were found at a high number of 3.76×10^7 CFU/g, which represented 95.4% of the total number of bacteria. Thereafter, 1000 isolates of these acid-forming bacteria were randomly replicated into NA-xylose and incubated at 37 °C for 24 h. Only 608 isolates (60.8%) showed the ability to grow and produce acid around the colonies. Of all isolates that were able to grow and produce acid in NA-xylose, only 51 isolates (10.2%) showed an acid-forming capability in the selective MRS-xylose agar and were presumptively identified as LAB. These results indicated that even xylose utilizing bacteria were found in significant numbers among acid-forming bacteria capable of growth in NA, but LAB was a minor component representing only 10.2% of the total number when tested on the LAB selecting medium such as MRS agar. These results were in agreement with the previous study, which suggested that the main population of culturable bacteria in Eri silkworm gut were non-LAB, including *Enterobacter* sp., *Pseudomonas* sp., *Citrobacter* sp., and *Bacillus* sp. [22]. This might be due to the pH condition of the insect gut, which is reported to be alkaline [10], which may limit the growth and number of LAB.

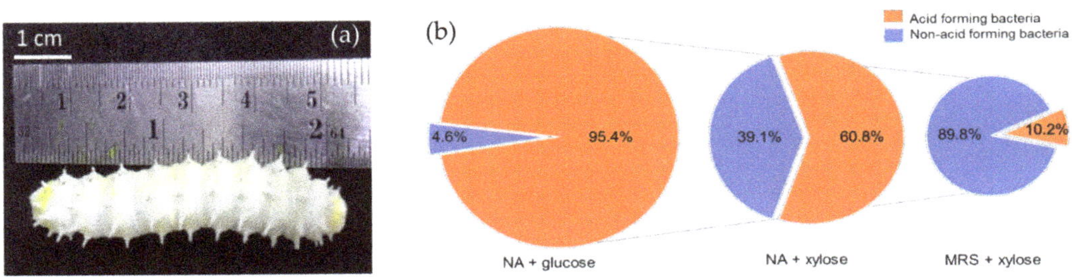

Figure 1. The mature Eri silkworm larvae cultivated at local farm in Chiang Mai, Thailand (**a**) and the pie diagram presenting the percentage of acid-forming bacteria from Eri silkworm midgut found in different agar medium (**b**).

3.2. Xylose Utilizing LAB Identification and Phylogenetic Analysis

The morphological study of 51 LAB isolates found that 47 isolates were Gram-positive with cocci shape, while the last four isolates were Gram-positive with irregular rod shape. All the xylose utilizing LAB isolates were identified using the sequence analysis of the 16S rRNA gene. The sequence was aligned with the related type strains, and the construction of phylogenetic trees was created and compared with the sequences of their closest relatives (Figure 2a). The full-length sequence of 16S rRNA gene of LAB isolates shared similarity to genera *Enterococcus* and *Weissella* with higher than 99%. Of 51 isolates, 35 were identified to be *Enterococcus mundtii*, which represented 68.6% of total isolates, followed by *Entercoccus faecalis* (15.7%), *Weissella cibaria* (7.8%), *Enterococcus hirae* (3.9%), *Enterococcus lactis* (2%), and *Enterococcus faecium* (2%) (Figure 2b). *Enterococcus* and

Lactobacillus have been reported to be the predominant LAB genera in mulberry silkworm midgut, where the relative abundance varied depending on silkworm species and their physiological activities [23]. In another study, the predominant gut microbiota from the healthy silkworm larvae were *Delftia* sp., *Ralstonia* sp., *Enterococcus* sp., *Staphylococcus* sp., and *Pelomonas* sp.; nevertheless, the observation of relative abundance change depended on the larval stage [24]. According to Chen et al. [25], approximately 10^7 CFU of bacteria were found in the whole gut of each sample. The identification using 16S rDNA sequence found that *Enterococcus* (62.1%) and *Clostridium* (35.4%) were the two main genera in each sample. Furthermore, the effect of bacteria including *Lactobacillus* spp., *Bifidobacterium*, *B. licheniformis*, and *B. niabensis* reflected in the improvement of growth and productivity characteristics of silkworm *Bombyx mori* L. [15,26,27]. Several studies recently have reported microbial abundance and diversity of Lepidopteran insect guts [22,23,28,29], and some xylose utilizing bacteria in Lepidoptera were also reported [16,30].

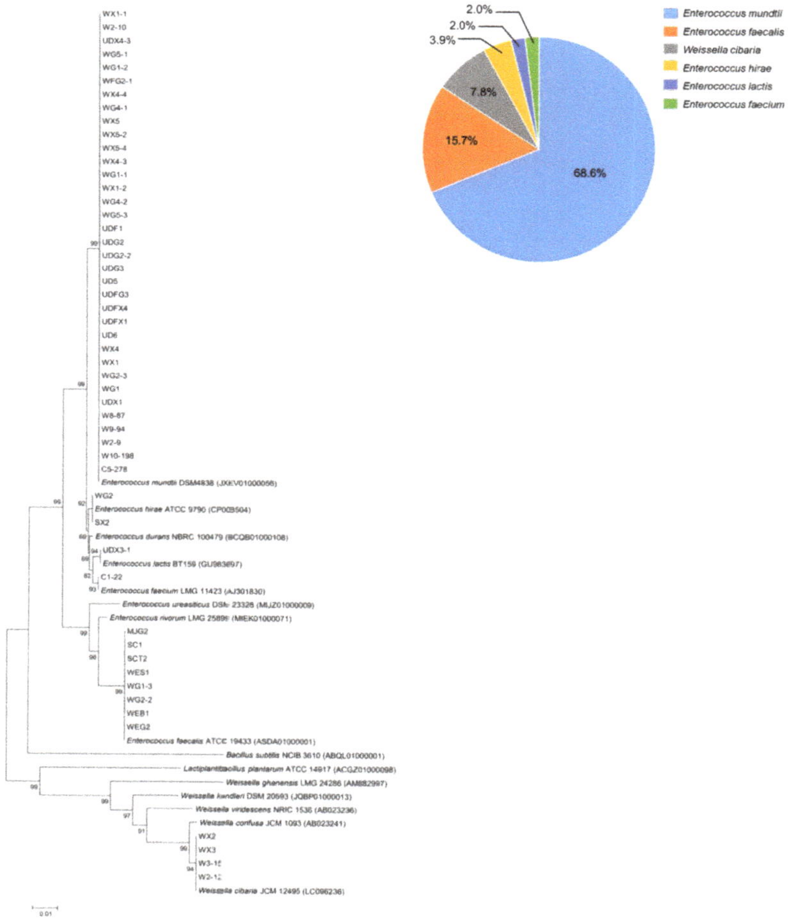

(a) (b)

Figure 2. Phylogenetic tree based on the 16S rRNA gene sequence analysis (**a**) and the proportion percentage of the identified species (**b**) of 51 selected LAB from Eri silkworm midgut.

3.3. Growth and Acid Formation on Polysaccharide Substrates

Silkworm does not code for cellulolytic genes; therefore, some of the cellulolytic-degrading enzymes might be produced by gut microbiota [31]. All selected xylose utilizing LAB do not show the extracellular polysaccharide degrading enzymes production on MRS agar supplemented with various polysaccharide substrates. However, all selected LAB showed positive growth, but the acid formation was found in varying levels when cultured on MRS agar supplemented with individual polysaccharides as the sole carbon source (data not shown). This means that these LAB may possess membrane-bound enzymes with the capability to convert these polysaccharides to free simple sugars and assimilate them into their metabolism via the lactic acid fermentation pathway. The results of this study revealed that some xylose utilizing LAB isolated in this study did not utilize various polysaccharide substrates by directly secreting the extracellular polysaccharide degrading enzymes, which is different from previous research where *Enterococcus* sp. was the host gut microbe with the capability of cellulolytic enzyme activity [14,32]. This evidence leads to a further suggestion that the main role of polysaccharides degradation in Eri silkworm may be undertaken by other non-LAB microbes such as the well-recognized polysaccharides degrading bacteria *Bacillus* spp., which is also reported to be the dominant bacterial population in the midgut of *Bombyx mori* L. silkworm [14,15].

3.4. Tannin-Tolerance Test

The results from these experiments found that all 51 LAB selected isolates showed tolerance against tannins supplemented in MRS-xylose agar at the final concentration of 1 g/L tannin with variation observed based on the extent of yellow clear zones around their colonies (Figure 3). The clear zones formed by 51 isolates on MRS-xylose with 1 g/L of tannin were found in the range of 4–15 mm, which was comparable to a lesser degree in size to their clear zone on the MRS-xylose agar without tannin supplementation. Interestingly, only three isolates (*E. faecalis* SC1, *E. faecalis* SCT2, and *E. hirae* SX2) were able to grow and form yellow clear zones on MRS-xylose supplemented with 5 g/L of tannin (Figure 3c), whereas the other isolates could not grow under this condition. Phenolic compounds of plants are secondary metabolites, including such as flavonoids, tannins, and phenolic acids, which exhibit many biological activities associated with plant growth and development [33]. Tannin is the polyphenolic compound that is commonly found in plant leaves, and this phenolic compound has been recognized for its antimicrobial activity effect as a result of binding either with the cellular enzymes in the cytoplasm or cell surface receptor proteins in the cell wall [34]. Thus, it seems likely that a tannin-rich environment provides a selective condition to the LAB that can survive in tannin-rich conditions. The capability of growth of *E. faecalis* SC1, *E. faecalis* SCT2, and *E. hirae* SX2 indicates the special beneficial characteristic for active survival of these LAB in the tannin-rich condition in Eri silkworm gut.

3.5. Antimicrobial Activity

One of the common benefits provided by probiotics is the ability to combat infection of pathogenic bacteria. In this study, *B. cereus*, *S. aureus*, and *P. vulgaris* were used as Gram-positive and Gram-negative pathogenic indicators, and the result is presented in Table 1. Overall, all 51 isolates exhibited inhibitory effects against at least one of the tested pathogens. Five isolates with the most excellent antagonistic activity against all three tested insect pathogenic bacteria were *E. faecalis* SCT2, *E. hirae* SX2, *E. mundtii* UDF1, *E. mundtii* UDFX1, and *E. mundtii* UDG3. *P. vulgaris* was found to be the most sensitive pathogenic strain based on the CFCS inhibitory effect produced by LAB isolates. The highest inhibitory effect was observed in *E. hirae* SX2 against all tested pathogenic bacteria (Figure 4).

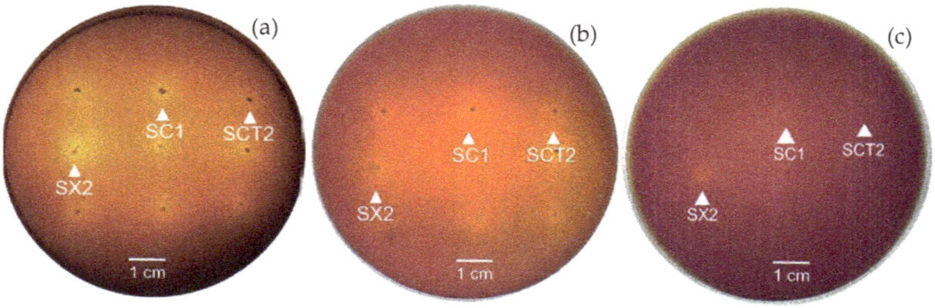

Figure 3. Growth and clear zone formation of xylose-utilizing LAB, *E. faecalis* SC1, *E. faecalis* SCT2, and *E. hirae* SX2, on MRS-xylose agar (**a**), MRS-xylose agar supplemented with 1 g/L tannin (**b**), and 5 g/L tannin (**c**), when cultivated at 37 °C for 24 h.

Table 1. Antimicrobial activity of LAB isolates against some pathogenic bacteria.

Isolates	Inhibition Zones against Pathogenic Bacteria (mm)					
	Proteus vulgaris		*Bacillus cereus*		*Streptococcus aureus*	
	Unneutralized CFCS	Neutralized CFCS	Unneutralized CFCS	Neutralized CFCS	Unneutralized CFCS	Neutralized CFCS
C1-22	++	++	+	+	ND	ND
C5-278	++	++	+	+	ND	ND
MJG2	++	++	+	+	ND	ND
SC1	++	++	+	+	ND	ND
SCT2	+++	+++	++	+	++	+
SX2	+++	+++	++	++	++	+
UD5	+++	+++	+	+	ND	ND
UD6	++	++	+	+	ND	ND
UDF1	+++	+++	++	+	+	ND
UDFG3	+	+	+	+	ND	ND
UDFX1	+++	+++	++	+	+	ND
UDFX4	++	++	+	+	ND	ND
UDG2	++	++	+	+	ND	ND
UDG2-2	++	++	+	+	ND	ND
UDG3	+++	+++	++	+	+	ND
UDX1	++	++	+	+	ND	ND
UDX3-1	++	++	+	+	ND	ND
UDX4-3	++	++	+	+	ND	ND
W10-198	++	++	+	+	ND	ND
W2-10	++	++	+	+	ND	ND
W2-12	++	++	+	+	ND	ND
W2-9	++	++	+	+	ND	ND
W3-15	++	++	+	+	ND	ND
W8-87	++	++	+	+	ND	ND
W9-94	++	++	+	+	ND	ND
WEB1	++	++	+	+	ND	ND
WEG2	++	++	+	+	ND	ND
WES1	+++	+++	+	+	ND	ND
WFG2-1	+	+	+	+	ND	ND
WG1	++	++	+	+	ND	ND
WG1-1	++	++	+	+	ND	ND
WG1-2	+	+	+	+	ND	ND
WG1-3	+	+	+	+	ND	ND
WG2	++	++	+	+	ND	ND
WG2-2	++	++	+	+	ND	ND
WG2-3	+	+	+	+	ND	ND
WG4-1	+	+	+	+	ND	ND
WG4-2	++	++	+	+	ND	ND
WG5-1	+	+	+	+	ND	ND
WG5-3	+	+	+	+	ND	ND
WX1	+	+	+	+	ND	ND
WX1-1	+	+	+	+	ND	ND
WX1-2	++	++	+	+	ND	ND
WX2	+++	+++	++	+	ND	ND
WX3	+++	+++	++	+	ND	ND
WX4	+++	+++	+	+	ND	ND
WX4-3	+++	+++	+	+	ND	ND
WX4-4	++	++	+	+	ND	ND
WX5	++	++	+	+	ND	ND
WX5-2	++	++	+	+	ND	ND
WX5-4	++	++	+	+	ND	ND

Note: clear zone around disc; +: 1–3 mm; ++: 3–5 mm; +++: >5 mm; ND: no inhibition zone was detected.

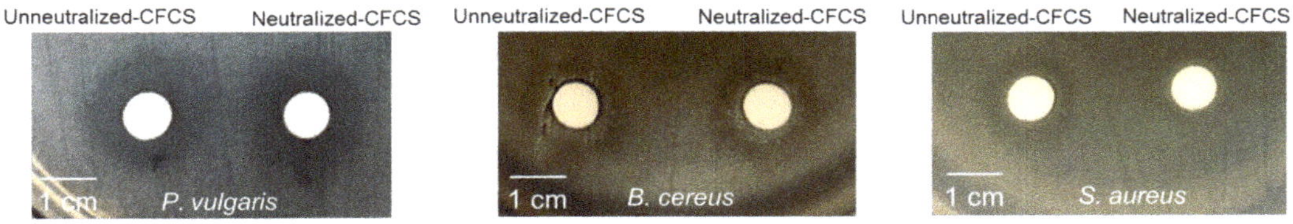

(a) (b) (c)

Figure 4. Antimicrobial activity of cell-free culture supernatant (CFCS) of *E. hirae* SX2 cultivated in MRS xylose against *Proteus vulgaris* (**a**), *Bacillus cereus* (**b**), and *Streptococcus aureus* (**c**).

Some previous reports confirmed that the silkworm gut is colonized with many bacteria [22,35]. *Streptococcus* sp. was found to be the most abundant pathogen in *Bombyx mori* larvae, whereas LAB from genus *Pediococcus*, *Leuconostoc*, and *Lactobacillus* did not cause any infection to silkworm [36]. However, the beneficial effect on silkworm host or interaction among the different microflora and the precise mechanism of action is unclear. Even though microbial supplementation as probiotics is commercially used for humans, ruminants, poultry, and fisheries, there are only few options of a probiotic formulation designed for silkworms [36]. Meanwhile, applying the commercial probiotic formulation was found to improve the larval body weight, effective rate of rearing, cocoon weight, pupal weight, shell ratio, and silk productivity [37–40]. Since synthetic antibiotics are widely applied in sericulture for disease control and some previous studies showed the antibiotic potential of LAB from silkworm against a range of both Gram-positive and Gram-negative pathogenic bacteria, considering the use of LAB isolated from silkworm gut with the antimicrobial activity has been proposed to be an alternative way for ecofriendly management of silkworm diseases [18]. Furthermore, some enterococci strains showed their ability to produce several bacteriocins [41], which confirm a competitive advantage toward the pathogenic microbes, and some *Enterococcus* spp. have been reported to be used as probiotics, such as *E. faecium*, *E. faecalis*, *E. lactis*, *E. hirae*, and *E. durans* [15,42,43]. However, the potential of enterococci used as probiotics supplement in Eri silkworm has not yet been reported. The dominant strains of xylose utilizing *Enterococcus* spp. isolated from the healthy silkworm of this study, and particularly *E. hirae* SX2 has the potential to contribute beneficial effects to the silkworm, and it is expected to be used as probiotic for Eri silkworm culture.

3.6. Effect of *E. hirae* SX2 Supplementation on Eri Silkworm Growth

The effects of presumptive probiotic *E. hirae* SX2 on the growth of Eri silkworm are shown in Table 2. The results revealed that the larvae fed on probiotics treated castor leaves exhibited a significant difference in weight when compared to the naturally fed control group. After 3 weeks of cultivation (fifth instar), larval weight of Eri silkworm reared with castor leaves supplemented with *E. hirae* SX2 were 4.53 ± 0.07 g, whereas larval weights of the control group were 4.20 ± 0.04 g. The results showed that larval weight of Eri silkworm reared with probiotic showed a significant increase of 7.85% ($p < 0.05$) when compared to the control leaves. In addition, the mortality rate of 3.3% was observed from the Eri silkworm reared with castor leaves supplemented with *E. hirae* SX2, which was 9.3% lower than that of the control. The culturable bacteria of Eri silkworm midgut reared on probiotic after 3 weeks of cultivation were enumerated (Figure 5). The results revealed that probiotic supplementation significantly affected total viable cells in the Eri silkworm gut. The high numbers of total viable cells and acid-forming bacteria in probiotic-treated larvae were detected when compared with untreated. Furthermore, the viable cells of LAB significantly increased when compared to the control treatment of castor leaves. These results indicated that probiotic *E. hirae* SX2 might be able to survive and grow in larval gut and could provide some benefits to the host larvae quality leading to the increase of the larval weight.

Table 2. Weight (g) of Eri silkworm reared with castor leaves and *E. hirae* SX2 supplemented castor leaves during 3 weeks of cultivation.

Treatments	0 Week (1st Instar Larva)	1 Week	2 Weeks	3 Weeks
Castor	0.03 ± 0.04	0.48 ± 0.03 [a]	1.53 ± 0.04 [b]	4.20 ± 0.04 [b]
Castor + SX2	0.02 ± 0.02	0.47 ± 0.01 [a]	1.69 ± 0.09 [a]	4.53 ± 0.07 [a]

Note: means in columns with different superscripts are statistically different at $p < 0.05$.

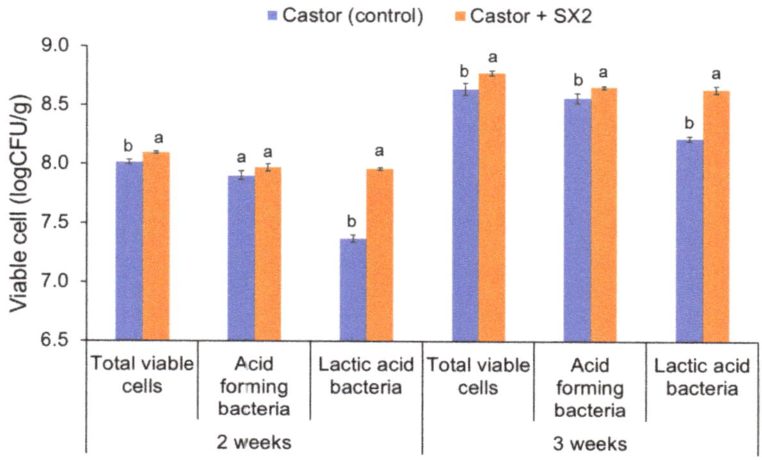

Figure 5. Total viable cells, acid-forming cells, and lactic acid-forming cells of probiotic-supplemented Eri silkworm midgut. Different small letters indicate significant differences in the viable cell (logCFU/g) in midgut between Eri silkworm reared with castor leaves and *E. hirae* SX2 treated castor ($p < 0.05$).

The probiotic microbial flora activity in the host gut might increase the efficiency of digestion and food assimilation, potentially leading to increase protein synthesis and assimilation in the host [44]. Moreover, probiotics have the ability to produce some vitamins and rapidly degrade the digestible compounds, potentially leading to nutritional assimilation. Based on the theory that beneficial microbial ecology is essential for eukaryotic metabolism, insects need gut microbial flora to provide a diversity of enzymes required for digestion of the nutritional components, which can contribute to the release of amino acids, fermentable sugars, and other molecules, which are beneficial to the growth of the insect [45]. Furthermore, the effect of probiotic bacteria including *Lactobacillus* spp., *Bifidobacterium*, *B. licheniformis*, and *B. niabensis* exhibited improvement in growth and quality characteristics of silkworm *Bombyx mori* L. [15,16,26,27]. Even some enterococci strains produce several bacteriocins, which have been reported with reference to the use of probiotics as mentioned previously, but there are no reports of the use of enterococci as probiotics supplement in Eri silkworm. The present study is the first report that indicates the potent beneficial effects of probiotic *Enterococcus* sp. on the Eri silkworm larval growth. However, more details concerning the influencing factors affecting the efficacy of probiotics, such as the probiotic dosage and probiotic formulation, are essential for advancing practical uses in Eri culture.

4. Conclusions

In this study, the xylose utilizing LAB from midgut of Eri silkworm larvae were isolated and characterized for evaluating probiotic characteristics such as tannin tolerance, production of cellulolytic enzymes, and antimicrobial activity against insect pathogenic bacteria, which potentially benefit Eri silkworm. Furthermore, among 51 LAB isolated from Eri silkworm, *E. hirae* SX2 was selected as the potential probiotic LAB based on beneficial characteristics such as tolerance to 0.5% (*w*/*v*) tannin and antimicrobial activity against insect pathogenic bacteria. The initial evaluation for targeting application of *E. hirae* SX2 in Eri silkworm rearing showed the beneficial and positive effect for improving larval quality, and this enterococci LAB is proposed as a new probiotic for Eri silk culture. Moreover, these findings will help to improve Eri silkworm economic traits and could also be used as the model in the further development of novel probiotics for silkworm (*B. mori*) or sericulture research.

Author Contributions: Conceptualization, K.U. and C.K.; methodology and formal analysis, K.U., A.K., P.K. and P.P.; investigation, K.U., A.K., P.K., P.P. and C.K.; writing—original draft preparation, K.U. and C.K.; writing—review and editing, K.U., K.S. and C.K.; supervision, C.K. All authors have read and agreed to the published version of the manuscript.

Funding: The authors are grateful to Chiang Mai University for financial support via postdoctoral fellowship and also funding from the Royal Golden Jubilee (RGJ) Ph.D. Program (Grant No. PHD/0137/2560 and PHD/0219/2561).

Institutional Review Board Statement: Not applicable.

Data Availability Statement: The data presented in this study are available in article.

Acknowledgments: The authors are grateful to Chiang Mai University for financial support via postdoctoral fellowship and Faculty of Agro-Industry for research facilities.

Conflicts of Interest: The authors declare no conflict of interest.

References

1. Pallavi, S.B. Commercial characters of selected eco-races of eri silkworm (*Samia cynthia ricini* Boisduval) reared on castor hybrid/variety. *Inter. J. Res. Anal. Rev.* **2018**, *5*, 513–518.
2. Charungkiattikul, S.; Joneurairatana, E. A revival for thailand's textile traditions: New value for local materials (eri silk) through art practice. *Textile* **2021**, 1–14. [CrossRef]
3. Leamkaew, V.; Jitjankarn, P.; Chairat, M. Polydopamine-dyed eri silk yarn for the improvement of wash and light fastness properties. *J. Text. Inst.* **2021**, *112*, 553–560. [CrossRef]
4. Prabhakar, C.S.; Sood, P.; Kanwar, S.S.; Sharma, P.N.; Kumar, A.; Mehta, P.K. Isolation and characterization of gut bacteria of fruit fly, *Bactrocera tau* (Walker). *Phytoparasitica* **2013**, *41*, 193–201. [CrossRef]
5. Paniagua Voirol, L.R.; Frago, E.; Kaltenpoth, M.; Hilker, M.; Fatouros, N.E. Bacterial symbionts in Lepidoptera: Their diversity, transmission, and impact on the host. *Front. Microbiol.* **2018**, *9*, 556. [CrossRef] [PubMed]
6. Nandy, G.; Chakraborti, M.; Shee, A.; Aditya, G.; Acharya, K. Gut microbiota from lower groups of animals: An upcoming source for cellulolytic enzymes with industrial potentials. *Biointerface Res. Appl. Chem.* **2021**, *11*, 13614–13637.
7. Ceja-Navarro, J.A.; Vega, F.E.; Karaoz, U.; Hao, Z.; Jenkins, S.; Lim, H.C.; Kosina, P.; Infante, F.; Northen, T.R.; Brodie, E.L. Gut microbiota mediate caffeine detoxification in the primary insect pest of coffee. *Nat. Commun.* **2015**, *6*, 7618. [CrossRef]
8. Kwong, W.K.; Moran, N.A. Gut microbial communities of social bees. *Nat. Rev. Microbiol.* **2016**, *14*, 374–384. [CrossRef]
9. Storelli, G.; Defaye, A.; Erkosar, B.; Hols, P.; Royet, J.; Leulier, F. *Lactobacillus plantarum* promotes Drosophila systemic growth by modulating hormonal signals through TOR-dependent nutrient sensing. *Cell Metab.* **2011**, *14*, 403–414. [CrossRef]
10. Engel, P.; Moran, N.A. The gut microbiota of insects—Diversity in structure and function. *FEMS Microbiol. Rev.* **2013**, *37*, 699–735. [CrossRef]
11. Naser, A.Z.; Deiab, I.; Darras, B.M. Polylactic acid (PLA) and polyhydroxyalkanoates (PHAs), green alternatives to petroleum-based plastics: A review. *RSC Adv.* **2021**, *11*, 17151–17196. [CrossRef]
12. Kim, J.-H.; Block, D.E.; Mills, D.A. Simultaneous consumption of pentose and hexose sugars: An optimal microbial phenotype for efficient fermentation of lignocellulosic biomass. *Appl. Microbiol. Biotechnol.* **2010**, *88*, 1077–1085. [CrossRef] [PubMed]
13. Bai, P.; Bai, M. Studies on the effect of aprobiotic and a neutraceutical agent on growth, development and commercial characteristics of silkworm, *Bombyx mori* L. *Indian J. Seric.* **2012**, *51*, 37–42.
14. Dantur, K.I.; Enrique, R.; Welin, B.; Castagnaro, A.P. Isolation of cellulolytic bacteria from the intestine of *Diatraea saccharalis* larvae and evaluation of their capacity to degrade sugarcane biomass. *AMB Express* **2015**, *5*, 1–11. [CrossRef] [PubMed]
15. Mala, M.; Vijila, K. Beneficial effects of *Bacillus licheniformis* and *Bacillus niabensis* on growth and economic characteristics of silkworm, *Bombyx mori* L. *Int. J. Chem. Stud.* **2018**, *6*, 1750–1754.

16. Prem Anand, A.A.; Vennison, S.J.; Sankar, S.G.; Gilwax Prabhu, D.I.; Vasan, P.T.; Raghuraman, T.; Jerome Geoffrey, C.; Vendan, S.E. Isolation and characterization of bacteria from the gut of *Bombyx mori* that degrade cellulose, xylan, pectin and starch and their impact on digestion. *J. Insect Sci.* **2010**, *10*, 107.
17. Quinto, E.J.; Jiménez, P.; Caro, I.; Tejero, J.; Mateo, J.; Girbés, T. Probiotic lactic acid bacteria: A review. *Food Nutr. Sci.* **2014**, *5*, 1765. [CrossRef]
18. Sonnenburg, J.L.; Chen, C.T.L.; Gordon, J.I. Genomic and metabolic studies of the impact of probiotics on a model gut symbiont and host. *PLoS Biol.* **2006**, *4*, e413. [CrossRef]
19. Unban, K.; Kodchasee, P.; Shetty, K.; Khanongnuch, C. Tannin-tolerant and extracellular tannase producing *Bacillus* isolated from traditional fermented tea leaves and their probiotic functional properties. *Foods* **2020**, *9*, 490. [CrossRef]
20. Tamura, K.; Dudley, J.; Nei, M.; Kumar, S. MEGA4: Molecular evolutionary genetics analysis (MEGA) software version 4.0. *Mol. Biol. Evol.* **2007**, *24*, 1596–1599. [CrossRef]
21. Fehlberg, L.C.C.; Nicoletti, A.G.; Ramos, A.C.; Rodrigues-Costa, F.; de Matos, A.P.; Girardello, R.; Marques, E.A.; Gales, A.C. *In vitro* susceptibility of *Burkholderia cepacia* complex isolates: Comparison of disk diffusion, Etest®, agar dilution, and broth microdilution methods. *Diagn. Microbiol. Infect. Dis.* **2016**, *86*, 422–427. [CrossRef] [PubMed]
22. MsangoSoko, K.; Gandotra, S.; Chandel, R.K.; Sharma, K.; Ramakrishinan, B.; Subramanian, S. Composition and diversity of gut bacteria associated with the eri silk moth, *Samia ricini*, (Lepidoptera: Saturniidae) as revealed by culture-dependent and metagenomics analysis. *J. Microbiol. Biotechnol.* **2020**, *30*, 1367–1378. [CrossRef] [PubMed]
23. Yeruva, T.; Vankadara, S.; Ramasamy, S.; Lingaiah, K. Identification of potential probiotics in the midgut of mulberry silkworm, *Bombyx mori* through metagenomic approach. *Probiotics Antimicro. Prot.* **2020**, *12*, 635–640. [CrossRef] [PubMed]
24. Sun, Z.; Lu, Y.; Zhang, H.; Kumar, D.; Liu, B.; Gong, Y.; Zhu, M.; Zhu, L.; Liang, Z.; Kuang, S. Effects of BmCPV infection on silkworm *Bombyx mori* intestinal bacteria. *PLoS ONE* **2016**, *11*, e0146313. [CrossRef]
25. Chen, B.; Teh, B.-S.; Sun, C.; Hu, S.; Lu, X.; Boland, W.; Shao, Y. Biodiversity and activity of the gut microbiota across the life history of the insect herbivore *Spodoptera littoralis*. *Sci. Rep.* **2016**, *6*, 29505. [CrossRef]
26. Masthan, K.; Rajkumar, T.; Narasimha Murthy, C.V. Studies on fortification of mulberry leaves with probiotics for improvement of silk quality. *Int. J. Biotechlol. Biochem.* **2017**, *13*, 73–80.
27. Moustafa, M.N.; Soliman, S. Nutritional efficiency and economic traits of silkworm *Bombyx mori*, L. reared on mulberry leaves fortified with synbiotics. *J. Plant Prot. Pathol.* **2019**, *10*, 671–675. [CrossRef]
28. Chen, B.; Yu, T.; Xie, S.; Du, K.; Liang, X.; Lan, Y.; Sun, C.; Lu, X.; Shao, Y. Comparative shotgun metagenomic data of the silkworm *Bombyx mori* gut microbiome. *Sci. Data* **2018**, *5*, 1–10. [CrossRef]
29. Chen, B.; Du, K.; Sun, C.; Vimalanathan, A.; Liang, X.; Li, Y.; Wang, B.; Lu, X.; Li, L.; Shao, Y. Gut bacterial and fungal communities of the domesticated silkworm (*Bombyx mori*) and wild mulberry-feeding relatives. *ISME J.* **2018**, *12*, 2252–2262. [CrossRef]
30. Klongklaew, A.; Unban, K.; Kanpiengjai, A.; Wongputtisin, P.; Pamueangmun, P.; Shetty, K.; Khanongnuch, C. Improvement of enantiomeric L-lactic acid production from mixed hexose-pentose sugars by coculture of *Enterococcus mundtii* WX1 and *Lactobacillus rhamnosus* SCJ9. *Fermentation* **2021**, *7*, 95. [CrossRef]
31. Pandiarajan, J.; Revathy, K. Cellulolytic potential of gut bacterial biomass in silkworm *Bombyx mori*. L. *Ecol. Genet. Genom.* **2020**, *14*, 100045. [CrossRef]
32. Shil, R.K.; Mojumder, S.; Sadida, F.F.; Uddin, M.; Sikdar, D. Isolation and identification of cellulolytic bacteria from the gut of three phytophagus insect species. *Braz. Arch. Biol. Technol.* **2014**, *57*, 927–932. [CrossRef]
33. Despres, L.; David, J.-P.; Gallet, C. The evolutionary ecology of insect resistance to plant chemicals. *Trends Ecol. Evol.* **2007**, *22*, 298–307. [CrossRef]
34. Buzzini, P.; Arapitsas, P.; Goretti, M.; Branda, E.; Turchetti, B.; Pinelli, P.; Ieri, F.; Romani, A. Antimicrobial and antiviral activity of hydrolysable tannins. *Mini-Rev. Med. Chem.* **2008**, *8*, 1179. [CrossRef] [PubMed]
35. Gandotra, S.; Kumar, A.; Naga, K.; Bhuyan, P.; Gogoi, D.; Sharma, K.; Subramanian, S. Bacterial community structure and diversity in the gut of the muga silkworm, *Antheraea assamensis* (Lepidoptera: Saturniidae), from India. *Insect Mol. Biol.* **2018**, *27*, 603–619. [CrossRef]
36. Bhalchandra, P.M.; Pathade, G. Antibacterial and cholesterol reducing lactic acid bacteria from silk worm (*Bombyx mori*) gut environment—A review. *Nat. Environ. Pollut. Technol.* **2011**, *10*, 319–326.
37. Anitha, J.; Sathish, J.; Sujatha, K. Nutritional efficiency and economic traits in *Samia ricini* donovan reared on castor leaves fortified with probiotic agent. *Int. J. Inn. Res. Sci. Eng. Tech* **2015**, *4*, 2319–8753.
38. Shruti, A.; Hadimani, D.; Sreenivas, A.; Beladhadi, R. Effect of probiotic feed supplements to mulberry silkworm, *Bombyx mori* L. for larval growth and development parameters. *Int. J. Chem. Stud.* **2019**, *7*, 3914–3919.
39. Singh, K.; Chauhan, R.; Pande, A.; Gokhale, S.; Hegde, N. Effect of use of *Lactobacillus plantarum* as a probiotics to improve cocoon production of mulberry silkworm *Bombyx mori* (L). *J. Basic Appl. Sci.* **2005**, *1*, 1–8.
40. Suraporn, S.; Sangsuk, W.; Chanhan, P.; Promma, S. Effects of probiotic bacteria on the growth parameters of the Thai silkworm, *Bombyx mori*. *Thai J. Agric. Sci.* **2015**, *48*, 29–33.
41. Shao, Y.; Chen, B.; Sun, C.; Ishida, K.; Hertweck, C.; Boland, W. Symbiont-derived antimicrobials contribute to the control of the Lepidopteran gut microbiota. *Cell Chem. Biol.* **2017**, *24*, 66–75. [CrossRef] [PubMed]

42. Li, B.; Zhan, M.; Evivie, S.E.; Jin, D.; Zhao, L.; Chowdhury, S.; Sarker, S.K.; Huo, G.; Liu, F. Evaluating the safety of potential probiotic *Enterococcus durans* KLDS6. 0930 using whole genome sequencing and oral toxicity study. *Front. Microbiol.* **2018**, *9*, 1943. [CrossRef] [PubMed]
43. Baccouri, O.; Boukerb, A.M.; Farhat, L.B.; Zébré, A.; Zimmermann, K.; Domann, E.; Cambronel, M.; Barreau, M.; Maillot, O.; Rincé, I. Probiotic potential and safety evaluation of *Enterococcus faecalis* OB14 and OB15, isolated from traditional tunisian testouri cheese and rigouta, using physiological and genomic analysis. *Front. Microbiol.* **2019**, *10*, 881. [CrossRef] [PubMed]
44. Rowland, I.; Gibson, G.; Heinken, A.; Scott, K.; Swann, J.; Thiele, I.; Tuohy, K. Gut microbiota functions: Metabolism of nutrients and other food components. *Eur. J. Nutr.* **2018**, *57*, 1–24. [CrossRef]
45. Liang, X.; Fu, Y.; Liu, H. Isolation and characterization of enzyme-producing bacteria of the silkworm larval gut in bioregenerative life support system. *Acta Astronaut.* **2015**, *116*, 247–253. [CrossRef]

insects

MDPI

Article

A Route to Translate a Silk-Based Medical Device from Lab to Clinic: The Silk Biomaterials Srl Experience

Giulia Alessandra Bassani, Valentina Vincoli, Marco Biagiotti, Elisa Valsecchi, Marta Virginia Zucca, Claudia Clavelli, Antonio Alessandrino and Giuliano Freddi *

Silk Biomaterials Srl, 22074 Lomazzo, Como, Italy; giulia.bassani@outlook.it (G.A.B.); valentina@silkbiomaterials.com (V.V.); marco@silkbiomaterials.com (M.B.); elisa@silkbiomaterials.com (E.V.); marta@silkbiomaterials.com (M.V.Z.); claudia@silkbiomaterials.com (C.C.); antonio@silkbiomaterials.com (A.A.)
* Correspondence: giuliano@silkbiomaterials.com

Citation: Bassani, G.A.; Vincoli, V.; Biagiotti, M.; Valsecchi, E.; Zucca, M.V.; Clavelli, C.; Alessandrino, A.; Freddi, G. A Route to Translate a Silk-Based Medical Device from Lab to Clinic: The Silk Biomaterials Srl Experience. *Insects* **2022**, *13*, 212. https://doi.org/10.3390/insects13020212

Academic Editors: Silvia Cappellozza, Morena Casartelli, Federica Sandrelli, Alessio Saviane, Gianluca Tettamanti and Brian T. Forschler

Received: 24 January 2022
Accepted: 17 February 2022
Published: 21 February 2022

Publisher's Note: MDPI stays neutral with regard to jurisdictional claims in published maps and institutional affiliations.

Simple Summary: Silk has always been a source of inspiration for textile designers to create very disruptive fashion products. More recently, silk has stimulated the creativity of scientists, as it emerged as a promising biomaterial for the development of next generation medical devices. Although silk has been clinically used as a suture for decades, only the development of novel processing approaches paved the way for the production of a plurality of regenerated silk-based materials, i.e., films, hydrogels, sponges, powder, nano- and microparticles, nanofibers, etc., which have been recognized as scaffolds-of-choice in different medical applications. However, the translation of research achievements into medical products used in clinical settings is complex from manufacturing, quality, and regulatory perspectives. The aim of this paper is to disclose how the clinical translation route works using, as a case study, a silk-based medical device recently developed by the Italian start up Silk Biomaterials srl. The results reported here will cover some fundamental aspects of the regulatory and quality path, from the demonstration of the robustness of the manufacturing process up to the evaluation of the biocompatibility, and of the functional performance of the device.

Abstract: The medical device is a nerve conduit entirely made of *Bombyx mori* silk fibroin. It is a tubular scaffold used for repairing peripheral nerve gaps, whose function is to protect the severed nerves and to favor their natural healing process. As any implantable medical device, the conduit must perform its function without causing adverse effects to the patient, meaning that it must be compliant with a range of regulations aimed at evaluating the risks related to the constituent materials and the manufacturing process, the toxicological impact of the processing aids, the biological safety, the functional performance, and the ability to sustain tissue regeneration processes. An exhaustive on-bench testing plan has been performed for the determination of the morphological, geometrical, physical, structural, and mechanical properties. For the toxicological analysis, the device was extracted with solvent and the number of leachable substances was determined by suitable chromatographic techniques. The biological safety was assessed by means of a set of tests, including cytotoxicity, delayed hypersensitivity, intracutaneous reactivity, pyrogen test, LAL (Limulus Amebocyte Lysate) test, acute systemic toxicity, and genotoxicity. Overall, the accumulated results demonstrated the suitability of the device for the intended use and supported the starting of a first-in-human clinical trial.

Keywords: silk fibroin; SILKBridge® nerve conduit; performance testing; biocompatibility; toxicology; regulatory requirements; quality requirements; control manufacturing

1. Introduction

Silk is a fibrous protein secreted by leaving organisms (insects, spiders, etc.) for housing, protection, or predation purposes. It is synthesized into the cells of specialized glands, accumulated into the lumen, and finally extruded to make extracorporeal filamentous structures (nets, cocoons, nests, etc.) [1,2]. The silks produced by *Lepidoptera* (*Bombyx*, *Antheraea*, *Phylosamia* genus) are an important source of starting materials for the textile industry [3]. The so-called domesticated or mulberry silk, obtained from rearing *Bombyx mori* silkworms and unraveling their cocoons, has been used for thousands of years to produce precious fabrics highly appreciated for their lustrous appearance, softness, and elegance. Silk production represents less than 0.1% of the global fiber market (estimated at about 110 million tons in the year 2020; source: https://textileexchange.org, accessed on 15 November 2021) but is still a significant billion-dollar industry because silk is the leading fiber in fashion design (https://fashionunited.com, accessed on 15 November 2021).

The first documented medical use of silk dates back to about 2200 years, when Claudius Galenus proposed to use silk as suture thread [4]. Other similar applications appeared throughout the centuries, but only in the 20th century have silk sutures become commercially available for routine clinical use. A key step for the development of a wider range of medical applications has been the dissolution of native silk fibroin fibers and the obtainment of a liquid feedstock from which different regenerated silk-based materials such as films, hydrogels, porous scaffolds, nanofibers, particles, etc., could be prepared [5]. Of the two main components of silk, fibroin and sericin, fibroin is the material that has attracted the most attention, although sericin has recently been carving out a significant space in the biomaterial field [6]. However, every time silk is mentioned in this article, reference will be made solely to the fibroin component.

From the 1980s onwards, the scientific community's interest in silk as a biomaterial has dramatically increased. Just to give an idea of the development of scientific interest in this area, a quick search in the Scopus database using "silk fibroin" as keyword returned about 50 papers published in the year 2000 and more than 800 in 2020, with an annual increase trend that is expected to continue in the future. While Japanese, European, and American research teams dominated the scientific production about silk at the beginning of the 21st century, the current panorama of the most prolific countries is led by China, whose scientific production is now three-times larger than that of America. It seems, therefore, that the millennial history of silk, which began in China and spread throughout the world via the Silk Road, has made a U-turn to return to where it originated, coupling the more traditional Chinese monopoly in silk production to the newly acquired scientific leadership.

The range of biomedical applications targeted by silk-based scaffolds is very wide, spanning from implantable devices for engineering and/or regeneration of soft and hard tissues to the development of tissue models, carriers for drug release, diagnostic devices, etc. [4,7,8]. The reason for such a wide interest in silk lies in its intrinsic properties, including biocompatibility, controllable degradability, absence of toxicity of degradation products, structural stability, mechanical performance, wide choice of manufacturing options which makes it possible to design scaffolds with desirable features for specific applications [5]. Different silk-based products have reached the pre-clinical stage and have been tested in small and large animal models, and in clinical trials with promising results [4]. Even if the number of silk-based devices approved for clinical use is still very low, the number of companies involved in the development of silk-based medical products has been increasing over the last decade; thus, it is likely that, quite soon, many of them will undertake the regulatory path that leads to the approval of devices for use in the clinical setting [4,9].

The translational process that allows a silk-based material to go through the regulatory pathway that transforms the initial idea and the starting material into a product to be used in the therapeutic field is particularly long and complex. Although there might be differences depending on the different rules put in place by European (CE mark) or American (Food and Drug Administration-FDA approval) regulatory agencies, many basic steps are common. These include the acquisition of raw materials in accordance with well-established quality assurance programs, the rigorous control of the manufacturing process that must be robust and compliant in terms of product specification, the implementation and execution of on-bench and in vitro testing programs, the thorough biocompatibility evaluation of the product according to guidelines such as ISO 10993-1 (International Organization for Standardization), the execution of relevant in vivo functional tests using appropriate animal models, the access to clinical trials to evaluate safety and performance of the device and the submission of the dossier to a regulatory body for device approval [9].

The aim of this paper is to disclose how the clinical translation route works using the experience of the Italian start up Silk Biomaterials srl in the development of a silk-based medical device, the SILKBridge® nerve conduit, as case study. The device is designed for the surgical repair of peripheral nerve discontinuities to support the regeneration of injured nerves [10]. It is intended to act as a bridge to guide and structurally support axonal growth across the gap during the healing process. The conduit is entirely made of silk fibroin obtained from *Bombyx mori* silkworms. During the manufacturing process, sericin is completely removed by degumming to leave pure silk fibroin. The conduit has a tubular structure with a three-layered wall comprising an intermediate textile layer intimately and firmly coupled with two electrospun layers, one in the inner and the other in the outer parts of the device wall [10]. This design allowed for optimizing both mechanical and biological properties, because the electrospun layers have biomimetic characteristics that enhance cell's adhesion and integration with the surrounding tissues, while the textile layer was designed to provide the required mechanical resistance during implantation (suture retention strength) and in vivo functioning (compression strength) [11]. The conduit displayed a slow degradation rate both in vitro [12] and in vivo [11]. As degradation occurred, the load bearing responsibility was transferred to the new tissue ingrowth such that mechanical integrity was maintained at the implantation site and a complete morphological and functional recovery of the transected nerve was achieved [11].

The results reported here complement those already published [10–12] about the synergic efficacy of the use of silk as biomaterial and of the design of a three-layered wall architecture for manufacturing the nerve conduit; moreover, they specifically tackle the following issues of the regulatory process: demonstration of the robustness of the production process through the evaluation of the functional parameters of the device defined in the design phase, verification of potential sources of risk that may emerge from the chemicals used for production through toxicological analysis, evaluation of biological safety as required by regulations in force for medical devices and start of the first phase of clinical trials.

2. Materials and Methods

2.1. Starting Materials and Device Manufacturing

The wall of the tubular device consists of three layers: two electrospun layers (ES, inner and outer) and an intermediate textile layer (TEX) [10]. Two different raw materials were used to produce the device: (i) a degummed silk yarn (organzine, 40 dtex), with which a tubular braid was manufactured by warp needle braiding technology; the braid forms the intermediate TEX layer; (ii) silk cocoons, which were degummed in an autoclave to remove sericin; the pure silk fibroin fibers were used as starting material to produce the inner and outer ES layers. Both silk cocoons and yarn were acquired from external suppliers, in compliance with traceability requirement. The incoming raw materials entered the device's manufacturing cycle after having passed control tests aimed at addressing quality and purity requirements.

The manufacturing process has been described in detail elsewhere [10]. Briefly, the TEX and ES layers comprising the wall of the device were coupled during electrospinning according to a patented process [13]. Two ES layers were assembled onto the inner and outer faces of a TEX textile braid. Coupling of the TEX layer with the two ES layers was made by means of two different welding media: (i) a solution of ionic liquid (1-ethyl-3-methylimidazolium acetate; EMIMAc; #51053, Sigma-Aldrich, Merk Life Science S.r.l., Milano, Italy) in water (EMIMAc/water 80/20% v/v); (ii) a solution of 15% w/w SF in EMIMAc. After electrospinning, the three-layered tubular structure was consolidated by immersion in 80% v/v ethanol for 30 min at room temperature, followed by overnight washing with distilled water and drying. Finally, the tubular structure was purified by extraction with ethanol (EtOH) in a microwave assisted extractor, washed with ultrapure water, dried, packaged in double pouches under laminar flow cabinet, and, finally, sterilized with ethylene oxide.

2.2. On-Bench Testing

2.2.1. Relaxed Internal Diameter

The cross-sections of the final finished devices were mounted on aluminum stubs and sputter coated with Au/Pd (Desk V, Denton Vacuum, LLC, Moorestown, NJ, USA). A Zeiss EVO MA10 scanning electron microscope (SEM) operating at 10 kV acceleration voltage, 100 μA beam current, and 35 mm working distance was used to acquire the image of the tubular cross-section of the device. The internal diameter was determined from the SEM image using the SEM software measuring tools.

2.2.2. Wall Thickness

Wall thickness was measured according to the ISO 7198:2016 standard method. The tubular device was cut longitudinally, flattened, and measured with a thickness tester MarCator 1075R (Mahr) equipped with a constant load thickness gauge of 0.3 cm^2 foot area that exerts a pressure of 1 kPa.

2.2.3. Linear Density (Mass Per Unit Length)

The length (L) of the sample was measured with a digital Vernier caliper (Metrica, Brighton, East Sussex, UK; Carbon Fiber Composite Digital Caliper), with 0.01 cm resolution. The mass (M) of the samples was measured with an analytical balance (Mettler Toledo, Milano, Italy, XRS105 Dual Range), with 0.01 mg resolution. The linear density (mass per unit length) of the sample, expressed in mg/cm, was calculated from the M/L ratio.

2.2.4. Porosity

Porosity (%P) was determined according to ISO 7198:2016, using the gravimetric test procedure, according to the following formula:

$$\%P = 100 \times [1 - (M/V\rho)]$$

where:

- M is the mass of the device (see: Section 2.2.3);
- V is the volume of the device, calculated from the values of wall thickness (see: Section 2.2.2) and relaxed internal diameter (see: Section 2.2.1);
- ρ is the density of silk fibroin (1.35 g/cm^3) [14].

2.2.5. Degree of Crystallinity

Attenuated total reflectance Fourier transform infrared spectroscopy (ATR-FTIR) was used to determine the degree of crystallinity. ATR-FTIR spectra were acquired with an ALPHA FTIR spectrometer (Bruker) equipped with an ATR platinum diamond accessory, at a resolution of 4 cm^{-1}, in the infrared range of 4000–400 cm^{-1}. The tubular specimen was cut longitudinally and the surface of the inner electrospun layer was pressed against the ATR crystal. The spectra recorded were corrected with a linear baseline and normalized to the CH_2 bending peak at about 1445 cm^{-1}. The crystallinity index was calculated by ratioing the intensity of the Amide III bands at 1260 cm^{-1} and 1230 cm^{-1} ($CI = A_{1260}/A_{1230}$) [15].

2.2.6. Thermal Properties

Thermal properties were determined by differential scanning calorimetry (DSC) using a calorimeter DSC 3500 Sirius (Netzsch, Selb, Germany). Samples (3–5 mg) were sealed in aluminum pans and subjected to a heating cycle from 50 °C to 400 °C, at a heating rate of 10 °C/min, under an N_2 atmosphere (flow rate: 20 mL/min). The following parameters were determined from the DSC curves: peak temperature of the endotherms associated with melting/degradation of the ES (T_{ES}) and TEX (T_{TEX}) layers; cumulative enthalpy ($DH_{ES/TEX}$) of the ES and TEX endotherms; ES:TEX weight ratio, calculated from the relative intensity of the respective DSC endotherms, using the DH values of −402 J/g and −307 J/g for the pure TEX and ES layers, respectively, to normalize the contribution of the two materials.

2.2.7. Compression Strength

Compression tests were performed under wet conditions, using an all-electric dynamic test instrument ElectroPuls E3000 (Instron, Norwood, MA, USA), equipped with a load cell of 10 N, a thermostatic bath (BioPuls), and appropriate grips submersed in water at 37 °C [10]. The sample of 10 mm length was mounted between the upper and lower plates and tested at a crosshead speed of 1.0 mm/min, with a preload of 0.1 N. The sample was conditioned in the thermostatic bath for 5 min before starting the test. The force was applied perpendicular to the longitudinal axis of the tubular sample. Compressive load values were recorded at strains of 20% and 40%. The modulus was calculated from the slope of the compression curve in the 0–40% load range.

2.2.8. Suture Retention Strength

The suture retention strength is the force necessary to pull a suture from the device. The ElectroPuls E3000 (Instron) tester equipped with a load cell of 10 N was used. The conduit was cut normal to the long axis and a suture was inserted 2 mm from the end of the sample through the wall to form a half loop. The sample was clamped in the lower fixed grip and the suture thread in the upper moving grip, which was pulled at the rate of 50 mm/min. The test was performed in the wet state with the sample submersed at 37 °C. The sample was conditioned in the thermostatic bath for 5 min before starting the test. The force required to pull the suture through the device was recorded.

2.2.9. Tensile Properties

Breaking load and elongation at break in uniaxial tensile test were determined under submersed conditions (in water at 37 °C), with the ElectroPuls E3000 (Instron) tester, equipped with a 250 N load cell. The test was performed in accordance with the provisions of ISO 7198:2016. The sample was conditioned in the thermostatic bath for 5 min before starting the test. The gauge length was 30 mm; a preload of 0.5 N was applied; the tests were run at 30 mm/min crossbar rate.

2.2.10. Statistical Analysis

Data from different production batches were compared using the one-way ANOVA statistical test to confirm the statistical equivalence between different production lots. Statistical analyses were conducted using Minitab statistical software.

2.3. Chemical Analysis

2.3.1. Determination of Bromide Ion

The sample was weighed into a crucible and reduced to ash. Ashes were recovered with water and the solution was filtered with a 0.45 µm filter before analysis by ion chromatography, with a Dionex™ ICS-4000 Integrated Capillary HPIC™ System, equipped with a Dionex IonPac AS11-HC-4 mm column, using a KOH solution as mobile phase in the gradient mode (0.02–80 mM in 15 min), at a flow rate of 15 mL/min. The injection volume was 0.4 mL; detection was made by a conductivity and QD charge detector. The results were expressed in mg/kg.

2.3.2. Determination of Lithium Ion

The sample was digested in a microwave oven (Anton Paar Multiwave 3000) with an acid mixture containing 4 mL HNO_3/2 mL H_2O_2/0.25 mL HF, with a power ramp from 0 to 1200 W in 25 min, a 15 min hold step at 1200 W and a cooling step for 20 min.

The analysis was performed by ICP-MS (inductively coupled plasma—mass spectrometry) employing a cross flow nebulizer with a Scott spray chamber (ICP-MS Perkin Elmer SCIEX mod ELAN 9000, autosampler AS90, Waltham, MA, USA). The element was quantified with an external calibration (AccuStandard ICP-MS-CAL2-1 multielement, New Haven, CT, USA), employing Germanium as the internal standard to compensate plasma fluctuations. The results were expressed in mg/kg.

2.3.3. Determination of Methyl and Ethyl Alcohols

The analysis was performed by head-space gas chromatography/mass spectrometry (HS-GC/MS) with a Shimadzu, mod. 2010 Plus, equipped with a MS detector QP2010 Ultra and an autosampler Shimadzu AOC 5000 Plus. Head space conditions were as follows: temperature 80 °C, time 60 min, injection volume 500 μL. A capillary column Agilent HP5-ms (30 m length, 0.25 mm inner diameter, 0.25 μm film thickness) was used for the separation. Other analytical conditions were as follows: carrier gas Helium at 1 mL/min; incubation temperature 80 °C; incubation time 60 min; splitless injector temperature 200 °C; temperature program hold 40 °C for 3 min—from 40 °C to 260 °C at 40 °C/min; injection volume 500 μL; transfer line temperature 280 °C; source temperature 200 °C; E 70 eV; SIM acquisition mode. Target ions were m/z 32 methyl alcohol (qualifier ion 31), m/z 31 ethyl alcohol (qualifier ion 45), and m/z 31 for 1-propanol (internal standard). The results were expressed in mg/kg.

2.3.4. Determination of 1-ethyl-3-methyl Imidazolium Acetate

The samples were finely chopped and extracted with methanol in an ultrasonic bath. The supernatant was filtered on a 0.22 μm PVDF syringe filter, dried, and recovered with 300 μL of eluent A. The analysis was performed by ultra high-pressure liquid chromatography-ion trap-mass spectrometry (UHPLC-IT-LTQ-MS/MS, FinniganMAT), with an Alltima HP C18 (3 μm, 150 mm × 2.0 mm) column, in the gradient mode (Eluent A: H_2O + 0.1% CH_3COOH; Eluent B: CH_3OH + 0.1% CH_3COOH; from 90% A to 100% B in 17 min), at a flow rate of 0.2 mL/min. The injection volume was 10 μL, the operating mode of the MS detector H-ESI(+), and the mass range m/z 50–300 uma. The results were expressed in mg/kg.

2.3.5. Determination of Formic Acid

The analysis was performed by head-space gas chromatography/mass spectrometry (HS-GC/MS), by using the same apparatus described in paragraph 2.3.3. for the determination of alcohols. Analytical conditions were the same, with exception of incubation time (60 min) and injection volume (500 μL). Target ions were formic acid (m/z 46) and acetic acid (m/z 60; internal standard). The results were expressed in mg/kg.

2.4. Biological Safety Evaluation

The ISO 10993 standard was used as benchmark for the biological safety evaluation of the nerve conduit. The following tests were performed: cytotoxicity (ISO 10993-5:2009); delayed hypersensitivity (ISO 10993-10:2010); irritation/intracutaneous reactivity (ISO 10993-10:2010); pyrogenicity (ISO 10993-11:2017; US Pharmacopoeia-USP 40<151>, and European Pharmacopoeia-EU 8.0 §2.6.8.); LAL test (USP <85>; USP <161>; EU par. 2.6.14 e par. 5.1.10; AAMI ST72); acute systemic toxicity (ISO 10993-11:2017); genotoxicity (ISO 10993-3:2014). Tests were carried out at Eurofins Biolab S.r.l. test facility (Good Laboratory Practice-GLP cert. n. 2017/16), Vimodrone (Milan), Italy.

3. Results

3.1. On-Bench Testing and Property/Performance Evaluation

Sources of possible variability may emerge at different points of the manufacturing process of a medical device and must be carefully monitored to avoid any negative impact on the properties and performance of the final device. Therefore, the design controls requirements according to USA (FDA, 21 CFR 820.30) and EU (MDR, 2017/745) directives were put in place for the design and manufacturing of the SILKBridge® nerve conduit, and an on-bench testing plan has been specifically developed for an in-depth property/performance evaluation. Three test categories, each one addressing specific functional endpoint, were defined. In particular:

1. the morphological and geometrical characteristics, which may affect the usability of the device during surgery, the performance at the site of implantation, and the biological response of surrounding tissues;
2. the physical and structural characteristics, which may impact the biological interaction with surrounding tissues and the rate and extent of degradation after implantation (the device is entirely made from a biodegradable natural protein polymer);
3. the mechanical characteristics, which must achieve a level high enough to withstand the mechanical stresses caused by suturing during implantation, as well as tensile, compression, and bending stresses caused by the surrounding tissues during in vivo functioning.

At the end of the manufacturing process, each device was visually inspected, cut to its final length (30 mm), and packaged for sterilization. The portions of the device exceeding the length of 30 mm were collected and analyzed for the determination of the morphological, geometrical, physical, and structural properties. This means that each device has undergone testing for the characteristics listed under points 1 and 2. On the other hand, the mechanical properties listed under point 3 were analyzed by sampling (1 device every 50 devices produced) because the tests require an entire device to be executed and are destructive. Figure 1 shows some representative experimental data for the three test categories analyzed on the finished device.

Tables 1–3 list the testing results obtained from the analysis of samples representative of three different production batches. For each batch, five samples were analyzed, and the results were reported as mean ± standard deviation. No statistically significant differences were observed among the production batches for each measured characteristic, indicating that the manufacturing process is under control and that the devices will perform as expected.

Table 1. On-bench testing results (mean ± Std. Dev.): morphological and geometrical characteristics (n = 5).

	Batch A	Batch B	Batch C	p Value [1]
Internal diameter (mm)	1.78 ± 0.10	1.76 ± 0.04	1.82 ± 0.07	>0.05
Wall thickness (mm)	0.43 ± 0.05	0.47 ± 0.04	0.47 ± 0.04	>0.05
Linear density (mg/cm)	5.7 ± 0.4	5.8 ± 0.2	6.2 ± 0.3	>0.05
Porosity (%)	86 ± 2	87 ± 2	86 ± 2	>0.05

[1] One-way ANOVA (confidence interval of the mean 95%). All data followed a normal distribution.

Table 2. On-bench testing results (mean ± Std. Dev.): physical and structural characteristics (n = 5).

	Batch A	Batch B	Batch C	p Value [1]
Degree of crystallinity	0.66 ± 0.03	0.66 ± 0.03	0.67 ± 0.01	>0.05
Thermal properties:				
T_{ES} (°C)	291 ± 1	291 ± 2	290 ± 1	>0.05
T_{TEX} (°C)	321 ± 1	322 ± 1	321 ± 2	>0.05
$\Delta H_{ES/TEX}$ (J/g)	411 ± 21	402 ± 32	410 ± 31	>0.05
ES:TEX ratio (%)	53 ± 1	53 ± 3	54 ± 2	>0.05

[1] One-way ANOVA (confidence interval of the mean 95%). All data followed a normal distribution.

Table 3. On-bench testing results (mean ± Std. Dev.): mechanical characteristics (n = 5).

	Batch A	Batch B	Batch C	p Value [1]
Suture retention (N)	4.6 ± 0.4	4.5 ± 0.8	4.7 ± 0.8	>0.05
Compression strength:				
Load at 20% strain (cN)	4.6 ± 1.2	4.2 ± 1.0	4.6 ± 1.0	>0.05
Load at 40% strain (cN)	9.2 ± 1.6	8.6 ± 1.9	9.7 ± 1.7	>0.05
Modulus (kPa)	93 ± 12	84 ± 17	97 ± 14	>0.05
Tensile strength:				
Breaking load (N)	24.8 ± 0.5	23.3 ± 0.9	23.1 ±1.8	>0.05
Elongation at break (%)	59.5 ± 2.7	60.5 ± 6.5	61.1 ± 3.6	>0.05

[1] One-way ANOVA (confidence interval of the mean 95%). All data followed a normal distribution.

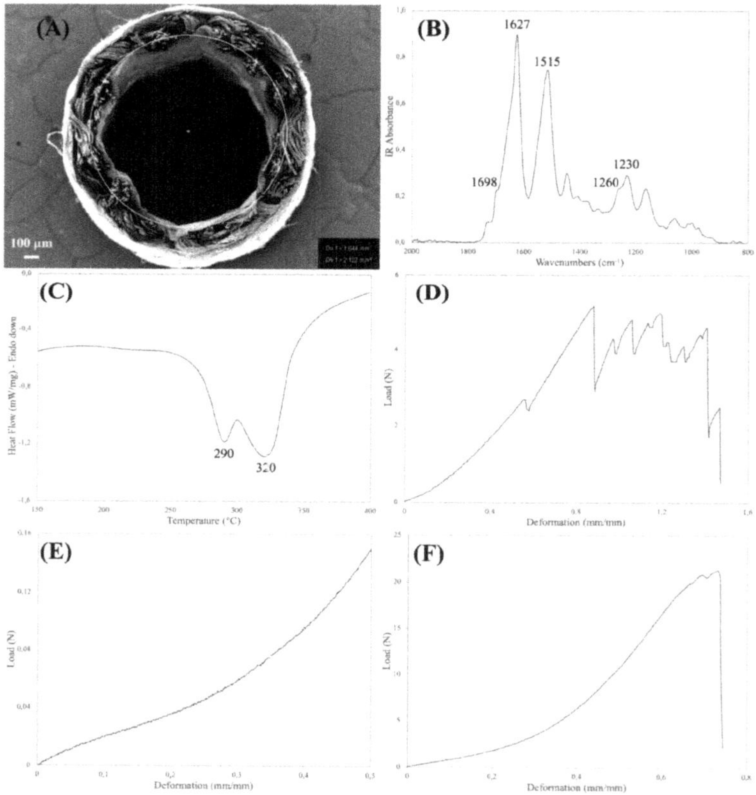

Figure 1. (**A**). SEM cross-section of the SILKBridge® nerve conduit showing the two inner and outer electrospun layers that enclose the intermediate textile layer. (**B**). ATR-FTIR spectrum of the inner electrospun layer in the 2000–800 cm^{-1} range showing the Amide bands characteristic of silk fibroin: Amide I, peak at 1627 cm^{-1} and shoulder at 1698 cm^{-1}; Amide II: peak at 1515 cm^{-1}; Amide III: peak at 1230 cm^{-1} and shoulder at 1260 cm^{-1}. The Degree of crystallinity is calculated from the I_{1260}/I_{1230} intensity ratio. (**C**). DSC thermograms of the device in the 150–400 °C temperature range. Peak at 290 °C: thermal degradation of the electrospun component. Peak at 320 °C: thermal degradation of the textile component. (**D**). Typical load/deformation curve of the suture retention test. The series of peaks recorded before ultimate failure represent the breakage of the yarns forming the textile layer. (**E**). Typical load/deformation curve of the compression strength test. (**F**). Typical load/deformation curve of the device analyzed in the uniaxial tensile mode.

3.2. Chemical Analysis and Toxicological Evaluation

For the manufacturing of medical devices, chemicals could be used as processing aids. The safety profile and the toxicological risks related to their use must be assessed. In the case of SILKBridge®, the processing aids used during manufacturing are: (i) methanol (MeOH), to clean starting materials; (ii) ethanol (EtOH), to consolidate the electrospun layers by solvent-induced random coil → β-sheet conformational transition and to clean the final device; (iii) lithium bromide (LiBr), to dissolve degummed cocoon silk fibroin fibers to produce SF films as a preparatory step of the electrospinning process; (iv) formic acid (FA), to dissolve SF films and prepare the electrospinning dope; (v) 1-ethyl-3-methyl imidazolium acetate (EMIMAc), the welding agent used to couple the TEX and ES layers. In terms of hazard classification (Regulation EC No 1272/2008), all these substances share an acute toxicity designation (oral, dermal, or by inhalation); some may cause eye/skin irritation (LiBr, EtOH, EMIMAc); EtOH and MeOH may show specific target organ toxicity, but none of them are reported as carcinogenic, mutagenic, toxic to reproduction or with endocrine-disrupting activity. Despite the thorough cleaning procedures adopted, residues of these chemicals may remain on the device and leach in the surrounding tissues during in vivo implantation. This might impact on the biocompatibility of the device and interfere with the smooth progress of tissue regeneration.

To evaluate the amount of residual processing aids that may remain in the final device, sterilized samples of nerve conduits were extracted, and the extracts were analyzed (Table 4, line 1). These results represent the starting point to perform the toxicological evaluation following the provisions of ISO 10993-17. As a first step, the concentration of each chemical expressed in mg/kg was transformed in μg/device to obtain the total amount of each compound released by one device (line 2). For this calculation, an average weight of 25 mg/device was considered. Then, assuming a worst-case scenario of multiple devices implanted at the same time in one patient (in this case: 10 devices), the value was multiplied by 10 to calculate the potential patient daily intake (PPDI, expressed in μg/day) (line 3), i.e., the total amount of chemical to which the patient is potentially exposed during the lifetime. A further worst case assumption was that the total amount of the chemical expressed by the PDDI value is released just in one day.

Table 4. Chemical and toxicological analysis of leachable processing aids.

		LiBr		EtOH	MeOH	EMIMAc	FA
		Li^+	Br^-				
1	Concentration (mg/kg)	4.0	324	2.7	1	4.1	200
2	Total Released Amount (μg/device) [1]	0.1	8.0	0.07	0.03	0.1	5.0
3	Potential Patient Daily Intake (PPDI; μg/day)	1.0	80	0.7	0.3	1.0	50
4	Tolerable Exposure (TE; μg/day)	250	1750	50,000	9000	10 [2]	35,400
5	Margin of Safety (MoS)	0.25×10^3	22	71×10^3	30×10^3	10	0.71×10^3

[1] An average weight of 25 mg has been considered for the device. [2] The TTC (threshold of toxicological concern) is used for EMIMAc instead of TE.

For each compound, the toxicological data were retrieved from specialized databases (TOXNET Databases, Pubmed, Pubchem, ECHA—European Chemical Agency, FDA—Food and Drug Administration, etc.). They allowed setting the values of tolerable exposure (TE, expressed in μg/day), i.e., the amount of chemical agent that doesn't pose concerns for human health (line 4). Since no relevant toxicological data were available for EMIMAc, the TE could not be calculated as recommended by ISO 10993-17. Instead, the value of TTC (threshold of toxicological concern) was estimated according to ISO/TS 21726. It is important to note that the TTC value assigned to EMIMAc (10 μg/day) derives from an extremely conservative and protective classification based on individual mutagenic impurities in pharmaceuticals and is presumed to be protective for both the potential carcinogenic and non-cancer effects that would occur following patient exposure to the medical device. Finally, the TE/PPDI ratio was calculated to obtain the margin of safety (MoS, line 5). For values of MoS ≤ 1, the compound is considered to pose a toxicological concern. For values of MoS > 1, the compound does not raise risks for human health. We adopted a more conservative approach, setting a threshold for MoS ≥ 10. As can be observed in Table 4, despite the worst-case assumptions made at different levels of the toxicological evaluation, i.e., total amount released in one day, multiple devices, MoS threshold higher than recommended by the reference standard, none of the compounds tested can be held responsible for causing toxicological concerns.

3.3. Biological Safety Evaluation

The ISO 10993-Part 1 provides a guidance for the assessment of medical devices according to the nature and duration of their contact with human tissues, and the other parts of the standard give indications on how to perform the tests required. SILKBridge® is categorized as an implant device in long term contact with tissue/bone (>30 days). Based on this categorization, as well as on the analysis of the risks associated with the constituent materials, their source, the manufacturing process, and the morphological, physical, chemical, and mechanical characteristics, the endpoints of the biological safety evaluation were established according to the whole ISO 10993 standard. Sterilized devices were then tested for each biological endpoint; the results are summarized in Table 5.

Table 5. Results of the biological safety evaluation.

Test	Evaluation	Final Score
Cytotoxicity	After 24 h of contact with the test sample extracts, discrete intracytoplasmic granules, no cell lysis, and no reduction of cell growth were observed (reactivity grade 0). The cell viability reduction was 4%.	Not cytotoxic
Delayed hypersensitivity	No abnormalities were observed in treated and control animals during the challenge phase (grade 0).	Not sensitizing
Intracutaneous reactivity	No abnormalities were observed immediately after injection. With polar extract all treated and control sites did not show signs of erythema, eschar or edema. With apolar extract, all treated and control sites showed a slight erythema, no eschar or edema. The Primary irritation index was 0.	Not irritating
Pyrogen test	The summed temperature rise of rabbits was +0.3 °C and the summed response did not exceed 2.8 °C. Only an individual temperature rise of one rabbit was higher than 0.5 °C and the summed temperature rise was lower than 3.3 °C.	Not pyrogenic
LAL test	The test sample solution had an endotoxin content <20 EU/device.	Not endotoxic
Acute systemic toxicity	None of the treated and control animals showed toxic signs, symptoms, mortality, or weight loss.	Not toxic
Genotoxicity	Bacterial Reverse Mutation Test: no toxic effects of the extracts were noted in any of the tester strains used; no biologically relevant increases in revertant colony numbers of any of the tester strains were observed following treatment with extracts. Mouse Lymphoma Test: no growth inhibition was observed for the polar and lipophilic extracts; no biologically relevant increase of mutants was found for the polar and lipophilic; the global evaluation factor (GEF) was not exceeded by the induced mutant frequency at any concentration; no dose-response relationship was observed.	Not mutagenic

4. Discussion

The development of an implantable medical device as the SILKBridge® nerve conduit must undergo a comprehensive series of control tests before being released for use to ensure that the manufacturing process is under control and that the device performs as expected. The on-bench testing framework complies with the design controls requirements set out according to FDA and EU/MDR regulations, and comprises a list of functional parameters and test methods inspired by the ISO 10993-18 standard, which can be used for the identification and evaluation of the physical, chemical, morphological, and topographical properties of materials in finished medical devices, all having a strong impact in terms of verification and validation of the target performance characteristics.

The results of the tests reported in Table 1 (morphological and geometrical characteristics), Table 2 (physical and structural characteristics), and Table 3 (mechanical characteristics) demonstrate that the comparison of three production lots gave satisfactory results. All the measured characteristics appeared highly reproducible, with a very low level of batch-to-batch variability, thus, ensuring the robustness of the manufacturing process. With reference to the suitability of the device for the intended use, the following discussion will provide deeper insights into the set of functional specifications that were defined during the design controls to comply with end-use requirements.

The inner diameter of the device is a geometrical characteristic determined by the foreseen clinical use. The transversal dimension of the conduit must be large enough to accommodate the distal and proximal extremities of the severed nerve and to allow their fixation by suturing during surgery. The size of the nerve conduits currently produced is 1.5–2.0 mm inner diameter, which is optimized for repairing transected digital nerves [16]. To repair larger nerves located in other anatomical positions, larger devices will be required. Usually, marketed nerve conduits cover a range of discrete sizes from 1 mm to 10 mm.

The wall thickness can be a more critical characteristic. A relationship between the formation of neuromas in regenerated nerves tissues and the thickness of the conduit wall has been reported [17]. The wall-thickness of SILKBridge® has been carefully designed considering the properties of the selected starting materials and the requirements in terms of mechanical performance, suturability, targeted degradation rate, and the permeability specifications [18]. In general, wall thickness values lower that 1 mm are highly suggested for nerve guide application, with 0.6 mm as the optimum target for the maximum value [19]. As shown in Table 1, the final device falls within this optimal range.

Linear density, i.e., the weight per unit length, allows for evaluating the amount of material implanted into the body. Silk, as any other non-autologous biomaterial, is likely to elicit a foreign body response at the site of implantation [7]. Moreover, it has been reported that silk-based devices implanted in soft tissues or with longer degradation times tend to induce a long-lasting inflammatory response than those with shorter degradation times or those located within hard tissues [8]. On this basis, it has been decided to keep the linear density as low as possible to reduce the impact of the device in terms of local reactions. The choice of a macro-porous texture of the textile braid for the central layer of the wall allowed for keeping the linear density of the final device largely below the threshold of 10 mg/cm, as defined in the design stage (Table 1).

The porosity of the wall of a nerve conduit is a key functional parameter. It must be optimized to maximize the influx of oxygen and nutrients from the interstitial fluid and to avoid the loss of neurotrophic factors secreted by Schwann cells at the distal stump. A wall porosity of about 80% or higher is considered ideal for peripheral nerve repair [17–19]. This target has been achieved with SILKBridge®, whose wall porosity higher than 80% results from the combination of the macro- and micro-porosity of the TEX and ES layers, respectively (Table 1). With reference to the pore size, this characteristic is entirely determined by the two ES layers and does not exceed the 5 μm threshold on average (data not shown). This allows for the permeation of small nutrient molecules, while preventing the migration of inflammatory cells and the infiltration of fibrous tissue inside the lumen of the conduit [18].

The physical and structural features of the device were explored by FTIR and DSC analyses (Table 2). Both the spectroscopic and thermal properties of silk fibroin-based materials have been extensively investigated [20,21]. The type, position, and intensity of the bands in the IR spectra and of the thermal transitions in the DSC thermograms not only reflect the chemical structure of the silk materials but also provide information about the molecular conformation taken by the silk fibroin chains in native and regenerated formats.

The aim of the FTIR analysis was to verify that the manufacturing process did not alter the chemical integrity of the constituent material, did not leave foreign matters or contaminants on the device, and allowed for achieving the desired degree of crystallinity of the regenerated electrospun layers. Loss of integrity and changes in the chemical and conformational structure of the constituent silk material may have a strong impact on the biodegradation rate of the device upon implantation. Moreover, the presence of contaminants may impact on the biocompatibility of the device. For the purpose of integrity and purity checks, the position and intensity of the IR bands found in the spectrum of the sample were compared with reference spectra, while the crystallinity index was calculated from the intensity ratio of the Amide III bands at 1260 cm^{-1} (crystalline phase) and 1230 cm^{-1} (amorphous phase) [15].

The major thermal events characteristics of silk fibroin fall in the 150–400 °C temperature range (glass transition and melting/degradation). As with FTIR, the aim of the DSC analysis was to verify the structural integrity of the constituent silk fibroin materials to avoid any impact on the performance of the device. The DSC analysis allowed us to also calculate the mass ratio between the electrospun and textile layers, which present distinct endothermic peaks [10]. This parameter is a very useful tool to further assess the robustness of the manufacturing process. The DSC and FTIR results listed in Table 2 allowed for confirming the structural integrity of the silk fibroin materials comprising the device.

Medical devices must be designed with an optimal set of mechanical properties fitting the requirements of the specific clinical need. Nerve conduits must provide a balanced combination of strength and elasticity to withstand clinical operation stresses, such as manipulation and suturing during implantation, to resist deformation caused by the biomechanical stresses generated in vivo, and to avoid channel collapse, since compression can result in damage to the growing axon [10,11]. To verify that the manufacturing process allows for reaching the desired mechanical characteristics, the devices were analyzed for determining the tensile, suture retention, and compression properties (Table 3).

To withstand physiological loads, the tensile properties of nerve conduits should at least approach the ones of natural nerves [18,22]. The device must be elastic enough to match the deformation of the natural nerve and strong enough to protect the growing axons from breaking. From the data reported in the literature, the stress and strain values of human nerves are about 7 MPa and 60%, respectively. Considering the geometry of the device, the stress value corresponds to a breaking load of about 15 N. As shown in Table 3, the SILKBridge® nerve conduit largely met the target values.

The suture retention test is intended to verify whether the device withstands the mechanical stresses applied by the surgeon during implantation and later during in vivo functioning. Suture failure may cause dramatic consequences to the patients, due to the loss of stability of the implant, lack of support for tissue regeneration, and need to re-operate. No threshold values for suture retention strength of nerve conduits have been reported in literature. On the contrary, limits were imposed for vascular grafts, a much more demanding application in terms of anastomotic strength, which requires that grafts are capable to exceed 2 N as suture retention strength [23–25]. Thanks to the strength of the textile braid, which is the load-bearing element of the wall of the device, the target threshold of suture retention strength was achieved by the final device (Table 3).

Finally, nerve conduits must withstand the mechanical compression stresses of surrounding tissues until complete nerve regeneration, avoiding collapse that may hinder the healing and cause pain to the patient [18]. To set acceptable threshold values for the compression performance, reference was made to the limit pressure values reported for the human median nerve exposed to carpal tunnel pressure [26]. In healthy individuals, carpal tunnel pressures are typically well below 10 mmHg, corresponding to a stress 1.33 kPa.

Considering the geometry of the SILKBridge®, this stress corresponds to a compression load of about 1.1 cN. The results listed in Table 3 indicate that the compression resistance threshold was easily achieved, and that the device can withstand compression stresses and remain open to allow for a smooth progression of the nerve regeneration.

The evaluation of the potential toxicological risks associated with leachable substances released by a medical device in the surrounding tissues is another important step for the identification and quantification of the biological hazards related to its use. The ISO 10993-12 standard provides provisions for the preparation of samples for analysis. The ISO 10993-18 standard specifies a framework for the identification of leachable substances and for the quantitative determination of their potential release in the human body. The ISO 10993-17 standard guides the manufacturer through the process of estimation and control of the toxicological risks associated with the medical device use. The chemical analysis addressed all the processing aids used along the manufacturing process. Their amount was determined by applying suitable laboratory extraction conditions and advanced analytical methods.

With reference to the mechanism by which the chemical compounds may be released into the tissues surrounding the implant area, a combination of contributions can play a role. Simple diffusion from the device to the tissue may occur in the first period after implantation, when the device is still intact. Afterwards, when the device starts to degrade, swelling and fragmentation of the polymer texture may open new ways for the chemical to leach outside. However, it must be considered that the release will never occur in a bursting way, but gradually over time and in a more physiologically compliant manner, so that the local load is diluted over time. This assumption is supported by the fact that SILKBridge® is made of silk fibroin materials characterized by a slow rate of degradation, i.e., from months to years for regenerated electrospun fibers and native microfibers, respectively [12].

As shown in Table 4, the results of the toxicological evaluation have demonstrated that the leachable substances coming from the manufacturing process were under control and that the cleaning procedures were effective in removing the greatest part of the processing aids from the final device. The residual amounts still present were largely below the threshold for toxicological concern, also considering the worst-case scenarios taken into account during the toxicological evaluation, i.e., the possibility of multiple implants (up to 10, one device for each finger), the ten-fold increase of the acceptable value of the Margin of Safety, and the assumption that the total amount is released in one day in a bursting way.

The targeted toxicological analysis here reported has been complemented with an untargeted one, which allowed to identify possible leachable compounds beyond those used for manufacturing, to include also unpredictable contaminants of environmental source, including laboratory materials, packaging, etc. (data not shown). It is worth noting that also this additional approach did not reveal the presence of unexpected contaminants likely to pose a health risk.

The evaluation of the toxicity profile of a degradable medical device such as SILKBridge® cannot disregard the possible toxicity of the degradation products. This issue has been addressed in a previous study where the in vitro degradation profile and the cytotoxicity of the degradation products were reported [12]. The bacterial protease type XIV from *Streptomyces griseus* was used as hydrolytic agent at three different enzyme/substrate ratios and for incubation times as long as 91 days. Degradation of the device occurred by surface erosion. The mass spectrometry analysis of the degradation products showed that the silk fibroin polypeptides recovered in the incubation buffers were representative of the aminoacidic sequence of the fibroin light and heavy chains, indicating that virtually the entire sequence of the fibroin protein was degraded. More important, the incubation buffers containing the soluble degradation products were tested with human HEK293 cells and mouse neuroblastoma N2a cells to assess cytotoxicity. No detrimental effects on cell viability were observed, suggesting that the degradation products, consisting of amino acids and small peptides, did not show any toxic property and their more likely fate was to enter the metabolic pathways of the host tissue cells [7]. Therefore, it is possible to conclude that the SILKBridge® nerve conduits will not elicit any toxic effect due to the constituent materials and/or to the manufacturing process.

Biocompatibility is an essential requirement for the medical devices and the large amount of scientific data on the safety of silk is not sufficient on a regulatory prospective to allow the use of silk-based device since the manufacturing process can change the characteristics of the raw material and have an impact on the biocompatibility of the finished device [8]. This perspective imposes that a comprehensive evaluation of the device biocompatibility is carried out to prevent any possible adverse effect for the patient's health. Different parts of the ISO 10993 standard provide a framework for the evaluation of the biological safety of the device. A plan has been designed and a set of tests, including cytotoxicity, delayed hypersensitivity, intracutaneous reactivity, pyrogen test, LAL test, acute systemic toxicity, and genotoxicity has been carried out. The overall results of the biological response allowed confirming that SILKBridge® is fully biocompatible (Table 5). This is in good agreement with the numerous data reported in the scientific literature about the biocompatibility of silk materials [8].

Finally, the evaluation of a medical device can't exempt from carrying out performance tests to verify that the product achieve it's intended use. In a previously published article [11] the implantation of the SILKBridge® in the median nerve of rats was discussed to demonstrate the regeneration of myelinated fibers along the conduit filling the gap previously created and leading to an effective morphological and functional recovery of the median nerve, like that observed with the reference autograft nerve reconstruction.

Altogether, the results reported here, and others previously published [10–12] represent an important achievement towards the implementation of a clinical study aimed at investigating the safety and efficacy of the SILKBridge® nerve conduit. The device has demonstrated an optimized balance of biomechanical and biological properties; it is intended as an "off-the-shelf" product ready to be used in the operating room as it is, without the need of adding neurotrophic and/or angiogenic factors or cells. The encouraging on-bench and pre-clinical results allowed us to proceed quickly towards the submission of a first in-human clinical study aimed at evaluating the reconstruction of digital nerve defects in humans (ClinicalTrials.gov identifier: NCT03673449). The study has already started at the Department of Plastic Surgery and Hand Surgery of the University Hospital of Zurich. Four out of 15 patients have been enrolled so far and implanted with SILKBridge® nerve conduit to repair a digital nerve gap, with very satisfactory outcome after one year in terms of functional recovery.

5. Conclusions

The path from silk-based starting materials (cocoons, yarns) to the production of a medical device and to its use in humans is long and tortuous. Patient safety and health are primary goals that must not be jeopardized in any kind of approach aimed at solving clinical problems. Of course, it is not surprising that regulatory agencies have put in place a set of very thorough and detailed directives to objectively assess all the risks associated with the use of implantable medical device.

As shown in this study, transforming a silk material into a medical device and taking it all the way to clinical application requires a network of expertise ranging from polymer science, chemistry, toxicology, biomedical engineering, medicine, biology, regulatory, and quality. It is also necessary to demonstrate that all manufacturing processes leading to the final device are scalable from laboratory or pilot scale to industrial scale. Finally, the support of clinicians is also of great importance from the beginning, with the identification of the clinical need to be solved, until the end, to evaluate the clinical results and to support product marketing.

Author Contributions: Conceptualization, A.A., G.F. and G.A.B.; Methodology, G.A.B., V.V. and M.B.; Formal Analysis, G.A.B.; Investigation, M.B., V.V., E.V., M.V.Z. and C.C.; Data Curation, G.A.B.; Writing—Original Draft Preparation, G.F.; Writing—Review & Editing, All; Supervision, G.F. and A.A.; Project Administration, A.A. Funding Acquisition, A.A. All authors have read and agreed to the published version of the manuscript.

Funding: This research did not receive any specific grant from funding agencies in the public, commercial, or not-for-profit sectors. It was entirely funded by Silk Biomaterials srl.

Institutional Review Board Statement: Animals were used in the following ISO 10993 tests for the biological safety evaluation: acute systemic toxicity (mice); pyrogenicity (rabbits); delayed hypersensitivity (guinea pigs); irritation/intracutaneous reactivity (rabbits), as recommended by the international standards of biocompatibility evaluation (ISO 10993, USP, EP), because no in vitro tests are available. Approval for the use of animals was under the responsibility of the accredited testing facility Eurofins Biolab S.r.l. (Good Laboratory Practice-GLP cert. n. 2017/16), Vimodrone (Milan), Italy.

Informed Consent Statement: Not applicable.

Data Availability Statement: The datasets generated for this study are available on reasonable request to the corresponding author.

Conflicts of Interest: A.A. is stock owners and employee of the funding organization Silk Biomaterials srl.; G.F. is stock owners and consultant of the funding organization; G.A.B., M.B., V.V., E.V., M.V.Z. and C.C. are employees of the funding organization. A.A., G.F., G.A.B., M.B., V.V., E.V., M.V.Z. and C.C. were involved in the collection, analyses or interpretation of data; in the writing of the manuscript, and in the decision to publish the results.

References

1. Vollrath, F.; Knight, D.P. Liquid crystalline spinning of spider silk. *Nature* **2001**, *410*, 541–548. [CrossRef]
2. Fedic, R.; Zurovec, M.; Sehnal, F. The Silk of Lepidoptera. *J. Insect Biotechnol. Sericol.* **2002**, *71*, 1–15. [CrossRef]
3. Babu, K.M. *Silk: Processing, Properties and Applications*; Woodhead Publishing: Sawston, UK, 2013; pp. 1–55.
4. Holland, C.; Numata, K.; Rnjak-Kovacina, J.; Seib, F.P. The Biomedical Use of Silk: Past, Present, Future. *Adv. Healthc. Mater.* **2019**, *8*, 1800465. [CrossRef]
5. Rockwood, D.N.; Preda, R.C.; Yücel, T.; Wang, X.; Lovett, M.L.; Kaplan, D.L. Materials fabrication from *Bombyx mori* silk fibroin. *Nat. Protoc.* **2011**, *6*, 1612–1631. [CrossRef]
6. Jo, Y.Y.; Kweon, H.Y.; Oh, J.H. Sericin for Tissue Engineering. *Appl. Sci.* **2020**, *10*, 8457. [CrossRef]
7. Altman, G.H.; Diaz, F.; Jakuba, C.; Calabro, T.; Horan, R.L.; Chen, J.; Lu, H.; Richmond, J.; Kaplan, D.L. Silk-based biomaterials. *Biomaterials* **2003**, *24*, 401–416. [CrossRef]
8. Thurber, A.E.; Omenetto, F.G.; Kaplan, D.L. In vivo bioresponses to silk proteins. *Biomaterials* **2015**, *71*, 145–157. [CrossRef]
9. Zhang, L.; Zhang, W.; Hu, Y.; Fei, Y.; Liu, H.; Huang, Z.; Wang, C.; Ruan, D.; Heng, B.C.; Chen, W.; et al. Systematic Review of Silk Scaffolds in Musculoskeletal Tissue Engineering Applications in the Recent Decade. *ACS Biomater. Sci. Eng.* **2021**, *7*, 817–840. [CrossRef]
10. Alessandrino, A.; Fregnan, F.; Biagiotti, M.; Muratori, L.; Bassani, G.A.; Ronchi, G.; Vincoli, V.; Pierimarchi, P.; Geuna, S.; Freddi, G. SILKBridge: A novel biomimetic and biocompatible silk-based nerve conduit. *Biomater. Sci.* **2019**, *7*, 4112–4130. [CrossRef] [PubMed]
11. Fregnan, F.; Muratori, L.; Bassani, G.A.; Crosio, A.; Biagiotti, M.; Vincoli, V.T.; Pierimarchi, P.; Geuna, S.; Alessandrino, A.; Freddi, G.; et al. Preclinical validation of SILKBridge for peripheral nerve regeneration. *Front. Bioeng. Biotechnol.* **2020**, *8*, 835. [CrossRef] [PubMed]
12. Biggi, S.; Bassani, G.A.; Vincoli, V.; Peroni, D.; Bonaldo, V.; Biagiotti, M.; Belli, R.; Alessandrino, A.; Biasini, E.; Freddi, G. Characterization of Physical, Mechanical, and Biological Properties of SILKBridge Nerve Conduit after Enzymatic Hydrolysis. *ACS Appl. Bio Mater.* **2020**, *3*, 8361–8374. [CrossRef] [PubMed]
13. Alessandrino, A. Process for the Production of a Hybrid Structure Consisting of Coupled Silk Fibroin Microfibers and Nanofibers, Hybrid Structure thus Obtained and Its Use as Implantable Medical Device. U.S. Patent 10,758,641, 1 September 2020.
14. Fossey, S.A.; Kaplan, D.L. Silk protein. In *Polymer Data Handbook*; Mark, J.E., Ed.; Oxford University Press, Inc.: Oxford, UK, 1999; pp. 970–974.
15. Bath, N.V.; Nadiger, G.S. Crystallinity in silk fibers: Partial acid hydrolysis and related studies. *J. Appl. Polym. Sci.* **1980**, *25*, 921–932. [CrossRef]
16. Ortiz, R.; Westenberg, R.F.; Langhammer, C.G.; Knaus, W.J.; Chen, N.C.; Eberlin, K.R. Nerve Diameter in the Hand: A Cadaveric Study. *Plast. Reconstr. Surg. Glob. Open* **2019**, *7*, e2155. [CrossRef]

17. Nectow, A.R.; Marra, K.G.; Kaplan, D.L. Biomaterials for the Development of Peripheral Nerve Guidance Conduits. *Tissue Eng. Part B* **2012**, *18*, 40–50. [CrossRef]
18. Chiono, V.; Tonda-Turo, C. Trends in the design of nerve guidance channels in peripheral nerve tissue engineering. *Prog. Neurobiol.* **2015**, *131*, 87–104. [CrossRef]
19. Lauren, E.; Kokai, L.E.; Lin, Y.C.; Oyster, N.M.; Marra, K.G. Diffusion of soluble factors through degradable polymer nerve guides: Controlling manufacturing parameters. *Acta Biomater.* **2009**, *5*, 2540–2550. [CrossRef]
20. Boulet-Audet, M.; Vollrath, F.; Chris Holland, C. Identification and classification of silks using infrared spectroscopy. *J. Exp. Biol.* **2015**, *218*, 3138–3149. [CrossRef]
21. Magoshi, J.; Nakamura, S. Studies on Physical Properties and Structure of Silk. Glass Transition and Crystallization of Silk Fibroin. *J. Appl. Polym. Sci.* **1975**, *19*, 1013–1015. [CrossRef]
22. Dumont, C.E.; Born, W. Stimulation of Neurite Outgrowth in a Human Nerve Scaffold Designed for Peripheral Nerve Reconstruction. *J. Biomed. Mater. Res. Part B Appl. Biomater.* **2005**, *73*, 194–202. [CrossRef]
23. Billiar, K.; Murray, J.; Laude, D.; Abraham, G.; Bachrach, N. Effects of carbodiimide crosslinking conditions on the physical properties of laminated intestinal submucosa. *J. Biomed. Mater. Res.* **2001**, *56*, 101–108. [CrossRef]
24. Mine, Y.; Mitsui, H.; Oshima, Y.; Noishiki, Y.; Nakai, M.; Sano, S. Suture Retention Strength of Expanded Polytetrafluoroethylene (ePTFE) Graft. *Acta Med. Okayama* **2010**, *64*, 121–128. [CrossRef]
25. Pensalfini, M.; Meneghello, S.; Lintas, V.; Bircher, K.; Ehret, A.E.; Mazza, E. The suture retention test revisited and revised. *J. Mech. Behav. Biomed. Mater.* **2018**, *77*, 711–717. [CrossRef]
26. Topp, K.S.; Boyd, B.S. Structure and biomechanics of peripheral nerves: Nerve responses to physical stresses and implications for physical therapist practice. *Phys. Ther.* **2006**, *86*, 92–109. [CrossRef]

insects

MDPI

Article

Transcriptome Analysis Reveals the Gene Expression Changes in the Silkworm (*Bombyx mori*) in Response to Hydrogen Sulfide Exposure

Rui Zhang [1,†], Yu-Yao Cao [1,†], Juan Du [2], Kiran Thakur [1,2], Shun-Ming Tang [3,4], Fei Hu [1,2,*] and Zhao-Jun Wei [1,2,*]

1 School of Food and Biological Engineering, Hefei University of Technology, Hefei 230009, China; 2020111323@mail.hfut.edu.cn (R.Z.); cyy199574@mail.hfut.edu.cn (Y.-Y.C.); kumarikiran@hfut.edu.cn (K.T.)
2 School of Biological Science and Engineering, North Minzu University, Yinchuan 750021, China; 20217528@stu.nun.edu.cn
3 Jiangsu Key Laboratory of Sericultural Biology and Biotechnology, School of Biotechnology, Jiangsu University of Science and Technology, Zhenjiang 212003, China; tangsm1971@just.edu.cn
4 Key Laboratory of Silkworm and Mulberry Genetic Improvement, Ministry of Agriculture, Sericultural Research Institute, Chinese Academy of Agricultural Sciences, Zhenjiang 212018, China
* Correspondence: hufei@hfut.edu.cn (F.H.); zjwei@hfut.edu.cn (Z.-J.W.)
† Co-first author.

Simple Summary: The fat body is one of the most important tissues in the body of insects due to its number of functions. Nowadays the new physiological function of H_2S has gained attention as a novel signaling molecule. H_2S performs crucial regulatory functions involving growth, the cardiovascular system, oxidative stress, and inflammation in many organisms. In this study, RNA-seq technology was used to investigate the fat body of the silkworm at the transcriptional level after H_2S exposure during the 5th larvae stage. A total of 1200 (DEGs) was identified after 7.5 μM H_2S treatment, of which 977 DEGs were up-regulated and 223 DEGs were down-regulated. DEGs were mainly involved in the transport pathway, cellular community, carbohydrate metabolism, and immune-associated signal transduction. Present research provides new insights on the gene expression changes in the fat body of silkworms after H_2S exposure.

Citation: Zhang, R.; Cao, Y.-Y.; Du, J.; Thakur, K.; Tang, S.-M.; Hu, F.; Wei, Z.-J. Transcriptome Analysis Reveals the Gene Expression Changes in the Silkworm (*Bombyx mori*) in Response to Hydrogen Sulfide Exposure. *Insects* **2021**, *12*, 1110. https://doi.org/10.3390/insects12121110

Academic Editors: Silvia Cappellozza, Morena Casartelli, Federica Sandrelli, Alessio Saviane and Gianluca Tettamanti

Received: 3 November 2021
Accepted: 9 December 2021
Published: 13 December 2021

Abstract: Hydrogen sulfide (H_2S) has been recognized for its beneficial influence on physiological alterations. The development (body weight) and economic characteristics (cocoon weight, cocoon shell ratio, and cocoon shell weight) of silkworms were increased after continuous 7.5 μM H_2S treatment. In the present study, gene expression changes in the fat body of silkworms at the 5th instar larvae in response to the H_2S were investigated through comparative transcriptome analysis. Moreover, the expression pattern of significant differentially expressed genes (DEGs) at the 5th instar larvae was confirmed by quantitative real-time PCR (qRT-PCR) after H_2S exposure. A total of 1200 (DEGs) was identified, of which 977 DEGs were up-regulated and 223 DEGs were down-regulated. Most of the DEGs were involved in the transport pathway, cellular community, carbohydrate metabolism, and immune-associated signal transduction. The up regulated genes under H_2S exposure were involved in endocytosis, glycolysis/gluconeogenesis, the citrate cycle (TCA cycle), and the synthesis of fibroin, while genes related to inflammation were down-regulated, indicating that H_2S could promote energy metabolism, the transport pathway, silk synthesis, and inhibit inflammation in the silkworm. In addition, the expression levels of these genes were increased or decreased in a time-dependent manner during the 5th instar larvae. These results provided insight into the effects of H_2S on silkworms at the transcriptional level and a substantial foundation for understanding H_2S function.

Keywords: silkworm; hydrogen sulfide; transcriptome; differentially expressed genes; expression pattern

1. Introduction

The silkworm, *Bombyx mori*, is a pivotal model of Lepidoptera, a holometabolous insect, with a considerable economic importance in the world. The silkworm has been considered a model animal because of its appropriate life cycle, cultivation, and fast reproduction [1,2]. The fat body for insects acts as the dynamic tissue and the metabolic organ, mainly involved in synthesis, nutrient storage, and energy metabolism [3], and it has been widely studied in various studies of silkworm to reflect various biological, physiological, and biochemical processes. The over expression of BmFoxO in the fat body could suppress protein translation in transgenic silkworms [4].

The toxic effects of hydrogen sulfide (H_2S) as an environmental pollutant have been studied in the past years. However, nowadays, the new physiological function of H_2S has gained attention as a novel signaling molecule. H_2S performs crucial regulatory functions involving growth, the cardiovascular system, oxidative stress, and inflammation in many organisms [5,6]. H_2S has been accepted as the third transmission gas followed by nitric oxide (NO) and carbon monoxide (CO) in past years with applications to various insects. Miller and Roth (2007) found that treatment with high concentrations of H_2S are toxic to *Caenorhabditis elegans*; However, exposure to low concentrations of H_2S increased the thermotolerance and lifespan via increasing the activity of SIR-2.1 [7]. Budde and Roth (2010) found that hif-1 is required when *C. elegans* were exposed to H_2S; in turn, H_2S could increase both HIF-1 protein concentration and nuclear localization [8]. H_2S is an endogenous regulator of oxidative damage, metabolism, and aging in *C. elegans* [9]. Wei and Kenyon (2016) found that removing germ cells in *C. elegans* triggers a level of hydrogen sulfide in nonreproductive somatic tissues. Hydrogen sulfide displayed the protective responses that slow aging [10]. H_2S treatment significantly increased the survival and lifespan of *Drosophila melanogaster* under arid and food-free conditions [11,12].

In our previous research, it was found that the development (body weight) and economic characteristics (cocoon weight, cocoon shell ratio, and cocoon shell weight) of silkworm (*Bombyx mori*) were increased after continuous treatment with 7.5 μM of H_2S. After exposure to H_2S, the level of metabolites related to inflammation ((6Z,9Z,12Z)-octadecatrienoic acid, hexadecanoic acid, choline phosphate, and malic acid, etc.) in the hemolymph of silkworms were significantly increased compared to control group [13]. However, when the concentrations of H_2S were higher than 7.5 μM, H_2S displayed toxic effects to the silkworm, including a decrease in body weight, cocoon weight, cocoon shell weight, etc. RNA-seq technology is widely used in the field of silkworms to investigate physiological and biochemical changes at the transcriptional level under diverse conditions [14]. However, there is no effective report for the effects of H_2S on silkworms at the transcriptional level.

The present study aimed to investigate the expression level of genes in the fat body of silkworms after exposure to H_2S. The results based on the transcriptomic and bioinformatic analysis showed significantly different responses to H_2S and provided useful insights into the role of H_2S on silkworms.

2. Materials and Methods

2.1. Silkworm Rearing and H2S Exposure

P50 strains of silkworm were obtained from the Sericulture Research Institute of the Chinese Academy of Agricultural Sciences, Zhenjiang, China. According to the previous study [13], from the 4th instar larvae to the 5th instar larvae, silkworms were divided into two groups (H_2S-treated and control) and reared in desiccators using fumigating treatment at 25 ± 1 °C and 12L:12D photoperiod conditions. Silkworms in the H_2S-treated group were treated by NaHS (Shandong West Asia Chemical Company, Jinan, China), a donor of H_2S. The experimental concentration of H_2S was set to 7.5 μM according to the report of Cao et al. (2020) [13]. There were triplicates in each group with 30 silkworms per replicate and the silkworms were fed abundant fresh mulberry leaves twice a day.

2.2. RNA Extraction

The total RNA of the silkworm fat body was extracted on ice from the 1st day (5L1D) to the 5th day (5L5D) at the 5th instar larvae using Trizol reagent (Invitrogen, CA, USA). The concentration and purity of total RNA were analyzed by NanoDrop ND-1000 (NanoDrop, Wilmington, DE, USA) and the ratios of A260/A280 and 260/230 were used to measure the quality of total RNA.

2.3. Sequencing, Data Processing, and Assembly

The total RNA of 5L3D was analyzed in the H_2S-treated and control groups by RNA-Sequencing. Total RNA (10 μg) of each sample was subjected to Poly(A) mRNA isolation with poly-T oligo-attached magnetic beads (Invitrogen, CA, USA). The poly(A) RNA was fragmented into small pieces using a Magnesium RNA Fragmentation Module (New England Biolabs (Beijing) LTD., cat. e6150, Beijing, China) under 94 °C for 5–7 min. The cleaved RNA fragments were then reverse-transcribed to create the cDNA by SuperScript^{TM} II Reverse Transcriptase (Invitrogen, CA, USA), which were next used to synthesize U-labeled second-stranded DNAs. An A-base was then added to the blunt ends of each strand, preparing them for ligation to the indexed adapters. Each adapter contained a T-base overhang for ligating the adapter to the A-tailed fragmented DNA. Single- or dual-index adapters were ligated to the fragments, and size selection was performed with AMPureXP beads. After the heat-labile UDG enzyme (New England Biolabs (Beijing) LTD., cat.m0280, Beijing, China) treatment of the U-labeled second-stranded DNAs, the ligated products were amplified with PCR by the following conditions: initial denaturation at 95 °C for 3 min; 8 cycles of denaturation at 98 °C for 15 s, annealing at 60 °C for 15 s, and extension at 72 °C for 30 s; and final extension at 72 °C for 5 min. The average insert size for the final cDNA library was 300 ± 50 bp. Lastly, we performed the 2 × 150 bp paired-end sequencing (PE150) on an Illumina Novaseq^{TM} 6000 (LC-Bio Technology CO., Ltd., Hangzhou, China) following the vendor's recommended protocol.

The transcriptome was sequenced on basis of the Illumina paired-end RNA-seq approach. The average insert size for the paired-end libraries was 300 bp (±50 bp). The low-quality reads with sequencing adaptors, sequencing primer, and nucleotide (q quality score (p value corrected by BH algorithm, also called FDR value, q value, p adj value) <20) were removed using Cutadapt software (https://cutadapt.readthedocs.io/en/stable/, accessed on 1 August 2021) [15] before assembly. We used HISAT2 software (https://daehwankimlab.github.io/hisat2/,version:hisat2-2.0.4, accessed on 1 August 2021) [16] to map reads to the genome. Three important parameters, such as Q20 (the percentage of bases with mass value ≥20), Q30 (mass value ≥30), and GC content were verified to evaluate all the reads using the FastQC online tool [17].

The *Bombyx mori* reference genome [18] was used for the alignment of the sample reads using the HISAT package [19]. StringTie (http://ccb.jhu.edu/software/stringtie/, accessed on 1 August 2021) was used for the assembly of the mapped reads of each sample and a comprehensive transcriptome was constructed combining all transcriptomes using Perl scripts [20]. The expression level of mRNAs was used to calculate the fragments per kilobase of exon model per million mapped reads (FPKM) by StringTie (http://www.bioconductor.org/packages/release/bioc/html/ballgown.html, accessed on 1 August 2021) [20]. The FPKM could evaluate the abundance of genes. The DEGs were selected with $\log_2$ (fold change) >1 or $\log_2$ (fold change) <−1 at p-value < 0.05, and q-value > p-value by R package-edgeR (https://bioconductor.org/packages/release/bioc/html/edgeR.html, accessed on 1 August 2021) [21].

2.4. Functional Annotation and Enrichment Analysis of DEGs

The reference sequences for all the silkworm genes were obtained from the NCBI database and BLASTX was used for annotation. All the sequences were aligned against Gene Ontology (GO) and Kyoto Encyclopedia of Genes and Genomes (KEGG) databases using DIAMOND (0.7.12) [22]. GO function or KEGG pathway significant enrichment analysis first mapped all the DEGs to each GO term in the GO database or each KEGG pathway in the KEGG database, and second calculated the number of genes in each GO term or KEGG pathway and finally applied a hypergeometric test to find a GO term or KEGG pathway that were significantly enriched in DEGs compared to the entire genome background.

2.5. Quantitative Real-Time PCR (qRT-PCR)

The qRT-PCR was used to validate the accuracy of DEGs in L5D3 at the transcriptome level and the expression pattern of DEGs in the 5th instar larvae was investigated. Reverse transcription was performed by HiScript II QRT SuperMix (Vazyme, Nanjing, China) as follows: total RNA (1 μg) (without genome contamination) and 4 × gDNA Wiper were mixed with 5 × HiScript II QRT SuperMix and then incubated at 25 °C for 10 min, 50 °C for 15 min, and 85 °C for 2 min [14]. The cDNA was used for real-time PCR detection using the UltraSYBR Mixture (Cwbio, Beijing, China) and qRT-PCR was executed using the Light Cycler® 480 System (Roche, Basel, Switzerland). The primer sequences for target genes were listed in Table S1 using primer 3.0 software (Premier Biosoft International, Palo Alto, CA, USA). The expression level of action A3, a silkworm housekeeping gene, was regarded as an internal reference in standardization [14]. Before Quantitative real-time PCR, the efficiency and specificity of primers were confirmed. The volume of the reaction system was 20 μL, containing 10 μL UltraSYBR Mixture, 1 μL cDNA, 0.5 μL forward and reverse primers and 8 μL ddH_2O. The qRT-PCR program was as follows: initial denaturation at 95 °C for 10 min; followed by 40 times of denaturation at 95 °C for 10 s, annealing at 60 °C for 30 s, and extension at 72 °C for 32 s; a melting curve program was set at 95 °C for 15 s, 60 °C for 1 min, 95 °C for 15 s, and a60 °C for 15 s with a final cooling step at 4 °C for 30 s [23]. The data were analyzed with the LightCycler®96 software (Roche Diagnostics, Indianapolis, IN, USA) using the $2^{-\Delta\Delta Ct}$ method.

2.6. Statistical Analysis

The data were analyzed using one-way ANOVA data analysis. All the data were obtained in triplicate and presented as mean ± standard error (SE) (n = 3). Statistical significance at $p < 0.05$ was measured using SPSS 20.0 software (IBM, Endicott, NY, USA).

3. Results

3.1. Transcriptome Sequencing and Assembly

The deep sequencing of RNA from the fat body (5L3D) treated with H_2S using the illumine sequencing and the data processing results were shown in Table 1. The raw data of H_2S-treated and control groups were approximately 7.75 GB and 7.66 GB, respectively. After filtering and removing the low-quality reads, adaptors and poly-N stretches, the valid data and valid reads ratio of each replicate in H_2S-treated group were obtained as follows: 7.55 Gb (99.12%), 7.87 Gb (98.91%), and 7.54 Gb (98.39%) and that of the control group were 7.42 Gb (94.00%), 7.37 Gb (98.99%), and 7.50 Gb (98.12%). The quality scores of Q30 exceeded 98% in each sample. Moreover, the mapped ratio in H_2S-treated and control groups were 97.35%, 97.43%, and 97.72% and 93.19%, 97.35%, and 97.52%, respectively, which highly matched with the silkworm genome (Table S2). The expression interval of genes was mainly focused on the 0.3~3.57 FPKM interval (FI) and the ratio was about 35% gene of the control group and 39% of H_2S-treated group (Table S3). These results revealed a good quality sequencing analysis and the genes were analyzed in the following section.

Table 1. The sequencing data of samples.

Samples	FB_Control_1	FB_Control_2	FB_Control_3	FB_H_2S_1	FB_H_2S_2	FB_H_2S_3
Raw reads	52,598,692 (7.89 Gb)	49,647,104 (7.45 Gb)	50,987,488 (7.65 Gb)	50,804,424 (7.62 Gb)	53,068,042 (7.96 Gb)	51,108,946 (7.67 Gb)
High quality reads	49,443,624 (7.42 Gb)	49,144,228 (7.37 Gb)	50,029,926 (7.50 Gb)	50,357,636 (7.55 Gb)	52,490,302 (7.87 Gb)	50,286,178 (7.54 Gb)
High quality reads ratio (%)	94.00	98.99	98.12	99.12	98.91	98.39
Q30 (%)	98.53	99.08	99.07	98.93	98.98	99.12
GC content (%)	50	48	49	48	48.50	48
Total mapped	46,077,049	47,843,354	48,788,614	49,023,730	51,140,817	49,141,008

3.2. DEG Identification and Analysis

DEGs in H_2S and control groups were calculated according to the FPKM value of the obtained genes and then a large number of genes were differentially expressed after H_2S exposure. The volcano plot of DEGs was shown in Figure 1 and the detailed information of DEGs was listed in Table S4. A total of 1200 DEGs were obtained between the H_2S-treated and control groups. The number of up-regulated and down-regulated DEGs was 977 and 223, and the up-regulated genes were higher than down-regulated genes.

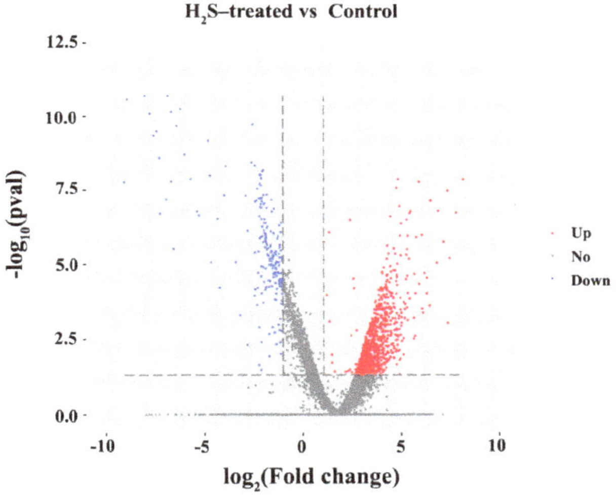

Figure 1. The volcano plot of DEGs. The red points represent the up-regulated genes, the blue points represent the down-regulated genes, and the grey points represent the genes without differentia expression.

3.3. GO Analysis of DEGs

The GO database can provide standard terms for describing the comprehensive properties of genes in organisms as an international classification system. In this study, DEGs were assigned to different GO categories such as biological processes, cellular components, and molecular functions. The major subcategories of DEGs for biological processes were border follicle cell migration, mitotic spindle elongation, myoblast fusion and microtubule-based processes, as well as others. (Figure 2 and Table S5). Through the results of the GO analysis, it was revealed that H_2S could influence these multiple biological processes by regulating the expression of genes in the fat body of silkworms at the molecular level.

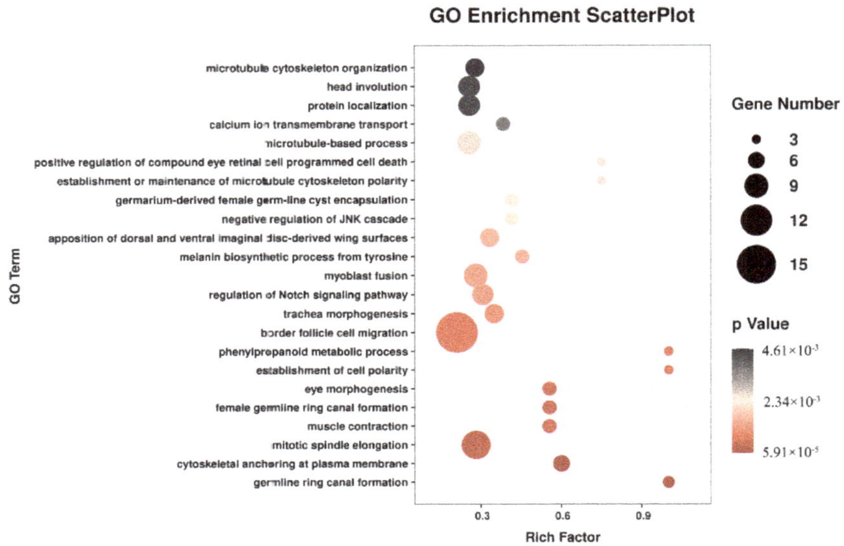

Figure 2. GO classification of DEGs ($p \leq 0.01$).

3.4. KEGG Pathway Analysis

To systematically analyze the biological function and metabolic pathways of DEGs, KEGG pathway annotations were obtained. The 561 genes of 1200 DEGs predicted to encode enzymes were mapped into 225 KEGG pathways. These annotated genes were assigned to six categories including organismal systems (49 pathways), metabolism (74 pathways), human diseases (39 pathways), genetic information processing (19 pathways), environmental information processing (28 pathways), and cellular processes (16 pathways) (Figure 3). In the organismal systems, vascular smooth muscle contraction (7 DEGs) is the most abundant metabolic pathway followed by the insulin signaling pathway (6 DEGs) and toll-like receptor signaling pathway (5 DEGs). The most abundant metabolic category in metabolism is inositol phosphate metabolism (12 DEGs) followed by glycolysis/gluconeogenesis (11 DEGs), glutathione metabolism (10 DEGs), and the citrate cycle (9 DEGs). In human disease, it is mainly related to insulin resistance (17 DEGs). In the category of genetic information processing, it is mainly enriched in ribosomes (33 DEGs) closely related to synthetic genetic material. In environmental information processing, the most abundant pathways are the Hippo signaling pathway (19 DEGs), MAPK signaling pathway (16 DEGs), and phosphatidylinositol signaling system (12 DEGs). In cellular processes, 36 DEGs were mainly enriched in endocytosis followed by focal adhesion (20 DEGs) and regulation of the actin cytoskeleton (15 DEGs).

A significant enrichment analysis was further performed on the metabolic pathways of DEGs, and a total of 26 significantly enriched metabolic pathways were obtained (Table 2). The terms were enriched to multiple molecular pathways such as the transport pathway and cellular community, including endocytosis, focal adhesion, regulation of the actin cytoskeleton, tight junction, and adherens junction; amino acid and carbohydrate metabolisms, such as glycolysis/gluconeogenesis, the citrate cycle (TCA cycle), phenylalanine metabolism, and histidine metabolism; immune-associated signal transduction such VEGF signaling pathway, the TNF signaling pathway, NOD-like receptor signaling, the MAPK signaling pathway, the Hippo signaling pathway-fly, the Toll-like receptor signaling pathway, and the NF-κB signaling pathway (Table S6). The top 10 metabolic pathways affected by H_2S were endocytosis, ribosome, focal adhesion, the Hippo signaling pathway, the MAPK signaling pathway, regulation of the actin cytoskeleton, tight junction, inositol phosphate metabolism, glycolysis/gluconeogenesis, and the TCA cycle (Table 2). These annotations indicated that H_2S could influence the multiple biological pathways in the fat body of silkworms and provide a new perspective to the effects of H_2S on silkworms.

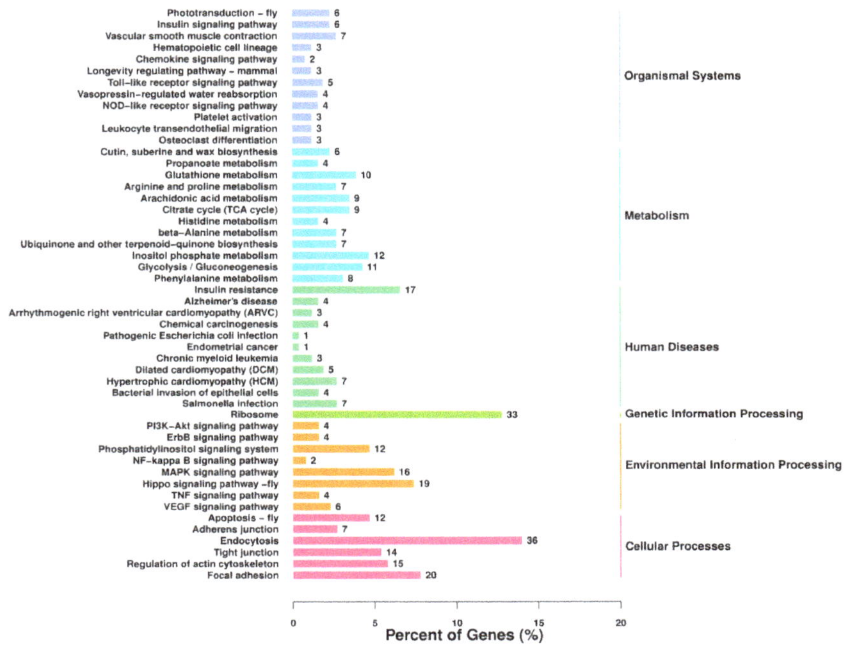

Figure 3. The KEGG classification of DEGs. The x-axis represents the number of genes annotated into the pathway and the proportion of the number of DEGs annotated to the total number of genes. The y-axis represents the name of the enriched KEGG pathways.

Table 2. The main enriched KEGG pathways ($p < 0.05$) in the H_2S-treated and the control groups.

Pathway ID	Pathway Name	Number of DEGs	*p*-Value
ko03010	Ribosome	33	<0.01
ko04510	Focal adhesion	20	<0.01
ko05132	Salmonella infection	7	<0.01
ko04370	VEGF signaling pathway	6	<0.01
ko04810	Regulation of actin cytoskeleton	15	<0.01
ko00360	Phenylalanine metabolism	7	<0.01
ko04530	Tight junction	13	<0.01
ko04668	TNF signaling pathway	4	0.01
ko00010	Glycolysis/Gluconeogenesis	11	0.01
ko04144	Endocytosis	35	0.01
ko00562	Inositol phosphate metabolism	12	0.01
ko00130	Ubiquinone and other terpenoid-quinone biosynthesis	6	0.01
ko04380	Osteoclast differentiation	3	0.01
ko04670	Leukocyte transendothelial migration	3	0.01
ko00410	beta-Alanine metabolism	7	0.01
ko05100	Bacterial invasion of epithelial cells	4	0.02
ko04611	Platelet activation	3	0.02
ko04520	Adherens junction	7	0.02
ko04621	NOD-like receptor signaling pathway	4	0.02
ko04391	Hippo signaling pathway-fly	19	0.02

Table 2. *Cont.*

Pathway ID	Pathway Name	Number of DEGs	*p*-Value
ko04010	MAPK signaling pathway	16	0.03
ko00340	Histidine metabolism	4	0.04
ko05410	Hypertrophic cardiomyopathy (HCM)	6	0.04
ko04064	NF-kappa B signaling pathway	2	0.04
ko00020	Citrate cycle (TCA cycle)	9	0.04
ko04962	Vasopressin-regulated water reabsorption	4	0.04

3.5. The Validation of DEGs by qRT-PCR

To validate the accuracy of DEGs obtained from the transcriptome between the H_2S-treated and control groups, 28 candidate genes were selected for qRT-PCR analysis. The selected DEGs were mainly annotated to glycolysis/gluconeogenesis, the TCA cycle, focal adhesion, endocytosis, adherens junction, the NF-κB signaling pathway, the MAPK signaling pathway, and the synthesis of fibroin. Good consistency between the results of qRT-PCR and the transcriptome confirmed the accuracy and reliability of the sequencing data and revealed the significant difference of these genes during H_2S exposure (Figure 4 and Table S3).

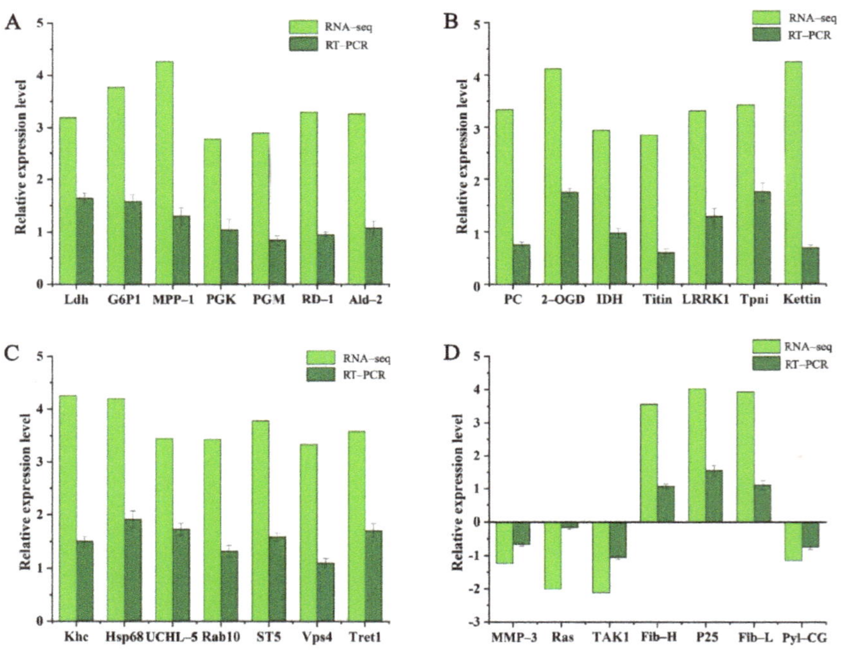

Figure 4. The qRT-PCR validation for DEGs expression after exposure to H_2S. DGEs related to the glycolysis pathway (**A**), the TCA cycle (**B**), endocytosis (**C**), and inflammation and silk protein synthesis (**D**) were analyzed. The full names of abbreviations of genes were listed in Table S1.

3.6. The Expression Pattern Analysis of DEGs

A total of 11 DEGs was representatively selected to analyze the expression level pattern in the 5th instar larvae using qRT-PCR. Genes (relating to the glycolytic pathway and TCA cycle) such as lactate phosphoglycerate kinase (PGK), phosphoglucomutase (PGM), NADP-dependent isocitrate dehydrogenase (IDH), and pyruvate carboxylase (PC) in the H_2S-treated group were up-regulated compared to control group (Figure 4A,B). Meanwhile, the levels of these genes were significantly increased at the 5L3D, 5L4D, and 5L5D comparing with 5L1D in the H_2S group (Figure 5A–D). In addition, the expression levels of the members of the Rab family of small GTPases (Rab10) and trehalose transporter 1 (Tret1) in endocytosis were up-regulated after H_2S treatment (Figure 4C). The expression levels of Rab10 and Tret1 were significantly increased in a time-dependent manner (Figure 5E,F). On the contrary, transforming growth factor-β activated kinase1 (Tak1) and matrix metalloproteinase-3 (MMP-3) were significantly decreased in the 5th instar larvae after H_2S exposure (Figure 5G,H). Moreover, the increased expression of fibroin synthesis genes, such as the fibroin heavy chain (Fib-H), fibroin light chain (Fib-L), and glycoprotein P25 (P25) was observed during the 5th larvae stage after H_2S exposure (Figure 4D). At the end of the 5th instar larvae, the expression levels were higher than the early 5th instar larvae (Figure 5I–K).

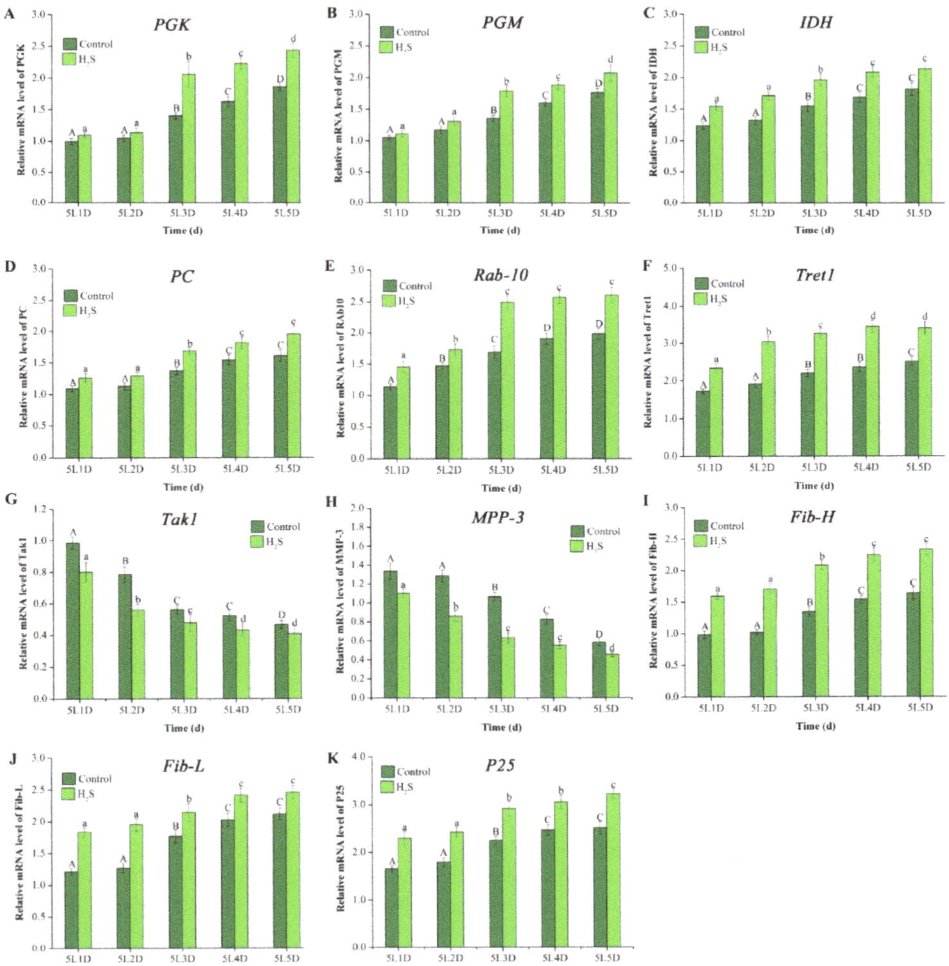

Figure 5. Expression changes of 10 DEGs after H_2S exposure at different time points in the 5th instar larvae (5L1D, 5L2D, 5L3D, 5L4D, and 5L5D). (**A**) PGK; (**B**) PGM; (**C**) IDH; (**D**) PC; (**E**) Rab-10; (**F**) Tret1; (**G**) Tak1; (**H**) MPP-3; (**I**) Fib-H; (**J**) Fib-L; and (**K**) P25. The lowercase letters (a, b, c, and d) represent the significant difference at $p \leq 0.05$ among the expression levels at different times of the H_2S treated group; the uppercase letters (A, B, C, and D) represent the significant difference at $p \leq 0.05$ among the expression levels at different times of the control group. The full names of abbreviations of genes are listed in Table S1.

4. Discussion

H_2S has been accepted as the third transmission gas followed by nitric oxide (NO) and carbon monoxide (CO) in recent years with applications to various insects, such as *C. elegans* [7–10], *D. melanogaster* [11,12], and *B. mori* [13]. RNA-seq technology is widely used in the field of silkworms to investigate the physiological and biochemical changes at the transcriptional level under diverse conditions [4,14,24]. In our previous research, it was found that the development (body weight) and economic characteristics (cocoon weight, cocoon shell ratio, and cocoon shell weight) were increased after continuous treatment with 7.5 μM of H_2S. However, there is no effective report for the effects of H_2S on the silkworm at the transcriptional level [13]. In the present study, the expression levels of genes in the fat body of silkworms after exposure to H_2S was investigated using transcriptomic and bioinformatic analysis. The results showed significantly different responses to H_2S and provided useful insights into the role of H_2S on silkworms.

Using transcriptome, a total of 1200 DEGs were identified in the fat body of silkworms at the 5th instar larvae. After treatment with H_2S, up-regulated DEGs (977) were higher than the down-regulated DEGs (223). These DEGs were mainly involved in glycolysis/gluconeogenesis, endocytosis, and the TCA cycle. The carbohydrate metabolism is one of the important energy metabolism pathways in organisms, incorporating glycolysis/gluconeogenesis, the TCA cycle, oxidative phosphorylation, and the pentose phosphate pathway [25]. Energy metabolism, especially the metabolic control of carbohydrates, is essential for insect growth and development [26]. Glucose is oxidized to pyruvate in glycolysis/gluconeogenesis [25]. Glycolytic activity is regulated by many enzymes in the glycolytic pathway such as PGK and PGM. PGM can convert glucose 1 phosphate into glucose 6 phosphate, providing the raw materials for the glycolytic pathway [27]. PGK is commonly known as a soluble and membrane-bound protein in different organisms [27]. In one of the glycolytic processes, phosphorylation of 1,3-diphosphoglycerate to 3-phosphoglycerate is catalyzed by PGK [28]. Like glycolysis/gluconeogenesis, the TCA cycle also provides the energy and intermediates for diverse biosynthetic pathways [25]. IDH, one of three rate-limiting enzymes in the TCA cycle, catalyzes the TCA cycle reaction wherein isocitrate is oxidized and decarboxylated to produce α-ketoglutarate, NADH, and CO_2 [29]. PC catalyzes the carboxylation of pyruvate into oxaloacetate for gluconeogenesis, the urea cycle, lipid synthesis, and other pathways which are the primary anaplerotic pathways for the TCA cycle where the intermediates are replenished by alanine and lactate [30]. In this study, the results showed that the levels of PGM, PGK, IDH, and PC were up-regulated after H_2S exposure. According to the expression pattern analysis of these genes, gene levels had gradually increased in a time-dependent manner. Especially for 5L4D and 5L5D, a significant difference was observed compared to 5L1D. These results indicated that the expression levels of the genes involved in carbohydrate metabolism were increased with H_2S exposure, suggesting H_2S can accelerate the carbohydrate metabolism of silkworms and provide much energy for silkworm metamorphosis.

The uptake and sorting of eukaryotes and the following recycling or degradation of fluid, membranes, membrane proteins, and macromolecules are regulated by endocytosis [31]. Endocytosis often depends on clathrin and occurs at clathrin-coated pits [32]. Utilizing endocytosis with clathrin-dependent or clathrin-independent uptake mechanisms, cargo proteins and adaptor molecules are delivered to early endosomes and sorted [33]. The Rab family of small GTPases can mediate membrane the trafficking processes of intracellular compartmentation, generation, and maintenance [34]. Rab10, which is closely related to clathrin-independent endocytosis is largely expressed in many model animals directing the trafficking of proteins. In *Caenorhabditis elegans*, Rab10 was required in interneuron postsynaptic glutamate receptor recycling and the transport between early endosomes and recycling endosomes [35]. Rab10 in the H_2S-treated group was up-regulated compared to the control group, which indicated that the H_2S-treated silkworms were more active in supporting endocytosis for the body. Tret1 participated in the transfer of trehalose, the major hemolymph sugar in most of the insect, which is synthesized in the fat body and released into the hemolymph. The transportation of trehalose with the up-regulated expression level of Tret1 after exposure to TiO_2 nanoparticles was accelerated in the silkworm, showing that TiO_2 nanoparticles could facilitate carbohydrate and the nutrition metabolism [36]. In this study, the level of Tret1 was up-regulated with the H_2S supplement to accelerate the transport rate of trehalose. Moreover, the expression levels of Rab10 and Tret1 were increased in a time-dependent manner during the 5th larvae. These results showed that H_2S could accelerate the transmission of mass in the silkworm fat body via up-regulating the expressions of Rab10 and Tret1 to provide material for larva–pupae metamorphosis.

Previous studies reported that H_2S exerted anti-inflammatory effects on T lymphocytes. Rats injected with NaHS inhibited inflammatory processes such as leukocyte infiltration and adherence to the vascular endothelium [37]. In this study, DEGs were annotated to various pathways related to inflammation by KEGG analysis, including the TNF signaling pathway, the MAPK signaling pathway, and the NF-κB signaling pathway.

Tak1 is a serine/threonine kinase and a member of the mitogen-activated protein kinase kinase (MAP3K) family which is activated by receptors such as TGF-β, TNF-α, and IL-1. MMP-3 is a matrix-soluble protein containing the binding site of NF-κB and can participate in the degradation of ECM and activate the precursors of MMPs. The content of MMP-3 was increased in the mouse arthritis model and the administration of baicalin can reduce the gene expression level of MMPs and alleviate inflammatory processes [38]. After H_2S exposure, levels of Tak1 and MMP-3 were down-regulated compared to the control group. Moreover, the expression levels of Tak1 and MMP-3 at the end of 5th instar larvae were significantly decreased compared to the early 5th instar larvae. These results indicated that H_2S might have an anti-inflammatory role in silkworm growth.

In addition, significant changes in the genes related to fibroin synthesis were observed in this study. The domestic silkworm has tremendous economic value and the most important manifestation of its value is silk production [39]. The function of the silk gland is to synthesize silk protein and during the 5th instar larvae it is the biggest organ of *B. mori* [39]. It was reported that Fib-H, Fib-L, and P25 encoding fibroin were expressed in all the larval instars. However, their highest expression was noticed before the last molting period [40]. The expression levels of Fib-L, P25, and Fib-H were up-regulated in response to H_2S compared to the control group. In addition, the expression of Fib-H at the end of the 5th instar larvae (5L4D, 5L5D) was rapidly increased and higher than on the other days. Similarly, the expression levels of P25 and Fib-L were also significantly increased. These results indicated that H_2S could promote the fibroin synthesis by up-regulating the expression levels of Fib-L, P25, and Fib-H, which was beneficial to silk production. These results were similar to the previous study which reported the up-regulated expression of Fib-L, P25, and Fib-H after treatment with titanium dioxide nanoparticles (TiO_2NPs) [41].

5. Conclusions

To summarize, this is the first-ever report to investigate the fat body of silkworms at the transcriptional level after H_2S exposure using the transcriptome analysis during the 5th larvae stage. Transcriptome analysis could detect numerous genes and signaling pathways and most of the DEGs were up-regulated compared to the control group. The expression of key genes related to energy metabolism, transport pathways, and fibroin synthesis in silkworms after H_2S exposure was up-regulated; conversely, the genes involved in inflammatory processes were down-regulated, which indicated that H_2S may lead to the promotion of biological processes such as energy and nutrition metabolism and inhibition of inflammatory processes. These results provide a novel approach to comprehending the molecular mechanisms underlying the H_2S effects in the silkworm fat body.

Supplementary Materials: The following are available online at https://www.mdpi.com/article/10.3390/insects12121110/s1, Tables S1–S6. Table S1. Primer's sequence of genes used in qRT-PCR analysis. Table S2. Statistical analysis of the RNAs Libraries. Table S3. Distribution of gene expression levels between H_2S-treated and the control group and the partial DEGs. Table S4. The list of DEGs comparison between H_2S treated group and control group. Table S5. Analysis of the enriched GO terms. Table S6. Analysis of the enriched KEGG pathway.

Author Contributions: Conceptualization, Z.-J.W. and F.H.; methodology, R.Z. and Y.-Y.C.; software, J.D.; validation, R.Z., J.D. and Y.-Y.C.; investigation, R.Z., J.D. and Y.-Y.C.; resources, S.-M.T.; data curation, K.T.; writing—original draft preparation, R.Z., K.T. and Y.-Y.C.; writing—review and editing, Z.-J.W., K.T. and F.H.; supervision, Z.-J.W. and F.H.; project administration, Z.-J.W. and F.H.; funding acquisition, Z.-J.W. All authors have read and agreed to the published version of the manuscript.

Funding: This research was funded by the National Natural Science Foundation of China (31772680, 32072795).

Institutional Review Board Statement: Not applicable.

Informed Consent Statement: Not applicable.

Data Availability Statement: The raw data in present paper was deposited in NCBI with the submission ID SUB10721255 (http://www.ncbi.nlm.nih.gov/bioproject/783962; 27 November 2021).

Conflicts of Interest: The authors declare no conflict of interest.

References

1. Goldsmith, M.R.; Shimada, T.; Abe, H. The genetics and genomics of the silkworm, *Bombyx mori*. *Annu. Rev. Entomol.* **2005**, *50*, 71–100. [CrossRef] [PubMed]
2. Liu, L.; Zhang, P.; Gao, Q.; Feng, X.; Han, L.; Zhang, F.; Bai, Y.; Han, M.; Hu, H.; Dai, F.; et al. Comparative transcriptome analysis reveals bmo-miR-6497-3p regulate circadian clock genes during the embryonic diapause induction process in bivoltine silkworm. *Insects* **2021**, *12*, 739. [CrossRef] [PubMed]
3. Zhu, Z.; Tan, Y.; Xiao, S.; Guan, Z.; Zhao, W.; Dai, Z.; Liu, G.; Zhang, Z. Solitary living brings a decreased weight and an increased agility to the domestic silkworm, Bombyx mori. *Insects* **2021**, *12*, 809. [CrossRef] [PubMed]
4. Lu, Z.Y.; Meng, Z.; Wen, M.Y.; Kang, X.; Zhang, Y.; Liu, Q.; Zhao, P.; Xia, Q. Overexpression of BmFoxO inhibited larval growth and promoted glucose synthesis and lipolysis in silkworm. *Mol. Genet. Genomics.* **2019**, *294*, 1375–1383. [CrossRef] [PubMed]
5. Zheng, S.F.; Jin, X.; Chen, M.H.; Shi, Q.; Zhang, H.; Xu, S. Hydrogen sulfide exposure induces jejunum injury via CYP450s/ROS pathway in broilers. *Chemosphere* **2019**, *214*, 25–34. [CrossRef]
6. Magierowski, M.; Magierowska, K.; Hubalewska-Mazgaj, M.; Surmiak, M.; Sliwowski, Z.; Wierdak, M.; Kwiecien, S.; Chmura, A.; Brzozowski, T. Cross-talk between hydrogen sulfide and carbon monoxide in the mechanism of experimental gastric ulcers healing, regulation of gastric blood flow and accompanying inflammation. *Biochem. Pharmacol.* **2018**, *149*, 131–142. [CrossRef] [PubMed]
7. Miller, D.L.; Roth, M.B. Hydrogen sulfide increases thermotolerance and lifespan in *Caenorhabditis elegans*. *Proc. Natl. Acad. Sci. USA* **2007**, *104*, 20618–20622. [CrossRef] [PubMed]
8. Budde, M.W.; Roth, M.B. Hydrogen sulfide increases hypoxia-inducible factor-1 activity independently of von Hippel-Lindau tumor suppressor-1 in *C. elegans*. *Mol. Biol. Cell* **2010**, *21*, 212–217. [CrossRef]
9. Qabazard, B.; Li, L.; Gruber, J.; Peh, M.T.; Ng, L.F.; Kumar, S.D.; Rose, P.; Tan, C.H.; Dymock, B.W.; Wei, F.; et al. Hydrogen sulfide is an endogenous regulator of aging in Caenorhabditis elegans. *Antioxid. Redox. Signal.* **2014**, *20*, 2621–2630. [CrossRef] [PubMed]
10. Wei, Y.; Kenyon, C. Roles for ROS and hydrogen sulfide in the longevity response to germline loss in *Caenorhabditis elegans*. *Proc. Natl. Acad. Sci. USA* **2016**, *113*, E2832–E2841. [CrossRef] [PubMed]
11. Kabil, H.; Kabil, O.; Banerjee, R.; Harshman, L.G.; Pletcher, S.D. Increased transsulfuration mediates longevity and dietary restriction in *Drosophila*. *Proc. Natl. Acad. Sci. USA* **2011**, *108*, 16831–16836. [CrossRef]
12. Zhong, J.F.; Wang, S.P.; Shi, X.X.; Mu, L.L.; Li, G.Q. Hydrogen sulfide exposure increases desiccation tolerance in *Drosophila melanogaster*. *J. Insect Physiol.* **2010**, *56*, 1777–1782. [CrossRef]
13. Cao, Y.Y.; Peng, L.L.; Jiang, L.; Thakur, K.; Hu, F.; Tang, S.M.; Wei, Z.J. Evaluation of the metabolic effects of hydrogen sulfide on the development of *Bombyx mori* (Lepidoptera: Bombycidae), using liquid chromatography-mass spectrometry-based metabolomics. *J. Insect Sci.* **2020**, *20*, 4. [CrossRef]
14. Jiang, L.; Peng, L.L.; Cao, Y.Y.; Thakur, K.; Tang, S.M.; Hu, F.; Wei, Z.J. Transcriptome analysis reveals gene expression changes of the fat body of silkworm (*Bombyx mori* L.) in response to selenium treatment. *Chemosphere* **2020**, *245*, 125660. [CrossRef]
15. Martin, M. Cutadapt removes adapter sequences from high-throughput sequencing reads. *EMBnet. J.* **2011**, *17*, 10–12. [CrossRef]
16. Pertea, M.; Kim, D.; Pertea, G.M.; Leek, J.T.; Salzberg, S.L. Transcript-level expression analysis of RNA-seq experiments with HISAT, StringTie and Ballgown. *Nat. Protoc.* **2016**, *11*, 1650–1667. [CrossRef]

17. Beekman, R.; Chapaprieta, V.; Russiñol, N.; Vilarrasa-Blasi, R.; Verdaguer-Dot, N.; Martens, J.H.A.; Duran-Ferrer, M.; Kulis, M.; Serra, F.; Javierre, B.M.; et al. The reference epigenome and regulatory chromatin landscape of chronic lymphocytic leukemia. *Nat. Med.* **2018**, *24*, 868–880. [CrossRef]
18. Xia, Q.; Zhou, Z.; Lu, C.; Cheng, D.; Dai, F.; Li, B.; Zhao, P.; Zha, X.; Cheng, T.; Chai, C.; et al. A draft sequence for the genome of the domesticated silkworm (*Bombyx mori*). *Science* **2004**, *306*, 1937–1940.
19. Kim, D.; Langmead, B.; Salzberg, S.L. HISAT: A fast spliced aligner with low memory requirements. *Nat. Methods* **2015**, *12*, 357–360. [CrossRef]
20. Pertea, M.; Pertea, G.M.; Antonescu, C.M.; Chang, T.C.; Mendell, J.T.; Salzberg, S.L. StringTie enables improved reconstruction of a transcriptome from RNA-seq reads. *Nat. Biotechnol.* **2015**, *33*, 290–295. [CrossRef]
21. Robinson, M.D.; McCarthy, D.J.; Smyth, G.K. edgeR: A Bioconductor package for differential expression analysis of digital gene expression data. *Bioinformatics.* **2010**, *26*, 139–140. [CrossRef]
22. Kanehisa, M.; Araki, M.; Goto, S.; Hattori, M.; Hirakawa, M.; Itoh, M.; Katayama, T.; Kawashima, S.; Okuda, S.; Tokimatsu, T.; et al. KEGG for linking genomes to life and the environment. *Nucleic Acids Res.* **2008**, *36*, D480–D484. [CrossRef]
23. Zhang, F.; Zhang, Y.Y.; Ma, R.H.; Thakur, K.; Han, J.; Hu, F.; Zhang, J.G.; Wei, Z.J. Multi-omics reveals the anticancer mechanism of asparagus saponin-asparanin A on endometrial cancer Ishikawa cells. *Food Funct.* **2021**, *12*, 614–632. [CrossRef]
24. Wang, J.J.; Bai, W.W.; Zhou, W.; Liu, J.; Chen, J.; Liu, X.Y.; Xiang, T.T.; Liu, R.H.; Wang, W.H.; Zhang, B.L.; et al. Transcriptomic analysis of two Beauveria bassiana strains grown on cuticle extracts of the silkworm uncovers their different metabolic response at early infection stage. *J. Invertebr. Pathol.* **2017**, *145*, 45–54. [CrossRef]
25. Adam, G.M.; Naama, K. NADH ties One-Carbon metabolism to cellular respiration. *Cell Press.* **2020**, *31*, 660–662.
26. Chen, Q.M.; Ma, Z.G.; Wang, X.; Li, Z.; Zhang, Y.; Ma, S.; Zhao, P.; Xia, Q. Comparative proteomic analysis of silkworm fat body after knocking out fibroin heavy chain gene: A novel insight into cross-talk between tissues. *Funct. Integr. Genomics.* **2015**, *15*, 611–637. [CrossRef]
27. Long, W.; Wu, J.; Shen, G.; Zhang, H.; Liu, H.; Xu, Y.; Gu, J.; Jia, L.; Lin, Y.; Xia, Q. Estrogen-related receptor participates in regulating glycolysis and influences embryonic development in silkworm Bombyx mori. *Inesct Mol. Biol.* **2020**, *29*, 160–169. [CrossRef]
28. Efe, S.; Duvernell, D.D.; Matzkin, M.L.; Duan, Y.H.; Zhu, C.T.; Verrelli, B.C.; Eanes, W.F. Single-Locus latitudinal clines and their relationship to temperate adaptation in metabolic genes and derived alleles in *Drosophila melanogaster*. *Genet. Soc. Am.* **2004**, *168*, 923–931.
29. Long, L.; Wang, C.C.; Zhao, X.Y.; Guan, J.X.; Lei, C.L.; Huang, Q.Y. Isocitrate dehydrogenase-mediated metabolic disorders disrupt active immunization against fungal pathogens in eusocial termites. *J. Pest. Sci.* **2020**, *93*, 291–301.
30. Meyer, S.; Gessner, D.K.; Wen, C.P.; Most, E.; Liebisch, G.; Zorn, H.; Ringseis, R.; Eder, K. The Antisteatotic and Hypolipidemic Effect of Insect Meal in Obese Zucker Rats is Accompanied by Profound Changes in Hepatic Phospholipid and 1-Carbon Metabolism. *Mol. Nutr. Food Res.* **2019**, *63*, e1801305. [CrossRef]
31. Shi, A.B.; Chen, C.C.; Banerjee, R.; Glodowski, D.; Audhya, A.; Rongo, C.; Grant, B.D. EHBP-1 functions with RAB-10 during endocytic recycling in Caenorhabditis elegans. *Mol. Biol. Cell.* **2010**, *21*, 2930–2943. [CrossRef]
32. Le Roy, C.; Wrana, J.L. Clathrin- and non-clathrin-mediated endocytic regulation of cell signaling. *Nat. Rev. Mol. Cell Biol.* **2005**, *6*, 112–126. [CrossRef]
33. Nichols, B. Caveosomes and endocytosis of lipid rafts. *J. Cell Sci.* **2003**, *116*, 4707–4714. [CrossRef]
34. Stenmark, H. Rab GTPases as coordinators of vesicle traffic. *Nat. Rev. Mol. Cell Biol.* **2009**, *10*, 513–525. [CrossRef]
35. Glodowski, D.R.; Chen, C.C.; Schaefer, H.; Grant, B.D.; Rongo, C. RAB-10 regulates glutamate receptor recycling in a cholesterol-dependent endocytosis pathway. *Mol. Biol. Cell.* **2007**, *18*, 4387–4396. [CrossRef]
36. Tian, J.H.; Xue, B.; Hu, J.H.; Li, J.X.; Cheng, X.Y.; Hu, J.S.; Li, F.C.; Chen, Y.H.; Li, B. Exogenous substances regulate silkworm fat body protein synthesis through MAPK and PI3K/Akt signaling pathways. *Chemosphere* **2017**, *171*, 202–207. [CrossRef]
37. Zanardo, R.C.; Brancaleone, V.; Distrutti, E.; Fiorucci, S.; Cirino, G.; Wallace, J.L. Hydrogen sulfide is an endogenous modulator of leukocyte-mediated inflammation. *FASEB J.* **2006**, *20*, 2118–2120. [CrossRef]
38. Jia, Q.; Wang, T.T.; Wang, X.Y.; Xu, H.; Liu, Y.; Wang, Y.; Shi, Q.; Liang, Q. Astragalin suppresses inflammatory responses and bone destruction in mice with collagen-induced arthritis and in human fibroblast-likesynoviocytes. *Front. Pharmacol.* **2019**, *10*, 94. [CrossRef]
39. Hu, W.; Wang, X.; Ma, S.; Peng, Z.; Cao, Y.; Xia, Q. CRISPR-Mediated endogenous activation of fibroin heavy chain gene triggers cellular stress responses in *Bombyx mori* Embryonic Cells. *Insects* **2021**, *12*, 552. [CrossRef]
40. Kunz, R.I.; Brancalhao, R.M.; Ribeiro, L.F.; Natali, M.R. Silkworm sericin: Properties and biomedical applications. *Biomed. Res. Int.* **2016**, *2016*, 8175701. [CrossRef]
41. Ni, M.; Zhang, H.; Li, F.C.; Wang, B.B.; Xu, K.Z.; Shen, W.D.; Li, B. Nanoparticulate anatase TiO_2 (TiO_2NPs) upregulates the expression of silkworm (*Bombyx mori*) neuropeptide receptor and promotes silkworm feeding, growth, and silking. *Peptides* **2015**, *68*, 64–71. [CrossRef]

insects MDPI

Article

Pathogenicity Detection and Genome Analysis of Two Different Geographic Strains of BmNPV

Huimin Guo [1,2], Benzheng Zhang [1], Xin Zheng [1], Juan Sun [1], Huiduo Guo [1], Gang Li [1], Guodong Zhao [1], Anying Xu [1] and Heying Qian [1,2,*]

[1] College of Biotechnology, Jiangsu University of Science and Technology, Zhenjiang 212018, China; 15735531904@163.com (H.G.); zbzheng3399@163.com (B.Z.); a18796017350@163.com (X.Z.); sj15062884263@163.com (J.S.); guohuiduo1991@126.com (H.G.); gangsri@just.edu.cn (G.L.); sdgdzhao@126.com (G.Z.); xaysri@126.com (A.X.)

[2] The Sericultural Research Institute, Chinese Academy of Agricultural Sciences, Zhenjiang 212018, China

* Correspondence: qhysri@just.edu.cn

Simple Summary: The hemolynamic septic disease in silkworms is caused by *Bombyx mori* nuclear polyhedrosis virus (BmNPV). It is the most severe viral disease that adversely affects the sericulture industry. Breeding BmNPV-resistant silkworm varieties is the most economic and effective solution. However, BmNPVs from different geographical strains have different pathogenicities. This brings the challenges of cultivating BmNPV-resistant silkworm varieties with wider adaptabilities. In this study, the genomes of two BmNPV strains (BmNPV ZJ and BmNPV YN) were sequenced and characterized to compare the difference in pathogenicity between the two strains. A total of 76 different genes in these two viruses were found with amino acid mutations. These included genes were associated with BmNPV replication and infection. In addition, the relative gene expression of the BmNPV YN strain was lower than that BmNPV ZJ. Thus, we speculate that the mutations in some genes may affect viral functions and may be the cause of the higher pathogenicity of BmNPV YN despite its lower proliferation rate. The present research provides new clues for further exploring the mechanism determining the difference in pathogenicity of different BmNPV strains.

Abstract: The pathogenicity of different concentrations of *Bombyx mori* nuclear polyhedrosis virus-Zhenjiang strain (BmNPV ZJ) and Yunnan strain (BmNPV YN) was assessed in Baiyu larvae. The structures of the two viral strains were observed by negative-staining electron microscopy, and their proliferation was examined by quantitative polymerase chain reaction (qPCR). The genomic sequences of these two viruses were obtained to investigate the differences in their pathogenicity. The lethal concentration 50 (LC_{50}) of BmNPV ZJ against Baiyu larvae was higher than that of BmNPV YN, indicating a relatively more robust pathogenicity in BmNPV YN. Electron microscopic images showed that the edges of BmNPV YN were clearer than those of BmNPV ZJ. The qPCR analysis demonstrated significantly higher relative expressions of immediately early 1 gene (*ie-1*), *p143*, *vp39*, and polyhedrin genes (*polh*) in BmNPV ZJ than in BmNPV YN at 12–96 h. The complete genomes of BmNPV ZJ and BmNPV YN were, respectively, 135,895 bp and 143,180 bp long, with 141 and 145 coding sequences and 40.93% and 39.71% GC content. Considering the BmNPV ZJ genome as a reference, 893 SNP loci and 132 InDel mutations were observed in the BmNPV YN genome, resulting in 106 differential gene sequences. Among these differential genes, 76 (including 22 hub genes and 35 non-hub genes) possessed amino acid mutations. Thirty genes may have been related to viral genome replication and transcription and five genes may have been associated with the viral oral infection. These results can help in understanding the mechanisms of pathogenicity of different strains of BmNPV in silkworms.

Keywords: *Bombyx mori* nuclear polyhedrosis virus; sequencing analysis; qPCR; pathogenicity difference

Citation: Guo, H.; Zhang, B.; Zheng, X.; Sun, J.; Guo, H.; Li, G.; Zhao, G.; Xu, A.; Qian, H. Pathogenicity Detection and Genome Analysis of Two Different Geographic Strains of BmNPV. *Insects* **2021**, *12*, 890. https://doi.org/10.3390/insects12100890

Academic Editors: Silvia Cappellozza, Morena Casartelli, Federica Sandrelli, Alessio Saviane and Gianluca Tettamanti

Received: 28 July 2021
Accepted: 27 September 2021
Published: 30 September 2021

1. Introduction

Bombyx mori nuclear polyhedrosis virus (BmNPV), a circular double-stranded DNA virus, is the first insect baculovirus to be discovered [1,2]. The *B. mori* hemolynamic septic disease caused by BmNPV is the most severe viral disease in silkworms that harms the sericulture industry [3]. It is a subacute disease that occurs over 3–5 days, and the onset is faster in summer and autumn. The typical characteristics of infected silkworms include manic creeping, the body becoming whitish and shiny with intersegmental swelling, the skin easily breaking and oozing a milky white body fluid, and, finally, death. The milky white body fluid contains the BmNPV virus particles, which contaminate the mulberry leaves on which the silkworms reside. The contaminated leaf is eaten by other silkworms, which contract oral infections or infections through the wound after coming into contact with other wounded silkworms. The outbreak of the disease in rural production areas usually occurs in the middle and late stages of the fifth instar of *B. mori*. It is difficult to prevent and leads to a significant reduction in the quantity of production and, sometimes, no harvest occurs. Cultivating silkworm varieties with high resistance to BmNPV is the most economical and effective means to reduce the loss in sericulture production. Over the last 10 years, various institutions in China have cultivated several silkworm varieties resistant to BmNPV [4–7]; for example, "Huakang No.2", which exhibits more than 100 times improved resistance to *B. mori* hemolynamic septic disease of the typical summer and autumn species "Qiufeng × Baiyu". The silk productivity of these varieties also indicates their remarkable disease resistance properties [8–11].

Like in other viruses, genetic variation has also been found in BmNPV [12–14] and characterizes the different geographical strains [15,16]. In addition, different geographic strains of BmNPV have differences in their morphology and pathogenicity [17,18]. Even the virulence of local strains of BmNPV from the same province also differs against *B. mori*. Bai et al. [19] and Tang et al. [12] found that the pathogenicity of eight strains of BmNPV in samples from various regions in Yunnan differed and exhibited varying rates of infectivity in *B. mori*. Wang et al. [20] reported that the pathogenicity of BmNPV from different origins in Guangxi Province was also distinct in different silkworm varieties, which brought a new challenge to the breeding of silkworms against BmNPV. It is therefore necessary to explore the pathogenic mechanisms of different mutant strains of BmNPV in relation to silkworms.

We isolated two BmNPVs from Zhenjiang, Jiangsu Province, and Luliang, Yunnan Province, with varying pathogenicities to the same silkworm varieties. In the present study, we applied Illumina second-generation sequencing technology and Pacbio third-generation sequencing technology to characterize their genomes and explore the underlying mechanism of the variation in pathogenicity of different BmNPV strains.

2. Materials and Methods

2.1. Materials

BmNPV strains were obtained from different geographic regions: the BmNPV Zhenjiang strain (BmNPV ZJ) was obtained from Zhenjiang, Jiangsu province, and the BmNPV Yunnan strain (BmNPV YN) was obtained from Luliang, Yunnan province. These two strains were isolated, purified, and preserved by our teams.

The silkworm variety "Baiyu", bred by the Institute of Sericulture of the Chinese Academy of Agricultural Sciences, was the parent of the control variety of silkworm varieties approved by the state for use in summer and autumn and was maintained in our laboratory.

2.2. Methods

2.2.1. Virus Collection and Purification

The budded virus (BV) of BmNPV ZJ was preserved and provided by the Pathology Laboratory of the Institute of Sericulture of the Chinese Academy of Agricultural Sciences. The BV of BmNPV YN was preserved and provided by the Institute of Sericulture of the Yunnan Academy of Agricultural Sciences. In the autumn of 2019, our team used the BV to puncture the fifth instar of silkworms. After the onset of the disease, the blood of the infected silkworm was collected, filtered, and centrifuged three to four times to obtain the purified virus solution. Then, it was diluted with an appropriate amount of double-distilled water and counted using a hemocytometer. The concentrations of BmNPV ZJ and BmNPV YN were 2.45×10^9 and 3.66×10^9, respectively; they were stored at 4 °C for further use.

2.2.2. Determination of Pathogenicity of Different BmNPV to *B. mori*

The original viral solutions of the two BmNPV strains were diluted with sterile water to a total of five concentrations, from 1×10^4–1×10^8, and smeared on mulberry leaves to feed the Baiyu larvae. The mulberry leaves with smooth surfaces were cut into small pieces (2.7 cm × 2.7 cm) using a punch, and 50 μL of different concentrations of polyhedral suspension was added dropwise on the upper surface of each leaf piece. After smearing evenly, these were fed to the second instar larvae of the silkworm, and leaves with the same volume of sterile water were used as blank controls. There were a total of 90 silkworm larvae per treatment, with three replicates (30 silkworms each), for each BmNPV concentration [20]. Four mulberry leaves carrying the virus were fed to each silkworm in each area at one time. After 24 h, the mulberry leaves carrying the virus were replaced with plain mulberry leaves with no virus. The growth and development of silkworms were observed every day, and the diseased silkworms were removed in time to avoid cross-infection. The number of dead silkworms that were infected with BmNPV was recorded after their third instar stage. The lethal concentration 50 (LC_{50}) of Baiyu was calculated using the Statistical Product and Service Solutions (SPSS, https://www.ibm.com/cn-zh/analytics/spss-statistics-software accessed on 26 September 2021), and the pathogenicities of BmNPV YN and BmNPV ZJ were compared.

2.2.3. Negative-Staining Electron Microscopy to Observe BmNPV Particles

The purified BmNPV ZJ and BmNPV YN were suspended on a copper mesh supported by a polyvinyl alcohol formaldehyde membrane (formvar membrane). The samples were stained with 2% phosphotungstic acid (pH 7.2) for 20 s. After washing and drying, the morphologies of the virus particles were observed with negative-staining electron microscopy.

2.2.4. Sequencing of BmNPV ZJ and BmNPV YN Genomes

DNA was extracted from BmNPV YN and BmNPV ZJ according to the alkaline lysis method [12]. The concentrations of the extracted viral DNA were measured using the Qubit quantitative detector. The DNA quality was assessed using 1% agarose gel electrophoresis. The high-quality viral DNA was used for genome sequencing using Illumina second-generation and PacBio third-generation sequencing technologies (Sangon Biotech Co., Ltd., Shanghai, China).

2.2.5. Analysis of BmNPV ZJ and BmNPV YN Genome Sequences

Using the genome of BmNPV T3 (L33180.1, 128,413 bp) as the reference, the sequencing data were assembled and corrected by SPAdes (https://cab.spbu.ru/software/spades/ accessed on 26 September 2021) and Primer-initiated Sequence Synthesis for Genomes (PrlnSeS-G, https://updeplasrv1.epfl.ch/prinses/ accessed on 26 September 2021) [21,22]. The Gene Ontology, Kyoto Encyclopedia of Genes and Genomes, Clusters of Orthologous Groups (COGs) of proteins, non-redundant proteins (NR), curated protein families (PFAM), Swiss-Prot, and TrEMBL databases were used for functional annotation of the virus genes. The genome sequences of these two BmNPVs were compared with other published BmNPV genome sequences. For multiple sequence alignment, the ClustalW parameters of Molecular Evolutionary Genetics Analysis (MEGA 7.1, https://megasoftware.net/ accessed on 26 September 2021) were used. The baculovirus repeated orfs gene (*bro*) of the BmNPV exists in diverse copies and nucleotide sequences in different ecological environments; therefore, this gene was also used for molecular identification of different strains of BmNPV [15,16,23]. The bootstrap statistical method was used for the calculation of

1000 replicates. The phylogenetic tree of BmNPV was constructed using the *bro-d* gene sequences of different BmNPV strains.

2.2.6. Detection of Gene-Relative Expression of BmNPV by qPCR

The two strains of BmNPV were diluted to 1×10^8 and fed to the fifth instar of the Baiyu silkworms, feeding 7 µL to each silkworm; there were five silkworms per treatment. The disease incidence rate among the silkworms was 100% at this concentration. At 12, 24, 48, 72, and 96 h, the silkworms were dissected and RNA was extracted from their midgut tissues. The very early gene *ie-1*, early gene *p143*, late gene *vp39*, and very late gene *polh* of BmNPV were chosen for quantitative polymerase chain reaction (qPCR) analysis. The primers for viral genes were designed in Primer 6.0 and their details are listed in Table 1. Reaction conditions for qPCR were: 95 °C, 30 s; 95 °C, 10 s; and 55 °C, 30 s, for a total of 40 cycles. Using *actin-3* (U49854) as the reference gene, the relative expression levels of the genes in the two BmNPV strains were calculated with the $2^{-\Delta\Delta CT}$ method.

Table 1. Primers used in qPCR.

Gene	Primer Sequences	Length/(bp)
actin-3	F: 5′-CGGCTACTCGTTCACTACC-3′ R: 5′-CCGTCGGGAAGTTCGTAAG-3′	147
ie-1	F: 5′-GGACGAATACTTGGACGAT-3′ R: 5′-GAGAACCTGTTGGAATTGTAG-3′	237
p143	F: 5′-GCACGGCAATACTTATCATC-3′ R: 5′-TGAGCACCAACAATAGTCC-3′	120
vp39	F: 5′-ACACGGAGGAATTGAGATT-3′ R: 5′-GATGTCACTGCTTCTTATCG-3′	116
polh	F: 5′-CTACAAGTTCCTCGCTCAA-3′ R: 5′-CTCGCTGTGGATGTTCAT-3′	163

2.2.7. BmNPV YN and BmNPV ZJ Differential Genes Analysis

Referring to the design process proposed by the Genome Analysis Toolkit (GATK) [24], the effective data for the BmNPV YN and BmNPV ZJ genomes were compared using the Burrows–Wheeler Aligner (BWA, http://bio-bwa.sourceforge.net/ accessed on 26 September 2021). To convert and sort the results, sequence alignment/map Tools (SAMtools, http://www.htslib.org/doc/samtools.html accessed on 26 September 2021) was used and the comparison results were statistically analyzed. The genotype differences between BmNPV YN and BmNPV ZJ were assessed using the HaplotypeCaller software, and the single nucleotide polymorphism (SNP) and insertion and deletion (InDel) information of the two strains were obtained. After quality control (Table 2), the SNP and InDel information were annotated using SNP effect software (SnpEff, https://pcingola.github.io/SnpEff/ accessed on 26 September 2021) [25]. According to the annotation results and BLAST analysis of the differential genes between BmNPV YN and BmNPV ZJ, candidate genes associated with the difference in pathogenicity of the BmNPV YN and BmNPV ZJ were screened.

Table 2. SNP and InDel quality control standards.

Quality Control Items	SNP	InDel
QualByDepth	≥2.0	≥2.0
RMSMappingQuality	≥40.0	-
FisherStrand	≤60.0	≤200.0
StrandOddsRatio	≤3.0	≤10.0
MappingQualityRankSumTest	≥−12.5	≥−12.5
ReadPosRankSumTest	≥−8.0	≥−8.0

3. Results

3.1. The Pathogenicity of BmNPV ZJ and BmNPV YN to B. mori

The mortality rate of Baiyu larvae increased with the increase in the concentration of BmNPV ZJ and BmNPV YN (Figure 1). At the same concentration, BmNPV YN exhibited stronger pathogenicity than BmNPV ZJ. In addition, the mortality rates at 1×10^7 and 1×10^8 of both BmNPVs differed significantly.

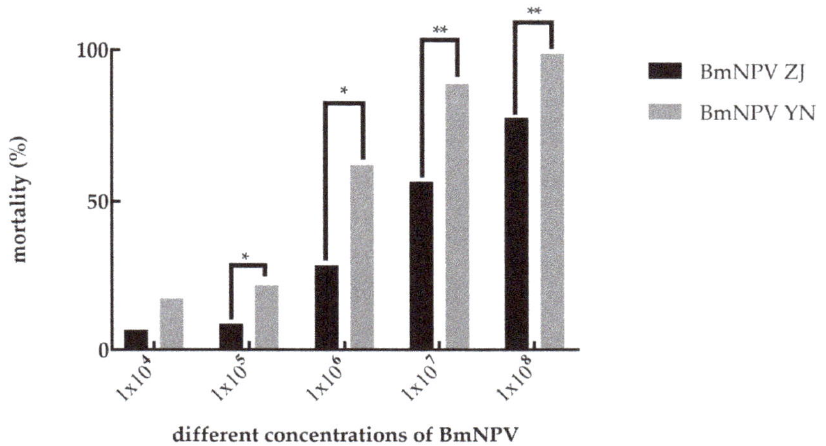

Figure 1. A comparison of the pathogenicities of different concentrations of the two BmNPV strains against Baiyu larvae. Note: "*" represents $p < 0.05$; "**" represents $p < 0.01$.

The median lethal dose of the two viruses in the Baiyu larvae was calculated by SPSS v21 (https://www.ibm.com/cn-zh/analytics/spss-statistics-software accessed on 26 September 2021) software. The LC_{50} concentration of BmNPV YN in Baiyu larvae was about 10 times less than that of BmNPV ZJ (Table 3), indicating that BmNPV YN could even cause death in half of these larvae at a low concentration.

Table 3. Comparison of virulence of BmNPV ZJ and BmNPV YN in Baiyu larvae.

Variety	Strain	Regression Equation	LC_{50}	95% Confidence Value	Slope of Regression Line/SE
Baiyu	BmNPV ZJ	y = −4.267 + 0.627x	6.45×10^6	3.88×10^6–1.14×10^7	0.056
Baiyu	BmNPV YN	y = −4.426 + 0.796x	3.62×10^5	1.98×10^5–6.45×10^5	0.063

3.2. The Morphology and Size of BmNPV ZJ and BmNPV YN

The purified BmNPV ZJ and BmNPV YN virus particle suspensions were observed by electron microscopy (Figure 2). The sizes of BmNPV ZJ and BmNPV YN appeared similar, with diameters of about 2.2–4.0 μm. The BmNPV polyhedra appeared mostly hexagonal, while a small number of particles appeared quadrilateral and irregular. The edges of BmNPV YN particles appeared sharper compared with those of BmNPV ZJ.

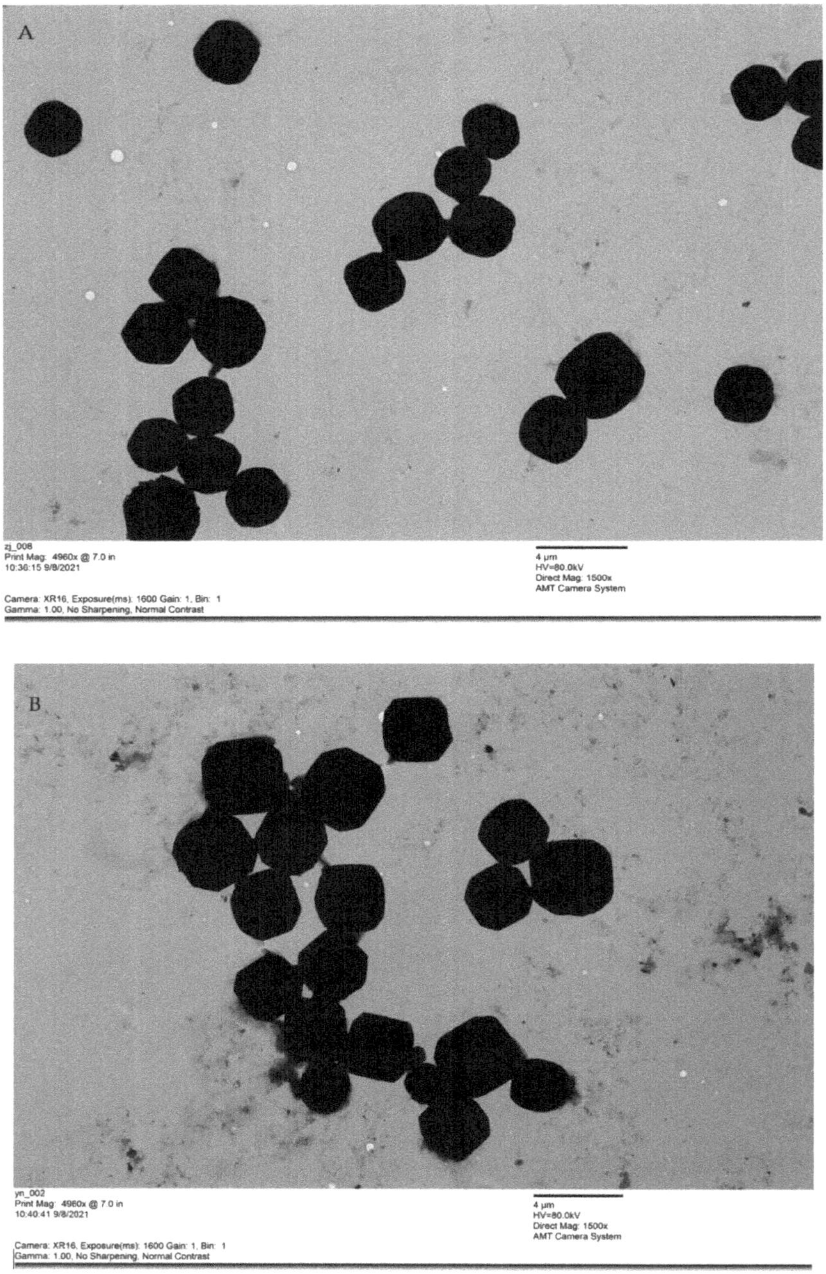

Figure 2. Morphological observation of BmNPV polyhedra by electron microscopy (×1500): (**A**) BmNPV ZJ and (**B**) Bm-NPV YN.

3.3. The Relative Expression of Genes of BmNPV ZJ and BmNPV YN in the Midgut of Baiyu

The qPCR analysis of the genes of BmNPV ZJ and BmNPV YN in the midgut of Baiyu larvae (Figure 3A) revealed that the relative expression of *ie-1* in BmNPV ZJ was higher than that in BmNPV YN at 12–96 h, with the highest level at 96 h. The expressions of *p143*, *vp39*, and *polh* in BmNPV ZJ were also higher than BmNPV YN before 96 h (Figure 3B–D).

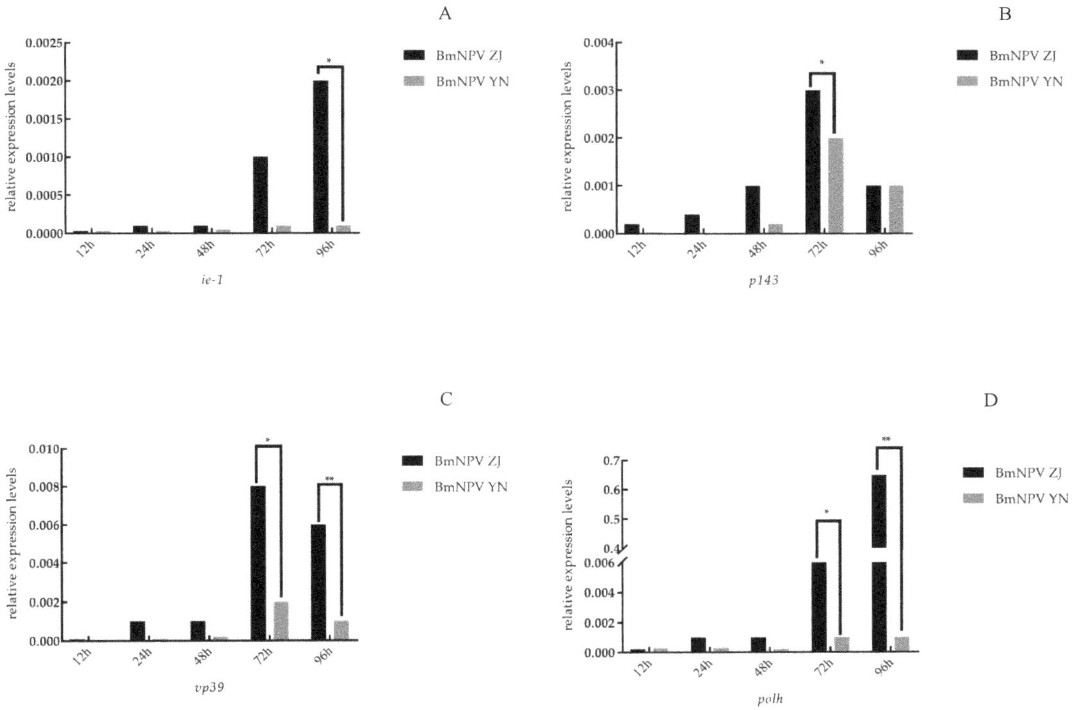

Figure 3. The relative expression of genes of the two BmNPV strains in the midgut of Baiyu larvae: (**A**) *ie-1*; (**B**) *p143*; (**C**) *vp39*; (**D**) *polh*. Note: "*" represents $p < 0.05$; "**" represents $p < 0.01$.

3.4. Structural Characteristics of Virus Genomes

3.4.1. Genome Characteristics of BmNPV ZJ

The genome of BmNPV ZJ was estimated to be 135,895 bp long, with 40.39% GC content. It comprised 141 predicted protein-coding sequences (CDSs), accounting for 83.86% of the entire genome sequence, of which 70 CDSs were on the positive strand and 71 on the negative strand. The average length of this genome was estimated to be 808 bp. Further analysis indicated 91 putative genes of 500 to 1000 bp lengths and 40 genes with ≥1000 bp lengths. Only 10 genes were less than 500 bp in length. In total, 139 genes were annotated. The annotated genes fibroblast growth factor (*fgf*), chi-tinase gene (*chi-a*), and alkaline exonuclease gene (*alk-exo*) were found in the BmNPV ZJ genome, but their homologous reading frames were absent in BmNPV YN. Further, late expression factor 12 (*lef-12*) and *AcMNPV 94K* were not annotated in the BmNPV T3 reference strain (Figure S1). Compared with the genome of BmNPV T3, some genes in these two viral genomes had SNP mutations, resulting in premature stop codons, and truncated or abnormal proteins. One putative gene with an unknown function was also found.

3.4.2. Genome Characteristics of BmNPV YN

The genome of BmNPV YN was estimated to be 143,180 bp long, with 39.71% GC content. It comprised 146 CDSs, accounting for 89.37% of the whole genome sequence, of which 76 CDSs were on the positive strand and 69 on the negative strand. The average length of this genome was estimated to be 882.52 bp, with 96 genes of 500 to 1000 bp lengths and 43 genes more than 1000 bp in length. Six genes were found to be of less than 500 bp in length and 137 genes were annotated. Further analysis revealed *Orf20*, late expression factor 7 (*lef-7*), and *p26* genes in the BmNPV YN genome, but the homologous genes were absent in the BmNPV ZJ genome. *Lef-12* was not annotated in the BmNPV T3 reference strain (Figure S2). In addition, seven putative genes with unknown functions were found.

3.4.3. Genome Comparison of BmNPV

The genome sequences of the two BmNPVs were compared with that of the BmNPV T3 strain as a reference, as shown in Table 4.

Table 4. General features of BmNPV genomes.

Items	BmNPV ZJ	BmNPV YN	BmNPV T3
Size (base)	135,895	143,180	128,413
G + C content (%)	40.39	39.71	40
Protein coding genes	141	145	136
Min length (base)	111	111	61
Max length (base)	2964	4050	1222
Average length (base)	808.23	882.52	852.154
Total coding gene (base)	113,961	127,965	115,893
Coding ratio (%)	83.86	89.37	90.25

The nucleotide sequences of the BmNPV ZJ and BmNPV YN genomes were analyzed by DNAMAN software (https://www.lynnon.com/dnaman.html accessed on 26 September 2021). The genes of BmNPV ZJ and BmNPV YN were 92.6–100.0% homologous. A comparison with the genome sequences of the other six BmNPV reference strains showed homologies of BmNPV ZJ and BmNPV YN in the range of 92.0–100.0% and 93.3–100.0%, respectively (Table S1). The *bro-d* gene sequence of BmNPV ZJ and BmNPV YN showed 95.9% homology, and the homologies with the other BmNPV reference strains were 92.1–99.6% and 93.3–97.3%, respectively (Table 5). The amino acid sequence homology of the BRO-D (encoded by the *bro-d*) of BmNPV ZJ and BmNPV YN was 96.6%, and the homologies with other BmNPVs were 92.0–99.7% and 92.2–97.1%, respectively (Table 5). The amino acid of the BRO-D protein of the two strains was mainly mutated at the N-terminal (Figure 4), which was consistent with the observation of Zhou et al. [26].

Table 5. Nucleotide and amino acid homology of *bro-d* gene sequences.

Reference Strain	Nucleotide Homology/(%)		Reference Strain	Amino Acid Sequence/(%)	
	BmNPV ZJ (N)	BmNPV YN (N)		BmNPV ZJ (AA)	BmNPV YN (AA)
Brazilian	94.3	94.2	Brazilian	93.7	92.5
Cubic	99.6	95.7	Cubic	99.7	96.3
Guangxi	92.1	93.3	Guangxi	91.7	92.2
India	96.8	97.3	India	96.6	97.1
T3	95.9	94.1	T3	92.0	92.5
Zhejiang	95.5	94.3	Zhejiang	94.0	92.8
Zhenjiang	-	95.9	Zhenjiang	-	95.9
Yunnan	96.9	-	Yunnan	96.6	-

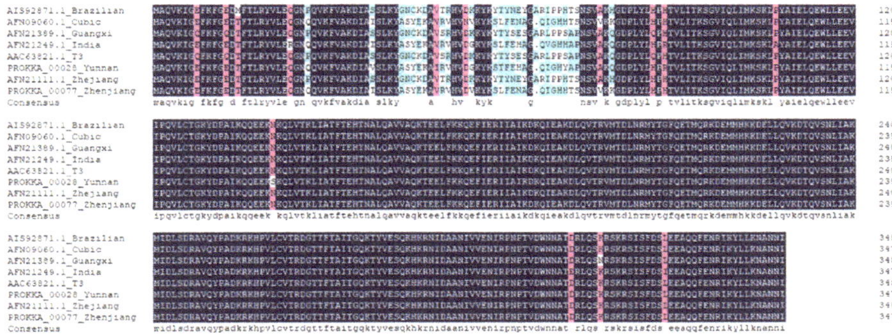

Figure 4. Alignment of amino acid sequences of BRO-D protein from different BmNPV strains.

The *bro-d* gene was chosen for the evolutionary analysis of the two BmNPVs. Using *Plutella xylostella* multiple nucleopolyhedrovirus (PxMNPV) as the exogenous reference, the phylogenetic tree was plotted using the MEGA 7.1 software (https://megasoftware.net/ accessed on 26 September 2021). The results are shown in Figure 5. The eight BmNPV strains were divided into two clades. Both BmNPV ZJ (Zhenjiang) and BmNPV YN (Yunnan) were grouped in clade I, and BmNPV ZJ was closer to the Cubic strain (IQ991009) while BmNPV YN was closer to the India strain (JQ991010). These finding are similar to those reported by Tang et al. [12].

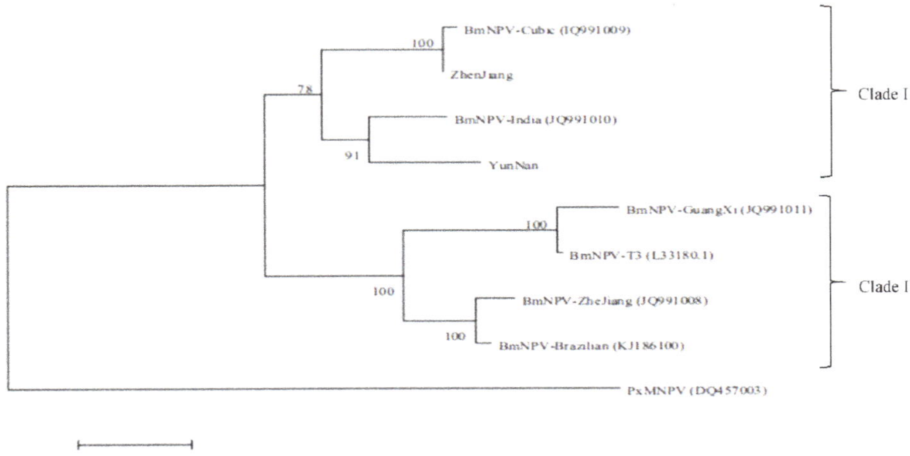

Figure 5. Phylogenetic tree constructed based on bro-d gene sequences of BmNPV isolates.

3.5. Detection and Analysis of Single Nucleotide Polymorphism and InDels in the Two Viral Genomes

Using BmNPV ZJ as the reference genome, a total of 893 SNP loci and 132 InDel variants were identified in BmNPV YN. The number of SNPs of transitions and transversions were 693 and 195, respectively. The SNP loci of the BmNPV YN strain were mainly enriched in the coding region, accounting for about 82.4% of the total number of SNPs. Among them, there were 487 synonymous mutations, 247 missense mutations, and 2 nonsense mutations, resulting in 102 different gene sequences (Table 6).

Table 6. Summary statistics of SNP base changes of BmNPV YN.

Type	SNP Number	Region
Transitions	693	
Transversions	195	
Mutations in coding region	736	
Intergenic mutation	151	
Synonymous mutation	487	CDS
Missense mutation	247	CDS
Nonsense mutation	2	CDS
Other mutations that could not be accurately judged	6	

Note: The reference genome was BmNPV ZJ.

Analysis of InDel mutations by SnpEff software showed that the numbers of InDel mutations were similar between coding and non-coding regions (Table 7). Further analysis revealed 38 in-frame mutations and 25 frameshift mutations, resulting in 30 differential gene sequences.

Table 7. InDel annotations of BmNPV YN.

Type		InDel Number	Region
Mutations in coding region		66	
Intergenic mutation		64	
Codon mutation	Code insertion	19	CDS
	Code deletion	19	CDS
Frameshift mutation		25	CDS
Other mutations that could not be accurately judged		6	

Note: The reference genome was BmNPV ZJ.

3.6. Differential Genes Analysis

The BmNPV ZJ and BmNPV YN strains were observed to have 106 different annotated genes sequences, which included non-synonymous mutations, synonymous mutations, and frameshift mutations. Among these, 76 differential nucleotide sequences led to differences in amino acid sequences in the two viruses (Table 8). The core genes [27] (shared genes of the baculovirus) accounted for 28.94% and non-core genes accounted for 47.37% of all differential genes, while some genes had not been reported earlier. Thirty annotated genes were associated with the viral genome replication and transcription, and five genes were related to the oral infection of viruses. These differences in genes may have been the cause of the difference in pathogenicity between the two virus strains.

Table 8. Viral functional genes with non-synonymous mutations and frameshift mutations.

Type	BmNPV ZJ Gene Number	BmNPV YN Gene Number	Gene Name	Mutation Type	Biological Function	Description
Core	PROKKA 00060	PROKKA 00011	*pif-2*	Transversions	Viral infection	The composition of the membrane, necessary for oral infection
	PROKKA 00082	PROKKA 00117	*p74 (pif-0)*	Transversions	Viral infection	It is related to the life cycle of the virus and participation in the oral infection of the virus
	PROKKA 00108	PROKKA 00135	*pif-1*	Transitions and transversions	Viral infection	The composition of the membrane, related to oral infections
	PROKKA 00113	PROKKA 00108	*pif-5 (odv-e56)*	InDel	Viral infection	Determines the virus host range, related to oral infections
	PROKKA 00009	PROKKA 00056	*Vlf-1*	Transversions and InDel	Replication, transcription	Late gene expression
	PROKKA 00020	PROKKA 00067	*dna pol*	Transitions and transversions	Replication, transcription	The catalytic activity, replication of the viral genome, and host DNA polymerase cannot replace viral enzymes in this process
	PROKKA 00035	PROKKA 00082	*lef-8*	Transitions	Replication, transcription	Late gene expression
	PROKKA 00043	PROKKA 00090	*p47*	Transitions	Replication, transcription	Regulation of viral transcription
	PROKKA 00067	PROKKA 00018	*lef-1*	InDel and transversions	Replication, transcription	Encoding DNA promoter and interacting with LEF-2
	PROKKA 00073	PROKKA 00024	*lef-2*	Transitions	Replication, transcription	Virus replication and late gene expression
	PROKKA 00092	PROKKA 00036	*lef-5*	Transitions	Replication, transcription	Regulation of viral transcription
	PROKKA 00095	PROKKA 00033	*AcMNPV orf103*	Transitions	Replication, transcription	Virus replication, influences virus particle formation
	PROKKA 00114	PROKKA 00109	*ie-1*	Transitions	Replication, transcription	The essential transactivated protein that initiates viral DNA replication and *bro* promoter transcription
	PROKKA 00123	PROKKA 00045	*lef-4*	Transitions and transversions	Replication, transcription	Regulation of viral transcription
	PROKKA 00128	PROKKA 00040	*dna hel/p143*	Transversions	Replication, transcription	DNA helicase activity, host domain determinant
	PROKKA 00132	PROKKA 00126	*p24*	Transitions	Replication, transcription	Regulation of viral transcription
	PROKKA 00003	PROKKA 00050	*p95*	Transitions	Structural protein	Composition of BV and ODV
	PROKKA 00006	PROKKA 00053	*gp41*	Transitions and transversions	Structural protein	Exists only in ODV, determining the manner and the ability of the virus to invade the host
	PROKKA 00008	PROKKA 00055	*AcMNPV orf78*	Transitions	Structural protein	Related to BV production and M-ODV formation
	PROKKA 00093	PROKKA 00035	*p40*	Transitions, transversions, and InDel	Structural protein	Includes body virus envelope components related to specific infection of host cells
	PROKKA 00100	PROKKA 00143	*AcMNPV orf109*	Transversions	Structural protein	Participates in viral nucleocapsid assembly
	PROKKA 00122	PROKKA 00046	*vp39*	Transitions and InDel	Structural protein	Related to virus infection
Non -core	PROKKA 00101	PROKKA 00142	*AcMNPV orf110*	Transitions	Viral infection	Related to oral infections
	PROKKA 00024	PROKKA 00071	*fp25K (ac61)*	InDel	Auxiliary function	Involved in BV and ODV formation, implicated in host degradation after death
	PROKKA 00058	PROKKA 00009	*pkip*	Transitions	Auxiliary function	Related to the BV nucleocapsid component
	PROKKA 00066	PROKKA 00017	*ecdysteroid UDP-glucosyl transferase (egt)*	Transversions	Auxiliary function	Hinders larvae molting and pupation

Table 8. *Cont.*

Type	BmNPV ZJ Gene Number	BmNPV YN Gene Number	Gene Name	Mutation Type	Biological Function	Description
	PROKKA 00070	PROKKA 00021	*pk1*	Transversions and InDel	Auxiliary function	Regulation of polyhedrin gene promoter activity
	PROKKA 00078	PROKKA 00029	*ptp*	Transitions	Auxiliary function	BV components, essential components for effective infection of larvae brain tissue
	PROKKA 00086	PROKKA 00129	*chitinase A*	Transitions	Auxiliary function	Related to virus transmission
	PROKKA 00140	PROKKA 00128	*viral cathepsin-like protein (v-cath)*	Transversions	Auxiliary function	Related to host liquefaction and degradation
	PROKKA 00016	PROKKA 00063	*AcMNPV orf69*	Transitions	Replication, transcription	Late gene expression
	PROKKA 00018	PROKKA 00065	*lef-3*	Transitions	Replication, transcription	SS-DNA binding and destruction of helical stability
	PROKKA 00031	PROKKA 00078	*lef-10*	Transitions	Replication, transcription	Regulation of viral transcription
	PROKKA 00042	PROKKA 00089	*lef-12*	Transitions	Replication, transcription	Late gene transcirition
	PROKKA 00052	PROKKA 00002	*bro-a*	Transitions and transversions	Replication, transcription	DNA binding protein, complementary to BRO-C
	PROKKA 00054	PROKKA 00005	*lef-6*	Transitions and transversions	Replication, transcription	Virus replication and late gene expression, affecting host cell apoptosis
	PROKKA 00057	PROKKA 00008	*dbp (DNA binding protein)*	InDel	Replication, transcription	SS-DNA binding protein co-localized with *ie-1* and *lef-3* in viral replication mechanism
	PROKKA 00069	PROKKA 00020	*bm (br) orf-4*	Transitions and transversions	Replication, transcription	Early gene expression of virus
	PROKKA 00071	PROKKA 00022	*orf1629*	Transitions and transversions	Replication, transcription	Virus replication
	PROKKA 00076	PROKKA 00027	*Bro-b*	Transitions, transversions, and InDel	Replication, transcription	DNA binding protein
	PROKKA 00077	PROKKA 00028	*Bro-d*	Transitions, transversions, and InDel	Replication, transcription	Virus replication and gene expression regulation
	PROKKA 00080	PROKKA 00105	*pe38*	Transitions and transversions	Replication, transcription	Virus replication and gene expression regulation
	PROKKA 00090	PROKKA 00038	*Bro-c*	Transitions, transversions, and InDel	Replication, transcription	DNA binding protein, complementary to BRO-A
	PROKKA 00091	PROKKA 00037	*39k*	InDel	Replication, transcription	Virus replication and gene expression regulation
	PROKKA 00097	PROKKA 00030	*he65*	Transitions and transversions	Replication, transcription	Virus replication
	PROKKA 00111	PROKKA 00106	*ie-2*	InDel	Replication, transcription	Virus replication and gene expression regulation
	PROKKA 00120	PROKKA 00115	*ie-0*	Transitions	Replication, transcription	Regulation of viral transcription
	PROKKA 00121	PROKKA 00116	*me53*	Transitions and transversions	Replication, transcription	Related to BV and ODV production
	PROKKA 00021	PROKKA 00068	*gp37*	Transitions	Structural protein	The formation of auxiliary components of polyhedra is involved in the transport of virus particles in the cell

Table 8. *Cont.*

Type	BmNPV ZJ Gene Number	BmNPV YN Gene Number	Gene Name	Mutation Type	Biological Function	Description
	PROKKA 00037	PROKKA 00084	*odv-e66*	Transitions and transversions	Structural protein	Participation in BV and ODV morphogenesis and oral infection
	PROKKA 00038	PROKKA 00085	*bmnpvcubigcp037*	InDel	Structural protein	Replication of the virus, regulation gof the transport of virus particles
	PROKKA 00065	PROKKA 00016	*Bv/odv-e26*	Transitions	Structural protein	Related to BV and ODV envelopes
	PROKKA 00096	PROKKA 00032	*vp80*	Transitions	Structural protein	Required for virus replication, BV production, and nucleocapsid maturation
	PROKKA 00135	PROKKA 00123	*AcMNPV orf132*	Transitions and transversions	Structural protein	Involved in BV and ODV formation
	PROKKA 00139	PROKKA 00118	*p10*	Transitions	Structural protein	Partcipates in the morphogenesis of viral polyhedra and promotes the release of polyhedra from infected insect cells
	PROKKA 00015	PROKKA 00062	*iap2*	Transitions	Apoptosis	Cell apoptosis inhibiting factor
	PROKKA 00055	PROKKA 00006	*iap1*	Transitions, transversions, and InDel	apoptosis	Induction of apoptosis

4. Discussion

BmNPV is a serious threat to silkworms and causes huge economic losses to the sericulture industry. Cultivating silkworm varieties resistant to a hemolymph-type septic diseases is the most effective and economical measure to minimize and mitigate these losses [28]. The pathogenicity of BmNPV varies in different geographic regions [18,19]. Furthermore, the underlying molecular mechanism that causes the variations in the pathogenicity of different strains of BmNPV remains unknown and poses a great challenge to the effective prevention of BmNPV. We compared the genomes of two BmNPV strains with different pathogenicities and found differences in some genes related to viral replication and infection.

In this study, the pathogenicities of BmNPV ZJ and BmNPV YN was determined and compared. The LC_{50} of BmNPV YN and BmNPV ZJ against Baiyu larvae were 3.62×10^5 and 6.45×10^6, respectively, indicating the semi-lethality of BmNPV YN in silkworms at a low concentration. Subsequently, the electron microscopic observation morphologies of virus particles revealed their hexagonal shapes with similar sizes (about 2.2–4.0 μm). These were typical silkworm nuclear polyhedrosis viruses, but the edges of BmNPV YN were clearer than those of BmNPV ZJ. The relative expressions of the *ie-1*, *p143*, *vp39*, and *polh* genes, expressed, respectively, at the very early, early, late, and very late stages of BmNPV ZJ, in the midgut of Baiyu larvae were higher than those of BmNPV YN within 12–96 h. This further indicates that the lethal concentration of BmNPV ZJ was higher than that of BmNPV YN. After a host is infected by a virus, in general, viruses with a high rate of proliferation tend to exhibit more robust pathogenicity. However, some more lethal virus strains may kill the host even with a lower rate of proliferation [29–31]. BmNPV YN might be such a type of lethal strain which, even in lower numbers, can kill the host. This inference is consistent with the finding of a tenfold lower LC_{50} for BmNPV YN than for BmNPV ZJ in the second instar of the Baiyu silkworm larvae. Therefore, when the virus was fed at a high concentration (10^8) to silkworms, the expression of the virus gene of BmNPV YN was lower, but it was enough to kill the host. Likewise, Ma [32] found that the mutation and recombination in genomes of different strains of *Helicoverpa armigera* nucleopolyhedrovirus might be the cause of the difference in their pathogenicity even when they are distributed in the same geographic location. Hence, the difference in pathogenicity of our strains might have been because of the variation in their genomes due to the differences in the climates of Zhenjiang, Jiangsu Province, and Luliang, Yunnan Province, which are located, respectively, along the southeast coast and the inland southwest of China.

To further explore the reasons for the difference in pathogenicity between BmNPV YN and BmNPV ZJ to *B. mori*, the whole genomes of the virus strains were sequenced and analyzed. The genome size of BmNPV ZJ was estimated to be 135,895 bp, with 40.39% GC content and encoding 141 genes, while the genome size of BmNPV YN was estimated to be 143,180 bp, with 39.71% GC content and encoding 145 genes. These data demonstrated that the genome sizes of the two viruses and the numbers of putative genes changed. For further analysis, we used the *bro-d* gene, which is highly conserved in different strains of BmNPV from varying geographical regions. BmNPV ZJ exhibited the highest homology with the Cubic strains published earlier in the GenBank, which might be due to the closer origin region of BmNPV. These findings suggest that BmNPV from different regions must be used for screening the BmNPV-resistant silkworm varieties for their wider adaptability.

A comparative analysis of BmNPV YN and BmNPV ZJ genomic sequences revealed a large number of SNP and InDel mutations in the two genomes. The nucleotide sequence homology of the two virus genomes was 92.6–100.0%, and a further BLAST analysis revealed 76 different genes between BmNPV YN and BmNPV ZJ. These genes included 22 core genes and 36 non-core genes, mainly the oral infection factor genes, viral replication genes, host molting, and pupation genes. The alterations of genes sequences affect the function of the encoded proteins. Therefore, we speculated that the difference in the genomes of the two viruses might have been the cause of the difference in their pathogenicity in silkworms.

Thus, the roles of viral genes and the mechanism of action of important proteins encoded by these genes need to be studied in the future.

5. Conclusions

To summarize, the BmNPV YN strain has a relatively lower rate of proliferation but stronger pathogenicity than BmNPV ZJ. The two viruses differ in terms of genome size and the number of coding genes. Among these, 76 genes are different, including genes associated with BmNPV virus replication and infection. These data illustrate that, due to different geographical environments, the genomes of different BmNPV strains mutate and rearrange, which finally leads to differential pathogenicity. The results of this study can provide references to further explore the molecular mechanism relating to the difference in pathogenicity among the viral strains and also provide some guidance in developing new insecticides.

Supplementary Materials: The following are available online at https://www.mdpi.com/article/10.3390/insects12100890/s1, Figure S1: BmNPV ZJ genome linear map, Figure S2: BmNPV YN genome linear map, Table S1: Nucleotide homology comparison of BmNPV ZJ and BmNPV YN with 6 reference strains.

Author Contributions: Conceptualization, H.G. (Huiduo Guo); methodology, H.Q. and G.Z.; software, G.L. and A.X.; validation, H.G. (Huimin Guo) and X.Z.; investigation, B.Z.; data curation, J.S.; writing—original draft preparation, H.G. (Huimin Guo); writing—review and editing, H.Q. and H.G. (Huiduo Guo). All authors have read and agreed to the published version of the manuscript.

Funding: This research was supported by the China Agriculture Research System of MOF and MARA and the Key R&D Plan of Jiangsu Province (Modern Agriculture) (BE2020418) and the Natural Science Foundation of Jiangsu Province (BK20181228, BK20201229).

Institutional Review Board Statement: Not applicable.

Informed Consent Statement: Not applicable.

Data Availability Statement: No new data were created or analyzed in this study. Data sharing is not applicable to this article.

Conflicts of Interest: The authors declare no conflict of interest.

References

1. Yuan, M.J.; Wu, W.B.; Liu, C.; Wang, Y.J.; Hu, Z.Y.; Yang, K.; Pang, Y. A highly conserved baculovirus gene *p48* (*ac103*) is essential for BV production and ODV envelopment. *Virology* **2008**, *379*, 87–96. [CrossRef] [PubMed]
2. Maeda, S.; Majima, K. Molecular cloning and physical mapping of the genome of *Bombyx mori* nuclear polyhedrosis virus. *J. Gen. Virol.* **1990**, *71*, 1851. [CrossRef] [PubMed]
3. Li, J.Q. Summary table of nationalsericulture production in 2019. *Chinses Seric.* **2020**, *41*, 70–72.
4. Xu, A.Y.; Qian, H.Y.; Sun, P.J.; Liu, M.Z.; Lin, C.Q.; Li, G.; Li, L.; Zhang, Y.H.; Zhao, G.D. Breeding of a new silkworm variety "Huakang 3" with resistance to *Bombyx mori* Nucleopolyhedrovirus disease. *Sci. Seric.* **2019**, *45*, 201–211.
5. Xu, A.Y.; Lin, C.Q.; Qian, H.Y.; Sun, P.J.; Zhang, Y.H.; Liu, M.Z.; Li, L. Breeding of a new silkworm variety "Huakang 2" with tolerance to *Bombyx mori* nucleopolyhedrovirus disease. *Sci Seric.* **2013**, *39*, 275–282.
6. Shi, M.N.; Bi, L.H.; Gu, J.D.; Fei, M.H.; Qi, G.J.; Wei, B.Y.; Huang, J.T.; Huang, L.L.; Su, H.M.; Meng, Y.Y. Breeding of a new highly resistant nuclear polyhedrosis disease silkworm variety Guican N2. *Guangxi Seric.* **2012**, *49*, 1–12.
7. Xu, A.Y.; Lin, C.Q.; Qian, H.Y.; Sun, P.J.; Zhang, Y.H.; Liu, M.Z.; Li, L. Brief introduction of new varieties of silkworm resistant to BmNPV. In Proceedings of the Tenth Domestic (Tussah) Silkworm Genetics and Breeding and Seed Breeding, Jilin, China, 19 July 2013; Chinese Silkworm Society: Jilin, China, 2014; pp. 184–185.
8. Zang, Y.L.; Hunang, L.; Wang, Y.W.; Wang, H.L.; Gao, H.J. A preliminary report on the feeding of silkworm varieties of "Huakang 1" and "Huakang 2" with tolerance to BmNPV disease in Taian. *Hebei J. For. Orchard. Res.* **2014**, *29*, 435–438.
9. Hu, S.Y.; Luo, C.B.; Sun, Y.P.; Yang, S.T.; Jiang, H.; Yang, B.; Yu, W.Z.; Huang, G.H. A brief report on feeding test of new silkworm variety Huakang 2 in Qingzhen city. *Chinses Seric.* **2016**, 24–27. [CrossRef]
10. Luo, P. Discussion on the prevention and control methods of silkworm nuclear polyhedrosis virus disease in subtropical regions. *Chinses Seric.* **2017**, *38*, 59–62,67.
11. Shi, M.N.; Tang, L.; Huang, H.Y.; Wei, T.X.; Pan, Z.X.; Tang, T.X.; Pan, Z.X.; Qi, G.J.; Pu, Y.X.; Mo, Y.X. Application of new silkworm variety 'Guican N2'. *Guangxi Seric.* **2015**, *52*, 27–31.

12. Tang, F.F.; Zhang, Y.H.; Shao, Y.L.; Zhu, F.; Bai, X.R. Virulence and phylogenetic analysis of *Bombyx mori* nucleopolyhedrovirus isolates from Yunnan, southwestern China. *Acta Entomol. Sin.* **2018**, *61*, 42–51.
13. Fu, J.Y.; Xi, Y.; Tang, M.J.; Yin, K.S. Study on the relationship between virulence and genetic structure of four wild isolates of *Euproctis pseudoconspersa* nuclear polyhedrosis virus. *J. Tea Sci.* **2011**, *31*, 289–294.
14. Fuxa, J.R. Ecology of insect nucleopolyhedroviruses. *Agric. Ecosyst. Environ.* **2004**, *103*, 27–43. [CrossRef]
15. Tang, F.F.; Shao, X.L.; Zhong, J.; Zhang, Y.H.; Huang, P.; Dong, Z.P.; Liao, P.F.; Bai, X.R. A preliminary study on molecular identification of *Bombyx mori* nucleopolyhedrovirus strains. *Sci. Seric.* **2014**, 1030–1035. [CrossRef]
16. Zhou, J.B. Identification of Thai Strain of *Bombyx mori* Nuclear Polyhedrosis Virus and Analysis of *bro* Gene Family. Master's Thesis, Anhui Agricultural University, Hefei, China, 2012.
17. Xu, Y.P.; Cheng, R.L.; Xi, Y.; Zhang, C.X. Genomic diversity of Bombyx mori nucleopolyhedrovirus strains. *Genomics* **2013**, *102*, 63–71. [CrossRef] [PubMed]
18. Sun, J.C.; Jin, F.L.; Xu, X.Y.; Tan, P.C. Comparative study on median lethal dose of BmNPV to *B. mori* in different regions. *Guangdong Seric.* **2001**, *035*, 35–37.
19. Bai, X.R.; Ran, R.F.; Dong, Z.P.; Dong, J.H.; Huang, P. Study on virulence of BmNPV to *Bombyx mori* in different areas of Yunnan. *Southwest China J. Agric. Sci.* **2010**, *23*, 2098–2101.
20. Wang, X.; Huang, X.H.; Jiang, M.G.; Tang, L.; Dong, G.Q., Huang, S.H.; Shi, M.N.; Pan, Z.X. Epidemic factors of *Bombyx mori* hemolymph-type septic disease in Guangxi and its molecular phylogenetic analysis. *J. South. Agric.* **2020**, *051*, 669–676.
21. Bankevich, A.; Nurk, S.; Antipov, D.; Gurevich, A.A.; Dvorkin, M.; Kulikov, A.S.; Lesin, V.M.; Nikolenko, S.I.; Pham, S.; Prjibelski, A.D. SPAdes: A new genome assembly algorithm and its applications to single-cell sequencing. *J. Bioinf. Comput. Biol.* **2012**, *19*, 455–477. [CrossRef] [PubMed]
22. Massouras, A.; Hens, K.; Gubelmann, C.; Uplekar, S.; Decouttere, F.; Rougemont, J.; Cole, S.T.; Deplancke, B. Primer-initiated sequence synthesis to detect and assemble structural variants. *Nat. Methods* **2010**, *7*, 485–486. [CrossRef] [PubMed]
23. Kang, W.; Suzuki, M.; Zemskov, E.; Okano, K.; Maeda, S. Characterization of Baculovirus Repeated Open Reading Frames (bro) in Bombyx moriNucleopolyhedrovirus. *J. Virol.* **1999**, *73*, 10339–10345. [CrossRef]
24. Mckenna, A.; Hanna, M.; Banks, E.; Sivachenko, A.; Cibulskis, K.; Kernytsky, A.; Garimella, K.; Altshuler, D.; Gabriel, S.; Daly, M. The Genome Analysis Toolkit: A MapReduce framework for analyzing next-generation DNA sequencing data. *Genome Res.* **2010**, *20*, 1297–1303. [CrossRef] [PubMed]
25. Cingolani, P. A program for annotating and predicting the effects of single nucleotide polymorphisms, SnpEff: SNPs in the genome of Drosophila melanogaster strain w1118; iso-2; iso-3. *Fly* **2012**, *6*, 1–13. [CrossRef]
26. Zhou, J.B.; Li, X.Q.; De-Eknamkul, W.; Suraporn, S.; Xu, J.P. Identification of a new *Bombyx mori* nucleopolyhedrovirus and analysis of its *bro* gene family. *Virus Genes* **2012**, *44*, 539–547. [CrossRef]
27. Herniou, W.A.; Olszewski, J.A.; Cory, J.S.; O'Reilly, D.R. The genome sequence and evolution of baculoviruses. *Annu. Rev. Entomol.* **2003**, *48*, 211–234. [CrossRef] [PubMed]
28. Qian, H.Y.; Li, G.; Zhao, G.D.; Liu, M.Z.; Sun, P.J.; Xu, A.Y. Identification of new silkworm vrieties resistant to nuclear polyhedrosis Virus to BmNPV. *Chinses Seric.* **2017**, *38*, 15–22.
29. Huang, C.; Fan, X.H.; Jiang, Y.H.; Song, D.Z.; Gao, L.Q.; Huang, Q.G.; Lai, Z.P. Anti-tumor effect of Newcastle disease virus strain D817 against nude mouse xenografts of human colon carcinoma. *Zhonghua Zhong Liu Za Zhi Chin. J. Oncol.* **2009**, *31*, 490.
30. Wang, Y.Y.; Lu, J.; Liu, H.; Ye, R. Comparison of the infectivity of three virus strains of the genus A-coronavirus to cells derived from different hosts. *China Sci.* **2013**. Available online: http://www.paper.edu.cn/releasepaper/content/201301-1114 (accessed on 26 September 2021).
31. Lu, W.Q.; Wei, P.W. Changes and Comparison of virus concentrations of six SMV strains in different soybean varieties. *Soybean Sci.* **1991**, *6*, 27–34.
32. Ma, X.C. Sequence Analysis of Genomes from Three Nucleopolyhedroviruses. Ph.D. Thesis, Zhejiang University, Hangzhou, China, 2006.

insects

MDPI

Article

Supercritical Carbon Dioxide Impregnation of Gold Nanoparticles Demonstrates a New Route for the Fabrication of Hybrid Silk Materials

Manish Singh [1], Estera S. Dey [1], Sunil Bhand [2] and Cedric Dicko [1,*]

[1] Pure and Applied Biochemistry, Chemistry Deptartment, Lund University, Naturvetarvägen 14, 22362 Lund, Sweden; manish.singh.itbhu06@gmail.com (M.S.); estera.dey1@gmail.com (E.S.D.)

[2] Deptartment of Chemistry, Birla Institute of Technology and Science, KK Birla Goa Campus, Pilani 403726, Zuarinagar, Goa, India; sunilbhand@goa.bits-pilani.ac.in

* Correspondence: cedric.dicko@tbiokem.lth.se

Simple Summary: The application of nanotechnology in textiles is limited by the difficulties of loading the fabrics with nanoparticles (NPs) and by their subsequent uncontrolled leakage. More fundamentally, there is a need to answer the question of the "space available" in textile fibers, and generally, other natural polymers for NPs loading. Due to these challenges, there is a risk that uncontrolled leakage of NPs from the textile industry could harm the environment and human health. Here, with a green and straightforward approach, using supercritical carbon dioxide ($scCO_2$) as a carrier fluid, we explored the impregnation in four types of Indian textile silks (Mulberry, Eri, Muga, and Tasar) with five standard sizes of gold NPs (5, 20, 50, 100 and 150 nm). The results suggested that all silks could be permanently impregnated with the gold nanoparticles (Au NPs) up to 150 nm. Knowing the available space in silk or other natural polymers can help us understand how and which natural polymers are suitable for use as catalysts, antimicrobial materials, UV-protective agents, and other valuable properties.

Abstract: How many nanoparticles can we load in a fiber? How much will leak? Underlying is the relatively new question of the "space available" in fibers for nanoparticle loading. Here, using supercritical carbon dioxide ($scCO_2$) as a carrier fluid, we explored the impregnation in four Indian silks (Mulberry, Eri, Muga, and Tasar) with five standard sizes of gold nanoparticles (5, 20, 50, 100 and 150 nm in diameter). All silks could be permanently impregnated with nanoparticles up to 150 nm in size under $scCO_2$ impregnation. Accompanying structural changes indicated that the amorphous silk domains reorganized to accommodate the gold NPs. The mechanism was studied in detail in degummed Mulberry silk fibers (i.e., without the sericin coating) with the 5 nm nanoparticle. The combined effects of concentration, time of impregnation, $scCO_2$ pressure, and temperature showed that only a narrow set of conditions allowed for permanent impregnation without deterioration of the properties of the silk fibers.

Keywords: silk; supercritical carbon dioxide impregnation; nanofiller

Citation: Singh, M.; Dey, E.S.; Bhand, S.; Dicko, C. Supercritical Carbon Dioxide Impregnation of Gold Nanoparticles Demonstrates a New Route for the Fabrication of Hybrid Silk Materials. *Insects* **2022**, *13*, 18. https://doi.org/10.3390/insects13010018

Academic Editors: Silvia Cappellozza, Morena Casartelli, Federica Sandrelli, Alessio Saviane and Gianluca Tettamanti

Received: 8 November 2021
Accepted: 20 December 2021
Published: 23 December 2021

1. Introduction

The primary purpose of impregnating nanoparticles (NPs) in fibers is to increase mechanical strength, improve physical properties, such as electrical conductivity and antistatic behavior, and add functionalities, such as antimicrobial, UV protection, flame retardance, and self-cleaning [1–3]. If homogeneously distributed, NPs in polymer matrices can increase the composite toughness and abrasive resistance. There are several ways to impregnate/synthesize nanoparticles, and most importantly, stabilize them on or in fibers. Now, standard methods include plasma treatment [4], in situ synthesis [5], sol-gel synthesis, chemical assembly [6], deposition [7], dip-coating [8,9], radiolysis [10], and sonochemical reduction [11].

The methods above have drawbacks, specifically a decrease in the tensile strength of the fibers; the process demands harsh chemical pretreatment, and leakage of nanoparticles occurs with time [12,13]. The severe chemical pretreatment step and leakage of NPs from fibers lead to serious environmental and health issues [12–14].

Within this context, the use of supercritical carbon dioxide ($scCO_2$) is an attractive alternative [15–19]. The synthesis/impregnation of NPs on a polymeric substrate under $scCO_2$ has several advantages due to some of its unique properties: low toxicity, non-flammable, inexpensive, low surface tension, and no residue in the treated medium after removal [15,20–23]. Typically, the exposure of natural and synthetic polymers to $scCO_2$ results in swelling and enhanced chain mobility of the polymers, which helps to load the additives [24–26]. Recent examples and applications using $scCO_2$ with dyes [25,27], conductive monomers [28,29], inorganic NPs [30], such as TiO_2 [31], Ag [32] into fibers have demonstrated the usefulness of the method [33]. However, the leakage, particle stabilization, and impregnation mechanism were unclear.

In 1995, NPs were synthesized on a polymer substrate for the first time using a supercritical fluid [34,35]. Afterward, substantial research led to the incorporation of metal NPs on various inorganic and organic substrates [20,32,35–38]. Silk has had a limited application with $scCO_2$. This is mainly within silk fibers dyeing [25,39], grafting [40,41], and controlled drug delivery application [42]. To the best of our knowledge, no report yet of metal/metal-oxide NPs impregnation in silk fibers using $scCO_2$.

Therefore, in the present work, we explored the impregnation in four different types of silks of standardized gold NPs. The choice of gold NPs was motivated by their colloidal stability, monodispersity and low chemical reactivity. For example, a careful evaluation of the plasmon peak of the gold NPs left in solution after impregnation provided a qualitative estimate for their chemical stability. Finally, the choice of size was limited to 150 nm since, at 200 nm, the colloidal stability of the gold NPs was poor.

The combined results demonstrated the usefulness of $scCO_2$ for impregnation and determined that the space available in silks was finite if no structural damage was the limiting factor.

2. Materials and Methods

Gold NPs of different diameters (5, 20, 50, 100, and 150 nm) were purchased from Sigma-Aldrich (Darmstadt, Germany). The NPs were phosphate stabilized and suspended in 0.1 mM phosphate buffer saline (PBS, Sigma-Aldrich, Darmstadt, Germany), having an optical density of 1 (O.D. = 1). At this O.D. the corresponding concentration were 5 nm (69 µg/mL, 3.5×10^{-7} M), 20 nm (53 µg/mL, 2.7×10^{-7} M), 50 nm (44 µg/mL, 2.2×10^{-7} M), 100 nm (38 µg/mL, 1.97×10^{-7} M) and 150 nm (63 µg/mL, $3.2\ 10^{-7}$ M). Details are in Table S1. The NPs were used without further treatment. Silk yarns from Mulberry (*Bombyx mori*), Eri (*Samia Cynthia ricini*), Muga (*Antheraea assamensis*), and Tasar (*Antheraea mylitta*) were purchased from Adarsha Traders, Davangere, Karnataka, India. Degummed (soap and sodium carbonate) silk fibers from Mulberry (*Bombyx mori*) were obtained from an online silk supplier (Wild Fibres, http://www.wildfibres.co.uk, accessed on 1 November 2021).

2.1. Preparation of Silks Fibers to Be Impregnated with Gold NPs

A bundle of fibers was weighed (0.051 ± 0.003 g), gently rinsed with 10 mL of double-distilled water twice and dried at room temperature for 12 h on the lab bench. The dry silk fibers were then immersed in 10 mL of reacting solution (9 mL of water and 1 mL of NPs) in a 20 mL glass vial. We used aluminum foil to cover the glass vial and magnetic bars to mix the solutions (see Figure 1). Total loading would yield the following µg of gold per mg of silk 1.4, 1, 0.9, 0.8 and 1.2 for the 5, 20, 50, 100 and 150 nm, respectively.

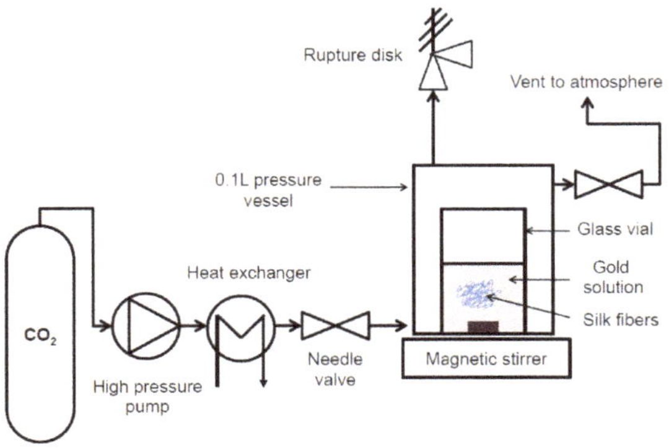

Figure 1. Schematic diagram of supercritical CO_2 set up. Briefly, when reaching supercritical conditions, the water and gold NPs become a ternary mixture of water/gold NPs/scCO_2. The new mixture has enhanced diffusion and interfacial properties allowing the gold NPs to be transported in the fibers.

2.2. Impregnation of Silk Fibers under Supercritical and Atmospheric Conditions

The scCO_2 impregnation was performed in a modified SFE-100 from Thar Technologies, Inc. (Pittsburg, PA, USA). Figure 1 illustrates the experimental apparatus. Although not intuitive, the CO_2 in the supercritical state permeates the whole reactor and mixes with the water/g NPs solution. The new mixture (water/gold NPs/scCO_2) has new diffusion and interfacial properties, transporting the gold NPs onto/into the fibers.

The setup consisted of a steel reaction vessel of 100 mL volume immersed in a thermostated water bath (see temperature details in the results section) with a magnetic stirrer.

The glass vial (20 mL) was inside the reaction vessel. The CO_2 was pumped in at 11 g per minute to reach the desired pressure. For the impregnation of the four silks, the temperature was kept at 40 °C and the pressure at 200 bars. For the detailed study of the impregnation using the fully degummed Mulberry silk fiber, the final pressure and temperature were adjusted according to the experimental design table (Table 1). The control impregnation happened at atmospheric pressure and at the corresponding temperature and time to match the supercritical conditions. The magnetic agitation speed of impregnation was kept constant at 300 rpm. Each experiment was run in triplicate.

Table 1. Factorial design parameters.

Factors	Levels	Range
A: Temperature (°C)	2	35, 50 and center point at 42.5
B: Pressure (bars)	2	100, 250 and center point at 175
C: Time (hours)	2	1, 3 and center point at 2
D: Mixing (rpm)	2	0 and 300

Note that direct attempts with the NPs resulted in the partial coloration of the fibers, thus incomplete and uneven impregnation (data not shown). Henceforth, the impregnation was carried on in the presence of water. The depressurizing effect was not studied, but the return to atmospheric conditions happened as slowly as the instrument allowed (i.e., 20 min). Another result is the known low pH under CO_2 conditions. Immediately after opening the reaction vessel, we found that the pH of the final solution had a pH of 3. Attempts to maintain the pH at 7 using a 1 M phosphate buffer did not significantly differ. Henceforth, the impregnation happened in water.

2.3. Fiber Post-Processing

Washing and fastness test: after the impregnation step, the fibers were recovered for post-processing, namely, washing and fastness treatment successively. The remaining liquid after impregnation was referred to as gold loading. Next, the washing step, the silk fibers, were transferred in a new glass vial and rinsed with 10 mL of water for 1 h. The remaining liquid is referred to as wash leakage. The final step is the fastness test; the silk fibers were transferred to a new glass vial containing 10 mL of water and heated to 50 °C for 1 h. This step is referred to as fastness leakage. The procedure is illustrated in Figure S1 in the Supplementary Information.

The liquid supernatants were measured by UV absorbance, and we calculated the gold loading, wash leakage, fastness leakage, and total efficiency as follows:

$$Gold\ loading\ (\%) = \left(1 - \frac{A_1}{A_0}\right) * 100 \quad (1)$$

$$Wash\ leakage\ (\%) = \frac{A_2}{A_0 - A_1} * 100 \quad (2)$$

$$Fast\ leakage\ (\%) = \frac{A_3}{(A_0 - (A_1 + A_2))} * 100 \quad (3)$$

$$Total\ efficiency\ (\%) = \frac{A_0 - (A_1 + A_2 + A_3)}{A_0} * 100 \quad (4)$$

where A_0 was the absorbance of gold suspension before impregnation, while A_1, A_2, and A_3 were the absorbance of the liquid supernatant after impregnation, water wash, and fastness treatment, respectively, the efficiencies were statistically compared using a general linear model on the arcsine transformed efficiencies (to avoid truncation) with a Tukey posthoc test for multiple comparisons. The analysis was performed using Minitab (Minitab, Inc., Philadelphia, PA, USA).

2.4. Characterization

2.4.1. UV-Visible Absorption Spectroscopy

The UV-Visible (UV-Vis) absorption measurements were performed with a Cary 60 UV-Vis spectrophotometer (Agilent) in the 200–800 nm wavelength range at a scan rate of 600 nm·min^{-1}; and a 1 cm plastic cuvette.

2.4.2. Fourier Transform Infrared Spectroscopy-Attenuated Total Reflectance (FTIR-ATR)

The Fourier transform infrared attenuated total reflectance (FTIR-ATR) spectra of silk fibers were measured at different steps: after the impregnation process (scCO_2 or atmospheric conditions), after washing and after fastness test using a Nicolet iS5 infrared spectrometer with an iD5 ATR accessory with diamond crystal (Thermo Scientific). Each spectrum was background corrected and collected between 550–4000 cm^{-1} (see supporting information Figure S2). Each spectrum was an average of 32 scans at 4 cm^{-1} resolution. The FTIR-ATR spectra were further processed to extract four structural parameters: the crystallinity degree (see Supplementary Materials Figure S3), the tyrosine ratio (see Supplementary Materials Figure S4), Amide I/II ratio and the hydrogen bond index. The silk fibers degree of crystallinity [43] was calculated by comparing the peaks' intensities at 1263 and 1230 cm^{-1} as follows:

$$Crystallinity\ degree\ (\%) = \frac{A_{1263}}{A_{1230} + A_{1263}} * 100 \quad (5)$$

A_{1263} and A_{1230} are the intensities of the peaks at 1263 and 1230 cm^{-1}, respectively.

The tyrosine ratio was calculated by estimating the area under the tyrosine peaks at ±830 and ±850 cm^{-1} and computing the ratio of intensities at 850/830 cm^{-1}. The weak features at ±850 and ±830 cm^{-1} make a doublet attributed to the Fermi resonance of the aromatic side chain of the tyrosine residue. The ratio is indicative of the local environment of tyrosine residues within the fibers and, by extension, the local environment of the amorphous regions of the silks [44].

Further, the ratio of the maximum intensities of Amide I (at around 1640 cm^{-1}) and II (at about 1510 cm^{-1}) peaks was calculated to estimate the total change in secondary structure upon treatment [45,46]. Typically, for silk fibers, the two peaks at ≈1620 cm^{-1} and ≈1510 cm^{-1} are mainly assigned to amide I (C=O and C–N) and amide II (N–H and C–N) β-sheet structures. The ratio will measure any changes due to water hydration and β-sheet structures. Additionally, the Amide I region (1600–1700 cm^{-1}) was deconvoluted to extract the secondary structure composition of the silks (see Supplementary Materials Figure S5).

An estimate of hydrogen bond intensity [47] was calculated using the ratio of intensities of N-H vibrations between 3200 cm^{-1} to 3500 cm^{-1}. In this region, a careful decomposition of the N-H stretching mode provides some information on the "free" (non-hydrogen bonded ±3400 cm^{-1}) and hydrogen-bonded N-H (±3320 cm^{-1}). The hydrogen bond index (HBI) was the ratio of bonded to free N-H intensities.

2.4.3. Scanning Electron Microscopy-Degummed Bombyx Mori

Scanning electron microscopy (SEM) images for degummed Bombyx mori were acquired on a JEOL JSM 6700F. The energy-dispersive X-ray spectroscopy (EDXS) analysis was performed using an Oxford X-MAX add-on. The spectra and images were reduced and analyzed using the Aztec software. Before SEM and EDXS, the fibers were carbon-coated.

2.4.4. Photographs

The silk fibers were captured using HP Scanjet G 4050 against a black background.

2.5. Factorial Design for Degummed Bombyx Mori Study

To test the effects of temperature, pressure, and time we designed a full factorial table. We measured the following responses: total efficiency and FTIR-ATR results (i.e., Amide I/II ratio, crystallinity, and tyrosine ratio). Table 1 summarizes the factorial table parameters. A total of 17 sample conditions were investigated.

3. Results

The four silks chosen are among the most common silks produced in India. Table 2 summarizes their most salient chemical and physical properties [48–50].

Table 2. Silk properties.

	Mulberry	Eri	Muga	Tasar
Average density (g/cm^3) [a]	1.357	1.288	1.34	1.323
Average moisture regain (%) [a]	7.04	8.03	7.61	8.52
Sericin content (%) [b]	10.4–24.4	6.5–10.1	8.6–12.7	8.2–14.4
Acid dye exhaustion (%) [c]	89.82	58.38	57.02	59.10
Disperse dye exhaustion (%) [d]	23.67	13.55	12.59	16.20
Elongation (%)	13.5	20.8	22.3	26.5
Tenacity (g/d) [e]	3.75	3.7	4.35	4.5
Initial modulus (g/d)	95	89	81	84
X-ray crystallinity (%)	38.2	32.6	35.0	35.2
Glass transition (°C) [f]	200–220	220–235	215–235	235–250
Basic/Acidic ratio [g]	0.65	1.3	1.24	0.97
Hydrophilic/hydrophobic ratio [g]	0.28	0.35	0.38	0.44
Bulky/non bulky side groups ratio [g]	0.17	0.24	0.28	0.33
Glycine/Alanine ratio [g]	1.58	0.8	0.82	0.81

[a] average of outer, middle and inner silk cocoons layers. [b] low and high sericin content in fibers (from reference [50]). [c] Texacid fast red A (acid dye). [d] Foron scarlet S-3GFL (disperse dye). [e] g/d: gram per denier. [f] from references [51,52]. [g] from reference [50].

3.1. The Efficiency of the Impregnation Process and Fibers Color Change

After correction for wash leakage and fastness leakage (see materials and methods), the total efficiencies are summarized in Figure 2. The total efficiency of gold loading was above 90% for the $scCO_2$ treated Mulberry silk fiber with gold NPs size 5, 20, 50, 100 nm, except for 150 nm, where the efficiency dropped to 75.9 ± 3.4%. In the controls, at atmospheric pressure, the total efficiency was constant from 5 to 100 nm at 24.0 ± 6.2%; and dropped to approximately 12% for the 150 nm particle sizes.

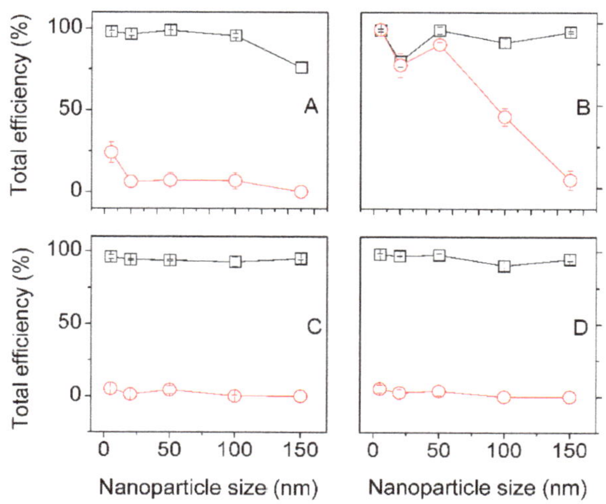

Figure 2. Comparison of total efficiency at supercritical and control impregnation of Indian silks: (**A**) Mulberry (**B**) Eri (**C**) Tasar (**D**) Muga with five different sizes of gold nanoparticle (5, 20, 50, 100 and 150 nm). Black squares (□) are $scCO_2$ samples, red circles (O) are control samples. The control and $scCO_2$ efficiencies were significantly different (N = 30, $p < 0$, $\alpha = 0.05$) for Mulberry, Tasar and Muga silks for all gold nanoparticles sizes. For Eri below 100 nm the efficiencies were not significant (N = 30, $p_{5nm} = 0.35$ $p_{20nm} = 0.12$ $p_{50nm} = 0.09$, $\alpha = 0.05$), above 100 nm the efficiencies were significantly different (N = 30, $p < 0$, $\alpha = 0.05$).

For the other three silks, namely: Eri, Tasar, and Muga, the total efficiency of gold loading was above 90% for $scCO_2$ impregnated regardless of the gold NPs' sizes. For those three silks, the $scCO_2$ treatment efficiencies were significantly larger (see caption Figure 2). However, Eri silks showed exceptionally high efficiency in the controls at 98.3 ± 0.01, 89.6 ± 2.9, and 94.6 ± 0.7% for 5, 20, and 50 nm size nanoparticles, respectively (not significantly different; see caption Figure 2). Beyond, the total efficiency dropped to 77.2 ± 2.1 and 61.9 ± 2.3% for 100 nm and 150 nm NPs (significantly different; see caption Figure 2).

Tasar and Muga, in the controls, showed a total efficiency of about 10% regardless of the size of the NPs. Increasing concentration, three times more gold at 5 nm, resulted in a decrease in total efficiency for all silks except Eri that stayed constant (see Supplementary Materials Figure S6).

Our observations of the total efficiencies for gold NPs in the four silks studies collectively showed an all-or-nothing effect, which meant that the process parameters have little influence. The four silks behaved identically under $scCO_2$ impregnation with no correlation to any of the silk properties shown in Table 2.

The two most remarkable observations were the loss of efficiency for Mulberry after 100 nm. The control sample from Eri silk displayed efficiencies comparable to the $scCO_2$ ones and up to 50 nm gold NP. We found, however, no significant properties of Eri that could explain this behavior. A possibility, therefore, was the sericin coating on each of the fibers. There is considerable variability in the sericin from each silk; even though we applied a pre-wash step before $scCO_2$ impregnation, we cannot fully control the sericin effect. The sericin is the natural target for dying silk fibers and would likely host the gold nanoparticles. The alternative is to remove the sericin and only use the remaining fibroin brins. The difficulty in sericin removal for Eri, Muga, and Tasar meant that only Mulberry silk fiber was further investigated.

Figure 3 shows the color changes in the silks from the gold plasmonic effect. Note that the color change for Muga and Tasar silks was not as evident as Mulberry and Eri since the formers are naturally colored.

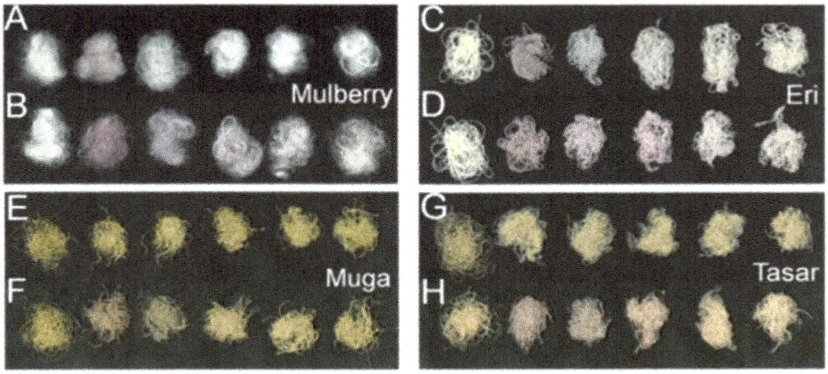

Figure 3. Photographs of the various silks after Gold NPs impregnation. Mulberry: $scCO_2$ (**A**), control (**B**); Eri: $scCO_2$ (**C**), control (**D**); Muga: $scCO_2$ (**E**), control (**F**); and Tasar: $scCO_2$ (**G**), control (**H**). The first fiber bundle from the left was the native untreated silk in all photographs. Then the fibers were treated with five different sizes of gold nanoparticles (5, 20, 50, 100, and 150 nm), respectively. The color change correlates with the initial gold solutions' color.

3.2. $ScCO_2$ Impregnation Induced Structural Changes

The evaluation of the impact of the impregnation procedure was principally conducted using FTIR-ATR and XRD. Figure 4 shows the effect of gold NPs size on a global structural parameter: amide I/II ratio. Changes in the amide I/II ratio indicate that a structural change has occurred. For Mulberry silk fiber, we observed no differences between $scCO_2$ treatment and control at different gold NPs sizes. Interestingly both traces overlapped with the Amide I /II ratio of the native silk (horizontal line in Figure 4A).

Muga silk behaved similarly. The Tasar silk presented a similar trend to Mulberry and Muga, except that the amide I/II ratio was consistently lower than the native Tasar silk. The Eri silk, on the other hand, the amide I/II ratio showed differences between treatments and departed from the native Eri silk amide I/II ratio.

The Amide I/II ratio changes can be resolved by fitting the Amide I peak with a sum of Gaussian contribution. The position and relative area of the Gaussian peaks were then interpreted in terms of secondary structure content and fraction (Table S2). Four secondary structures were extracted for the four silks, namely, a peak centered at around 1620 cm^{-1} representing intermolecular β-sheets; a peak centered at 1655 cm^{-1} representing a mixture of random coils and α-helices; a peak centered at about 1678 cm^{-1} representing β-turns and a weak peak centered at around 1695 cm^{-1} representing β-sheets structures (see supporting information Figure S5).

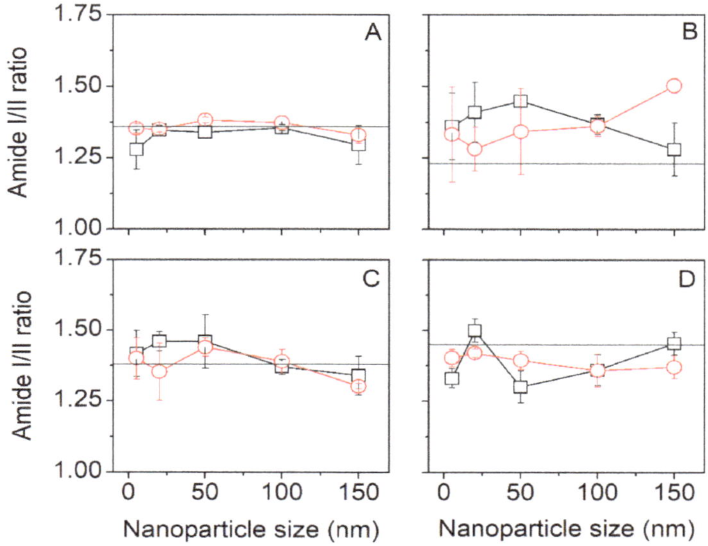

Figure 4. FTIR-ATR change in amide I/ II ratio for Mulberry (**A**), Eri (**B**), Muga (**C**), and Tasar (**D**) silks. Black squares (□) are $scCO_2$ samples, red circles (O) are control samples. The horizontal line is the amide I/II ratio in native silk. We found that for $scCO_2$ treatment alone, the amide I/II ratio was not significantly different from the native silks.

In all silks, the β-sheets content appeared constant regardless of the impregnated gold NPs (see Supplementary Materials Figure S7). In Mulberry only, we found an interconversion from the random coil/α-helical structures to β-turns with increasing gold NP sizes (see Supplementary Materials Figure S7). Noteworthy is the overlapping signal from water in the Amide I region that may bias the decomposition of secondary structures. For example, $scCO_2$ drying followed by washing and high-temperature treatment may affect the water signal contribution in the Amide I differently. Additionally, the contribution of the sericin coating may not be constant throughout the process. These, among others, were the reason for the next section study with the degummed silk.

Besides, we found that the crystallinity index (similar to the β-sheets structure content) did vary marginally from the native silks' crystallinity, only for Eri, Muga, and Tasar. In Eri, the crystallinity index from the gold impregnated silks was higher than the native Eri fibers but similar to the controls. For Muga and Tasar, we observed the opposite trend, a lower crystallinity (see Supplementary Materials Figure S8). The crystallinity index was independent of the gold NP size in all fibers.

The relative intensity of the tyrosine doublet (Intensity at 850 cm^{-1}/Intensity at 830 cm^{-1}) was used as a spectral marker of the environment of the hydroxyl groups and the strength of hydrogen bonds involving these groups. The tyrosine residues usually exist in the amorphous regions, containing most amino acids with bulky and polar side chains. An increase in the tyrosine ratio led to the conclusion that the hydrogen bonds involving the tyrosine residues were weak, and consequently, the mobility of the tyrosine residues was higher [44].

For Mulberry, Muga, and Tasar, we found that the tyrosine ratio was constant and closed to the value from the respective native silks (see Supplementary Materials Figure S9). On the other hand, Eri silk showed a constant tyrosine ratio with increasing gold NP size. Still, a systematically lower ratio than native Eri silk, suggesting the amorphous region in Eri was stiffer after the treatment.

The XRD confirmed the gold NPs with an increasing diffraction peak at 38° (Figure 5). The silks diffraction peaks were conserved regardless of the gold NP size used. No further attempt at analyzing the XRD was deemed necessary.

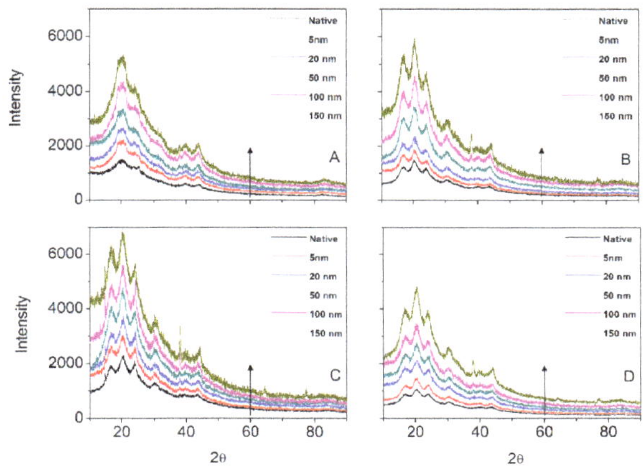

Figure 5. X-ray diffractogram as a function of gold NP size for Mulberry (**A**), Eri (**B**), Muga (**C**), and Tasar (**D**) silks. The arrow indicates an increasing gold NP size. Note at around 37° the (111) reflection for the gold.

In the next section, we focus only on the Mulberry silk fiber and the 5 nm gold NP to unravel the processes underlying the gold impregnation capacity of silk fibers.

3.3. *Effect of Degumming on Bombyx Mori and 5 nm Gold NPs*

3.3.1. Facile Gold Impregnation in Supercritical Carbon Dioxide at 40 °C and 200 Bars

The supercritical treatment of silk fibers resulted in a high gold loading (95.5%—Figure 6A), while the control treatment yielded a poor gold loading (6.3%—Figure 6A). The high percentage of wash leakage (68.4%—Figure 6B) and fastness leakage (58.3%—Figure 6C) for the control silk suggested that the gold NPs were weakly attached to the surface of silk fibers.

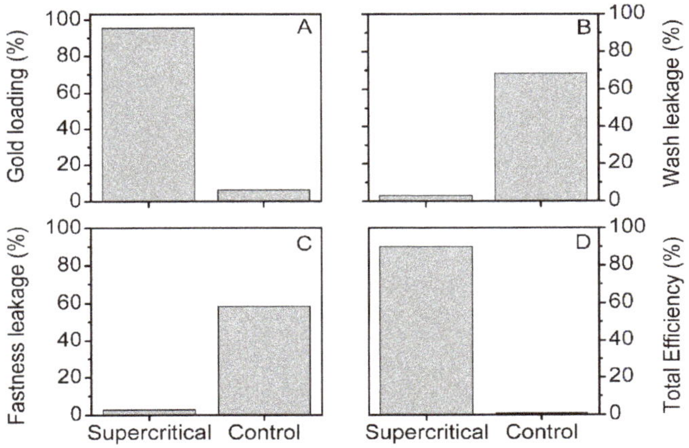

Figure 6. Comparative gold loading (%)—(**A**), wash leakage (%)—(**B**), fastness leakage (%)—(**C**), and total efficiency (%)—(**D**), for supercritical impregnated and control degummed silk fibers. See the materials and methods for details.

The $scCO_2$ impregnated silk was, on the other hand, in sharp contrast with the control. We found a low percentage of wash leakage (3.0%—Figure 6B) and fastness leakage (2.8%—Figure 6C), thus a small amount of weakly attached gold NPs to silk fibers. The substantial difference in the total efficiency of $scCO_2$ and control-treated silk fibers stressed the effectiveness of the former over the latter (Figure 6D).

3.3.2. Effect of Initial Gold NPs Concentration and Time of Impregnation

An essential set of parameters is gold NP concentration and time. Figure 7 summarizes our three FTIR-ATR markers and the total impregnation efficiency. Figure 7B–D shows that the first point at t = 0 min represents native degummed silk.

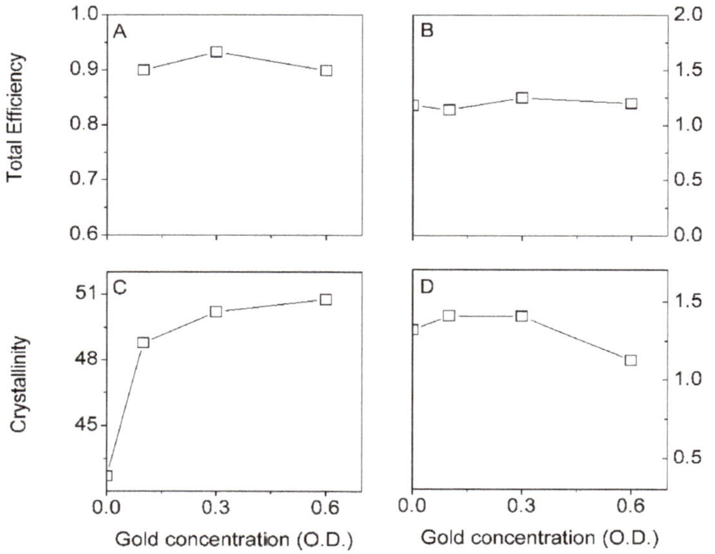

Figure 7. Effect of increasing gold concentration (in optical density O.D.) on total efficiency (**A**), Amide I/II ratio (**B**), Crystallinity (**C**), and Tyrosine ratio (**D**). Interestingly, even though we observed no changes in the amide I/II ratio, we found that the crystallinity and tyrosine ratio were affected. The O.D. concentrations correspond to approximately 6.9, 20.7 and 41.3 μg/mL for the 0.1, 0.3 and 0.6 O.D.

We found that the total efficiency and Amide I/II ratio were independent of the gold NP concentration. On the other hand, the silk crystallinity increased with concentration, whereas the tyrosine ratio decreased. It is important to note that the crystallinity and tyrosine ratio's positive and negative changes would result in a zero net change in structure, as shown in the amide I/II ratio plot (Figure 7B). The result suggested that some amorphous silk was converted in [-sheet structures (crystallinity)] the typical Silk I to Silk II conversion with increasing gold NP concentration. The existing β-sheets structures were becoming larger through interchain crystallization.

The contribution from the gold NPs to the FTIR-ATR spectra was measured to be at around 1734, 1599, 1448, and 1245 cm^{-1} (data not shown). However, we did not observe significant peaks in those regions, suggesting that silk most likely covered the gold signal.

A photograph (Figure 8) illustrates the sharp color change with increasing concentration.

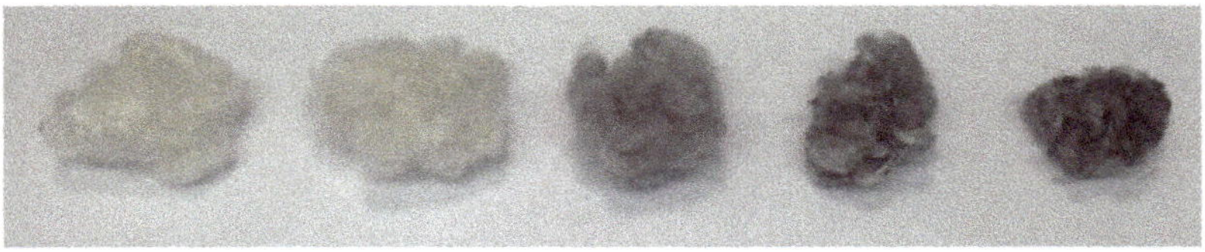

Figure 8. Photograph of impregnated silks, from **left** to **right**: native silk, control impregnation (no $scCO_2$), $scCO_2$ impregnation 0.1, 0.3, and 0.6 O.D, respectively.

Figure 9 illustrates the effect of $scCO_2$ impregnation time. Similarly to the gold NP concentration, we found that the efficiency was constant at around 90%. Interestingly, the amide I/II ratio decreased sharply with increasing times, whereas the crystallinity appeared steady, and the tyrosine ratio showed a slight decrease with time.

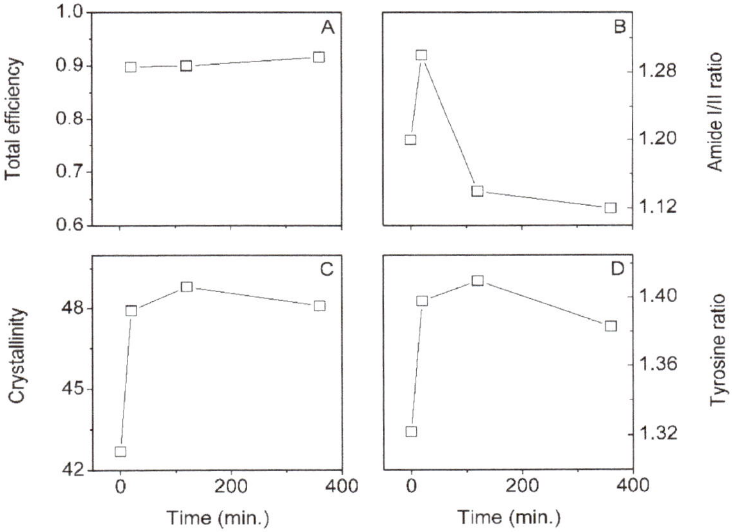

Figure 9. Effect of increasing $scCO_2$ impregnation time on total efficiency (**A**), Amide I/II ratio (**B**), Crystallinity (**C**), and Tyrosine ratio (**D**). The gold NP concentration was kept constant at 0.1 O.D.

Figure S10 shows the SEM images of a native degummed and $scCO_2$ impregnated silk (panel A and C, respectively). EDX spectra of selected features on the silk surface showed no gold for the control and traces of gold for the $scCO_2$ (panel B and D, respectively).

Taken together, the low amount of gold NPs on the surface and no gold NPs in the solution, we could suggest that most of the gold NPs were inside the silk, as expected from the efficiencies. A more precise quantification was obtained by XRD (See Figure 10) with the characteristic gold peak at 38° (note that the other peaks at 65° and 78° were barely visible for the highest gold NP concentration used). The distinct silk peaks [48] at 25°, 40°, and 42° appeared unchanged with increasing gold NPs concentrations.

The detailed peak deconvolution of the amide I (see Supplementary Materials Figure S11) showed that the β-sheets structures were constant with gold concentration and impregnation time. We observed, however, a conversion from the random coil/α-helical structures to β-turns with increasing gold concentration and time.

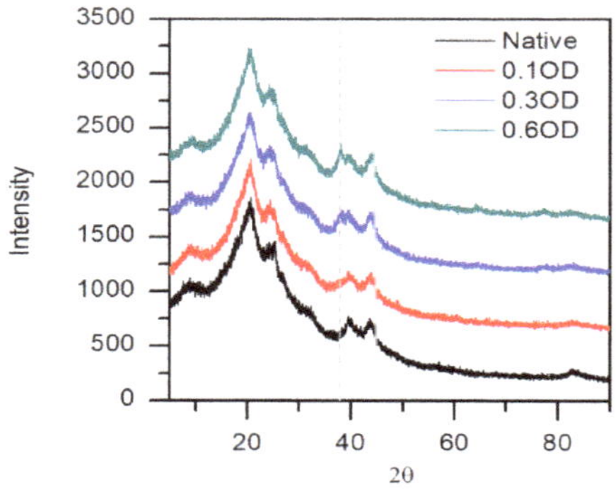

Figure 10. Intensities versus diffraction angle (2θ) as derived by our X-ray diffraction for silk fibers with different amounts of gold. For three times and six times gold addition, the gold peaks were at 2θ = 38°, 65° and 78°. No gold was detected for one-time gold addition. The silk characteristic peaks were at 20°, 25°, 40° and 42°. The grey dotted line indicates the gold reflection at 38°.

Throughout the analysis so far, we found that changes happened in the amorphous region of the silk. A complementary analysis of the interactions within the silk structures was the hydrogen bond index (HBI) from N-H vibration (ratio of intensities at 3320 to 3400 cm^{-1}). Alternatively, one could use the carbonyl signal C=O between 1600 to 1700 cm^{-1}. The presence of other strongly overlapping bands in the C=O region precluded a correct estimation of the HBI.

The results for the HBI of the three samples (Figure 11) suggested that the hydrogen bond index increased relative to the native silk but was not different from the two $scCO_2$ treated silks. Overall, the structural effects observed were predominantly coming from $scCO_2$ treatment.

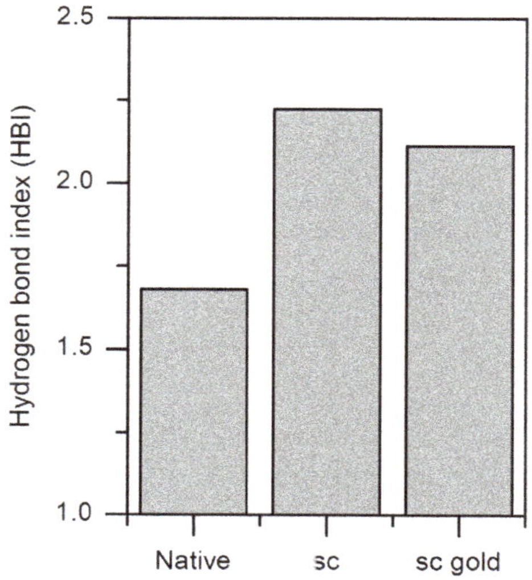

Figure 11. Change in Hydrogen bond index for N-H vibration for native silk, $scCO_2$ only treated silk, and $scCO_2$ plus gold NPs impregnated silk.

To help understand further the effect of the $scCO_2$ process parameters (temperature, time, pressure, and mixing), we used a factorial design.

3.4. Factorial Design—Relationships between Process Parameters and Multiple Responses

In the factorial design table (see Table 1), we combined the various factors of interest at different levels. We explored our four typical responses: total efficiency, crystallinity, Amide I/II ratio, and tyrosine ratio.

We found for the total efficiency that only pressure and time contributed significantly (significance level $\alpha = 0.1$) to explain the total efficiency variance. We also noted that both pressure and time contributed equally. For the Amide I/II ratio, we found that pressure and time were significant, except that pressure had a more substantial effect than time. For the crystallinity, only the temperature contributed significantly to the effect. Eventually, only the time had a more significant effect on the tyrosine ratio, and the pressure was marginally significant (see Supplementary Materials Figure S12).

One advantage of the factorial design is that one can seek the optimal conditions for a set of parameters. Optimally, we wished to maximize efficiency while maintaining the crystallinity at the lowest level possible. We found an optimal set of conditions: 35 °C for temperature, 250 bar for pressure, 1 h for the time, and 300 rpm for the mixing speed.

4. Discussion

Mechanism of Impregnation and Gold Nanoparticles Location

The permeation of $scCO_2$ into a polymer causes it to swell. Aided by its zero-surface tension, the addition of $scCO_2$ into the polymer phase gives the chains higher mobility. The CO_2 molecules act as lubricants, which reduce chain-chain interactions by increasing the polymer's inter-chain distance and free volume [53–56], also known as plasticization. The physical properties of the polymer are changed dramatically, including the depression of the glass transition temperature (Tg), the lowering of interfacial tension and a reduction of the viscosity of the polymer melt. $ScCO_2$ may increase the crystallinity of the polymers because the polymer chains are freer to align themselves with a more favorable order [57]. The above phenomena describe well the impregnation of soluble molecules into polymers. In the case of NPs, however, little is known.

Nevertheless, the studied silks could host gold NPs with sizes up to 150 nm using SCO_2. However, it is unclear as to the location specificity of the gold NPs with size. For example, would the smaller NPs be preferentially located more in-depth, and, as the size increases, the NPs would be nearer the silk surface.

The details of the degummed Mulberry silk fiber impregnation mechanism suggested that the gold NPs were limited to the silk inter-fibrillar space and, more specifically, around the amorphous regions. Figure 12 illustrates our findings.

Further research on ternary systems comprising "CO_2 + Nonsoluble NPs + fibers (solid substrates)" is required for a detailed understanding of mass transfer and diffusion in the substrate and of influences on the properties of the bulk material as crystallinity, morphology, anisotropy, and reactivity [58].

Collectively our results lead to the hypothesis that, under the supercritical conditions used in our experiment, the transport of the NPs would happen due to a gradient in surface tension. Park et al. [59] found, for example, that polystyrene (a polymer close to silks) surface tension decreases with CO_2 increased solubility in the polymer (i.e., higher pressures and higher temperatures). They also found that the polymer surface tension was independent of the polymer conformational entropy; in other words, its internal organization.

One could envision a transport phenomenon akin to the Marangoni flow [60]. The nature of the gradient remains unclear.

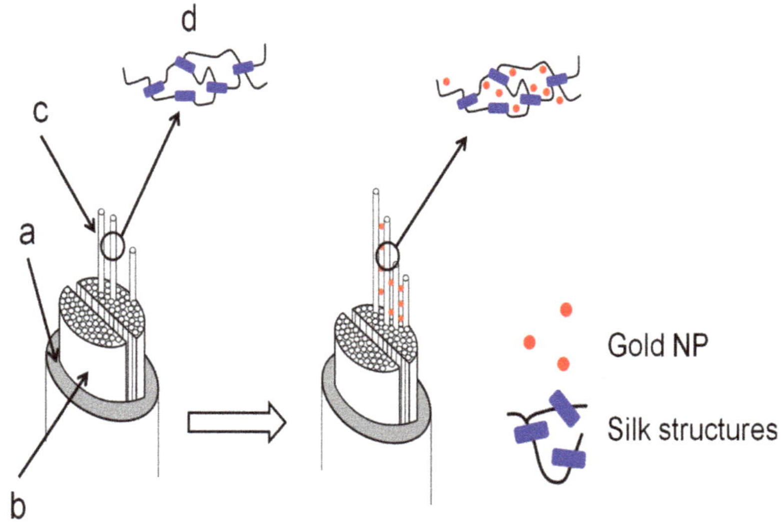

Figure 12. Gold NPs impregnation in silk. The typical silk fiber consists of (**a**) sericin coating, (**b**) fibroin brins, (**c**) fibrils, and (**d**) the secondary structure (idealized here the β-sheet as blue rectangles and the α-helices, random coils and turns as curvy black lines). After impregnation, the gold NPs were found in the inter-fibrillar spaces.

5. Conclusions

In conclusion, we demonstrated the permanent loading of gold NPs in four types of silk (Bombyx, Eri, Tasar, and Muga) using green and scalable technology like $scCO_2$. This work addressed the existing problem of uncontrolled leakage of the loaded particles from fibers, which is a significant concern for the environment and a hurdle in nanotechnology applications for synthetic and natural fibers' functionalization.

Further, we reported space availability in silk fibers by $scCO_2$ assisted impregnation at low temperatures. The four silks (Mulberry, Eri, Muga, and Tasar) displayed a remarkable capacity for the size of gold NPs (up to 150 nm). The detailed study of the impregnation mechanism in degummed mulberry silk fiber suggested a narrow window of process parameters with no detrimental effect on the fiber. The mechanism of impregnation of NPs into a solid fiber substrate is yet to be resolved; we hypothesized that the transport of the NPs was possible because of a surface tension gradient at the liquid-solid interface.

This developed approach is scalable, environmentally friendly. The results could help predict the application of natural fiber loaded with NPs as catalysts, self-cleaning, antimicrobial materials, UV-protective agents, and other valuable properties.

Supplementary Materials: The following are available online at https://www.mdpi.com/article/10.3390/insects13010018/s1, Figure S1: wash procedure and leakage calculation. Figure S2: Typical FTIR-ATR spectra of Mulberry, eri, muga, and tasar silks., Figure S3: Crystallinity index. The spectrum is truncated between 1180 and 1300 cm^{-1}, and a linear baseline is subtracted. The crystallinity index is computed using the intensities of the peaks at 1263 and 1230 cm^{-1} (red stars). Figure S4: Tyrosine ratio estimate. The spectra are truncated between 790 and 870 cm^{-1}, and a linear baseline is subtracted. The ratio is computed from the peak maxima at around 830 and 850 cm^{-1} (red stars in the figure). Figure S5: Secondary structure decomposition and analysis. Typical peak deconvolution of the Amide I and Amide II (top panel) and associated residuals (bottom panel). Briefly, a series of Gaussian peaks are fitted simultaneously to the Amide I and II. The initial position of the peaks is determined by secondary derivative analysis [37]. The final number of peaks is determined using an F-test to compare different models. The Amide I/II ratio is computed directly from the two main peaks in the upper panel. Figure S6: Comparison of total efficiency of 5 nm gold NP impregnation at 0.1 and 0.3 O.D. Figure S7: FTIR-ATR changes in secondary structure content as a function of Au NP size for Mulberry (A), Eri (B), Muga (C), and Tasar (D) silks. Black squares (□) are inter β-sheets, red circles (O) are β-turns, blue triangles (Δ) are α-helices and random coils, inverted green triangles (∇) are intra β-sheets. Figure S8: FTIR-ATR change in crystallinity for Mulberry (A), Eri (B), Muga (C), and Tasar (D) silks. Black squares (□) are $scCO_2$ samples, red circles (O) are control samples. The horizontal line is the crystallinity in native silk. We found that for $scCO_2$ treatment alone, the crystallinity was not significantly different from the native silks. Figure S9: FTIR-ATR changes in tyrosine ratio for Mulberry (A), Eri (B), Muga (C), and Tasar (D) silks. Black squares (□) are $scCO_2$ samples, red circles (O) are control samples. The horizontal line is the tyrosine ratio in native silk. We found that for $scCO_2$ treatment alone, the tyrosine ratio was not significantly different from the native silks. Figure S10: SEM images of native (A, scale bar 5 μ) and impregnated (C, scale bar 10 μ) silk fibers. The white squares represent the areas where the EDX spectra were collected. EDX spectra of native (B) and impregnated (C) silk fibers. The spectra showed only weak gold signals at 2.5, 7, 9.5, 10.5, 11.5, and 13.5 keV (D). Figure S11: FTIR-ATR changes in secondary structure content as a function of Au NP concentration. Black squares (□) are inter β-sheets, red circles (O) are β-turns, blue triangles (Δ) are α-helices and random coils, inverted green triangles (∇) are intra β-sheets. Figure S12: Pareto charts of the standardized effect from the full factorial analysis. The red dotted line is the standardized value above which a factor or combination of factors is considered significant (α level = 0.1). Table S1: Gold nanoparticles concentration and mass. Table S2. Consensus assignment for silk secondary structures determination by FTIR.

Author Contributions: Conceptualization, M.S., E.S.D., S.B. and C.D.; methodology, M.S.; validation, M.S., E.S.D. and C.D.; investigation, M.S.; data curation, C.D.; writing—original draft preparation, M.S.; writing—review and editing, M.S., E.S.D., S.B. and C.D. All authors have read and agreed to the published version of the manuscript.

Funding: The Swedish Science Research Council funded this research under grant number 348-2013-6218.

Institutional Review Board Statement: Not applicable.

Informed Consent Statement: Not applicable.

Data Availability Statement: The data presented in this study are available in the manuscript.

Acknowledgments: The authors acknowledge the Swedish Science Research Council for support (VR-grant 348-2013-6218).

Conflicts of Interest: The authors declare no conflict of interest.

References

1. El-Shishtawy, R.M.; Asiri, A.M.; Abdelwahed, N.A.M.; Al-Otaibi, M.M. In situ production of silver nanoparticle on cotton fabric and its antimicrobial evaluation. *Cellulose* **2011**, *18*, 75–82. [CrossRef]
2. Alongi, J.; Malucelli, G. Cotton flame retardancy: State of the art and future perspectives. *RSC Adv.* **2015**, *5*, 24239–24263. [CrossRef]
3. Onar, N.; Aksit, A.C.; Sen, Y.; Mutlu, M. Antimicrobial, UV-protective and self-cleaning properties of cotton fabrics coated by dip-coating and solvothermal coating methods. *Fiber Polym.* **2011**, *12*, 461–470. [CrossRef]
4. Alongi, J.; Carosio, F.; Malucelli, G. Current emerging techniques to impart flame retardancy to fabrics: An overview. *Polym. Degrad. Stab.* **2014**, *106*, 138–149. [CrossRef]
5. Dong, B.H.; Hinestroza, J.P. Metal nanoparticles on natural cellulose fibers: Electrostatic assembly and in situ synthesis. *ACS Appl. Mater. Interfaces* **2009**, *1*, 797–803. [CrossRef]
6. Li, G.; Liu, H.; Zhao, H.; Gao, Y.; Wang, J.; Jiang, H.; Boughton, R.I. Chemical assembly of TiO_2 and TiO_2@Ag nanoparticles on silk fiber to produce multifunctional fabrics. *J. Colloid Interface Sci.* **2011**, *358*, 307–315. [CrossRef]
7. Shateri-Khalilabad, M.; Yazdanshenas, M.E.; Etemadifar, A. Fabricating multifunctional silver nanoparticles-coated cotton fabric. *Arab. J. Chem.* **2017**, *10*, S2355–S2362. [CrossRef]
8. Tang, B.; Sun, L.; Kaur, J.; Yu, Y.; Wang, X. In-situ synthesis of gold nanoparticles for multifunctionalization of silk fabrics. *Dye. Pigm.* **2014**, *103*, 183–190. [CrossRef]
9. Tang, B.; Li, J.; Hou, X.; Afrin, T.; Sun, L.; Wang, X. Colorful and antibacterial silk fiber from anisotropic silver nanoparticles. *Ind. Eng. Chem. Res.* **2013**, *52*, 4556–4563. [CrossRef]
10. Grand, J.; Ferreira, S.R.; de Waele, V.; Mintova, S.; Nenoff, T.M. Nanoparticle alloy formation by radiolysis. *J. Phys. Chem. C* **2018**, *122*, 12573–12588. [CrossRef]

11. Park, J.-E.; Atobe, M.; Fuchigami, T. Sonochemical synthesis of conducting polymer–metal nanoparticles nanocomposite. *Electrochim. Acta* **2005**, *51*, 849–854. [CrossRef]
12. Mitrano, D.M.; Rimmele, E.; Wichser, A.; Erni, R.; Height, M.; Nowack, B. Presence of nanoparticles in wash water from conventional silver and nano-silver textiles. *ACS Nano* **2014**, *8*, 7208–7219. [CrossRef]
13. Hristozov, D.; Malsch, I. Hazards and risks of engineered nanoparticles for the environment and human health. *Sustainability* **2009**, *1*, 1161–1194. [CrossRef]
14. Rastogi, A.; Zivcak, M.; Sytar, O.; Kalaji, H.M.; He, X.; Mbarki, S.; Brestic, M. Impact of metal and metal oxide nanoparticles on plant: A critical review. *Front. Chem.* **2017**, *5*, 78. [CrossRef] [PubMed]
15. Wakayama, H.; Fukushima, Y. Supercritical CO_2 for making nanoscale materials. *Ind. Eng. Chem. Res.* **2006**, *45*, 3328–3331. [CrossRef]
16. Zhang, X.; Heinonen, S.; Levänen, E. Applications of supercritical carbon dioxide in materials processing and synthesis. *RSC Adv.* **2014**, *4*, 61137–61152. [CrossRef]
17. Cansell, F.; Chevalier, B.; Demourgues, A.; Etourneau, J.; Even, C.; Pessey, V.; Petit, S.; Tressaud, A.; Weill, F. Supercritical fluid processing: A new route for materials synthesis. *J. Mater. Chem.* **1999**, *9*, 67–75. [CrossRef]
18. Xu, Y.; Musumeci, V.; Aymonier, C. Chemistry in supercritical fluids for the synthesis of metal nanomaterials. *React. Chem. Eng.* **2019**, *4*, 2030–2054. [CrossRef]
19. Lane, M.K.M.; Zimmerman, J.B. Controlling metal oxide nanoparticle size and shape with supercritical fluid synthesis. *Green Chem.* **2019**, *21*, 3769–3781. [CrossRef]
20. Chih, Y.-W.; Cheng, W.-T. Supercritical carbon dioxide-assisted synthesis of silver nanoparticles in polyol process. *Mater. Sci. Eng. B* **2007**, *145*, 67–75. [CrossRef]
21. Parandhaman, T.; Dey, M.D.; Das, S.K. Biofabrication of supported metal nanoparticles: Exploring the bioinspiration strategy to mitigate the environmental challenges. *Green Chem.* **2019**, *21*, 5469–5500. [CrossRef]
22. Hua, M.; Hua, X. polymer nanoparticles prepared by supercritical carbon dioxide for in vivo anti-cancer drug delivery. *Nano-Micro Lett.* **2014**, *6*, 20–23. [CrossRef]
23. Wang, Y.; Dave, R.N.; Pfeffer, R. Polymer coating/encapsulation of nanoparticles using a supercritical anti-solvent process. *J. Supercrit. Fluids* **2004**, *28*, 85–99. [CrossRef]
24. Liu, Z.-T.; Zhang, L.; Liu, Z.; Gao, Z.; Dong, W.; Xiong, H.; Peng, Y.; Tang, S. Supercritical CO_2 dyeing of ramie fiber with disperse dye. *Ind. Eng. Chem. Res.* **2006**, *45*, 8932–8938. [CrossRef]
25. van der Kraan, M.; Fernandez Cid, M.V.; Woerlee, G.F.; Veugelers, W.J.T.; Witkamp, G.J. Dyeing of natural and synthetic textiles in supercritical carbon dioxide with disperse reactive dyes. *J. Supercrit. Fluids* **2007**, *40*, 470–476. [CrossRef]
26. Guney, O.; Akgerman, A. Synthesis of controlled-release products in supercritical medium. *AIChE J.* **2002**, *48*, 856–866. [CrossRef]
27. Fernandez Cid, M.V.; van Spronsen, J.; van der Kraan, M.; Veugelers, W.J.T.; Woerlee, G.F.; Witkamp, G.J. Excellent dye fixation on cotton dyed in supercritical carbon dioxide using fluorotriazine reactive dyes. *Green Chem.* **2005**, *7*, 609–616. [CrossRef]
28. Singh, M.; Bollella, P.; Gorton, L.; Dey, E.; Dicko, C. Conductive and enzyme-like silk fibers for soft sensing application. *Biosens. Bioelectron.* **2019**, *150*, 111859. [CrossRef]
29. Cooper, A.I. Polymer synthesis and processing using supercritical carbon dioxide. *J. Mater. Chem.* **2000**, *10*, 207–234. [CrossRef]
30. Jiang, N.; Wang, Y.; Li, D.; Niu, J.; Wang, S.; Chen, A. Carbon-doped metal oxide nanoparticles prepared from metal nitrates in supercritical CO_2-enabled polymer nanoreactors. *Part. Part. Syst. Charact.* **2019**, *36*, 1900016. [CrossRef]
31. Yu, Q.; Wu, P.; Xu, P.; Li, L.; Liu, T.; Zhao, L. Synthesis of cellulose/titanium dioxide hybrids in supercritical carbon dioxide. *Green Chem.* **2008**, *10*, 1061–1067. [CrossRef]
32. Gittard, S.D.; Hojo, D.; Hyde, G.K.; Scarel, G.; Narayan, R.J.; Parsons, G.N. Antifungal textiles formed using silver deposition in supercritical carbon dioxide. *J. Mater. Eng. Perform.* **2009**, *19*, 368–373. [CrossRef]
33. Ding, X.; Yu, M.; Wang, Z.; Zhang, B.; Li, L.; Li, J. A promising clean way to textile colouration: Cotton fabric covalently-bonded with carbon black, cobalt blue, cobalt green, and iron oxide red nanoparticles. *Green Chem.* **2019**, *21*, 6611–6621. [CrossRef]
34. Watkins, J.J.; McCarthy, T.J. Polymer/metal nanocomposite synthesis in supercritical CO_2. *Chem. Mater.* **1995**, *7*, 1991–1994. [CrossRef]
35. Zhang, Y.; Kang, D.; Saquing, C.; Aindow, M.; Erkey, C. Supported Platinum Nanoparticles by Supercritical Deposition. *Ind. Eng. Chem. Res.* **2005**, *44*, 4161–4164. [CrossRef]
36. Ye, X.-R.; Lin, Y.; Wang, C.; Engelhard, M.H.; Wang, Y.; Wai, C.M. Supercritical fluid synthesis and characterization of catalytic metal nanoparticles on carbon nanotubes. *J. Mater. Chem.* **2004**, *14*, 908–913. [CrossRef]
37. Hasell, T.; Thurecht, K.J.; Jones, R.D.W.; Brown, P.D.; Howdle, S.M. Novel one pot synthesis of silver nanoparticle–polymer composites by supercritical CO_2 polymerization in the presence of a RAFT agent. *Chem. Commun.* **2007**, *38*, 3933–3935. [CrossRef]
38. Wong, B.; Yoda, S.; Howdle, S.M. The preparation of gold nanoparticle composites using supercritical carbon dioxide. *J. Supercrit. Fluids* **2007**, *42*, 282–287. [CrossRef]
39. Fan, Y.; Zhang, Y.-Q.; Yan, K.; Long, J.-J. Synthesis of a novel disperse reactive dye involving a versatile bridge group for the sustainable coloration of natural fibers in supercritical Carbon Dioxide. *Adv. Sci.* **2019**, *6*, 1801368. [CrossRef] [PubMed]
40. Peng, Q.; Xu, Q.; Sun, D.; Shao, Z. Grafting of methyl methacrylate onto Antheraea pernyi silk fiber with the assistance of supercritical CO_2. *J. Appl. Polym. Sci.* **2006**, *100*, 1299–1305. [CrossRef]

41. Peng, L.; Guo, R.; Lan, J.; Jiang, S.; Wang, X.; Lin, S.; Li, C. Silver nanoparticles coating on silk fabric with pretreatment of 3-aminopropyltrimethoxysilane in supercritical carbon dioxide. *J. Ind. Text.* **2016**, *47*, 883–896. [CrossRef]
42. Marin, M.A.; Mallepally, R.R.; McHugh, M.A. Silk fibroin aerogels for drug delivery applications. *J. Supercrit. Fluids* **2014**, *91*, 84–89. [CrossRef]
43. Bhat, N.V.; Nadiger, G.S. Crystallinity in silk fibers: Partial acid hydrolysis and related studies. *J. Supercrit. Fluids* **1980**, *25*, 921–932. [CrossRef]
44. Monti, P.; Freddi, G.; Bertoluzza, A.; Kasai, N.; Tsukada, M. Raman spectroscopic studies of silk fibroin from bombyx mori. *J. Raman Spectrosc.* **1998**, *29*, 297–304. [CrossRef]
45. Boulet-Audet, M.; Vollrath, F.; Holland, C. Identification and classification of silks using infrared spectroscopy. *J. Exp. Biol.* **2015**, *218*, 3138–3149. [CrossRef]
46. Ishida, K.P.; Griffiths, P.R. Comparison of the amide I/II intensity ratio of solution and solid-state proteins samples by transmission, attenuated total reflectance, and diffuse reflectance spectrometry. *Appl. Spectrosc.* **1993**, *47*, 584–589. [CrossRef]
47. Chen, Z.; Huang, W.; Fang, P.F.; Yu, W.; Wang, S.J.; Xiong, J.; Xu, Y.S. The hydrogen bond and free volume property of poly(ether-urethane) irradiated by neutron. *J. Polym. Sci. B Polym. Phys.* **2010**, *48*, 381–388. [CrossRef]
48. Sen, K.; Babu, K.M. Studies on Indian silk. II. Structure–property correlations. *J. Appl. Polym. Sci.* **2004**, *92*, 1098–1115. [CrossRef]
49. Sen, K.; Babu, M.K.; Babu, K.M. Studies on Indian silk. III. Effect of structure on dyeing behavior. *J. Appl. Polym. Sci.* **2004**, *92*, 1116–1123. [CrossRef]
50. Sen, K.; Babu, K.M. Studies on Indian silk. I. macrocharacterization and analysis of amino acid composition. *Polymer* **2003**, *92*, 1080–1097. [CrossRef]
51. Guan, J.; Wang, Y.; Mortimer, B.; Holland, C.; Shao, Z.; Porter, D.; Vollrath, F. Glass transitions in native silk fibres studied by dynamic mechanical thermal analysis. *Soft Matter* **2016**, *12*, 5926–5936. [CrossRef] [PubMed]
52. Mazzi, S.; Zulker, E.; Buchicchio, J.; Anderson, B.; Hu, X. Comparative thermal analysis of Eri, Mori, Muga, and Tussar silk cocoons and fibroin fibers. *J. Therm. Anal. Calorim.* **2014**, *116*, 1337–1343. [CrossRef]
53. Aymonier, C.; Loppinet-Serani, A.; Reverón, H.; Garrabos, Y.; Cansell, F. Review of supercritical fluids in inorganic materials science. *J. Supercrit. Fluids* **2006**, *38*, 242–251. [CrossRef]
54. Brunner, G.J. Application of Supercritical Fluids. *Annu. Rev. Chem. Biomol. Enginering* **2010**, *1*, 321–342. [CrossRef] [PubMed]
55. Kikic, I.; Vecchione, F. Supercritical impregnation of polymers. *Curr. Opin. Solid State Mater. Sci.* **2003**, *7*, 399–405. [CrossRef]
56. Zhao, X.; Lv, L.; Pan, B.; Zhang, W.; Zhang, S.; Zhang, Q. Polymer-supported nanocomposites for environmental application: A review. *Chem. Eng. J.* **2011**, *170*, 381–394. [CrossRef]
57. Kazarian, S.G. Polymer processing with supercritical fluids. *Polym. Sci.* **2000**, *42*, 78–101.
58. Weidner, E. Impregnation via supercritical CO_2—What we know and what we need to know. *J. Supercrit. Fluids* **2018**, *134*, 220–227. [CrossRef]
59. Park, H.; Thompson, R.B.; Lanson, N.; Tzoganakis, C.; Park, C.B.; Chen, P. Effect of temperature and pressure on surface tension of polystyrene in supercritical carbon dioxide. *J. Phys. Chem. B* **2007**, *111*, 3859–3868. [CrossRef]
60. Rengasamy, R.S. Wetting phenomena in fibrous materials. In *Thermal and Moisture Transport in Fibrous Materials*; Pan, N., Gibson, P., Eds.; Woodhead: Melbourne, Australia, 2006.

insects

MDPI

Article

Hemolymph Ecdysteroid Titer Affects Maternal mRNAs during *Bombyx mori* Oogenesis

Meirong Zhang [1,2,*], Pingzhen Xu [1,2] and Tao Chen [1,2,*]

[1] Jiangsu Key Laboratory of Sericultural Biology and Biotechnology, School of Biotechnology, Jiangsu University of Science and Technology, Zhenjiang 212028, China; xpz198249@just.edu.cn

[2] Key Laboratory of Silkworm and Mulberry Genetic Improvement, Ministry of Agriculture, Sericultural Research Institute, Chinese Academy of Agricultural Sciences, Zhenjiang 212028, China

* Correspondence: zhangmr@just.edu.cn (M.Z.); 199000001909@just.edu.cn (T.C.)

Simple Summary: Both maternal genes and ecdysteroids play important roles during embryonic development. In this study, we aimed to characterize the dynamic landscape of maternal mRNAs and the relationship between maternal genes and ecdysteroids during silkworm oogenesis. For the first time, we determined the start of the accumulation of maternal mRNAs in the ovary at the wandering stage during the larval period. We detected the developmental expression profiles of each gene in the ovary or ovariole. We finally confirmed the role of 20-hydroxyecdysone in regulating maternal gene expression. Taken together, our findings expand the understanding of insect oogenesis and provide a perspective on the embryonic development of the silkworm.

Abstract: Silkworm larval–pupal metamorphosis and the first half of pupal–adult development occur during oogenesis from previtellogenesis to vitellogenesis and include two peaks of the hemolymph ecdysteroid titer. Moreover, a rise in 20-hydroxyecdysone titer in early pupae can trigger the first major transition from previtellogenesis to vitellogenesis in silkworm oogenesis. In this study, we first investigated the expression patterns of 66 maternal genes in the ovary at the wandering stage. We then examined the developmental expression profiles in six time-series samples of ovaries or ovarioles by reverse transcription–quantitative PCR. We found that the transcripts of 22 maternal genes were regulated by 20-hydroxyecdysone in the isolated abdomens of the pupae following a single injection of 20-hydroxyecdysone. This study is the first to determine the relationship between 20-hydroxyecdysone and maternal genes during silkworm oogenesis. These findings provide a basis for further research into the embryonic development of *Bombyx mori*.

Keywords: *Bombyx mori*; maternal mRNAs; 20-hydroxyecdysone; pupa; wandering

Citation: Zhang, M.; Xu, P.; Chen, T. Hemolymph Ecdysteroid Titer Affects Maternal mRNAs during *Bombyx mori* Oogenesis. *Insects* **2021**, *12*, 969. https://doi.org/10.3390/insects12110969

Academic Editors: Silvia Cappellozza, Morena Casartelli, Federica Sandrelli, Alessio Saviane and Gianluca Tettamanti

Received: 7 September 2021
Accepted: 25 October 2021
Published: 27 October 2021

1. Introduction

Early embryonic development occurs in the absence of de novo transcription and is maternally regulated [1,2]. Maternal mRNAs and proteins stored in the egg during oogenesis are activated to initiate and regulate embryonic development [2–4]. Maternal mRNAs are responsible for early embryogenesis by driving cellular division until zygotic genome activation [2]. The handoff of developmental control from the maternal to the zygotic genome is known as the maternal to zygotic transition (MZT). Most striking during the MZT is the elimination of maternal transcripts and the onset of transcription from the newly activated zygote's genome [2,5,6]. This transition has been well studied in model organisms, and many early developmental processes are highly conserved and are critical for organism survival [7]. RNA-binding proteins (RBPs) regulate the translation, stability, and localization of maternal mRNAs [8]; Smaug (SMG) RBP is essential to maternal mRNAs degradation in *Drosophila melanogaster* [9]. Another type of RBP comprises the AU-rich element-binding proteins (ARE-BPs), which have a role in maternal mRNA clearance during the MZT in *Caenorhabditis elegans* [10], zebrafish (*Danio rerio*) [11], *Xenopus laevis* [12], and mice (*Mus musculus*) [13]. The transcription factors critical to the activation of the zygotic genome following maternal mRNA degradation have been identified in *Drosophila* (Zelda) [14], zebrafish (Nanog, Pou5f1 and SoxB1) [15], mice (Dux, Nfy, Dppa2, and Dppa4) [16–18], and humans (OCT4 and DUX4) [16,19,20].

The domestic silkworm, *Bombyx mori* (*B. mori*), is not only economically prized for its silk production, but is also as a model of Lepidoptera [21,22]. Therefore, it is widely used in basic and applied research. The ovary structure of the silkworm is polytrophic meroistic. In silkworms, oogenesis can be subdivided into the 3 distinct developmental periods of previtellogenesis, vitellogenesis, and choriogenesis, while the development of the follicles is divided into 12 different stages according to various morphological criteria [23]. The oogenesis developmental period of previtellogenesis consists of stages 1–3 during the larval period. After larval–pupal ecdysis, the ovarioles emerge gradually via elongation into the abdominal cavity, which is part of the oogenesis developmental periods of vitellogenesis and choriogenesis [23].

Throughout the latter half of the fifth instar, a small rise in ecdysteroid titer occurs upon commencement of wandering, and this is followed by a plateau. The next day, the titer starts to increase gradually and then elevates steeply to form a peak 1 day later [24]. After this peak, the ecdysteroid titer decreases rapidly to a very low level, but then increases again after pupation up to a very high titer (forming the second peak) 2 days after pupation. The titer then decreases irregularly to reach a minimum at 6 days after pupation [24]. There is a strong parallel between the overall changes in follicular development and the ecdysteroid titers during larval–pupal and the first half of pupal–adult development. The increase in 20-hydroxyecdysone (20E) titer in the early pupa stage can trigger previtellogenic development and vitellogenesis in the silkworm [25–27]. The ecdysteroids in silkworm ovaries are synthesized in follicle cells and transferred into the oocyte where they are phosphorylated by EcKinase, the mRNA of which originates from nurse cells and the oocyte itself [28,29].

The silkworm ovariole has unique features in the context of physiological and biochemical studies, which provide an excellent model for studies on changes in gene expression [30–32]. We previously reported that 66 maternal genes have been successfully identified via orthologous comparison and expression detection, and their mRNAs form three clusters of degradation patterns during the MZT period in the silkworm [32,33]. Although the roles of maternal genes and ecdysteroids in insect embryonic development are well known, their relationships remain to be determined. Therefore, in the present study, we first investigated the expression characteristics of 66 maternal genes in the ovary of the wandering silkworm and then examined their developmental expression profiles in the ovaries or ovarioles of silkworms at stages 3 to 5 of oogenesis by reverse transcription–quantitative PCR (RT-qPCR). We found that the transcripts of 22 maternal genes were regulated in the isolated abdomens of the pupae by a single injection of 20E. Our data also demonstrate the start of the accumulation of maternal mRNAs during stage 3 of the oogenesis developmental period of previtellogenesis in the larva.

2. Materials and Methods

2.1. Experimental Animals and 20-Hydroxyecdysone

B. mori (Dazao) larvae were reared under standard conditions (25 °C and 70% humidity). The larvae–pupae, pupae, and ligation pupae were maintained under a 12 h light/12 h dark photoperiod at 25 °C and 70% humidity. 20-hydroxyecdysone (20E), an ecdysteroid (A506554), was purchased from Sangon Biotech Co., Ltd. (Shanghai, China) and was dissolved in anhydrous ethanol and diluted to 2 g L^{-1} using sterile distilled water.

2.2. Ligature and 20-Hydroxyecdysone Treatment

Immediately after pupation, female pupae of similar sizes were ligated between the thorax and the abdomen using cotton thread. The isolated abdomens of the pupae (20E-depleted abdomens) were treated with 5 μL 20E (10.0 μg) of solution or 5 μL of sterile distilled water (control) four days later.

2.3. Sample Preparation

The ovaries or ovarioles were collected from the wandering larvae, pre-pupae, day-0 pupae, day-1 pupae, day-2 pupae, and day-3 pupae and were injected with 20E at 24 h and 48 h after depletion of the abdomens. The dissection was conducted on ice for purposes of anaesthetization.

2.4. Reverse-Transcription PCR and Reverse Transcription–Quantitative PCR Analysis

Total RNA was extracted using TRIzol reagent (Invitrogen, Carlsbad, CA, USA). The total RNA concentrations were quantified, and a fraction of the RNA was treated with DNase. The RNA was used to synthesize first-strand cDNA using the PrimeScript™ RT Master Mix (Perfect Real Time; TaKaRa, Dalian, China) according to the manufacturer's instructions. The expression patterns of maternal genes in the ovaries of the wandering larvae were analyzed using reverse-transcription PCR (RT-PCR) and reverse-transcription–quantitative PCR (RT-qPCR). The ovaries or ovarioles were collected from the wandering larvae, pre-pupae, day-0 pupae, day-1 pupae, day-2 pupae, and day-3 pupae and were used to determine the developmental expression profiles of each gene via RT-qPCR. The ovarioles of pupae injected with 20E at 24 h and 48 h after depletion of the abdomens were used for maternal gene RT-qPCR analysis. The thermal cycling conditions of reverse-transcription PCR (RT-PCR) were 94 °C for 5 min, 35 cycles at 94 °C for 30 s, 58 °C for 30 s, 72 °C for 30 s, and a final extension at 72 °C for 10 min before storing at 12 °C. The RT-PCR products of each gene were separated by 1.2% agarose gel electrophoresis. RT-qPCR was carried out in an ABI PRISM® 7500 (Applied Biosystems, USA) using SYBR Green Supermix (TaKaRa, China). The thermal program of RT-qPCR consisted of an initial denaturation at 95 °C for 3 min, 40 cycles at 95 °C for 15 s, and 60 °C for 31 s, and melting from 60 °C to 95 °C. Translation initiation factor 4a (*TIF-4A*) was used as a reference gene [34]. The cycle threshold (Ct) values were converted to linear values using the comparative CT method [35]. Student's *t*-tests were used to analyze the data. The specific primers for each gene are shown in Table S1.

2.5. Kyoto Encyclopedia of Genes and Genomes Annotation

Kyoto Encyclopedia of Genes and Genomes (KEGG) annotation of the maternal genes was performed using the KEGG web service [36]. Protein sequences in fasta format were blasted in the database.

3. Results

3.1. Gene Expression Analysis of the Ovary at the Wandering Stage

We previously identified 66 maternal genes in *B. mori* [32,33]. At stage 3, a small rise in ecdysteroid titer occurs at the start of wandering, and this is followed by a plateau [24]; then, glycogen synthesis begins in the oocyte, and the follicle cells near the oocyte–nurse cells interface begin to migrate centripetally between the oocyte and the nurse cells [23]. The results of the transcriptional analysis of the 66 maternal genes in the ovary of the wandering silkworm showed transcriptional signals for all the genes (Figures 1 and S1). This is contrary to their low expression in the ovary of the day-3 fifth instar [32]. This result implies the start of the accumulation of maternal mRNAs during stage 3 of the oogenesis developmental period of previtellogenesis, which takes place in the larval period. KEGG ontology assignments of the 66 maternal genes were performed (Table S2). These genes were mainly found to be involved in genetic information processing and environmental information processing (Figure S2). The genes *eIF4AIII*, *Bin1*, *Pabn2*, *Csp(DnaJ-7)*, *wbl*, *me31B*, and *Mat89Ba* were involved in genetic information processing and were further analyzed. *eIF4AIII*, *Bin1*, and *Pabn2* were involved in the mRNA surveillance pathway, and *Csp(DnaJ-7)*, *wbl*, and *me31B* were involved in RNA degradation (Table S3).

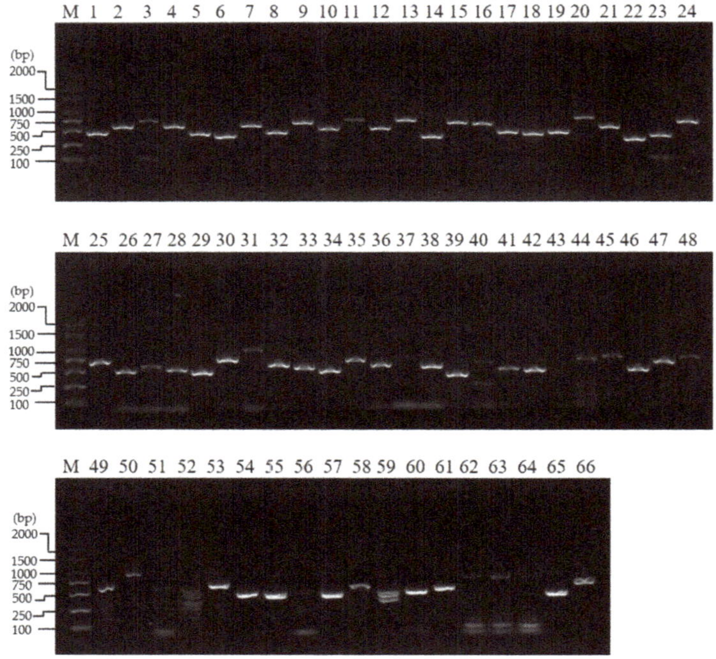

Figure 1. Expression patterns of maternal genes in the ovary of the wandering silkworm by reverse transcription (RT) PCR. M: DL2000 Plus DNA Maker; 1: *me31B*; 2: *lok*; 3: *vri*; 4: *Egfr*; 5: *Su(var) 205*; 6: *Hp1b-l*; 7: *spz*; 8: *tkv*; 9: *CycB*; 10: *proPPAE*; 11: *asp*; 12: *PAH*; 13: *aub*; 14: *Csp(DnaJ-7)*; 15: *SPE*; 16: *BAEE*; 17: *PPAE*; 18: *Sod2*; 19: *esc*; 20: *Src42A*; 21: *Smg*; 22: *Eif-4a*; 23: *eIF4AIII*; 24: *rod*; 25: *vfl*; 26: *bai*; 27: *Nelf-E*; 28: *Pabn2*; 29: *Bin1*; 30: *tud*; 31: *Moe*; 32: *Sel(cnpy1)*; 33: *Hip14(ZDHHC17)*; 34: *mamo*; 35: *sax*; 36: *babo*; 37: *h*; 38: *Chc*; 39: *Snap25*; 40: *SPE-like*; 41: *Src64B*; 42: *wbl*; 43: *Mat89Ba*; 44: *Dif*; 45: *ndl(osp)*; 46: *Nelf-A*; 47: *tld*; 48: *proSP7*; 49: *gammaTub*; 50: *Th*; 51: *pie*; 52: *gro*; 53: *hb*; 54: *pip*; 55: *spoon(AKAP1)*; 56: *snk*; 57: *Btk29A*; 58: *dpp*; 59: *Msp300(nesprin-1)*; 60: *KCNQ*; 61: *shot*; 62: *sog*; 63: *Pc*; 64: *Dst*; 65: *TPH1*; 66: *glo(hnRNPF)*.

3.2. Expression Profile of Each Gene during Stages 3–5 of Oogenesis

From stages 3 to 5 of silkworm oogenesis, the ovarioles emerged gradually via elongation into the abdominal cavity. The oocyte grew rapidly and was found to contain small protein yolk spheres from stage 4 (Figure S3). To identify the expression profile of each gene during stages 3–5, six time-series samples of ovary or ovariole were collected from the wandering larvae, pre-pupae, day-0 pupae, day-1 pupae, day-2 pupae, and day-3 pupae and were analyzed by RT-qPCR. The thresholds for the up- and downregulation of fold change (≥1.5 and ≤0.67) were found to be differentially reached in the ovarioles of the five other developmental stages in comparison with the wandering stage. In total, the temporal control of the expressional profiles gave rise to three different trends during stages 3–5 (Figures 2–4 and Table 1). The transcript levels of 18 maternal genes were increased (Figure 2 and Table 1). In contrast, the transcripts of nine genes were significantly decreased (Figure 3 and Table 1). Meanwhile, the transcript levels of 39 genes showed no change (Figure 4 and Table 1). A small rise in ecdysteroid titer occurs at the start of wandering, and this is followed by a plateau. The next day, the titer starts to increase gradually and elevates steeply to form a peak 1 day later [24]. After this peak, the ecdysteroid titer decreases rapidly to a very low level, but then increases again after pupation and reaches a very high titer (forming the second peak) 2 days after pupation [24]. Between the two peaks of the ecdysteroid titer, the silkworm gradually commences larval–pupal metamorphosis, and the organs of the silkworm are reconstructed; specifically, the oocyte grows rapidly. The developmental expression profiles of maternal genes may be closely related to ecdysteroid fluctuation during larval–pupal and the first half of pupal–adult development.

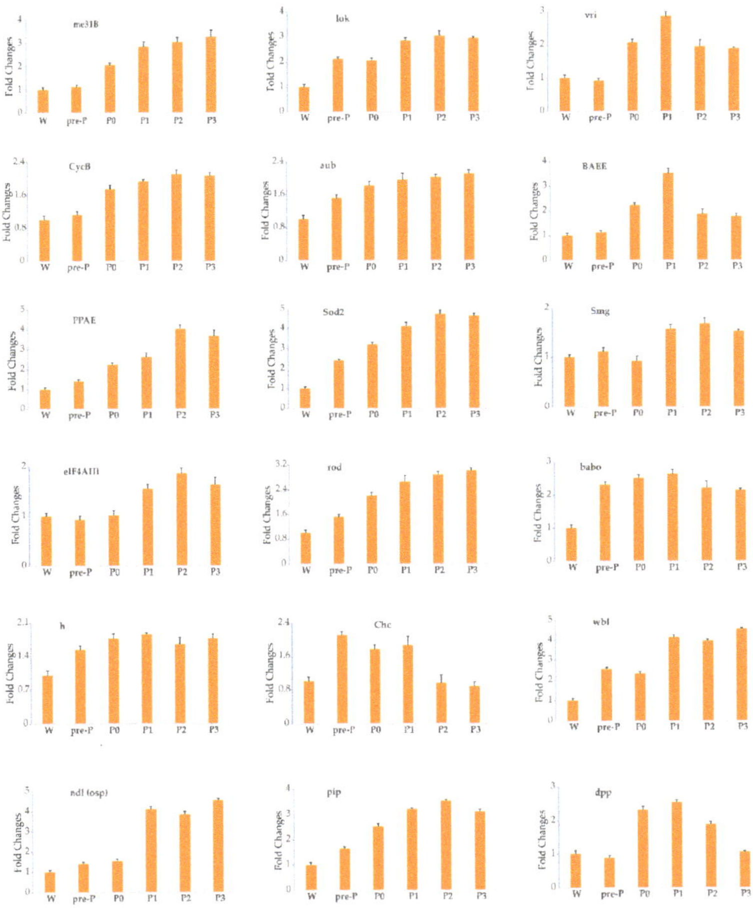

Figure 2. The expression profiles of 18 maternal genes trended upward, as determined by RT-qPCR during stages 3–5 of silkworm oogenesis. Six time-series samples of ovary or ovariole were collected. Each time point was assessed thrice. *TIF-4A* was used as the internal control, and the relative quantities for each gene at the wandering stage were set to 1. Bars indicate the standard deviation. W, wandering larvae; pre-P, pre-pupae; P0, P1, P2 and P3, day-0, day-1, day-2, and day-3 pupae.

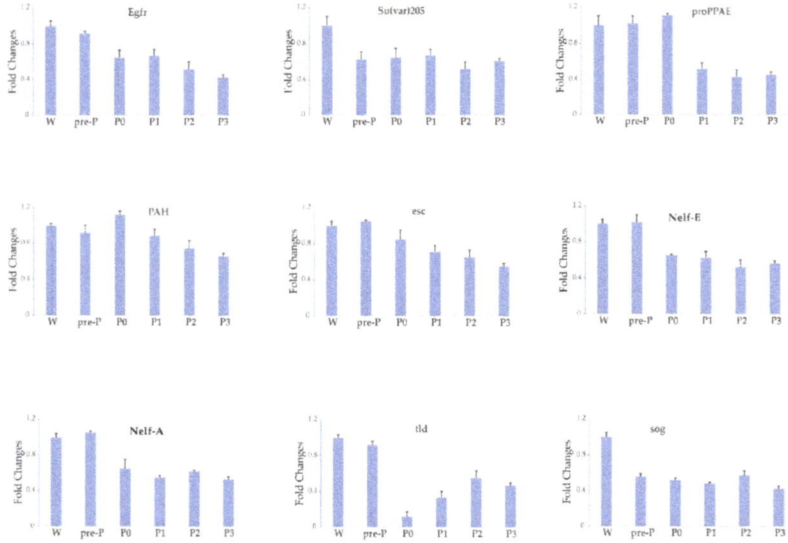

Figure 3. The expression profiles of nine maternal genes trended downward, as determined by RT-qPCR during stages 3–5 of silkworm oogenesis. Six time-series samples of ovaries or ovarioles were collected. Each time point was assessed thrice. *TIF-4A* was used as the internal control, and the relative quantities for each gene at the wandering stage were set to 1. Bars indicate the standard deviation. W, wandering larvae; pre-P, pre-pupae; P0, P1, P2 and P3, day-0, day-1, day-2, and day-3 pupae.

3.3. 20-Hydroxyecdysone Regulates the Expression of Select Genes

To confirm the regulatory function of 20E in the expression of maternal genes, an experiment in which exogenous 20E was injected into the 20E-depleted abdomens was performed. The level of ecdysteroid titer is very low in day-0 pupae [24]. When a day-0 pupa is ligated between the thorax and the abdomen, it is found that the abdomen contains no sources of the ecdysteroid that is synthesized and released by the prothoracic glands, which are entwined in pairs in the tracheal bush of the first spiracle in the thorax [37]. This allowed us to examine the effect of 20E. The injection of 20E into 20E-depleted abdomens initially caused significant morphological changes in the ovarian structure, and while the ovarioles developed normally, the development of the ovaries in control individuals was arrested (Figure S4). Our ligation experiments suggest that 20E plays a central role in silkworm oogenesis. We therefore examined the gene expression profiles after 24 and 48 h of 20E treatment. The transcripts of 17 genes (*me31B*, *lok*, *vri*, *CycB*, *aub*, *BAEE*, *PPAE*, *Sod2*, *Smg*, *eIF4AIII*, *rod*, *babo*, *Chc*, *wbl*, *ndl(osp)*, *pip*, and *tld*) were increased, while the transcripts of 5 genes (*proPPAE*, *PAH*, *esc*, *Nelf-E*, and *Nelf-A*) were decreased, and the transcripts of another 5 genes (*h*, *dpp*, *Egfr*, *Su(var)205*, and *sog*) showed no change (Figure 5). This indicates that the transcript levels of the maternal genes were regulated by 20E. They may have regulatory function during early silkworm oogenesis.

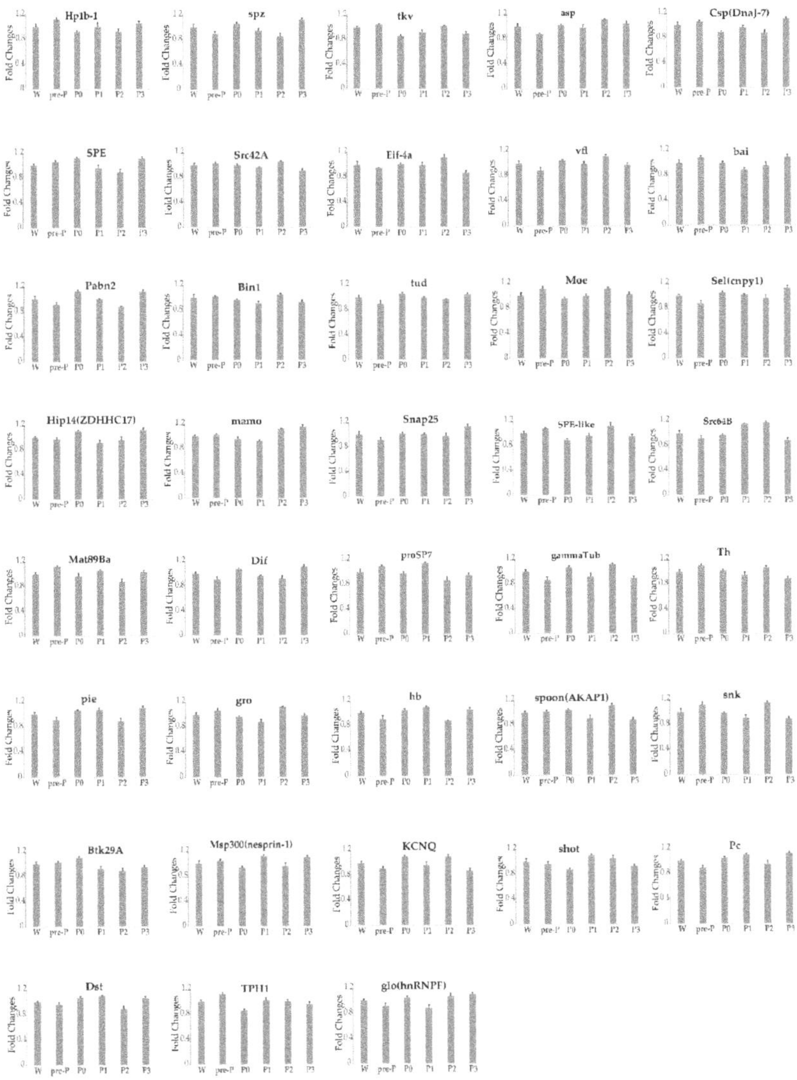

Figure 4. The expression profiles of 39 maternal genes showed a flattened trend, as determined by RT-qPCR during stages 3–5 of silkworm oogenesis. Six time-series samples of ovary or ovariole were collected. Each time point was assessed thrice. *TIF-4A* was used as the internal control, and the relative quantities for each gene at the wandering stage were set to 1. Bars indicate standard deviation. W, wandering larvae; pre-P, pre-pupae; P0, P1, P2 and P3, day-0, day-1, day-2, and day-3 pupae.

Table 1. Summary of the expression profile of maternal genes during stages 3–5 of oogenesis.

	No. of Maternal Genes	Name of Maternal Genes
Upward trend	18	*me31B, lok, vri, CycB, aub, BAEE, PPAE, Sod2, Smg, eIF4AIII, rod, babo, h, Chc, wbl, ndl(osp), pip, dpp*
Downward trend	9	*Egfr, Su(var)205, proPPAE, PAH, esc, Nelf-E, Nelf-A, tld, sog*
Flattening trend	39	*Pabn2, Eif-4a, Bin1, bai, tud, gammaTub, Mat89Ba, Pc, Btk29A, Src64B, shot, spoon(AKAP1), Msp300(nesprin-1), TPH1, vfl, KCNQ, Sel(cnpy1), Hip14(ZDHHC17), Hp1b-l, spz, mamo, tkv, hb, snk, asp, proSP7, glo(hnRNPF), Csp(DnaJ-7), pie, SPE, sax, gro, SPE-like, Dif, Th, Src42A, Dst, Snap25, Moe*

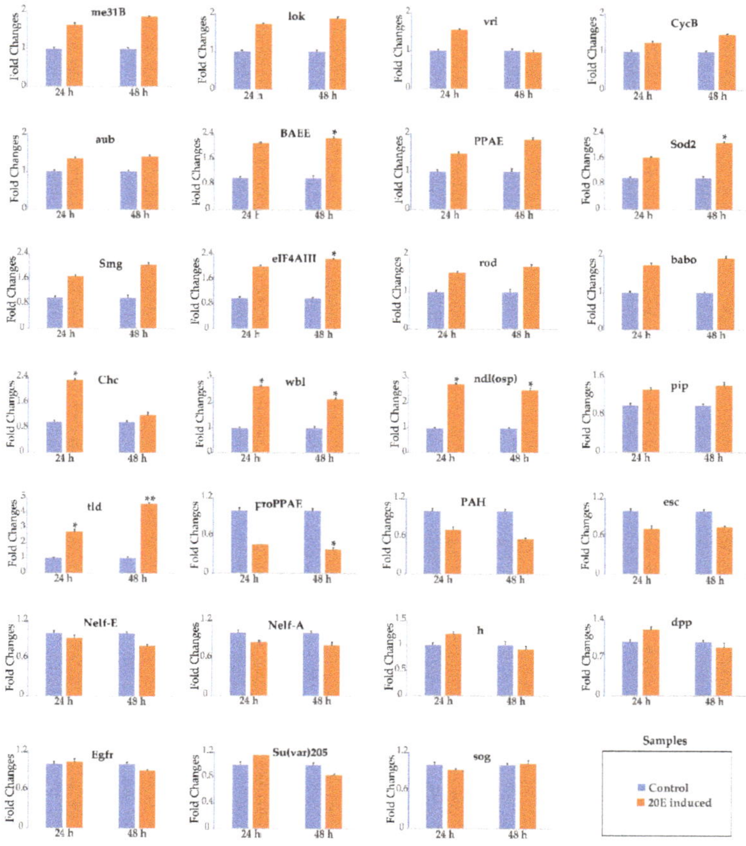

Figure 5. 20-hydroxyecdysone (20E) regulation of the expression of select genes determined by RT-qPCR. The isolated abdomens of the pupae (20E-depleted abdomens) were injected with 5 μL of a 20E (10.0 μg) solution or 5 μL of sterile distilled water (control). The samples of ovary and ovariole were collected at 24 h and 48 h after 20E treatment. Each time point was assessed thrice. *TIF-4A* was used as the internal control. Bars indicate the standard deviation. Significant analysis: * $p < 0.05$ or ** $p < 0.01$.

4. Discussion

In insects, ecdysteroids play an important role in molting and metamorphosis. It has been established that insect eggs (including those of the silkworm) contain various ecdysteroids, and the amounts of these ecdysteroids fluctuate during embryonic development [38,39]. Ecdysteroids are essential for embryonic development [40,41] and are synthesized de novo, while free ecdysteroids participate in morphogenesis at an early embryonic stage when the prothoracic glands have not yet differentiated. Thus, ecdysteroids in insect eggs may be of maternal origin [42,43]. Moreover, the increase in 20E titer in young pupae can trigger the first major transition from previtellogenesis to vitellogenesis in silkworm oogenesis [25–27]. Meanwhile, the development of animal embryos is initially directed and controlled by the maternal gene products loaded into the egg during oogenesis [1,2]. We previously successfully identified 66 maternal genes in the silkworm [32,33]. In the present study, we first investigated the expression characteristics of the 66 maternal genes in the ovary of the wandering silkworm, and then examined the developmental expression profiles of each gene in the ovary or ovariole from larvae to day-3 pupae, finally confirming the role of 20E in regulating maternal gene expression.

The maternal gene products that are loaded into the egg drive the earliest stages of development, until the zygotic genome can be transcribed. This tightly regulated process is known as the MZT and has been well-studied within model species [2,7,44–46]. Nevertheless, we have little knowledge of the time required for this process to begin or of the dynamic landscape of maternal mRNAs during oogenesis in insects. The 66 maternal genes' transcripts were detected in the ovary of the wandering silkworm. This is contrary to their low expression levels found in the ovaries of day-3 fifth instar insects in our previous study [32]. The wandering silkworm period is stage 3 of oogenesis, at which point the oocyte–nurse cell clusters are arranged in single file in the ovariole, and the follicle cells begin to migrate centripetally between the oocyte and the nurse cells. Glycogen synthesis begins in the oocyte [23]. This indicates the start of the accumulation of maternal mRNAs during stage 3 of the oogenesis developmental period of previtellogenesis in larvae. This obviously differs from the hypothesis that the genes expressed continually during choriogenesis are likely to be maternal genes, because at this stage, the follicle cells gradually experience apoptosis, and the oocyte is uniquely reserved in the mature egg [31,43].

The ovarioles were still enclosed in a protective capsule in day-0 pupae and 20E-depleted abdomens. The protective capsule ruptured, and the ovarioles gradually emerged into the abdominal cavity, shortly after the injection of 20E. In the 20E-depleted abdomens, the ovarioles were still enclosed in a protective capsule following the injection of juvenile hormone (JH) (data not shown). Ovarian development is induced by the injection of tebufenozide into 20E-depleted abdomens, while this is arrested at the stage of mid-vitellogenesis [47]. Ovarian development can be induced secondarily in 20E-depleted abdomens via a single injection of 20E, indicating that 20E plays a central role in silkworm oogenesis and is essential for the initial transition from previtellogenesis to vitellogenesis. In addition, our data show that the dynamic landscape of the 18 genes (incremental expression) and the 9 genes (decremental expression) cohere with the 20E titer changes, along with two peaks during stages 3–5 of silkworm oogenesis. Indeed, we found that the transcripts of 22 genes were regulated in the 20E-depleted abdomens by a single injection of 20E. This finding suggests that 20E plays a central role and regulates the transcription of maternal genes in stages 3–5 of silkworm oogenesis.

eIF4AIII (ATP-dependent RNA helicase) is involved in the mRNA surveillance pathway, and *me31B* is involved in RNA degradation, as determined by the KEGG ontology assignments. *eIF4AIII* is essential to the translation of nuclear cap-binding complex (CBC)-associated mRNAs [48,49]. *Me31B* mediates the translational silencing of maternal mRNAs and is an essential regulatory mechanism during early *Drosophila* oogenesis [50,51]. The transcripts of *eIF4AIII* and *me31B* were increased, which may indicate their essential regulatory action during early silkworm oogenesis.

Genes performing essential housekeeping functions are required at all stages of animal development and are transcribed by both the mother and the zygote [7]. The maternal and zygotic genomes are coordinated during early embryonic development to such a degree that the transcript levels of these genes remain relatively constant [7,32], despite the transition between the genomes of origin of these transcripts [7]. The stable mRNAs of 39 maternal genes perform essential housekeeping functions required during stages 3–5 of silkworm oogenesis.

5. Conclusions

In summary, this study has taken a much-needed first step towards determining the relationship between 20E and maternal genes during stages 3–5 of silkworm oogenesis. Our findings reveal the role of 20E in regulating the transcription of maternal genes and, accordingly, elucidate the dynamic landscape of gene regulation. Our data also demonstrate the start of the accumulation of maternal mRNAs during stage 3 of the oogenesis developmental period of previtellogenesis in larvae. Our findings expand the understanding of insect oogenesis and provide a perspective on embryo development in *B. mori*.

Supplementary Materials: The following are available online at https://www.mdpi.com/article/10.3390/insects12110969/s1, Figure S1. Expression patterns of maternal genes in the ovary of the wandering silkworm by reverse-transcription quantitative PCR (RT-qPCR). TIF-4A was used as the internal control, and the relative quantity of TIF-4A was set to 1. Bars represent the standard deviation. Figure S2. Kyoto Encyclopedia of Genes and Genomes KEGG analysis of maternal genes. Figure S3. View of the ovary or ovariole of the silkworm from the wandering larvae to day-3 pupae. Figure S4. View of the 20E-depleted abdomens and the ovary or ovariole of the 20E-depleted abdomens injected with 5 μL of a 20E (10.0 μg) solution or 5 μL of sterile distilled water. Table S1. Primers of genes for RT-PCR and RT-qPCR. Table S2. KEGG annotation. Table S3. Genetic information processing.

Author Contributions: M.Z. drafted the manuscript. P.X. performed the literature review. T.C. analyzed the data. All authors have read and agreed to the published version of the manuscript.

Funding: This work was financially supported by the National Natural Science Foundation of China (grant No. 31302035) and the Natural Science Foundation of Jiangsu Province (grant No. BK2012273).

Institutional Review Board Statement: Not applicable.

Informed Consent Statement: Not applicable.

Data Availability Statement: Data are contained within the article and Supplementary Materials.

Conflicts of Interest: The authors declare no conflict of interest.

References

1. Langley, A.R.; Smith, J.C.; Stemple, D.L.; Harvey, S.A. New insights into the maternal to zygotic transition. *Development* **2014**, *141*, 3834–3841. [CrossRef] [PubMed]
2. Vastenhouw, N.L.; Cao, W.X.; Lipshitz, H.D. The maternal-to-zygotic transition revisited. *Development* **2019**, 146. [CrossRef]
3. Horner, V.L.; Wolfner, M.F. Mechanical stimulation by osmotic and hydrostatic pressure activates *Drosophila* oocytes in vitro in a calcium-dependent manner. *Dev. Biol.* **2008**, *316*, 100–109. [CrossRef] [PubMed]
4. Tadros, W.; Lipshitz, H.D. The maternal-to-zygotic transition: A play in two acts. *Development* **2009**, *136*, 3033–3042. [CrossRef]
5. Chang, H.; Yeo, J.; Kim, J.G.; Kim, H.; Lim, J.; Lee, M.; Kim, H.H.; Ohk, J.; Jeon, H.Y.; Lee, H.; et al. Terminal Uridylyltransferases Execute Programmed Clearance of Maternal Transcriptome in Vertebrate Embryos. *Mol. Cell* **2018**, *70*, 72–82.e77. [CrossRef] [PubMed]
6. Zhang, B.; Wu, X.; Zhang, W.; Shen, W.; Sun, Q.; Liu, K.; Zhang, Y.; Wang, Q.; Li, Y.; Meng, A.; et al. Widespread Enhancer Dememorization and Promoter Priming during Parental-to-Zygotic Transition. *Mol. Cell* **2018**, *72*, 673–686.e676. [CrossRef]
7. Atallah, J.; Lott, S.E. Evolution of maternal and zygotic mRNA complements in the early *Drosophila* embryo. *PLoS Genet.* **2018**, *14*, e1007838. [CrossRef] [PubMed]
8. Cao, W.X.; Kabelitz, S.; Gupta, M.; Yeung, E.; Lin, S.C.; Rammelt, C.; Ihling, C.; Pekovic, F.; Low, T.C.H.; Siddiqui, N.U.; et al. Precise Temporal Regulation of Post-transcriptional Repressors Is Required for an Orderly *Drosophila* Maternal-to-Zygotic Transition. *Cell Rep.* **2020**, *31*, 107783. [CrossRef]

9. Benoit, B.; He, C.H.; Zhang, F.; Votruba, S.M.; Tadros, W.; Westwood, J.T.; Smibert, C.A.; Lipshitz, H.D.; Theurkauf, W.E. An essential role for the RNA-binding protein Smaug during the *Drosophila* maternal-to-zygotic transition. *Development* **2009**, *136*, 923–932. [CrossRef]
10. Gallo, C.M.; Munro, E.; Rasoloson, D.; Merritt, C.; Seydoux, G. Processing bodies and germ granules are distinct RNA granules that interact in *C. elegans* embryos. *Dev. Biol.* **2008**, *323*, 76–87. [CrossRef]
11. Rabani, M.; Pieper, L.; Chew, G.L.; Schier, A.F. A Massively Parallel Reporter Assay of 3 ′ UTR Sequences Identifies In Vivo Rules for mRNA Degradation. *Mol. Cell* **2017**, *68*, 1083–1094. [CrossRef] [PubMed]
12. Giraldez, A.J.; Mishima, Y.; Rihel, J.; Grocock, R.J.; Van Dongen, S.; Inoue, K.; Enright, A.J.; Schier, A.F. Zebrafish MiR-430 promotes deadenylation and clearance of maternal mRNAs. *Science* **2006**, *312*, 75–79. [CrossRef] [PubMed]
13. Ramos, S.B.; Stumpo, D.J.; Kennington, E.A.; Phillips, R.S.; Bock, C.B.; Ribeiro-Neto, F.; Blackshear, P.J. The CCCH tandem zinc-finger protein Zfp36l2 is crucial for female fertility and early embryonic development. *Development* **2004**, *131*, 4883–4893. [CrossRef] [PubMed]
14. Liang, H.L.; Nien, C.Y.; Liu, H.Y.; Metzstein, M.M.; Kirov, N.; Rushlow, C. The zinc-finger protein Zelda is a key activator of the early zygotic genome in *Drosophila*. *Nature* **2008**, *456*, 400–403. [CrossRef]
15. Lee, M.T.; Bonneau, A.R.; Takacs, C.M.; Bazzini, A.A.; DiVito, K.R.; Fleming, E.S.; Giraldez, A.J. Nanog, Pou5f1 and SoxB1 activate zygotic gene expression during the maternal-to-zygotic transition. *Nature* **2013**, *503*, 360–364. [CrossRef]
16. De Iaco, A.; Planet, E.; Coluccio, A.; Verp, S.; Duc, J.; Trono, D. DUX-family transcription factors regulate zygotic genome activation in placental mammals. *Nat. Genet.* **2017**, *49*, 941–945. [CrossRef]
17. De Iaco, A.; Coudray, A.; Duc, J.; Trono, D. DPPA2 and DPPA4 are necessary to establish a 2C-like state in mouse embryonic stem cells. *Embo Rep.* **2019**, *20*, e47382. [CrossRef]
18. Eckersley-Maslin, M.; Alda-Catalinas, C.; Blotenburg, M.; Kreibich, E.; Krueger, C.; Reik, W. Dppa2 and Dppa4 directly regulate the Dux-driven zygotic transcriptional program. *Genes Dev.* **2019**, *33*, 194–208. [CrossRef]
19. Hendrickson, P.G.; Dorais, J.A.; Grow, E.J.; Whiddon, J.L.; Lim, J.W.; Wike, C.L.; Weaver, B.D.; Pflueger, C.; Emery, B.R.; Wilcox, A.L.; et al. Conserved roles of mouse DUX and human DUX4 in activating cleavage-stage genes and MERVL/HERVL retrotransposons. *Nat. Genet.* **2017**, *49*, 925–934. [CrossRef]
20. Gao, L.; Wu, K.L.; Liu, Z.B.; Yao, X.L.; Yuan, S.L.; Tao, W.R.; Yi, L.Z.; Yu, G.L.; Hou, Z.Z.; Fan, D.D.; et al. Chromatin Accessibility Landscape in Human Early Embryos and Its Association with Evolution. *Cell* **2018**, *173*, 248–259. [CrossRef]
21. Goldsmith, M.R.; Shimada, T.; Abe, H. The genetics and genomics of the silkworm, *Bombyx mori*. *Annu. Rev. Entomol.* **2005**, *50*, 71–100. [CrossRef] [PubMed]
22. Xia, Q.; Cheng, D.; Duan, J.; Wang, G.; Cheng, T.; Zha, X.; Liu, C.; Zhao, P.; Dai, F.; Zhang, Z.; et al. Microarray-based gene expression profiles in multiple tissues of the domesticated silkworm, *Bombyx mori*. *Genome Biol.* **2007**, *8*, R162. [CrossRef] [PubMed]
23. Yamauchi, H.; Yoshitake, N. Developmental stages of ovarian follicles of the silkworm, *Bombyx mori* L. *J. Morphol.* **1984**, *179*, 21–31. [CrossRef]
24. Mizoguchi, A.; Ohashi, Y.; Hosoda, K.; Ishibashi, J.; Kataoka, H. Developmental profile of the changes in the prothoracicotropic hormone titer in hemolymph of the silkworm *Bombyx mori*: Correlation with ecdysteroid secretion. *Insect Biochem. Mol. Biol.* **2001**, *31*, 349–358. [CrossRef]
25. Swevers, L.; Eystathioy, T.; Iatrou, K. The orphan nuclear receptors BmE75A and BmE75C of the silkmoth *Bombyx mori*: Hornmonal control and ovarian expression. *Insect Biochem. Mol. Biol.* **2002**, *32*, 1643–1652. [CrossRef]
26. Swevers, L.; Iatrou, K. The ecdysone regulatory cascade and ovarian development in lepidopteran insects: Insights from the silkmoth paradigm. *Insect Biochem. Mol. Biol.* **2003**, *33*, 1285–1297. [CrossRef]
27. Sonobe, H.; Yamada, R. Ecdysteroids during early embryonic development in silkworm *Bombyx mori*: Metabolism and functions. *Zool. Sci.* **2004**, *21*, 503–516. [CrossRef] [PubMed]
28. Ito, Y.; Yasuda, A.; Sonobe, H. Synthesis and phosphorylation of ecdysteroids during ovarian development in the silkworm, *Bombyx mori*. *Zool. Sci.* **2008**, *25*, 721–727. [CrossRef]
29. Ito, Y.; Sonobe, H. The role of ecdysteroid 22-kinase in the accumulation of ecdysteroids in ovary of silkworm *Bombyx mori*. *Ann. NY Acad. Sci.* **2009**, *1163*, 421–424. [CrossRef]
30. Niwa, Y.S.; Niwa, R. Transcriptional regulation of insect steroid hormone biosynthesis and its role in controlling timing of molting and metamorphosis. *Dev. Growth Differ.* **2016**, *58*, 94–105. [CrossRef]
31. Zhang, Q.; Sun, W.; Sun, B.Y.; Xiao, Y.; Zhang, Z. The dynamic landscape of gene regulation during *Bombyx mori* oogenesis. *BMC Genom.* **2017**, *18*, 714. [CrossRef]
32. Zhang, M.R.; Xu, P.Z.; Pang, H.L.; Chen, T.; Zhang, G.Z. Expression Analysis of mRNA Decay of Maternal Genes during *Bombyx mori* Maternal-to-Zygotic Transition. *Int. J. Mol. Sci.* **2019**, *20*, 5651. [CrossRef]
33. Zhang, M.R.; Qin, S.; Xu, P.Z.; Zhang, G.Z. Identifying potential maternal genes of *Bombyx mori* using digital gene expression profiling. *PLoS ONE* **2018**, *13*, e0192745. [CrossRef]
34. Guo, H.; Jiang, L.; Xia, Q. Selection of reference genes for analysis of stress-responsive genes after challenge with viruses and temperature changes in the silkworm *Bombyx mori*. *Mol. Genet. Genom.* **2016**, *291*, 999–1004. [CrossRef]
35. Livak, K.J.; Schmittgen, T.D. Analysis of relative gene expression data using real-time quantitative PCR and the 2(-Delta Delta C(T)) Method. *Methods* **2001**, *25*, 402–408. [CrossRef]

36. Kanehisa, M.; Sato, Y.; Kawashima, M. KEGG mapping tools for uncovering hidden features in biological data. *Protein Sci.* **2021**. [CrossRef] [PubMed]
37. Xu, P.Z.; Zhang, M.R.; Wang, X.Y.; Wu, Y.C. Precocious Metamorphosis of Silkworm Larvae Infected by BmNPV in the Latter Half of the Fifth Instar. *Front. Physiol.* **2021**, *12*, 650972. [CrossRef]
38. Ohnishi, E.; Hiramoto, M.; Fujimoto, Y.; Kakinuma, K.; Ikekawa, N. Isolation and identification of major ecdysteroid conjugates from the ovaries of *Bombyx mori*. *Insect Biochem.* **1989**, *19*, 95–101. [CrossRef]
39. Makka, T.; Seino, A.; Tomita, S.; Fujiwara, H.; Sonobe, H. A possible role of 20-hydroxyecdysone in embryonic development of the silkworm *Bombyx mori*. *Arch. Insect Biochem. Physiol.* **2002**, *51*, 111–120. [CrossRef] [PubMed]
40. Kozlova, T.; Thummel, C.S. Essential roles for ecdysone signaling during *Drosophila* mid-embryonic development. *Science* **2003**, *301*, 1911–1914. [CrossRef] [PubMed]
41. Wang, C.F.; Zhang, Z.; Sun, W. Ecdysone oxidase and 3-dehydroecdysone-3beta-reductase contribute to the synthesis of ecdysone during early embryonic development of the silkworm. *Int. J. Biol. Sci.* **2018**, *14*, 1472–1482. [CrossRef]
42. Hanaoka, K.; Onishi, E. Changes in ecdysone titre during pupal-adult development in the silkworm, *Bombyx mori*. *J. Insect Physiol.* **1974**, *20*, 2375–2384. [CrossRef]
43. Carter, J.M.; Baker, S.C.; Pink, R.; Carter, D.R.; Collins, A.; Tomlin, J.; Gibbs, M.; Breuker, C.J. Unscrambling butterfly oogenesis. *BMC Genom.* **2013**, *14*, 283. [CrossRef]
44. Blythe, S.A.; Wieschaus, E.F. Zygotic Genome Activation Triggers the DNA Replication Checkpoint at the Midblastula Transition. *Cell* **2015**, *160*, 1169–1181. [CrossRef]
45. Briggs, J.A.; Weinreb, C.; Wagner, D.E.; Megason, S.; Peshkin, L.; Kirschner, M.W.; Klein, A.M. The dynamics of gene expression in vertebrate embryogenesis at single-cell resolution. *Science* **2018**, *360*, aar5780. [CrossRef]
46. Sankar, A.; Lerdrup, M.; Manaf, A.; Johansen, J.V.; Gonzalez, J.M.; Borup, R.; Blanshard, R.; Klungland, A.; Hansen, K.; Andersen, C.Y.; et al. KDM4A regulates the maternal-to-zygotic transition by protecting broad H3K4me3 domains from H3K9me3 invasion in oocytes. *Nat. Cell Biol.* **2020**, *22*, 380–388. [CrossRef] [PubMed]
47. Swevers, L.; Iatrou, K. The ecdysone agonist tebufenozide (RH-5992) blocks the progression into the ecdysteroid-induced regulatory cascade and arrests silkmoth oogenesis at mid-vitellogenesis. *Insect Biochem. Molec.* **1999**, *29*, 955–963. [CrossRef]
48. Ballut, L.; Marchadier, B.; Baguet, A.; Tomasetto, C.; Seraphin, B.; Le Hir, H. The exon junction core complex is locked onto RNA by inhibition of eIF4AIII ATPase activity. *Nat. Struct. Mol. Biol.* **2005**, *12*, 861–869. [CrossRef]
49. Choe, J.; Ryu, I.; Park, O.H.; Park, J.; Cho, H.; Yoo, J.S.; Chi, S W.; Kim, M.K.; Song, H.K.; Kim, Y.K. eIF4AIII enhances translation of nuclear cap-binding complex-bound mRNAs by promoting disruption of secondary structures in 5′UTR. *Proc. Natl. Acad. Sci. USA* **2014**, *111*, E4577–E4586. [CrossRef]
50. Nakamura, A.; Amikura, R.; Hanyu, K.; Kobayashi, S. Me31B silences translation of oocyte-localizing RNAs through the formation of cytoplasmic RNP complex during *Drosophila* oogenesis. *Development* **2001**, *128*, 3233–3242. [CrossRef] [PubMed]
51. Gotze, M.; Dufourt, J.; Ihling, C.; Rammelt, C.; Pierson, S.; Sambrani, N.; Temme, C.; Sinz, A.; Simonelig, M.; Wahle, E. Translational repression of the *Drosophila* nanos mRNA involves the RNA helicase Belle and RNA coating by Me31B and Trailer hitch. *RNA* **2017**, *23*, 1552–1568. [CrossRef] [PubMed]

insects

MDPI

Article

Function of Polyamines in Regulating Cell Cycle Progression of Cultured Silkworm Cells

Li Chang [1,2], Zhiqing Li [1,2,*], Hao Guo [1,2], Wenchang Zhang [1,2], Weiqun Lan [1,2], Jue Wang [3], Guanwang Shen [1,2], Qingyou Xia [1,2] and Ping Zhao [1,2]

1 State Key Laboratory of Silkworm Genome Biology, Biological Science Research Center, Southwest University, Chongqing 400715, China; cl17783721226@163.com (L.C.); guohaosf001@163.com (H.G.); zwczmu@163.com (W.Z.); sparkle2021lwq@163.com (W.L.); gwshen@swu.edu.cn (G.S.); xiaqy@swu.edu.cn (Q.X.); zhaop@swu.edu.cn (P.Z.)

2 Chongqing Key Laboratory of Sericultural Science, Chongqing Engineering and Technology Research Center for Novel Silk Materials, Southwest University, Chongqing 400715, China

3 College of Sericulture, Textile and Biomass Sciences, Southwest University, Chongqing 400715, China; 15123540502@163.com

* Correspondence: lizhiqing@swu.edu.cn

Citation: Chang, L.; Li, Z.; Guo, H.; Zhang, W.; Lan, W.; Wang, J.; Shen, G.; Xia, Q.; Zhao, P. Function of Polyamines in Regulating Cell Cycle Progression of Cultured Silkworm Cells. *Insects* **2021**, *12*, 624. https://doi.org/10.3390/insects12070624

Academic Editors: Silvia Cappellozza, Morena Casartelli, Federica Sandrelli, Alessio Saviane and Gianluca Tettamanti

Received: 20 May 2021
Accepted: 8 July 2021
Published: 8 July 2021

Publisher's Note: MDPI stays neutral with regard to jurisdictional claims in published maps and institutional affiliations.

Simple Summary: The mechanism of the polyamine pathway in the lepidopteran silkworm is largely unknown. In the current study, we aimed to characterize the function of polyamines and polyamine pathway genes in silkworm cells as a regulator of cell cycle progression. For the first time, we identified the homologous genes of the polyamine pathway in silkworm, and analyzed their expression characteristics in different tissues and their subcellular localizations in cultured silkworm cells. We measured the abundant levels of polyamines in silkworm cells by HPLC analysis. We found that exogenous supplementation of spermidine in cells promoted DNA replication and cell cycle progression and, in contrast, treatment with polyamine biosynthesis inhibitors DFMO and MGBG prevented DNA replication and cell cycle progression. Indeed, the mechanism studies indicated that spermidine increased the expression of cell cycle-related genes, whereas this increase could be abolished by treatment with inhibitors. Taken together, our findings highlight that appropriate levels of polyamines have beneficial effects on the progression of the cell cycle by regulating cell cycle genes in silkworm.

Abstract: Background: Putrescine, spermidine, and spermine are polyamines that are ubiquitously distributed in prokaryotic and eukaryotic cells, which play important roles in cell proliferation and differentiation. Methods: We investigated the expression profiles of polyamine pathway genes by qRT-PCR in different tissues of the lepidopteran silkworm. The polyamine levels in cultured silkworm cells were measured by HPLC. Spermidine and polyamine biosynthetic inhibitors were used for treating the cultured silkworm cells in order to clarify their effects on cell cycle progression. Results: We identified the anabolic and catabolic enzymes that are involved in the polyamine biosynthetic pathway in silkworm. Transcriptional expression showed at least seven genes that were expressed in different silkworm tissues. Treatments of the cultured silkworm cells with spermidine or inhibitor mixtures of DFMO and MGBG induced or inhibited the expression of cell cycle-related genes, respectively, and thus led to changed progression of the cell cycle. Conclusions: The present study is the first to identify the polyamine pathway genes and to demonstrate the roles of polyamines on cell cycle progression via regulation of the expression of cell cycle genes in silkworm.

Keywords: *Bombyx mori*; polyamines; cell cycle progression

1. Introduction

Polyamines are low-molecular-weight polycationic aliphatic amines derived from the metabolism of arginine, which mainly include putrescine, spermidine, and spermine, and are found in prokaryotic and eukaryotic cells [1,2]. Due to their polycationic nature, polyamines are fully protonated under physiological pH and ionic strength conditions. Therefore, they can easily interact with negatively charged macromolecules, such as DNA, RNA, ATP, phospholipids, or proteins [3–5], and thereby participate in a number of biological processes, such as modulation of gene expression and enzyme activities, activation of DNA synthesis, transcriptional processes, and regulation of cell proliferation and differentiation [6–8].

The biosynthesis of polyamines is regulated by multiple feedback loops that maintain the homeostasis of their metabolites in cells [7]. It begins with the conversion of putrescine from the amino acid ornithine by the enzyme ornithine decarboxylase (ODC), a rate-limiting enzyme in polyamine synthesis [9,10]. Putrescine is sequentially converted to the higher polyamines spermidine and spermine via spermidine synthase (SPDS) and spermine synthase (SPMS), respectively, together with S-adenosyl-methionine decarboxylase (SAMDC), the second rate-limiting enzyme [7,11]. The synthesis of spermine and spermidine is reversible by the action of spermine oxidase (SMO) and acetylpolyamine oxidase (APAO), and requires the spermidine/spermine acetyltransferase (SSAT) enzyme [7].

In addition to the de novo biosynthesis of polyamines in cells, polyamines are also natural components of our diets. For example, spermidine and spermine are found in foods of both plant and animal origin, such as soybeans, mushrooms, cowpeas, fruits, shellfish, and meat products. These dietary polyamines can be absorbed by our gut and transported into different tissue cells, and further regulate the intracellular homeostasis of polyamines [12].

It has been shown that a deficiency of polyamines in cells delays cell cycle progression, with most cells arrest at the G1 to S phase, and the rate of DNA synthesis is also decreased [13,14]. The basic functions of polyamines have been related to cellular DNA protection against exogenous agents and radiation injuries [15,16]. To date, studies on polyamines have mainly focused on spermidine. The evidence has shown that spermidine can promote the growth and development of cells, as well as accelerate the occurrence of autophagy and delay aging [17]. The addition of spermidine in micromolar concentrations into diets has been found to promote human hair growth, and prolong the mean lifespan of *Drosophila melanogaster* and *Caenorhabditis elegans* [17,18]. Additionally, excessive application of spermidine can cause serious diseases, such as respiratory symptoms, nephrotoxicity, and even cancers [19,20].

To investigate the functions of polyamines in cells, a number of inhibitors have been developed to disturb the biosynthesis and homeostasis of polyamines. Among of these inhibitors, difluoromethylornithine (DFMO) and methylglyoxal-bis(guanylhydrazone) (MGBG) are two critical inhibitors that can specifically interfere with the key enzyme activities of ODC and SAMDC, respectively [21–24]. A combination treatment with DFMO and MGBG can be used to produce a profound anti-proliferative effect and the cell cycle of the cells is arrested at the G1 to S phase, which has demonstrated significant therapeutic effects [7,13].

The silkworm *Bombyx mori* is an economically important insect for silk production and a lepidopteran model for investigating gene functions [25,26]. Although previous studies have demonstrated the positive effects of spermidine supplementation on the development of silk glands, as well as economic parameters [27,28], the enzymes involved in the polyamine pathway and the function of polyamines in the silkworm remain unexplored. Therefore, in the present study, we identified the genes in the polyamine pathway, detected their expression characteristics in different tissues, and analyzed their subcellular localization in cultured silkworm cells. We also determined the levels of various polyamines in cultured silkworm cells by HPLC, and further investigated the cellular function of polyamines by exogenous addition of spermidine and polyamine biosynthetic inhibitors.

2. Materials and Methods

2.1. Cell Culture

The BmN cell line derived from the silkworm ovary was used in the present work [29]. BmN cells were grown in TC-100 medium (Sigma, St. Louis, MO, USA) supplemented with 10% fetal bovine serum (FBS, Hyclone, Logan, UT, USA) and penicillin-streptomycin (Thermo Fisher Scientific, Waltham, MA, USA) at a temperature of 27 °C. Spermidine was purchased from Sigma (St. Louis, MO, USA), DFMO inhibitor was obtained from Tocris Bioscience (Bristol, UK), and MGBG inhibitor was obtained from Beijing Zongheng Technology Co, Ltd. (Beijing, China).

2.2. Identification of Polyamine Pathway Genes in Silkworm

The annotated polyamine pathway genes from *D. melanogaster* and humans were downloaded from NCBI (Accessed date: 19 September 2018, http://www.ncbi.nlm.nih.gov/). We used these protein sequences as queries to perform a BLASTP search against the silkworm genome database (Accessed date: 23 October 2019, https://silkdb.bioinfotoolkits.net/). The identification for each gene is listed in Figure 1B.

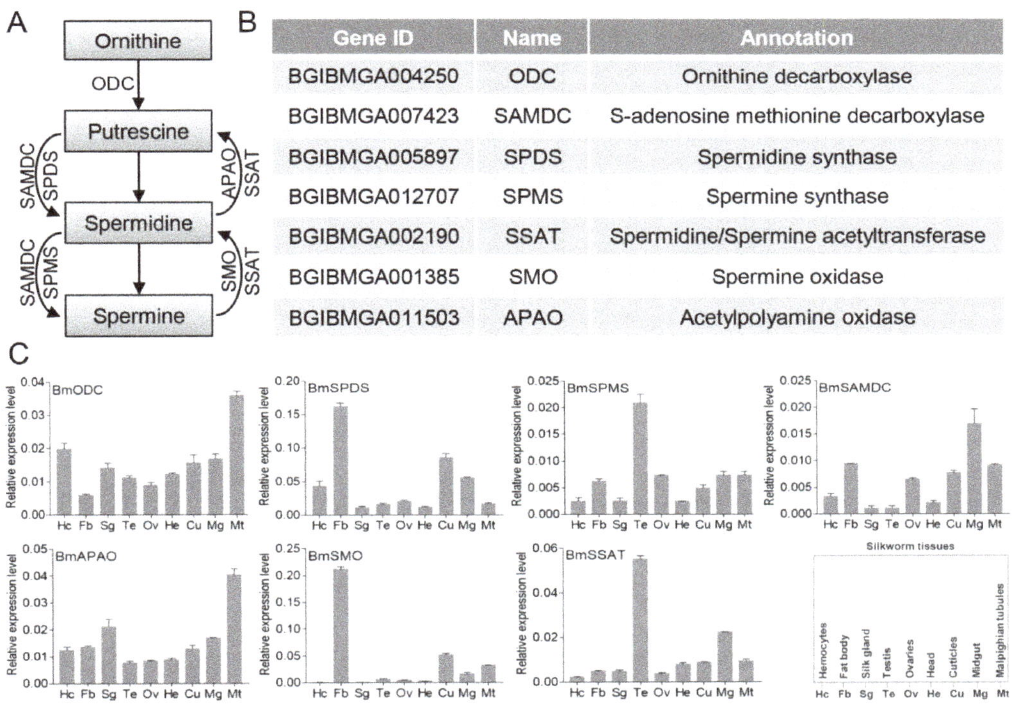

Gene ID	Name	Annotation
BGIBMGA004250	ODC	Ornithine decarboxylase
BGIBMGA007423	SAMDC	S-adenosine methionine decarboxylase
BGIBMGA005897	SPDS	Spermidine synthase
BGIBMGA012707	SPMS	Spermine synthase
BGIBMGA002190	SSAT	Spermidine/Spermine acetyltransferase
BGIBMGA001385	SMO	Spermine oxidase
BGIBMGA011503	APAO	Acetylpolyamine oxidase

Figure 1. Identification and expression profiles of polyamine pathway genes in the silkworm. (**A**) A schematic representation of the polyamine biosynthesis pathway. (**B**) Identification and annotation of polyamine pathway genes. (**C**) Tissue expression profile of polyamine pathway genes in the silkworm.

2.3. Plasmids

Expression constructs for polyamine pathway genes were amplified from the cDNA library of cultured silkworm cells using the primers listed in Table S1 and further cloned into an NcoI–XhoI/NotI site of pENTR11 (Invitrogen, Carlsbad, CA, USA) vector [30]. All plasmids were verified by sequencing. The pENTR11-BmODC, pENTR11-BmSAMDC, pENTR11-BmSPMS, pENTR11-BmSPDS, pENTR11-BmSSAT, pENTR11-BmAPAO, and pENTR11-BmSMO were respectively inserted into the expression vectors of pPBO_ie2GW (containing N-terminal EGFP tag) by gateway reaction to construct the expression plasmids [31].

2.4. EdU Staining

EdU staining was performed by using a commercial BeyoClick™ EdU-488 Kit (Beyotime, Shanghai, China). BmN cells were treated with spermidine at 10 μM or inhibitors at 50 μM (DFMO and MGBG mixtures in a molar ratio of 1:1) for 48 h and stained with 10 μM EdU for 2 h at 27 °C. The cells were fixed with stationary liquid for 15 min and then incubated with Click Additive Solution for 30 min behind the scenes. Subsequently, cells were stained with Hoechst 33,342 (Beyotime, Shanghai, China) and immersed in 1× PBS buffer. Finally, fluorescence signals were captured by a fluorescent microscope (Z16, Leica, Wetzlar, Germany). To quantify the EdU-positive cells (green), cells were analyzed by selecting three different fluorescent fields and counted by the counting function of Photoshop CS6 software (Adobe Systems Incorporated, San Jose, CA, USA).

2.5. Polyamine Measurements

Polyamine levels were measured by HPLC as described previously [32]. Briefly, sample pretreatment and derivatization: 1×10^7 cells treated with inhibitors at 50 μM for 48 h were collected at 1000 rpm and immersed in 0.7 mL of 1× PBS buffer, ultrasonicated for 10 min, and centrifuged at 8000× *g* for 10 min. Then, 0.2 mL of supernatant were added into a brown centrifuge tube, and then 40 μL of 2 M NaOH, 60 μL saturated $NaHCO_3$ and 0.2 mL of acetone solution of dansyl chloride (10 mg/mL) were added in order into a brown centrifuge tube (Light protection) (Axygen, San Francisco, CA, USA) and reacted in a water bath at 50 °C for 40 min in darkness. The reaction solution was cooled at room temperature, and 100 μL of ammonia were added into a brown centrifuge tube, and incubated for 30 min at room temperature. Finally, the total amount of 1 mL was filled using methanol and filtered by a pinhead filter. The experimental conditions were as follows: Chromatograph: agilent1100 high-performance liquid chromatograph with a wavelength of 254 nm; Column: Kromasil C18 reversed-phase column (250 mm × 4.6 mm, 5 μm); Column temperature: 30 °C; Flow rate: 1 mL/min; Injection volume: 10 μL; Mobile phase gradient: A, methanol; B, water. All assays were performed in triplicate.

2.6. Quantitative Real-Time PCR Analysis

To analyze the tissue expression characteristics of polyamine pathway genes and evaluate the mRNA levels of cell cycle-related genes after exogenous addition of spermidine or inhibitors, total RNA samples were prepared from BmN cells and silkworm tissues (Dazao strain) dissected on the third day of the fifth instar larvae of silkworm individuals by using Trizol reagent (Invitrogen, Carlsbad, CA, USA). According to the manufacturer's protocol of the M-MLV Reverse Transcriptase Kit (Promega, Madison, WI, USA), 1 μg of total RNA was used for cDNA synthesis. qRT-PCR was performed with a SYBR Premix ExTaq Kit (Takara, Kyoto, Japan) and a qTower 2.2 Real-time PCR Detection System. The BmEif-4a gene was used as the internal control. All experiments were independently performed with three biological replicates, and the relative mRNA expression levels were calculated using the $2^{-\triangle\triangle CT}$ method. All primers used for qRT-PCR are listed in Table S1.

2.7. Subcellular Localization Assay

The analysis of fluorescently tagged proteins followed our previous protocol [30]. Briefly, cells were cultured on glass coverslips and transfected with 500 ng of expression plasmids. Then, 72 h post-transfection, cells were washed once with PBS and fixed with 4% paraformaldehyde in PBS for 10 min. The nuclei DNA were counterstained by 4′,6-diamidino-2-phenylindole (DAPI) (Invitrogen, Carlsbad, CA, USA). Fluorescence signals were captured by a fluorescent microscope (Z16, Leica, Wetzlar, Germany).

2.8. Flow Cytometry Analysis

Cell cycle distributions were monitored by flow cytometer analysis (CytoFLEX S, Beckman, Brea, CA, USA) of cellular DNA stained with propidium iodide (Sangon Biotech, Shanghai, China) according to the procedure described previously [30].

2.9. Statistical Analysis

Data are presented as the mean ± standard deviation (SD) of three independent biological replicates. Statistical significance (p-value) was analyzed by the Student's t-test. Statistical significance is denoted as follows: * $p < 0.05$, ** $p < 0.01$, and *** $p < 0.001$.

3. Results

3.1. Identification and Expression Profiles of Polyamine Pathway Genes in Silkworm

Polyamines are widely found in prokaryotic and eukaryotic cells, and at least seven enzymes have been reported to be involved in the pathway of polyamine metabolism [7] (Figure 1A). In order to explore the function of this polyamine pathway in silkworm, we identified the genes in silkworm. We referred to proteins from the annotated species, including *D. melanogaster* and humans, and performed BLASTP analysis against the silkworm genome database, and named BmODC, BmSAMDC, BmSPDS, BmSPMS, BmSSAT, BmSMO, and BmAPAO (Figure 1B). Next, by qRT-PCR analysis, we analyzed the expression profiles of the polyamine pathway genes in silkworm tissues on the third day of the fifth instar larvae (Figure 1C), which showed that the polyamine pathway genes were expressed in all the examined tissues, although the expression levels of different genes in the same tissue were varied. The expression patterns of BmODC and BmAPAO were similar, both of which were highly expressed in Malpighian tubules. The expression pattern of BmSPDS was similar to BmSMO, and both were abundantly expressed in the fat body. The expression pattern of BmSPMS was similar to BmSSAT, both of which were predominantly expressed in the testis. These different expression profiles may indicate that polyamine pathway genes contribute to the regulation of the dynamic levels of polyamines in different tissues of the silkworm.

3.2. Subcellular Localizations of Polyamine Pathway Genes in Cultured Silkworm Cells

Subcellular localization of a target protein is frequently used to exhibit its function in cells. Enhanced green fluorescent protein (EGFP) expressed in cells is capable of producing green fluorescence when excited by illumination with ultraviolet light, which makes it an excellent reporter for in vivo observation [33]. Therefore, fusion expression of fluorescent protein with a target protein has been extensively used for direct visualization of target proteins in various cells and tissues by using confocal microscopy. To further analyze the potential functions of the polyamine pathway genes in cultured silkworm cells, polyamine pathway genes were constructed into the EGFP fusion expression vector, and then transfected into the silkworm cells for fluorescence observation. Subcellular localizations showed that BmODC, BmSAMDC, BmSPDS, BmSPMS, BmSSAT, BmSMO, and BmAPAO were predominantly localized to the cytoplasm and especially concentrated in the nuclear membrane as compared with the EGFP control (Figure 2). It was interesting that BmSAMDC and BmSSAT, both involved in the biosynthesis and metabolism of spermidine/spermine, respectively, also showed identical subcellular localization in the nucleolus, which may indicate their critical roles in regulating homeostasis of spermidine/spermine in whole cells. All these results revealed the specific localization of polyamine pathway genes in cultured silkworm cells.

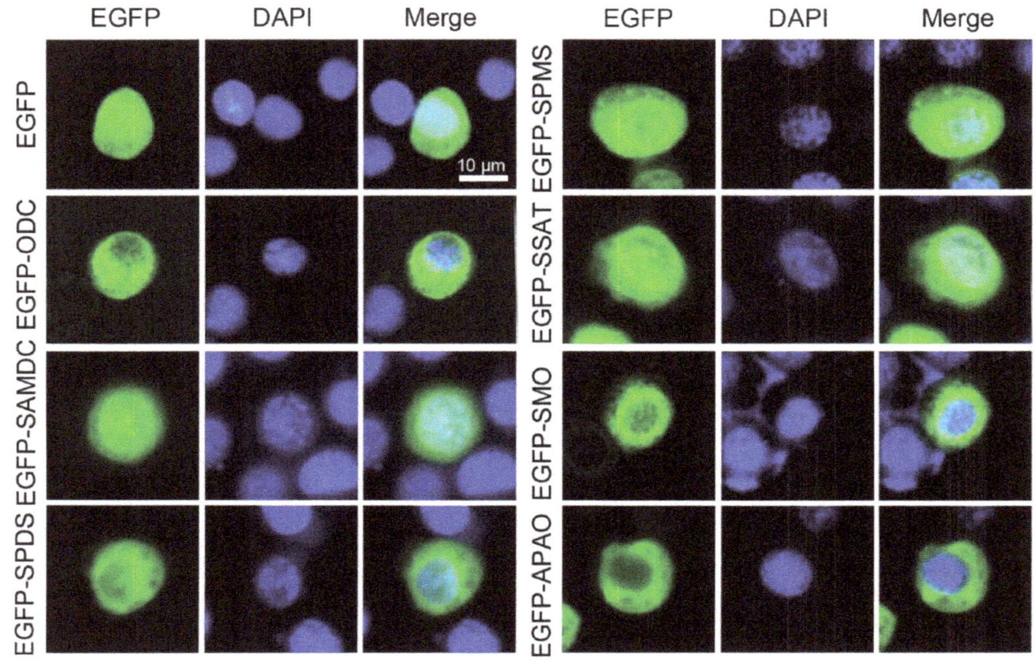

Figure 2. Subcellular localizations of polyamine pathway genes in cultured silkworm cells. Subcellular localization of transiently expressed EGFP-fused polyamine pathway genes in the silkworm BmN cells was determined by fluorescence (green) and the nuclei DNA was counterstained with DAPI (blue). Scale bar, 10 μm.

3.3. Spermidine Supplementation Promotes DNA Replication in Cultured Silkworm Cells

Polyamines have been reported to promote cell cycle progression. To evaluate whether polyamines also contribute to the progression of silkworm cells, we added spermidine to the cells and measured its effects on the cell cycle. Firstly, the toxicity of spermidine was detected in silkworm cells and the results showed that spermidine, in the range of 0–200 μM, had no obvious toxicity effect on silkworm cells (Figure 3A). Then, we selected 10 μM of spermidine to treat the cells for further analysis. Ongoing DNA synthesis was detected in cells by using EdU staining, and it was shown that the treatment with spermidine clearly promoted DNA replication (Figure 3B). The proportional analysis showed that the number of EdU-stained cells was also significantly increased in spermidine-treated cells, which was consistent with the fluorescence observations of EdU staining (Figure 3C). Further, flow cytometry analysis was used to examine the phase of the cell cycle after spermidine treatment. It was shown that spermidine was able to increase the percentage of cells at the S and G2/M phases and decrease G1 phase, which suggests that spermidine may promote cell cycle progression (Figure 3D). Interestingly, spermidine supplementation markedly induced the expression of cell cycle-related genes, such as BmCyclinA, BmCyclinB, BmCyclinB3, BmCyclinD, and BmCyclinE (Figure 3E). Taken together, it was shown that spermidine supplementation in silkworm cells can promote DNA replication and cell cycle progression by inducing the expression of cell cycle genes.

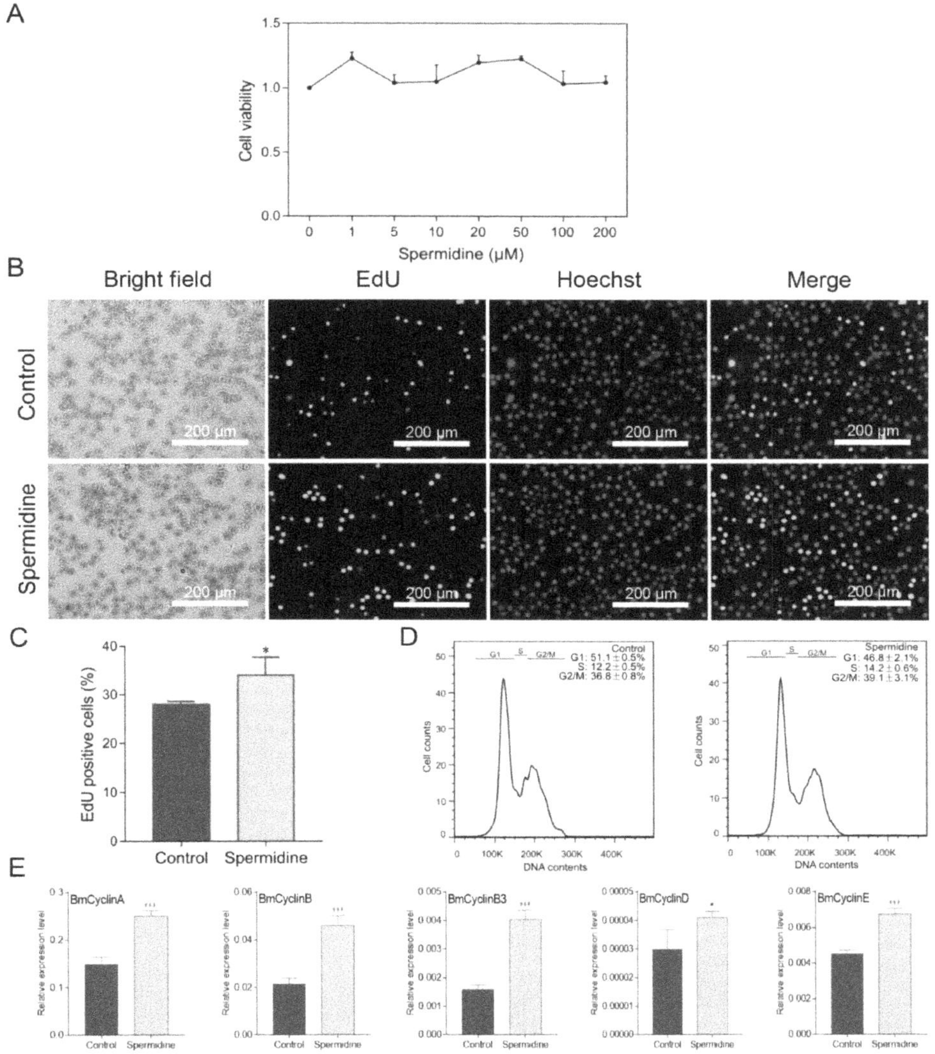

Figure 3. Spermidine supplementation promotes DNA replication in cultured silkworm cells. (**A**) Cytotoxicity test of different concentrations of spermidine on silkworm cells. (**B**) BeyoClick™ EdU-488 staining was used to detect DNA replication. Scale bar, 200 μm. (**C**) EdU-positive cells (green) were analyzed by selecting three different fluorescent fields and counted by Photoshop software. The data presented are the means ± SD (n = 3). (**D**) Cell cycle distribution was analyzed by using a flow cytometry and cells in G1, S, or G2/M were quantitated by FlowJo software. Three repeats were done for each treatment and the percentage of cells are the means ± SD (n = 3). (**E**) Expression of cell cycle-related genes was detected by qRT-PCR. For significant analysis: * $p < 0.05$ and *** $p < 0.001$.

3.4. Inhibition of Polyamine Pathway Activity Reduces Polyamine Levels of Cultured Silkworm Cells

To further explore whether a decrease of intracellular polyamine levels would affect the cell cycle activity of silkworm cells, ODC inhibitor (DFMO) and SAMDC inhibitor (MGBG) were used to inhibit the activity of polyamine biosynthesis. To determine the toxicity of inhibitors on cultured silkworm cells, first, we treated the cells with DFMO and MGBG mixed at a molar ratio of 1:1. The results showed that treatment with the inhibitors had no obvious toxicity in the range of 0–200 μM on cells (Figure 4A). Then, cells were treated with 50 μM concentrations of inhibitors for further experiments. After the treatment with the inhibitors, we collected the cells and measured different polyamine levels by HPLC analysis. As shown in Figure 4B,C, it clearly showed that the levels of putrescine, spermidine, and spermine were significantly decreased in cells. These results indicated that the treatment with inhibitors successfully reduces polyamine levels in silkworm cells.

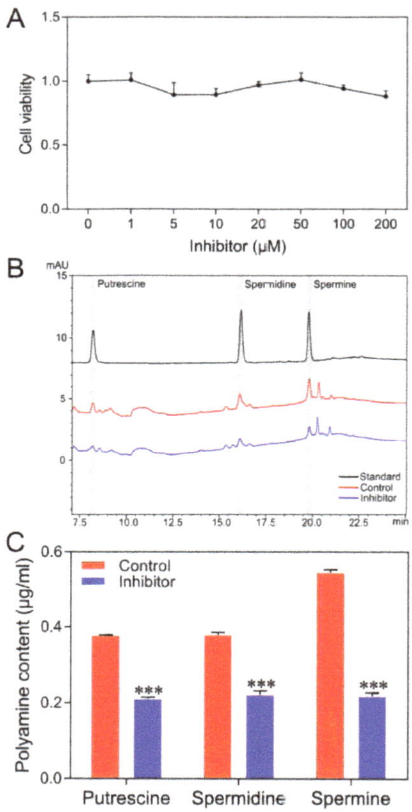

Figure 4. Inhibition of polyamine pathway activity decreases the levels of polyamines in cultured silkworm cells. (**A**) Cytotoxicity test of different concentrations of spermidine on the silkworm cells. (**B**) The standards of various polyamines and intracellular polyamines were determined by HPLC and the representative figure is shown. (**C**) Three independent experiments for polyamine content were analyzed, and the data presented are the means ± SD ($n = 3$). For significant analysis: *** $p < 0.001$.

3.5. Inhibition of Polyamine Pathway Activity Decreases DNA Replication in Cultured Silkworm Cells

Next, we investigated DNA replication ability by EdU staining when the levels of intracellular polyamines were reduced by treatment with inhibitors. The result showed that the DNA replication ability was attenuated (Figure 5A), and the proportion of the EdU-positive cells was also significantly decreased (Figure 5B). The effect of inhibitors on the progression of the cell cycle was also evaluated. As shown in Figure 5C, a delay of the transition from G1 to S phase was observed, and the percentage of cells at the G1 phase was higher in cells treated by inhibitors than that in control cells. Interestingly, the G1 arrest of cells by inhibitors could be partially recovered by the addition of spermidine (Figure 5C), which further confirmed the role of spermidine in regulating cell cycle progression. In contrast to spermidine supplementation, the qRT-PCR analysis of the expression for cell cycle genes revealed that the treatment with inhibitors decreased the expression of all the genes (Figure 5D). Taken together, these results showed that the dynamic levels of polyamines in cultured silkworm cells play important roles in regulating cell cycle progression, and the traditional inhibitors of the polyamine pathway can be used to investigate polyamine functions in the silkworm.

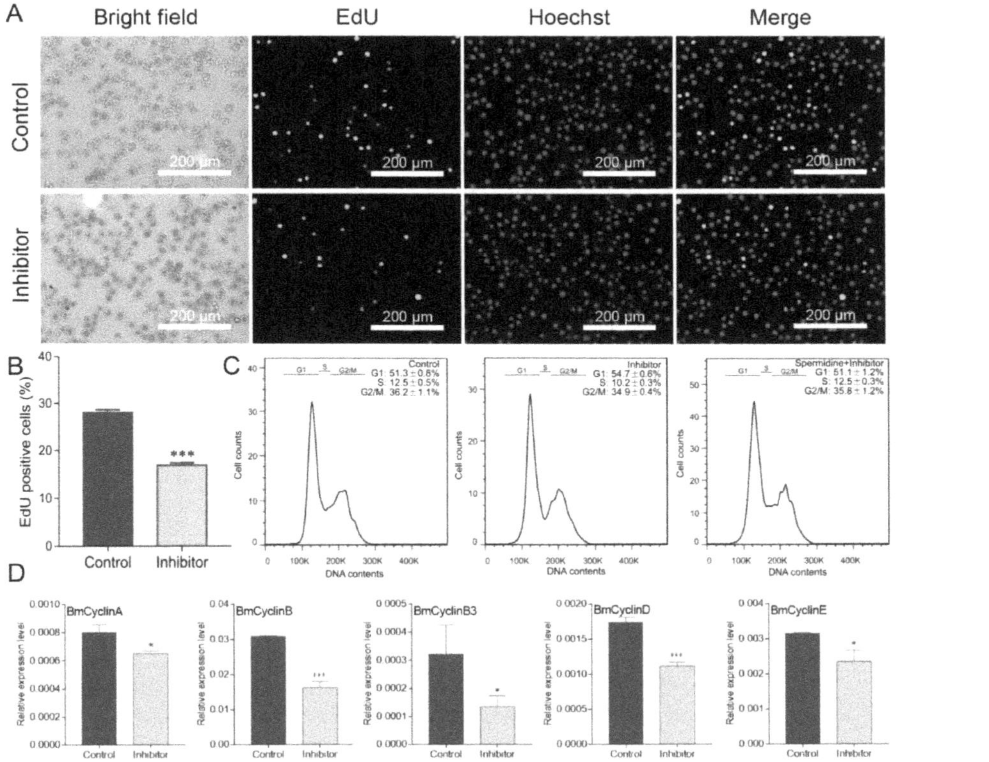

Figure 5. Inhibition of polyamine pathway activity decreased DNA replication in cultured silkworm cells. (**A**) BeyoClick™ EdU-488 staining was used to detect DNA replication. Scale bar, 200 μm. (**B**) EdU-positive cells (green) were analyzed by selecting three different fluorescent fields and counted by Photoshop software. The data presented are the means ± SD (n = 3). (**C**) Cell cycle distribution was analyzed by using a flow cytometry and cells in G1, S, or G2/M were quantitated by FlowJo software. Three repeats were done for each treatment and the percentage of cells are the means ± SD (n = 3). (**D**) Expression of cell cycle-related genes was detected by qRT-PCR. For significant analysis: * $p < 0.05$ and *** $p < 0.001$.

4. Discussion

In the current study, we identified the anabolic and catabolic enzymes that are involved in the polyamine biosynthetic pathway in the silkworm, and analyzed their expression characteristics in different tissues, as well as their subcellular localization patterns in cultured silkworm cells. Moreover, we investigated the role of the polyamine pathway on DNA replication and cell cycle progression by the exogenous additions of spermidine and inhibitors. The obtained data expanded the functional insights of polyamines in the lepidopteran model insect.

Biosynthesis of polyamines is regulated by a series of enzymes [7]. For example, ODC and APAO are involved in the synthesis of putrescine, SPDS and SMO co-regulate the synthesis of spermidine, while SPMS and SSAT regulate the homeostasis of spermidine and spermine. It was speculated that the enzymes regulating the synthesis of the same substance may have similar expression characteristics in a given tissue. Therefore, we analyzed the tissue expression profiles of polyamine pathway genes in the silkworm individuate. Although the tissue expression characteristics showed that polyamine pathway genes were expressed in all analyzed silkworm tissues, their expression levels were varied. It was interesting that the expression patterns of BmODC/BmAPAO, BmSPDS/BmSMO, and BmSPMS/BmSSAT are similar in different tissues, and have very high expression in a specific tissue. These observations showed that the varied expression abundances of polyamine pathway genes in different tissues contribute to different polyamine levels, and suggested that different levels of polyamines are important in different tissues.

Polyamines are involved in a number of biological processes, including promoting cell proliferation and cell cycle progression [6,13]. Consistent with this, exogenous addition of spermidine promotes DNA replication and cell cycle progression of cultured silkworm cells. Meanwhile, treatment of cells with polyamine pathway inhibitors DFMO and MGBG prevents DNA replication and the transition from G1 to S phase is also delayed, and importantly, the levels of putrescine, spermidine, and spermine are significantly decreased by inhibitors. Therefore, the efficacy of DFMO and MGBG for inhibiting polyamine levels in silkworm cells provides the potential to investigate polyamine functions by inhibitor treatment in silkworm individuals.

Furthermore, we found that the cell cycle-related genes, including BmCyclinA, BmCyclinB, BmCyclinB3, BmCyclinD, and BmCyclinE, were all upregulated by spermidine supplementation and downregulated by inhibitor treatment, respectively, which implied that spermidine may affect cell cycle progression by participating in the regulation of cell cycle gene expression. This is consistent with the view that polyamines can regulate gene expression by binding to negatively charged nucleic acids and altering chromatin and RNA structure or increasing translation efficiency [34]. Whether spermidine can directly regulate the cell cycle gene expression or not needs to be explored in the future.

Indeed, it has been reported that silkworm fed with spermidine-treated leaves enhanced the expression of BmFib-H, a gene encoding for fibroin heavy chain that is the largest and most abundant silk protein in the posterior silk glands of the silkworm [35]. Moreover, the quantity and quality of silk were improved by the spermidine treatment [27,28], which indicates the potential application of polyamines for better silk production. We also examined the silk properties after the injection of spermidine into the fifth instar larva of the silkworm strain stocked in our laboratory; however, no significant changes were observed in the cocoon weight and shell ratio of the spermidine-treated silkworm as compared with the control (data not shown). The difference in our findings might be due to the different silkworm strain or different treatment of spermidine supplement. However, we will detect whether spermidine could enhance the expression of silk proteins and regulate endoreplication of silk gland cells in the next work. Because of the critical role of ODC, a rate-limiting enzyme in polyamine synthesis, we are currently establishing the stable expression of ODC in the silkworm in order to increase the levels of polyamines in the silk glands. It is expected that this strategy could provide us novel insights into regulating the expressions of silk proteins and increasing the yields of silk in silkworm.

5. Conclusions

In summary, this study is the first to analyze the polyamine biosynthetic pathway genes in the lepidopteran silkworm and to investigate the role of spermidine on DNA replication and cell cycle progression of cultured silkworm cells, which demonstrated the conserved function of the polyamine pathway among species. Our data also supported that commercial polyamines and inhibitors can be used to investigate the roles of the polyamine pathway in silkworm cells and individuals. Future studies need to clarify the molecular mechanism by which polyamines regulate the expression of target genes in silkworm.

Supplementary Materials: The following are available online at https://www.mdpi.com/article/10.3390/insects12070624/s1, Table S1. List of primers used in this study.

Author Contributions: Conceptualization, Z.L.; Data curation, L.C. and Z.L.; Funding acquisition, Z.L., Q.X. and P.Z.; Investigation, L.C. and Z.L.; Methodology, H.G., W.Z., W.L., J.W. and G.S.; Project administration, Z.L.; Resources, Q.X. and P.Z.; Supervision, Z.L.; Writing—original draft preparation, L.C. and Z.L.; Writing—review and editing, Z.L. All authors have read and agreed to the published version of the manuscript.

Funding: This work was supported by National Natural Science Foundation of China (Nos. 32030103 and 31772532), Fundamental Research Funds for the Central Universities (No. XDJK2019B007), and Innovation and Entrepreneurship Training Program of Southwest University (No. S202010635009).

Institutional Review Board Statement: Not applicable.

Informed Consent Statement: Not applicable.

Data Availability Statement: Data are contained within the article and Supplementary Materials.

Conflicts of Interest: The authors declare no conflict of interest.

References

1. Tabor, C.W.; Tabor, H. Polyamines. *Annu. Rev. Biochem.* **1984**, *53*, 749–790. [CrossRef]
2. Tabor, C.W.; Tabor, H. Polyamines in microorganisms. *Microbiol. Rev.* **1985**, *49*, 81–99. [CrossRef]
3. Van Dam, L.; Korolev, N.; Nordenskiold, L. Polyamine-nucleic acid interactions and the effects on structure in oriented DNA fibers. *Nucleic Acids Res.* **2002**, *30*, 419–428. [CrossRef]
4. Igarashi, K.; Kashiwagi, K. Modulation of cellular function by polyamines. *Int. J. Biochem. Cell Biol.* **2010**, *42*, 39–51. [CrossRef] [PubMed]
5. Schuster, I.; Bernhardt, R. Interactions of natural polyamines with mammalian proteins. *Biomol. Concepts* **2011**, *2*, 79–94. [CrossRef]
6. Thomas, T.J. Polyamines in cell growth and cell death: Molecular mechanisms and therapeutic applications. *Cell. Mol. Life Sci.* **2001**, *58*, 244–258. [CrossRef] [PubMed]
7. Casero, R.A., Jr.; Stewart, T.M.; Pegg, A.E. Polyamine metabolism and cancer: Treatments, challenges and opportunities. *Nat. Rev. Cancer* **2018**, *18*, 681–695. [CrossRef] [PubMed]
8. Gerner, E.W.; Meyskens, F.L., Jr. Polyamines and cancer: Old molecules, new understanding. *Nat. Rev. Cancer* **2004**, *4*, 781–792. [CrossRef] [PubMed]
9. Coffino, P. Regulation of cellular polyamines by antizyme. *Nat. Rev. Mol. Cell Biol.* **2001**, *2*, 188–194. [CrossRef] [PubMed]
10. Yuan, Q.; Ray, R.M.; Viar, M.J.; Johnson, L.R. Polyamine regulation of ornithine decarboxylase and its antizyme in intestinal epithelial cells. *Am. J. Physiol. Gastrointest. Liver Physiol.* **2001**, *280*, G130–G138. [CrossRef]
11. Wallace, H.M. The physiological role of the polyamines. *Eur. J. Clin. Investig.* **2000**, *30*, 1–3. [CrossRef]
12. Del Rio, B.; Redruello, B.; Linares, D.M.; Ladero, V.; Ruas-Madiedo, P.; Fernandez, M.; Martin, M.C.; Alvarez, M.A. Spermine and spermidine are cytotoxic towards intestinal cell cultures, but are they a health hazard at concentrations found in foods? *Food Chem.* **2018**, *269*, 321–326. [CrossRef]
13. Yamashita, T.; Nishimura, K.; Saiki, R.; Okudaira, H.; Tome, M.; Higashi, K.; Nakamura, M.; Terui, Y.; Fujiwara, K.; Kashiwagi, K.; et al. Role of polyamines at the G1/S boundary and G2/M phase of the cell cycle. *Int. J. Biochem. Cell Biol.* **2013**, *45*, 1042–1050. [CrossRef]
14. Oredsson, S.M. Polyamine dependence of normal cell-cycle progression. *Biochem. Soc. Trans.* **2003**, *31*, 366–370. [CrossRef] [PubMed]
15. Rider, J.E.; Hacker, A.; Mackintosh, C.A.; Pegg, A.E.; Woster, P.M.; Casero, R.A., Jr. Spermine and spermidine mediate protection against oxidative damage caused by hydrogen peroxide. *Amino Acids* **2007**, *33*, 231–240. [CrossRef]
16. Warters, R.L.; Newton, G.L.; Olive, P.L.; Fahey, R.C. Radioprotection of human cell nuclear DNA by polyamines: Radiosensitivity of chromatin is influenced by tightly bound spermine. *Radiat. Res.* **1999**, *151*, 354–362. [CrossRef] [PubMed]
17. Eisenberg, T.; Knauer, H.; Schauer, A.; Buttner, S.; Ruckenstuhl, C.; Carmona-Gutierrez, D.; Ring, J.; Schroeder, S.; Magnes, C.; Antonacci, L.; et al. Induction of autophagy by spermidine promotes longevity. *Nat. Cell Biol.* **2009**, *11*, 1305–1314. [CrossRef] [PubMed]
18. Madeo, F.; Eisenberg, T.; Büttner, S.; Ruckenstuhl, C.; Kroemer, G. Spermidine: A novel autophagy inducer and longevity elixir. *Autophagy* **2010**, *6*, 160–162. [CrossRef]

19. Til, H.; Falke, H.; Prinsen, M.; Willems, M. Acute and subacute toxicity of tyramine, spermidine, spermine, putrescine and cadaverine in rats. *Food Chem. Toxicol.* **1997**, *35*, 337–348. [CrossRef]
20. Pegg, A.E. Toxicity of Polyamines and Their Metabolic Products. *Chem. Res. Toxicol.* **2013**, *26*, 1782–1800. [CrossRef]
21. Metcalf, B.W.; Bey, P.; Danzin, C.; Jung, M.J.; Casara, P.; Vevert, J.P. Catalytic irreversible inhibition of mammalian ornithine decarboxylase (E.C.4.1.1.17) by substrate and product analogs. *J. Am. Chem. Soc.* **1978**, *100*, 2551–2553. [CrossRef]
22. Seppänen, P.; Alhonen-Hongisto, L.; Jänne, J. Relation of the Antiproliferative Action of Methylglyoxal-bis(guanylhydrazone) to the Natural Polyamines. *Eur. J. Biochem.* **1980**, *110*, 7–12. [CrossRef] [PubMed]
23. Qu, N.; Ignatenko, N.A.; Yamauchi, P.; Stringer, D.E.; Levenson, C.; Shannon, P.; Perrin, S.; Gerner, E.W. Inhibition of human ornithine decarboxylase activity by enantiomers of difluoromethylornithine. *Biochem. J.* **2003**, *375*, 465–470. [CrossRef] [PubMed]
24. Pegg, A.E. S-Adenosylmethionine decarboxylase. *Essays Biochem.* **2009**, *46*, 25–45. [CrossRef]
25. Xia, Q.; Li, S.; Feng, Q. Advances in Silkworm Studies Accelerated by the Genome Sequencing of *Bombyx mori*. *Annu. Rev. Entomol.* **2014**, *59*, 513–536. [CrossRef] [PubMed]
26. International Silkworm Genome Consortium. The genome of a lepidopteran model insect, the silkworm *Bombyx mori*. *Insect Biochem. Mol. Biol.* **2008**, *38*, 1036–1045. [CrossRef]
27. Lattala, G.M.; Kandukuru, K.; Gangupantula, S.; Mamillapalli, A. Spermidine enhances the silk production by mulberry silkworm *J. Insect Sci.* **2014**, *14*, 207. [CrossRef] [PubMed]
28. Yerra, A.; Mysarla, D.K.; Siripurapu, P.; Jha, A.; Valluri, S.V.; Mamillapalli, A. Effect of polyamines on mechanical and structural properties of *Bombyx mori* silk. *Biopolymers* **2017**, *107*, 20–27. [CrossRef]
29. Pan, M.-H.; Cai, X.-J.; Liu, M.; Lv, J.; Tang, H.; Tan, J.; Lu, C. Establishment and characterization of an ovarian cell line of the silkworm, *Bombyx mori*. *Tissue Cell* **2010**, *42*, 42–46. [CrossRef]
30. Li, Z.; Mon, H.; Xu, J.; Zhu, L.; Lee, J.M.; Kusakabe, T. A conserved SUMOylation signaling for cell cycle control in a holocentric species *Bombyx mori*. *Insect Biochem. Mol. Biol.* **2014**, *51*, 71–79. [CrossRef]
31. Li, Z.; Cui, Q.; Xu, J.; Cheng, D.; Wang, X.; Li, B.; Lee, J.M.; Xia, Q.; Kusakabe, T.; Zhao, P. SUMOylation regulates the localization and activity of Polo-like kinase 1 during cell cycle in the silkworm, *Bombyx mori*. *Sci. Rep.* **2017**, *7*, 15536. [CrossRef] [PubMed]
32. Igarashi, K.; Kashiwagi, K.; Hamasaki, H.; Miura, A.; Kakegawa, T.; Hirose, S.; Matsuzaki, S. Formation of a compensatory polyamine by Escherichia coli polyamine-requiring mutants during growth in the absence of polyamines. *J. Bacteriol.* **1986**, *166*, 128–134. [CrossRef] [PubMed]
33. Liu, L.-P.; Deng, Z.-N.; Qu, J.-W.; Yan, J.-W.; Catara, V.; Li, D.-Z.; Long, G.-Y.; Li, N. Construction of EGFP-Labeling System for Visualizing the Infection Process of *Xanthomonas axonopodis* pv. *citri in planta*. *Curr. Microbiol.* **2012**, *65*, 304–312. [CrossRef]
34. Miller-Fleming, L.; Olin-Sandoval, V.; Campbell, K.; Ralser, M. Remaining Mysteries of Molecular Biology: The Role of Polyamines in the Cell. *J. Mol. Biol.* **2015**, *427*, 3389–3406. [CrossRef] [PubMed]
35. Ma, S.; Shi, R.; Wang, X.; Liu, Y.; Chang, J.; Gao, J.; Lu, W.; Zhang, J.; Zhao, P.; Xia, Q. Genome editing of *BmFib-H* gene provides an empty *Bombyx mori* silk gland for a highly efficient bioreactor. *Sci. Rep.* **2014**, *4*, 6867. [CrossRef]

insects

MDPI

Article

Reference Transcriptome Data in Silkworm *Bombyx mori*

Kakeru Yokoi [1,2,*,†], Takuya Tsubota [3,†], Akiya Jouraku [1], Hideki Sezutsu [3] and Hidemasa Bono [4,5]

1 Insect Genome Research and Engineering Unit, Division of Applied Genetics, Institute of Agrobiological Sciences (NIAS), National Agriculture and Food Research Organization (NARO), 1-2 Owashi, Tsukuba, Ibaraki 305-8634, Japan; joraku@affrc.go.jp
2 Research Center for Agricultural Information Technology (RCAIT), National Agriculture and Food Research Organization (NARO), Kintetsu Kasumigaseki Building Kasumigaseki 3-5-1, Chiyoda-ku, Tokyo 100-0013, Japan
3 Transgenic Silkworm Research Unit, Division of Biotechnology, Institute of Agrobiological Sciences (NIAS), National Agriculture and Food Research Organization (NARO), 1-2 Owashi, Tsukuba, Ibaraki 305-8634, Japan; tsubota@affrc.go.jp (T.T.); hsezutsu@affrc.go.jp (H.S.)
4 Database Center for Life Science (DBCLS), Joint Support-Center for Data Science Research, Research Organization of Information and Systems, 1111 Yata, Mishima, Shizuoka 411-8540, Japan; bonohu@hiroshima-u.ac.jp
5 Program of Biomedical Science, Graduate School of Integrated Sciences for Life, Hiroshima University, 3-10-23 Kagamiyama, Higashi-Hiroshima City, Hiroshima 739-0046, Japan
* Correspondence: yokoi123@affrc.go.jp; Tel.: +81-29-838-6129
† These authors equally contributed to this work.

Simple Summary: The silkworm *Bombyx mori* is a lepidopteran model insect with biological and industrial importance. Its high-quality genome sequence has recently become available and the utilization of this information, in combination with extensive transcriptomic analyses, is expected to provide an elaborate gene model. It will also be possible to clarify the gene expression in detail using this approach. In the present study, we established reference transcriptome data for the silkworm and performed a detailed examination of the gene expression in silkworm tissues. The results obtained will contribute to our understanding of silkworm biology and further promote the industrial application of the silkworm and other insects.

Abstract: Herein, we performed RNA-seq analysis of ten major tissues/subparts of silkworm larvae. The sequences were mapped onto the reference genome assembly and the reference transcriptome data were successfully constructed. The reference data provided a nearly complete sequence for *sericin-1*, a major silk gene with a complex structure. We also markedly improved the gene model for other genes. The transcriptomic expression was investigated in each tissue and a number of transcripts were identified that were exclusively expressed in tissues such as the testis. Transcripts strongly expressed in the midgut formed tight genomic clusters, suggesting that they originated from tandem gene duplication. Transcriptional factor genes expressed in specific tissues or the silk gland subparts were also identified. We successfully constructed reference transcriptome data in the silkworm and found that a number of transcripts showed unique expression profiles. These results will facilitate basic studies on the silkworm and accelerate its applications, which will contribute to further advances in lepidopteran and entomological research as well as the practical use of these insects.

Keywords: silkworm; *Bombyx mori*; transcriptome analysis; RNA-seq; gene model transcriptional factor

Citation: Yokoi, K.; Tsubota, T.; Jouraku, A.; Sezutsu, H.; Bono, H. Reference Transcriptome Data in Silkworm *Bombyx mori*. *Insects* **2021**, *12*, 519. https://doi.org/10.3390/insects12060519

Academic Editors: Silvia Cappellozza, Morena Casartelli, Federica Sandrelli, Alessio Saviane and Gianluca Tettamanti

Received: 22 April 2021
Accepted: 28 May 2021
Published: 3 June 2021

1. Introduction

The silkworm *Bombyx mori* is a lepidopteran insect that has been utilized in studies of physiology, genetics, molecular biology, and pathology. Functional analyses of genes related to hormone synthesis/degradation, pheromone reception, larval marking formation, and virus resistance have been performed using this silkworm [1–5], and the findings obtained have contributed to the promotion of insect science. The silkworm has the ability to produce large amounts of silk proteins, which is one of the most prominent characteristics of this species. Silk proteins are mainly composed of the fibrous protein Fibroin and aqueous protein sericin, which are produced in the larval tissue silk gland (SG) [6]. A transgenic technique has been applied to the silkworm [7] that has enabled the production of a large amount of recombinant proteins through the introduction of transgenes for overexpression in SG [8]. The silkworm can be utilized as a significant bioreactor through this approach.

Based on its biological and industrial importance, the whole genome sequence of the silkworm was reported in 2004 by two research groups [9,10]. This was the first lepidopteran genomic analysis and has served as a fundamental basis for genomic studies on Lepidoptera. This silkworm genome data were updated in 2008 [11], and related data have since become available, including microarray-based gene expression profiles, a BAC-based linkage map, and full-length cDNA data [12–14]. These data have strongly promoted studies on *B. mori* and other lepidopteran insects in the past few decades.

A new and high-quality reference genome assembly of the silkworm p50T (*daizo*) strain using PacBio long-read and Illumina short-read sequencers was reported in 2019 [15]. The new genome assembly consists of 696 scaffolds with N50 of 16.8 Mb and only 30 gaps. A predicted new gene model based on this novel genome assembly, using cDNA, protein, and RNA-seq data as hints, was constructed and was more precise that the previous model made via the old genome assembly [15]. The next step in the establishment of genome-related data is transcriptome data, which contain reference transcriptome sequence data and gene expression profiles in major tissues. These data will significantly contribute to advances in research on the silkworm and other lepidopterans.

In the present study, we constructed a reference transcriptome dataset using RNA-seq data obtained from ten major tissues/subparts of silkworm larvae (Figure 1). RNA-seq data were mapped on the new genome assembly and reference transcriptome sequence data were successfully constructed. We also performed functional annotation of the reference transcriptome using human and *Drosophila* protein datasets. in addition to NCBI-nr data. The established reference transcriptome sequence data provided a nearly complete structure for *sericin-1* (*Ser1*), a major silk gene with a complex sequence. The expression of the transcriptome was investigated in each tissue, and the expression of a number of transcripts was found to be confined to tissues such as the testis (TT). Among them, genes with transcripts that were strongly expressed in the midgut (MG) formed tight genomic clusters, suggesting that they originated via gene duplication. The transcripts of transcriptional factor (TF) genes expressed in specific tissues or SG subparts were also detected, and we speculate that these genes play key roles in the major biological process of these tissues/territories. The present results will accelerate molecular biological studies on the silkworm as well as other related species, and this is an essential milestone to promote entomological research as well as the practical use of insects.

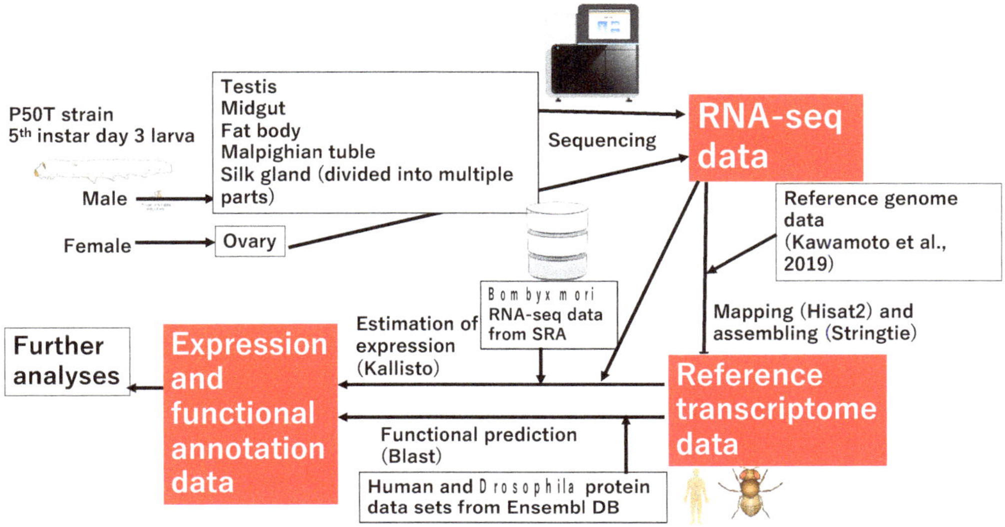

Figure 1. Workflow of the data analysis performed in the present study. To obtain reference transcriptome sequences, Fastq data of 10 tissues/subparts from fifth instar larvae were mapped to the new reference genome [15]. Kallisto software was used to estimate the expression abundance of each transcript in these tissues. Regarding RNA-seq data obtained from the public database (o751 strain; see Table 1). We performed a Blast search against human and *Drosophila* genome data to perform functional annotations of the reference transcriptome. The images in Figure 1 is from TogoTV (© 2016 DBCLS TogoTV).

2. Materials and Methods

2.1. Silkworm Rearing, RNA Extraction, and Sequencing

The silkworm p50T strain was reared on an artificial diet (Nihon Nosan Kogyo, Yokohama, Japan) at 25 °C under a 12-h light/dark photoperiod. The SG, fat body (FB), MG, Malpighian tuble (MT), TT, and ovary (OV) were dissected on the third day of the fifth instar larvae. The SG was further subdivided into the anterior silk gland (ASG), anterior part of the middle silk gland (MSG_A), middle part of the middle silk gland (MSG_M), posterior part of the middle silk gland (MSG_P), and posterior silk gland (PSG). Each tissue/subpart was dissected from one individual, and three biological replicates were obtained and separately analyzed (see Table 1). The tissues were homogenized using ISOGEN (NIPPON GENE, Tokyo, Japan) and the SV Total RNA Isolation System (Promega, Madison, WI, Tara) was used for the RNA extraction. The total RNA samples extracted were sequenced by Illumina NovaSeq6000 (Macrogen Japan Corp., Kyoto, Japan).

2.2. Construction of RTD and Estimation of the Expression of Each Transcript

The raw RNA-seq data of 30 samples were trimmed by Trimmomatic version 0.36 [16]. The trimmed RNA-seq data of each tissue were mapped to the new reference genome with a new gene model using HISAT2 version 2.1.0 [15,17]. The mapped data were each assembled to transcriptome data by StringTie version 1.3.3 [18]. The 30 transcriptome datasets were merged into one transcriptome dataset, referred to as "a reference transcriptome" by StringTie. GffCompare version 0.10.6 was used (URL: https://ccb.jhu.edu/software/stringtie/gffcompare.shtml: Accessed on 1 June 2021) for comparisons with the reference transcriptome and previously reported gene sets [15]. The transcripts detected at the newly identified loci were categorized into the "new loci" group, those newly detected at the previously identified loci into the "new isoform" group, and other genes and transcripts into the "identified in gene models".

To estimate the expression of the reference transcriptome in 30 samples, the raw fastq data of each sample and the reference transcript data were used with Kallisto version 0.44.0 [19]. The raw RNA-seq data of multiple tissues in B. mori strain o751 from the Sequence Read Archive (SRA) and reference transcript data were used in comparisons of the transcriptome data; the accession numbers of the raw RNA-seq data are DRA005094 DRA005878, and DRA005094 [20–22].

We used TIBCO Spotfire Desktop (version 7.6.0) software with the "Better World" program license (TIBCO, Inc., Palo Alto, CA, USA; Tara, URL: http://spotfire.tibco.com/better-world-donation-program/: Accessed on 1 June 2021) for the classification of differentially expressed samples in silkworm tissues in HC using Ward's method. Morpheus was also used for HC (https://software.broadinstitute.org/morpheus: Accessed on 1 June 2021). R (version 3.6.0) was used in the PCA analysis. Regarding the Venn diagram construction and the scatter plot analysis, R (version 4.0.2) was used. The relationships among the gene expression profiles in the SG territories were evaluated using Spearman's rank correlation.

2.3. Annotation for the Reference Transcriptome and Functional Enrichment Analyses

Transcoder (version 5.5.0: URL: https://transdecoder.github.io/: Accessed on 1 June 2021) was used to identify the coding regions within the transcript sequences and to convert transcript sequences to amino acid sequences. Transcriptome sequence sets were compared at the predicted amino acid sequence level by the successive execution of the blastp program in the NCBI BLAST software package (version 2.9.0+), with default parameters and an e-value cut-off of 1e-10 [23]. Regarding the reference database sets to be blasted, human and fruit fly (*D. melanogaster*) protein datasets in the Ensembl database (version 97) were used because the sequences of these organisms were functionally well-annotated and amenable to computational methods, such as a pathway analysis [24]. The names of the top-hit genes in the human and fruit fly datasets were annotated to *B. mori* transcripts utilizing Ensembl Biomart [25] and Spotfire Desktop software under TIBCO Spotfire's "Better World" program license. Functional enrichment analyses (FEA) were performed using the metascape portal site (version 3.5 on 10 May 2021) [26] with annotation results against the human gene set.

2.4. Comparison of Gene Structures among Different Models

The gene structures among the reference transcriptome data (RTD), gene model data (GMD), and cDNA-based data were compared in the silkBase (URL: http://silkbase.ab.a.u-tokyo.ac.jp/cgi-bin/index.cgi: Accessed on 1 June 2021) or KAIKObase [27]. Amino acid sequences were aligned using CLC Genomics Workbench 20.0.04 (QIAGEN, Aarhus, Denmark)

Table 1. Samples for RNA-seq and run accession IDs.

Sample	SRA Run ID	Strain	Sex	Reference
ASG-1,2,3	DRR186474,DRR186475,DRR186476	p50T	Male	This study
MSG_A-1,2,3	DRR186477,DRR186478,DRR186479	p50T	Male	This study
MSG_M-1,2,3	DRR186480,DRR186481,DRR186482	p50T	Male	This study
MSG_P-1,2,3	DRR186483,DRR186484,DRR186485	p50T	Male	This study
PSG-1,2,3	DRR186486,DRR186487,DRR186488	p50T	Male	This study
FB-1,2,3	DRR186489,DRR186490,DRR186491	p50T	Male	This study
MG-1,2,3	DRR186492,DRR186493,DRR186494	p50T	Male	This study
MT-1,2,3	DRR186495,DRR186496,DRR186497	p50T	Male	This study
TT-1,2,3	DRR186498,DRR186499,DRR186500	p50T	Male	This study
OV-1,2,3	DRR186501,DRR186502,DRR186503	p50T	Female	This study
BN_TT-1,2,3	DRR068893,DRR068894,DRR068895	o751	Male	[25]
BN_FB-1,2,3	DRR095105,DRR095106,DRR095107	o751	Male	[27]
BN_MG-1,2,3	DRR095108,DRR095109,DRR095110	o751	Male	[26]
BN_MT-1,2,3	DRR095111,DRR095112,DRR095113	o751	Male	[27]
BN_SG-1,2,3	DRR095114,DRR095115,DRR095116	o751	Male	[27]

2.5. RT-PCR

cDNA was synthesized using Superscript IV (Thermo Fisher Scientific Inc., Waltham, MA, USA) according to the manufacturer's instructions. Five hundred nanograms of the total RNAs extracted from the ASG, MSG_A, MSG_M, MSG_P, and PSG were used for the cDNA synthesis. KOD FX neo polymerase (Toyobo, Osaka, Japan) was used for RT-PCR. The PCR conditions were as follows: 95 °C for 1 min, followed by 22 cycles (for rp49) or 30 cycles (for TF genes) of 95 °C for 30 s, 58 °C for 30 s, followed by 68 °C for 1 min, and additional 68 °C for 1 min after the cyclic phase. The primer sequences are listed in Table S1.

3. Results

3.1. Reference Transcriptome Data

We performed a transcriptomic analysis of the major silkworm larval tissues, namely, the SG, FB, MG, MT, TT, and OV, to acquire more expanded RNA-seq data (Figure 2 and Table 1). The gene expressions were clearly differentiated among subregions in the SG [28,29], and, thus, we subdivided the SG into five subparts and investigated the gene expression in each region (ASG, MSG_A, MSG_M, MSG_P, and PSG; Figure 2 and Table 1). In total, ten tissues/subparts were dissected from the fifth instar third day larvae of p50T strain with three biological replicates (Table 1). Thirty sets of RNAs were used for RNA-seq. We mapped the RNA-seq data on the reference genome assembly using information from the previously established gene model [15] (hereafter referred to as gene model data (GMD)) and constructed reference transcriptome data (RTD). Hereafter in this article, we defined "gene" and "transcript" as a representative sequence producing a protein of single or multiple isoforms in a single loci shown in GMD data and a isoform derived from a gene shown in RTD data. RTD comprises 51,926 transcripts in 24,236 loci (Figure 3A), the numbers of which are higher than those of GMD (24,236 vs. 16,845 loci and 51,926 vs. 16,880 transcripts; see Figure 3A,B). Therefore, RTD is an extension of GMD. To perform functional annotations, coding sequence (CDS) regions and amino acid sequence data were constructed using RTD (Data S1), and it was found that 39,619 transcripts, derived from 16,632 loci, had at least one CDS in RTD. The predicted amino acid sequences were used for gene functional annotations through a homology search against human and Drosophila gene sets. This analysis revealed that 26,698 transcripts showed homology to human transcripts and 29,177 to fruit fly transcripts (Data S2). We also performed a blastp analysis using the NCBI nr database and found that 43,358 amino acids had homological proteins in this database (Data S3).

3.2. Comparison between Constructed Reference Transcriptome Data and Previous Gene Model Data

RTD represents a marked improvement over GMD. Several misassembled genes are present in GMD, as represented by the KWMTBOMO00087-88, KWMTBOMO00196-197, or KWMTBOMO00222-223 genes [15]. These genes are split into two structures, even though full-length cDNA data define them as single genes: BMgn002111, BMgn000626, and BMgn000572 (full-length cDNA data ID) covers the KWMTBOMO00087-88, 00196-19,7 and 00222-223 gene regions, respectively (Figure S1) [14,15]. We investigated their structures in our model and found that all were accurately predicted (MSTRG.494.1, MSTRG.649.1-2, and MSTRG.704.1-3; Figure S1). The elucidated gene structure was attributed to the extensive RNA-seq analysis in the present study; the previous study lacked gene expression data from tissues such as TT, OV, MT, and PSG, and these genes were all strongly expressed in these tissues (Table S2).

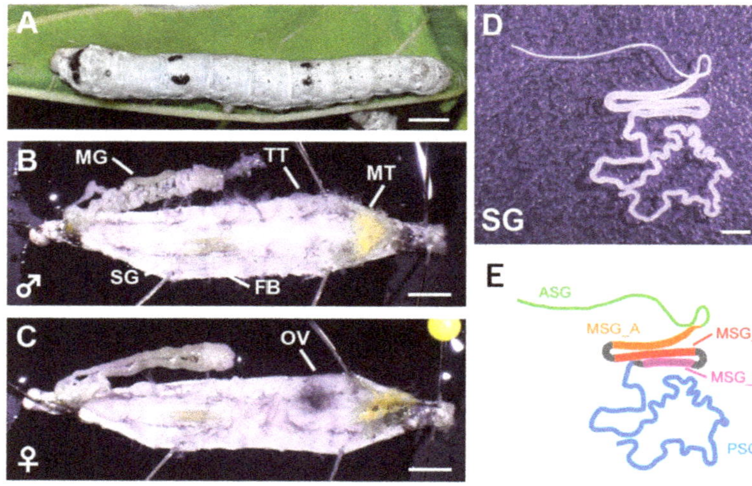

Figure 2. Tissues used in this study. (**A**) Final (fifth) instar larva of the silkworm p50T (daizo) strain. Scale bar = 5 mm. (**B**,**C**) Male (**B**) or (**C**) female individuals dissected on the third day of fifth instar larvae. Scale bar = 5 mm. MG—mid gut; TT—testis; MT—Malpighian tubules; SG—silk gland; FB—fat body; OV—ovary. (**D**,**E**) Image (**D**) and schematic (**E**) of the silk gland. ASG—anterior silk gland; MSG_A—anterior part of the middle silk gland; MSG_M—middle part of the middle silk gland; MSG_P—posterior part of the middle silk gland; PSG—posterior silk gland. Scale bar = 2.5 mm.

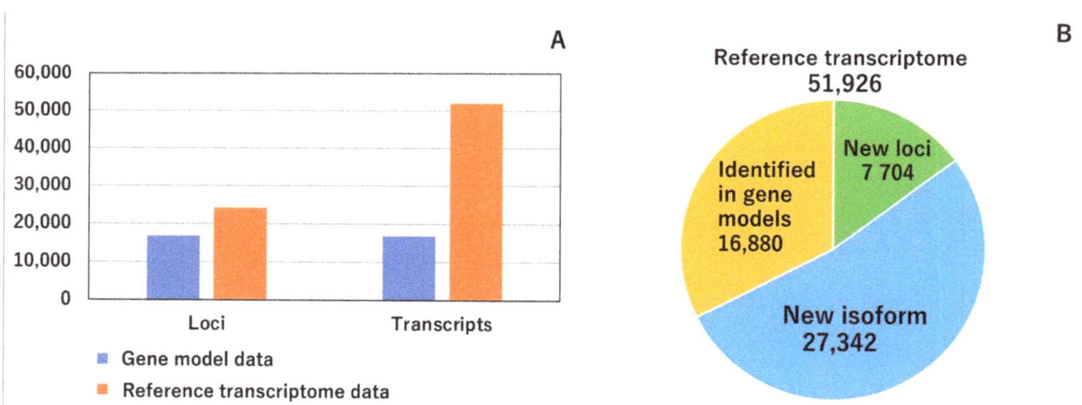

Figure 3. Basal characteristics of the reference transcriptome. (**A**) Comparison of the gene model data [15] and the reference transcriptome data of the present study. The numbers of the loci and transcripts are shown. These numbers were calculated from gff files of the two data sets. (**B**) Classification of 51,926 transcripts. Each transcript was classified into three categories, and the numbers of the three categories are shown in a pie chart. Definitions of the three categories are described in the main text.

RTD also provided a number of novel genes/isoforms. Comparisons of RTD with GMD revealed that among the 51,926 transcripts identified, 7704 belonged to the new loci group, whereas 27,342 were categorized as new isoforms (Figure 3B, see Materials and Methods). Among the 7704 new loci group transcripts, a number of transcripts were also present in the previously established full-length cDNA-based gene model [14]. However, 2324 transcripts did not hit this gene set and, thus, the genes to which these transcripts belong were perceived to be novel genes. An expression analysis revealed that many of these transcripts belonging to these genes were commonly expressed in all of the tissues investigated herein, whereas the other transcripts were exclusively expressed in specific tissues, such as TT (Figure S2). The functional annotation analysis revealed that newly identified transcripts included a trypsin inhibitor (MSTRG.14562.2), carboxypeptidase (MSTRG.16874.1-3), and pyruvate kinase (MSTRG.18651.2). Therefore, our RTD represents a significant improvement over GMD (Additional file 4).

3.3. Elaborated Structures of Silk Genes

Silk production is one of the most prominent characteristics of the silkworm. Silk genes are strongly expressed in the silk-producing tissue SG, and our extensive RNA-seq data are expected to show highly elaborated models for these genes. *Ser1* is one of the major silk genes, is strongly expressed in the MSG, and encodes a >400-kDa serine-rich protein [6,30]. *Ser1* is composed of nine exons, among which exon 6 has a long repetitive sequence with a length of ~6500 bp [6,31]. The full-length sequence of this exon is yet to be elucidated because of its complexity. We demonstrate herein that our model MSTRG.2477.1 provided an almost complete sequence for this exon (6234 bp; Figure S3). Exon 6 encodes serine, glycine, threonine, asparagine, and aspartic acid-rich residues (Figure S4), which is consistent with previous findings showing that *Ser1* comprises large numbers of these residues (Table S3) [32]. The detailed structural analysis revealed that the long repetitive motif identified here comprised 53 repeats of a 38-amino acid unit (Figure S5). Each unit had serine-rich residues and a slight difference was observed in the sequences among units (Figure S5) [6]. The 38-amino acid-based repeat unit was also observed in exon 8 of *Ser1* (Figures S4 and S5) or in the sericins of saturniid species [6,33], and, thus, the repeat unit of this length is expected to have a structural function in a number of sericin proteins.

We also observed significant improvements for other sericin genes. *Sericin-3* (*Ser3*) is another major silk protein that has a relatively soft texture and possesses serine-rich residues [34,35]. In GMD, a 73-bp deletion was detected in exon 3, and because of this structural error, a frame shift was present in the predicted amino acid sequence (KWMTBOMO06311; Figure S6). In contrast, our RTD (MSTRG.2595.1) successfully provided an accurate gene structure (Figure S6). *Sericin-4* (*Ser4*) is another sericin protein that is composed of 34 exons [36]. This gene is split into three distinct structures in GMD (KWMTBOMO06324, KWMTBOMO06325, and KWMTBOMO06326), whereas RTD provided an exact model (MSTRG.2610.1; Figure S7). Collectively, these results suggest that our RTD provided highly defined structures, even for complex silk genes.

3.4. Estimating the Abundance of the Reference Transcriptome in Multiple Tissues

Our extensive transcriptomic analysis provided fundamental insights into the transcriptomic expression profiles in multiple silkworm tissues. The expression abundance of each transcript was calculated as transcripts per million (tpm; Additional file 14) and the transcriptomic expression was compared among tissues through two independent methods, one for hierarchical clustering (HC) and another for a principal component analysis (PCA). To avoid the effects of low expression transcripts, we performed these analyses using transcripts with tpm values > 30 in at least one sample. HC using all transcript tpm data were also performed for comparison. These analyses revealed that the biological replicate samples collected herein were highly reproducible, because they were derived from the same tissues forming tight clusters (Figure 4; Figure S8). In the HC analysis, a single cluster was formed for the MSG_M and MSG_P in each sample (Figure 4; Figure S8B), and we speculated that this was as a result of the highly conserved transcriptomic expression between these SG territories. A correlation analysis of the transcriptomic expression using SG transcriptome data supported this hypothesis (Table S4). A previous study performed an RNA-seq analysis of multiple silkworm larval tissues in another silkworm strain, o751 [20–22], and these RNA-seq data were added to our analysis (Figure 4, Table 1, and Figure S8; these samples are referred to as BN_MG, BN_FB, BN_MT, BN_SG, and BN_TT). We found that the samples collected from the same tissues clearly formed clusters (Figure 4). In the o751 strain, the SG was collected as a whole tissue. Although all of the SG subparts (MSG_A, MSG_M, MSG_P, ASG, and PSG) were closely located in the HC analysis of transcripts with >30 (tpm) in at least one sample, BN_SG formed a single cluster distant from those of the SG subparts (Figure 4). Therefore, our transcriptomic data are a robust platform for analyzing and comparing the gene expressions in multiple tissues

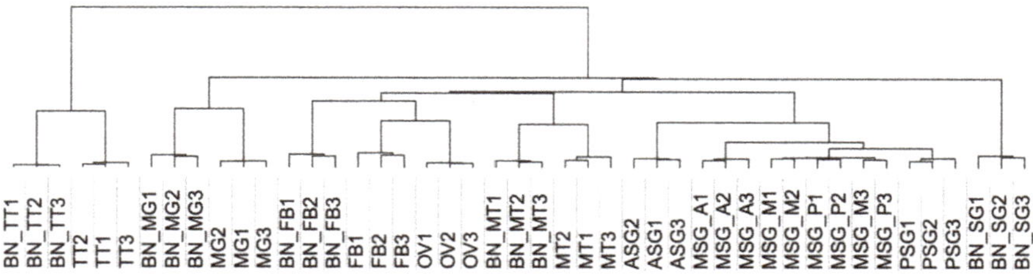

Figure 4. Hierarchical clustering of expression data in 45 samples of transcripts showing a tpm value > 30 in at least one sample. Abbreviations that start with "BN" indicate samples collected in a previous study [20–22]. The numbers added to the abbreviations mean biological replicates.

3.5. Transcript Abundance in Each Tissue and in the Silk Gland

Using the data described above, we investigated the transcriptomic expression in detail in each tissue. We focused on transcripts with a tpm value > 30, which accounted for approximately the top 5% of the most strongly expressed transcripts, and regarded such transcripts as being expressed in each tissue. To reveal the whole profile of these strongly expressed transcripts, we investigated which tissues these transcripts were expressed in, and counted the number of transcripts expressed in a single or multiple tissues (Figure 5). We found that 711 transcripts were expressed in all of the tissues (Figure 5), suggesting ubiquitous functions. We also detected transcripts expressed in specific tissues (Figure 5 and Table S5). Among them, transcripts solely expressed in TT were the most abundant (1882), followed by those expressed in OV (799), MT (499), and MG (440) (Figure 5 and Table S5). We also identified transcripts expressed in more than two tissues, such as TT and OV (397; Figure 5). FEA was performed on transcripts with a tissue-restricted expression, and the functional clusters enriched in each tissue were very diversified; for example, transcripts exclusively expressed in the TT were strongly enriched for "cilium organization", "Huntington's disease", and "cilium or flagellum-dependent cell motility" (Figure 6A), whereas "Metabolism of RNA", "regulation of mRNA metabolic process", and "ribonucleoprotein complex biogenesis" were enriched in the OV (Figure 6B). Comparisons of the transcript expression levels among the tissues revealed that the expression levels of the strongly expressed transcripts were very high in the ASG (Figure 7). These transcripts comprised fungal protease inhibitors, cuticular protein genes, and others (Table S6). In contrast, the levels of strongly expressed transcripts were lower in the MSG-M/MSG-P (Figure 7 and Table S6). We also investigated the expression profiles of the transcripts strongly expressed in each tissue, and found tissue-restricted expression for these transcripts in the MG, FB, and MSG_A and a ubiquitous expression for those in the other tissues examined (Table S6 and Figure S9). The genomic positions of the tissue-enriched transcripts were examined, and the transcripts strongly expressed in the MG formed tight genomic clusters (Figure 8 and Table S6). We also found clusters for strongly expressed transcripts in other tissues (Table S6 and Figure S10).

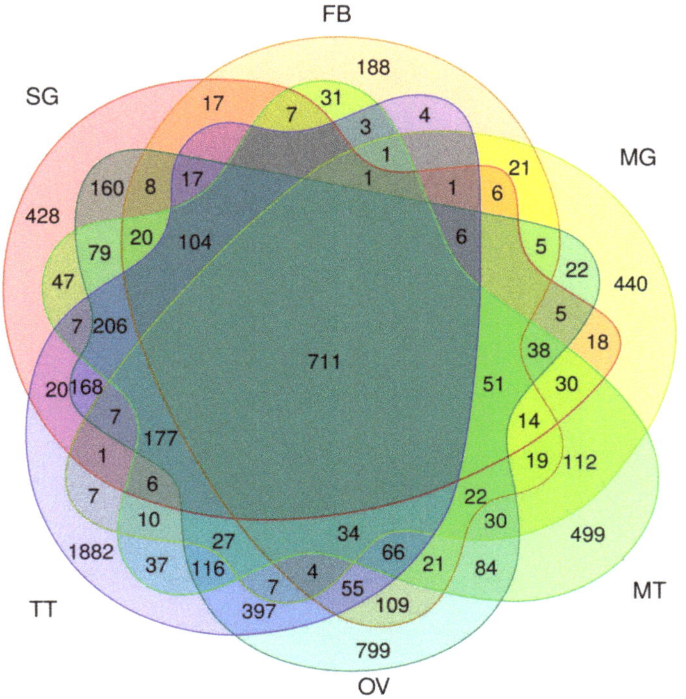

Figure 5. Venn diagram showing the transcripts expressed in each tissue. The number of transcripts with a tpm value > 30 is shown. SG—silk gland; FB—fat body; MG—mid gut; MT—Malpighian tubules; OV—ovary; TT—testis.

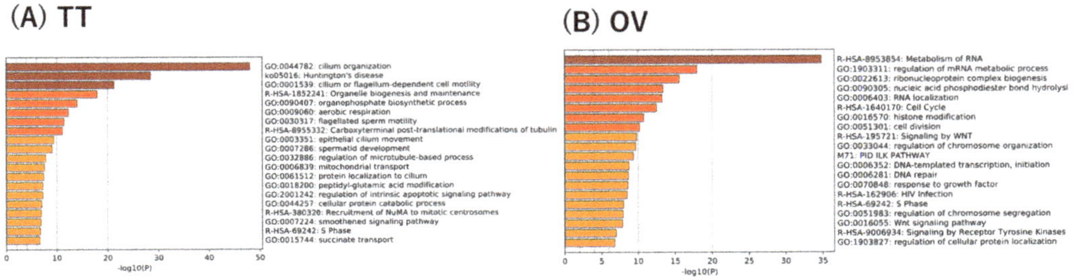

Figure 6. Results of the enrichment analysis by Metascape in the testis (TT) (**A**) and ovary (OV) (**B**). An enrichment analysis was performed using annotation data against the human gene set of the reference transcripts expressed in specific tissues. $-\log 10$ (P) represents $-\log 10$ (p-value). For example, $-\log 10$ (p) = 5 represents p-value = 10^{-5}.

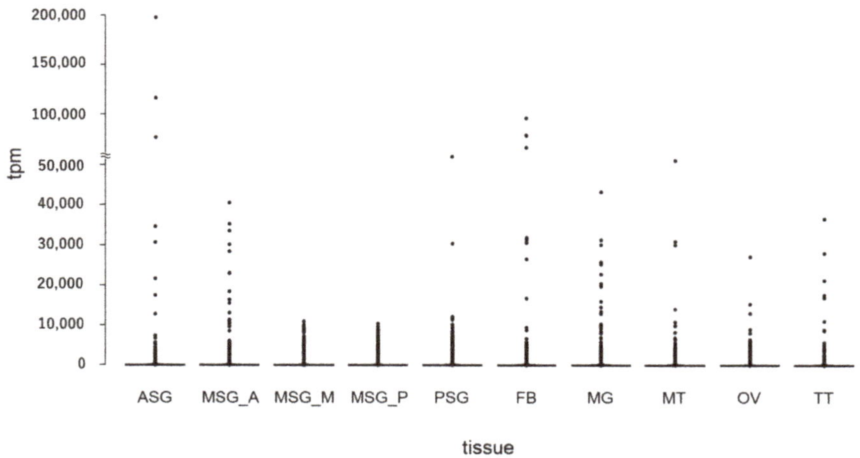

Figure 7. A scatter plot of the transcript expression in each tissue. Each spot shows the tpm value.

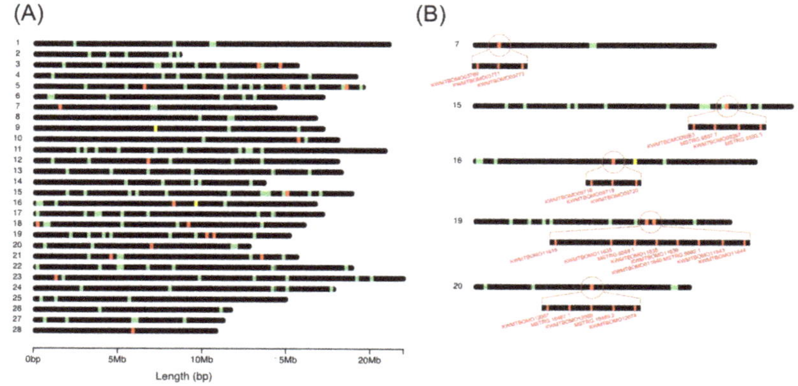

Figure 8. Genomic position of the transcripts strongly expressed in the MG only (red bar), in other than the MG (green bar), and commonly in the MG and other tissues (yellow bar). The top 50 strongly expressed transcripts are shown. The black bar indicates the chromosome. The number at the left side of each chromosome indicates the chromosomal number. (**A**) The genomic positions of all chromosomes. (**B**) The genomic positions of chromosomes 7, 15, 16, 19, and 20, in which tight clusters of strongly expressed transcripts in the MG are present.

We then investigated the transcript expression in the SG in more detail. Previous studies revealed that a number of transcripts showed a territory-specific expression in the SG [28,29]. However, the overall transcript expression in each territory remained unclear. Herein, we demonstrated that >1000 transcripts were commonly expressed in all SG subparts, and also that a number of transcripts were expressed in specific territories (Figure 9). They included 351 ASG-restricted, 180 MSG_A-restricted, 99 MSG_M-restricted, 71 MSG_P-restricted, and 100 PSG-restricted transcripts, respectively (Figure 9 and Table S7). Furthermore, we identified transcripts that were commonly expressed in more than two territories (Figure 9). They included transcripts expressed in MSG_M and MSG_P (Figure 9), and combined with the results of HC and the correlation analysis (Figure 4 and Table S4), we speculate that gene expression is highly conserved between MSG_M and MSG_P. This result was supported by the presence of a smaller number of transcripts expressed solely in the MSG_M or MSG_P (Figure 9). We also found that the transcripts that were exclusively expressed in ASG were more abundant than in other territories, and fewer transcripts were commonly expressed in the ASG and other subparts (Figure 9). We speculate that this reflects the functional diversification of the ASG, because the numbers of transcripts exclusively expressed in ASG (351) and those expressed in the MSG and PSG (395) were comparable, suggesting similar diversified functions (Figure 9). This may also be the case for MSG_A, based on the presence of similar characteristics (Figure 9). FEA revealed that the functional clusters enriched in each SG subpart were largely diversified (Figure S11).

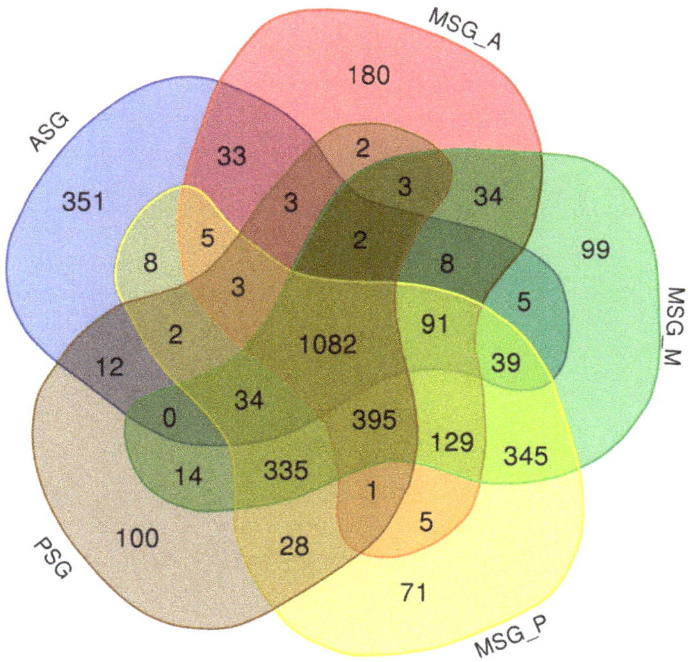

Figure 9. Venn diagram showing the number of transcripts with a tpm value > 30 in each silk gland part.

3.6. Expression Analysis of Transcriptional Factor Genes in the Silkworm

A transcriptomic analysis is a powerful tool for identifying genes with low levels of expression. TF genes are considered to show low expression levels, even though they have many important functions in developmental, physiological, and other major biological processes. Therefore, an expression analysis of TF genes will contribute to a more detailed understanding of the silkworm biology. Silkworm TF genes have recently been catalogued [37] and we investigated their expression levels in various silkworm tissues using this information. According to the low level of expression of TF genes, herein, we perceived TF transcripts with a tpm value > 5 as those expressed in each tissue. This analysis revealed that a number of TF transcripts were exclusively expressed in the OV and/or TT (Figure 10A, Table S8). These transcripts included KWMTBOMO02002 (traffic jam), KWMTBOMO002212 (mirror), KWMTBOMO01693 (vismay), and KWMTBOMO06584 (Sox100B), all of which play significant roles in gonad morphogenesis, oogenesis, spermatogenesis, as well as TT differentiation in *Drosophila melanogaster* (Table S8) [38–41]. We also found that KWMTBOMO09369/KWMTBOMO10218, the silkworm counterparts of human GATA4, were expressed in the OV, similar to that in humans (Table S8) [42]. We speculate that these TF genes have conserved functions in a wide variety of organisms, possibly in the development, differentiation, or homeostasis of reproductive tissues.

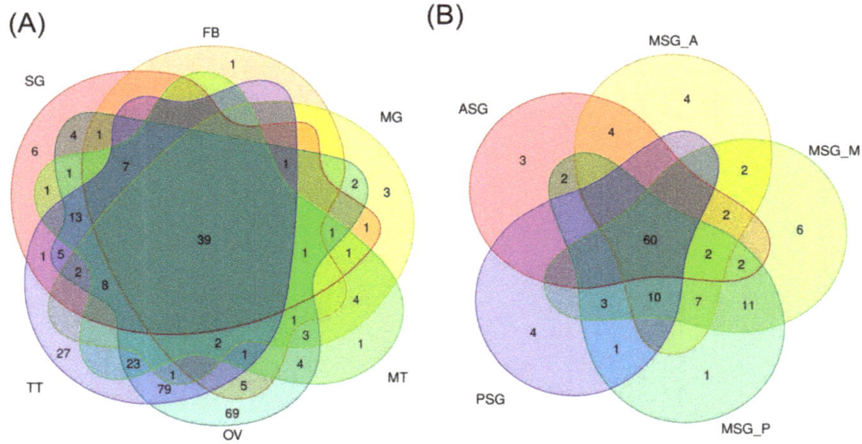

Figure 10. Venn diagrams showing the number of transcriptional factor (TF) transcripts with a tpm value > 5 (**A**) in each tissue and (**B**) in the silk gland subparts.

We also investigated TF gene expression in the SG. Previous studies revealed the central roles of the TF genes forkhead (fkh), Antennapedia (Antp), and Arrowhead (Awh) in the regulation of the silk gene expression [43–47]. Other TF genes that function in the regulation of silk gene expression may also exist because silk genes are strongly expressed and a number of silk genes are expressed in specific territories in the SG (Figure 9). In addition to previously identified TF genes, the expression of novel TF genes, such as those belonging to the High Mobility Group (HMG), Zinc Finger (ZF), Paired box (PAX), or forkhead family, was confined to the SG (Figure 10B and Table 2). The tpm values of these transcripts are shown in Table S8, and we identified TF genes expressed solely in the ASG or MSG_A. Transcripts expressed in more than two subparts were also detected (Table S9). We confirmed their expression in an RT-PCR analysis and the experimental results were consistent with the tpm values (Figure 11). The TF genes identified herein include those with indispensable roles in the development of *D. melanogaster*; for example, Dichaete (D), gooseberry-neuro (gsb-n), and sloppy paired 2 (slp2) are essential for embryonic segmentation [48–50], enhancer of split mβ-HLH (E(spl)mβ-HLH) for neurogenesis [51], and twin of eyegone (toe) for eye development [52]. We consider that it is very interesting if these genes have significant roles in the regulation of the silk gene expression, a function that is independent from the developmental regulation, in the silkworm.

Table 2. List of transcription factors expressed in the silk gland (more detailed data was in Table S9). human_gname and fly_gname indicate functional annotations against human and fly genomes, respectively, shown in Data S3.

Transcript ID	Family	Human_Gname	Fly_Gname
KWMTBOMO02291.mrna1	HMG	SOX14	D
KWMTBOMO02290.mrna1	HMG	SOX21	Sox21b
KWMTBOMO07638.mrna1	ETS	SPDEF	Ets98B
KWMTBOMO08935.mrna1	zf-LITAF-like		CG30273
KWMTBOMO08827.mrna1	zf-C2H2	GLIS2	sug
KWMTBOMO10947.mrna1	zf-CCCH	ZFP36L1	Tis11
KWMTBOMO09826.mrna1	zf-C2H2	ZBTB49	Clamp
KWMTBOMO12968.mrna1	Fork_head	FOXN1	jumu
KWMTBOMO15603.mrna1	zf-C2H2	ZNF606	CG9215
KWMTBOMO01121.mrna1	zf-C2H2	NAF1	CG10341
KWMTBOMO00301.mrna1	zf-LITAF-like		CG32280
KWMTBOMO07252.mrna1	zf-C2H2	PRDM10	
KWMTBOMO03284.mrna1	zf-BED		
KWMTBOMO09501.mrna1	TRAM_LAG1_CLN8	CERS5	schlank
KWMTBOMO00651.mrna1	Homeobox	LHX8	Awh
KWMTBOMO07825.mrna1	MYB	SMARCA1	Iswi
KWMTBOMO08651.mrna1	Homeobox	PBX1	exd
KWMTBOMO11294.mrna1	zf-C2H2	ZNF891	CG17328
KWMTBOMO07831.mrna1	PAX	PAX2	toe
KWMTBOMO01266.mrna1	PAX	PAX3	gsb-n
KWMTBOMO02915.mrna1	Fork_head	FOXB1	fd96Ca
KWMTBOMO07945.mrna1	zf-C2H2	OVOL1	ovo
KWMTBOMO01265.mrna1	PAX	PAX3	gsb-n
KWMTBOMO13459.mrna1	Pou	POU3F4	vvl
KWMTBOMO07731.mrna1	zf-LITAF-like	LITAF	CG13510
KWMTBOMO15317.mrna1	TF_bZIP	FOSL1	kay
KWMTBOMO07734.mrna1	zf-LITAF-like		CG13510
KWMTBOMO10990.mrna1	TF_bZIP	ATF3	Atf3
KWMTBOMO07931.mrna1	ETS	ETV6	aop
KWMTBOMO01705.mrna1	zf-C2H2	KLF10	cbt
KWMTBOMO16597.mrna1	zf-C2H2	KLF18	
KWMTBOMO02077.mrna1	zf-C2H2	GFI1B	sens

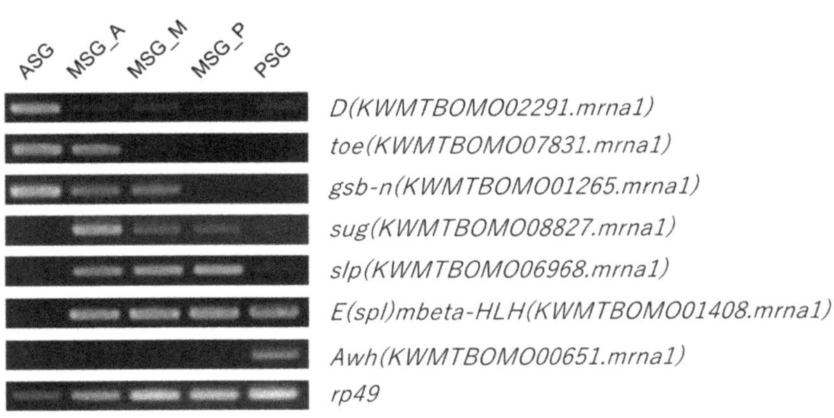

Figure 11. RT-PCR of TF transcripts showing territory-specific expression in the SG.

4. Discussion

In the present study, we performed RNA-seq analysis of multiple larval tissues from the silkworm *B. mori*. We established RTD using a recently reported high-quality reference genome assembly [15] and RNA-seq data newly obtained herein. RTD showed marked improvements over GMD, most notably the establishment of a nearly complete structure for *Ser1* (Figures S3 and S4); its full sequence has never been entirely elucidated because of its complexity. Our results indicate that an extensive RNA-seq analysis in combination with high-quality reference genome data provides highly refined gene structures, even for complex genes. The cost of performing a deep sequencing analysis has recently decreased and, thus, it has become affordable for every researcher to conduct not only a short-read RNA-seq analysis, but also long-read genome sequencing, using their own species. The present results are a significant proof-of-concept that highly refined gene structures may be established using a combination of these data, even for non-model organisms. Furthermore, more elaborate gene structures may be constructed using the RNA-seq data derived from other tissues and/or stages.

Herein, we found that a number of transcripts showed a tissue-restricted expression in the silkworm (Figure 5). Among them, transcripts exclusively expressed in the TT were the most abundant (Figure 5). A previous study identified a number of TT-specific genes in the silkworm [12], which is consistent with the present results. The presence of a number of TT-specific genes was also demonstrated in the jewel wasp, *Nasonia vitripennis* [53], indicating that this is a common feature in insects. Recent studies on *Drosophila* revealed that newly emerging genes were strongly biased for expression in the male reproductive system [54]. Therefore, the TT-specific transcripts identified in the present study may have similar traits. This issue may be confirmed by investigating the evolutionary ages of these genes, and, if this is the case, addressing the question of why the TT is a tissue that is permissive for new gene birth, a phenomenon observed not only in insects, but also in vertebrates [55], which will become possible using the silkworm. Another important result obtained from our cross-tissue gene expression analysis is that genes strongly expressed in the MG showed a strong tissue-restricted expression and also formed tight genomic clusters (Figure 8, Table S6, and Figure S9). Comparisons of the sequences of these transcripts revealed that the transcripts in each cluster encoded homological proteins; chr.7 transcripts encoded trypsins, chr.15 juvenile hormone-binding proteins (JHBPs), chr.16 fatty acid-binding proteins, chr.19 actin cytoskeleton-regulatory complex proteins, and chr.20 multiprotein bridging factor 2 (MBF2; Table S6). The predicted amino acid sequences of the transcripts within each cluster annotated with the same functions are similar to each other. Among these genes, a strong expression in the MG has already been demonstrated for *jhbps* [56], and Trypsin proteins were present in the digestive juices [57]. Homological genes that cluster in the genome are generally considered to have originated via tandem gene duplication, and to the best of our knowledge, few studies have investigated clustered genes that are expressed in the MG. In one case study on *Drosophila*, neutral lipase genes expressed in the MG clustered in the genome and were presumably under positive selection to retain different substrate specificities towards new lipid components of the diet [58]. Based on these findings, the transcripts identified in the present study may also have advantages in the silkworm MG, such as enhancing the activity for digestion and/or xenobiotic detoxification. The present study provides valuable insights into gene evolution and neofunctionalization in insects, which may be validated in more detail in future studies. In addition, the results obtained herein will facilitate the practical application of the silkworm; targeted gene integration into the clusters identified in the present study will enable strong gene expression in the MG, which will contribute to the establishment of strains with valuable properties, such as increased antiviral or antibiotic activities. The latest genome editing technologies should promote the establishment of useful silkworm strains.

Our detailed transcript expression analysis provides fundamental information on the traits of the SG, particularly each subpart. The SG is a tissue that arises from a single embryonic segment and has a long tubular structure [59]. Transcriptomic expression in each SG subpart is largely diversified, as demonstrated herein (Figure 9 and Table S7) and in previous studies [28,29]. The most important result of the present study is that the transcriptomic expression in the ASG was diversified the most within the SG, as demonstrated by the presence of a number of transcripts exclusively expressed in the ASG (351; Figure 9), a number of transcripts expressed both in the MSG and PSG (395; Figure 9), and the location of the ASG at the outermost site in the SG cluster in the HC analysis (Figure 4). We speculate that these results indicate the functional diversification of the ASG, which is consistent with previous findings showing that the ASG functions in silk fiber processing and the MSG/PSG in silk protein production [60]. FEA revealed that functional clusters enriched in the ASG were the "carbohydrate metabolic process" and "transport of small molecules" (Figure S11A), while those enriched in the MSG/PSG were "ribonucleoprotein complex biogenesis" and "translation" (Figure S11B–E), further supporting this concept. In this context, another result showing that the transcript expression in the MSG_A also showed diversification is of interest. We demonstrated that transcripts specifically expressed in the MSG_A were abundant (180; Figure 9), a number of transcripts were commonly expressed in the MSG_M/MSG_P/PSG (335; Figure 9), and the MSG_A was located at the outermost site in the MSG/PSG cluster in the HC analysis (Figure 4). MSG_A is a part of the MSG and functions in the production of the sericin proteins *Ser2* and *Ser3* [35,61], similar to the function of the MSG_M/MSG_P in the production of *Ser1* [32]. Nevertheless, our results indicate that the transcript expression in the MSG_A was more diverse than that in the MSG_M/MSG_P compared with that of the PSG; PSG is the territory that produces fibroin and not sericin [62], whereas the transcript expression appeared to be more conserved between the PSG and MSG_M/MSG_P than between the MSG_A and MSG_M/MSG_P (Figures 4 and 9). These results may be attributed to the presence of a number of transcripts that are strongly and specifically expressed in the MSG_A, including ecdysone oxidase, fatty acid hydroperoxide dehydratase, and other transcripts (Figure 9 and Table S7). We found that transcripts strongly expressed in the MSG_A were enriched for the functional clusters of "metabolism of vitamins and cofactors" and "organic hydroxy compound transport" (Figure S11B), and speculated that these clusters define the biological functions that are unique to this territory. Therefore, our extensive RNA-seq analysis provides fundamental insights into the functions of the SG, as well as its evolution, which has not yet been elucidated in detail. In the *Lepidoptera*, the morphology of the SG is largely diversified among species [63] and the *Saturniidae*, a family that is phylogenetically close to the *Bombycidae*, have MSGs with one territory and no morphological separation [6]. Therefore, the differentiation of the MSG gene expression observed herein may be specific to *B. mori* and/or other closely related species. Further studies are needed to clarify whether the differences in the gene expression among species are a driving force that generates diversity in cocoon properties, including shape, size, and physical activity. The TF genes identified in the present study may be one of the key factors responsible for the differences in the gene expression within the SG or among species.

5. Conclusions

We carried out RNA-seq analysis on the major larval tissues of the silkworm. Using these data, we successfully improved the gene model greatly as well as clarified gene expression in detail in each tissue. Our result should be a fundamental basis for the further promotion of the silkworm study as well as contribute to the practical application of the insects.

Supplementary Materials: Additional data are available in The Life Science Database Archive. The title in the Archive is "KAIKO—Metadata of reference transcriptome data" (http://doi.org/doi:10.18908/lsdba.nbdc02443-000.V001; Accessed on 21 April 2021) and figshare (http://doi.org/doi:10.6084/m9.figshare.c.5333894; Accessed on 28 May 2021). Table S1: Primer sequences used for RT-PCR (http://doi.org/doi:10.6084/m9.figshare.14217308; Accessed on 21 April 2021). Table S2: Tpm values in MSTRG.494.1, MSTRG.649.1-2, and MSTRG.704.1-3 (http://doi.org/doi:10.6084/m9.figshare.14206034; Accessed on 21 April 2021). Table S3: Comparisons of the *sericin-1* amino acid composition elucidated by an amino acid analysis and gene model. Mole% is shown (http://doi.org/doi:10.6084/m9.figshare.14206748; Accessed on 22 April 2021). Table S4: Spearman's rank correlation coefficient among silk gland territories (http://doi.org/doi:10.6084/m9.figshare.14217071; Accessed on 22 April 2021). Table S5: List of transcripts expressed in specific tissues (http://doi.org/doi:10.6084/m9.figshare.14217206; Accessed on 22 April 2021). Table S6: List of transcripts strongly expressed in each tissue or in the silk gland subparts. The top 50 strongly expressed transcripts are shown (http://doi.org/doi:10.6084/m9.figshare.14217218; Accessed on 22 April 2021). Table S7: List of territory-specific transcripts in the silk gland (http://doi.org/doi:10.6084/m9.figshare.14217242; Accessed on 22 April 2021). Table S8: List of tissue-specific TF genes (http://doi.org/doi:10.6084/m9.figshare.14217272; Accessed on 22 April 2021). Table S9: List of territory-specific TF genes in the silk gland, (http://doi.org/doi:10.6084/m9.figshare.14217296; Accessed on 12 May 2021). Data S1: Predicted amino acid sequences of reference transcriptome (http://doi.org/doi:10.18908/lsdba.nbdc02443-004; Accessed on 21 April 2021). Data S2: Functional annotations of the reference transcriptome (blast against human and *Drosophila* gene sets; http://doi.org/doi:10.18908/lsdba.nbdc02443-003; Accessed on 21 April 2021). Data S3: Functional annotations of the reference transcriptome (blast against the NCBI nr database; http://doi.org/doi:10.6084/m9.figshare.14192741; Accessed on 21 April 2021). Data S4: Expression data of each transcript in multiple tissues (http://doi.org/doi:10.18908/lsdba.nbdc02443-002.V001; Accessed on 2 April 2021). Figure S1: Comparison of gene structures among gene model data, cDNA-based data, and reference transcriptomic data. (A) Locus around KWMTBOMO00087, (B) KWMTBOMO00196, and (C) KWMTBOMO00222. GMD—gene model data; CBD—cDNA-based data; RTD—reference transcriptomic data (http://doi.org/doi:10.6084/m9.figshare.14205785; Accessed on 22 April 2021). Figure S2: Expression of new loci transcripts that did not hit the cDNA-based gene model. Numbers indicate transcripts with a tpm value > 0.01 in each tissue. SG—silk gland (using average tpm values in the five SG subparts); FB—fat body; MG—mid gut; MT—malpighian tubules; OV—ovary; TT—testis (http://doi.org/doi:10.6084/m9.figshare.14217335; Accessed on 22 April 2021). Figure S3: Structure of the *sericin-1* gene. MSTRG.2477.1 has a long exon 6. GMD—gene model data; CBD—cDNA-based data; RTD—reference transcriptomic data (http://doi.org/doi:10.6084/m9.figshare.14206166; Accessed on 22 April 2021). Figure S4: The presumptive full-length amino acid sequence of *sericin-1* deduced by the reference transcriptomic data (MSTRG.2477.1). The orange characters show the amino acids encoded by exon 6 and blue characters by exon 8 (http://doi.org/doi:10.6084/m9.figshare.14206736; Accessed on 22 April 2021). Figure S5: Sequence of the 38-amino acid-based repeat unit encoded by exon 6 and exon 8 in *sericin-1*. Exon 6 comprises 53 repeats and exon 8 comprises 13 repeats (http://doi.org/doi:10.6084/m9.figshare.14212673; Accessed on 22 April 2021). Figure S6: Comparison of the *sericin-3* sequence among different gene models. The amino acid sequences identified in the previous study (NM_001114644; [23]), derived from cDNA-based data (BMgn014348), reference transcriptomic data (MSTRG.2595.1), and gene model data (KWMTBOMO06311), are shown. Amino acids that differ in KWMTBOMO06311 are shown as white letters. Note that a frame shift occurs in the gene model data (http://doi.org/doi:10.6084/m9.figshare.14216927; Accessed on 22 April 2021). Figure S7: Comparison of the *sericin-4* gene structure. The gene structures elucidated by the previous study (*sericin-4*; [24]) and modeled by the gene model data (GMD) and reference transcriptomic data (RTD) are shown (http://doi.org/doi:10.6084/m9.figshare.14216951; Accessed on 22 April 2021). Figure S8: (A) Principal component analysis (PCA) results with expression profiles in 45 samples of transcripts showing a tpm value > 30 in at least one sample. Abbreviations and the numbers of samples are the same as in Figure 4. The X axis and Y axis are the principal components 1 (PC1) and PC2, respectively. (B) Hierarchical clustering of the expression data in 45 samples using all transcript tpm values (http://doi.org/doi:10.6084/m9.figshare.14216993; Accessed on 22 April 2021). Figure S9: A scatter plot of the transcript expression in each tissue. Each spot shows the tpm value. The top ten ranked strongly expressed transcripts in (A) ASG, (B) MSG_A, (C) MSG_M, (D) MSG_P, (E) PSG, (F) FB, (G) MG, (H) MT, (I) OV, and (J) TT are marked in red (http://doi.org/doi:10.6084/m9.figshare.14217227; Accessed on 22 April 2021). Figure S10: Genomic position of the genes strongly expressed in each tissue. In (A), the genes strongly expressed in the ASG only are shown with a red bar, in subparts other than the ASG with a green bar, and those commonly expressed in the ASG and other tissues with a yellow bar. The same applies to MSG_A (B), MSG_M (C), MSG_P (D), PSG (E), FB (F), MT (G), OV (T), and TT (I). (A') shows the chr. 11 of the ASG and (F') shows chr. 20 of the FB. Regarding MG, see Figure 8. The top 50 strongly expressed transcripts are shown. The black bar indicates the chromosome, and the number at the left side of each chromosome indicates the chromosomal number (http://doi.org/doi:10.6084/m9.figshare.14217233; Accessed on 22 April 2021). Figure S11: Results of the enrichment analysis for territory-specific transcripts in the silk gland. The number of transcripts used for the analysis is shown in the bracket. −log10 (P) represents −log10 (p-value). For example, −log10 (p) = 5 represents p-value = 10^{-5}, (http://doi.org/doi:10.6084/m9.figshare.14217257; Accessed on 22 April 2021).

Author Contributions: Conceived and designed experiments, K.Y., T.T., and H.S.; performed experiments, T.T.; contributed reagents/materials/analysis tools, H.S.; analyzed data, K.Y., T.T., A.J., and H.B.; contributed to the writing of the manuscript under the draft version, K.Y., T.T., and H.B. All of the authors discussed the data and helped with the manuscript preparation. K.Y. supervised the project. All authors have read and agreed to the published version of the manuscript.

Funding: This work was supported by the National Bioscience Database Center of the Japan Science and Technology Agency (JST) to H.B. This work was supported by the Cabinet Office, Government of Japan, Cross-ministerial Strategic Innovation Promotion Program (SIP), "Technologies for Smart Bio-industry and Agriculture" (funding agency Bio-oriented Technology Research Advancement Institution, NARO) to K.Y., T.T., and H.S. This work was also supported by a grant from the Ministry of Agriculture, Forestry, and Fisheries of Japan (Research Project for Sericultural Revolution) to K.Y., T.T., and H.S. The computing resource was partly provided by the super computer system at the National Institute of Genetics (NIG), Research Organization of Information and Systems (ROIS), Japan.

Institutional Review Board Statement: Not applicable.

Data Availability Statement: The RNA-seq reads supporting the conclusions of the present study are available in the SRA with accession number DRA008737 (the accession number of the RNA-seq data of each sample is shown in Table 1). Reference transcriptome data are available at the Transcriptome Shotgun Assembly (TSA) database under accession IDs ICPK01000001-ICPK01051926, and the Gff file of the transcriptome is available via doi:10.18908/lsdba.nbdc02443-001.V001. The estimated abundance of transcripts is available from the Gene Expression Archive (GEA) in DDBJ under accession ID E-GEAD-315 or from The Life Science Database Archive (doi:10.18908/lsdba.nbdc02443-002.V001).

Acknowledgments: The authors thank Kaoru Nakamura and Toshihiko Misawa for rearing the silkworms, Satoko Kawamoto for technical assistance, and Jianqiang Sun for data analysis advice.

Conflicts of Interest: The authors declare no conflict of interest. The funders had no role in the design of the study; in the collection, analyses, or interpretation of data; in the writing of the manuscript, or in the decision to publish the results.

References

1. Tan, A.; Tanaka, H.; Tamura, T.; Shiotsuki, T. Precocious metamorphosis in transgenic silkworms overexpressing juvenile hormone esterase. *Proc. Natl. Acad. Sci. USA* **2005**, *102*, 11751–11756. [CrossRef] [PubMed]
2. Ito, K.; Kidokoro, K.; Sezutsu, H.; Nohata, J.; Yamamoto, K.; Kobayashi, I.; Uchino, K.; Kalyebi, A.; Eguchi, R.; Hara, W. De-letion of a Gene Encoding an Amino Acid Transporter in the Midgut Membrane Causes Resistance to a Bombyx Parvo-like Virus. *Proc. Natl. Acad. Sci. USA* **2008**, *105*, 7523–7527. [CrossRef]
3. Sakurai, T.; Mitsuno, H.; Haupt, S.S.; Uchino, K.; Yokohari, F.; Nishioka, T.; Kobayashi, I.; Sezutsu, H.; Tamura, T.; Kanzaki, R. A Single Sex Pheromone Receptor Determines Chemical Response Specificity of Sexual Behavior in the Silkmoth *Bombyx mori*. *PLoS Genet.* **2011**, *7*, e1002115. [CrossRef] [PubMed]
4. Daimon, T.; Uchibori, M.; Nakao, H.; Sezutsu, H.; Shinoda, T. Knockout silkworms reveal a dispensable role for juvenile hormones in holometabolous life cycle. *Proc. Natl. Acad. Sci. USA* **2015**, *112*, E4226–E4235. [CrossRef]
5. KonDo, Y.; Yoda, S.; Mizoguchi, T.; Ando, T.; Yamaguchi, J.; Yamamoto, K.; Banno, Y.; Fujiwara, H. Toll ligand Spätzle3 controls melanization in the stripe pattern formation in caterpillars. *Proc. Natl. Acad. Sci. USA* **2017**, *114*, 8336–8341. [CrossRef]
6. Yukuhiro, K.; Sezutsu, H.; Tsubota, T.; Takasu, Y.; Kameda, T.; Yonemura, N. Insect Silks and Cocoons: Structural and Mo-Lecular Aspects. In *Extracellular Composite Matrices in Arthropods*; Springer: Berlin, Germany, 2016; pp. 515–555.
7. Tamura, T.; Thibert, C.; Royer, C.; Kanda, T.; Eappen, A.; Kamba, M.; Kômoto, N.; Thomas, J.-L.; Mauchamp, B.; Chavancy, G.; et al. Germline transformation of the silkworm *Bombyx mori* L. using a piggyBac transposon-derived vector. *Nat. Biotechnol.* **2000**, *18*, 81–84. [CrossRef] [PubMed]
8. Tatematsu, K.-I.; Kobayashi, I.; Uchino, K.; Sezutsu, H.; Iizuka, T.; Yonemura, N.; Tamura, T. Construction of a binary transgenic gene expression system for recombinant protein production in the middle silk gland of the silkworm *Bombyx mori*. *Transgenic Res.* **2009**, *19*, 473–487. [CrossRef] [PubMed]

9. Mita, K.; Kasahara, M.; Sasaki, S.; Nagayasu, Y.; Yamada, T.; Kanamori, H.; Namiki, N.; Kitagawa, M.; Yamashita, H.; Yasukochi, Y.; et al. The Genome Sequence of Silkworm, *Bombyx mori*. *DNA Res.* **2004**, *11*, 27–35. [CrossRef]
10. Biology Analysis Group; Xia, Q.; Zhou, Z.; Lu, C.; Cheng, D.; Dai, F.-Y.; Liu, B.; Zhao, P.; Zha, X.; Cheng, T.; et al. A Draft Sequence for the Genome of the Domesticated Silkworm (*Bombyx mori*). *Science* **2004**, *306*, 1937–1940. [CrossRef]
11. Consortium, I.S.G. The Genome of a Lepidopteran Model Insect, the Silkworm *Bombyx mori*. *Insect Biochem. Mol. Biol.* **2008**, *38*, 1036–1045.
12. Xia, Q.; Cheng, D.; Duan, J.; Wang, G.; Cheng, T.; Zha, X.; Liu, C.; Zhao, P.; Dai, F.; Zhang, Z. Microarray-Based Gene Ex-pression Profiles in Multiple Tissues of the Domesticated Silkworm, *Bombyx mori*. *Genome Biol.* **2007**, *8*, 1–13. [CrossRef]
13. Yamamoto, K.; Nohata, J.; Kadono-Okuda, K.; Narukawa, J.; Sasanuma, M.; Sasanuma, S.-I.; Minami, H.; Shimomura, M.; Suetsugu, Y.; Banno, Y.; et al. A BAC-based integrated linkage map of the silkworm *Bombyx mori*. *Genome Biol.* **2008**, *9*, R21. [CrossRef]
14. Suetsugu, Y.; Futahashi, R.; Kanamori, H.; Kadono-Okuda, K.; Sasanuma, S.-I.; Narukawa, J.; Ajimura, M.; Jouraku, A.; Namiki, N.; Shimomura, M.; et al. Large Scale Full-Length cDNA Sequencing Reveals a Unique Genomic Landscape in a Lepidopteran Model Insect, *Bombyx mori*. *G3 Genes Genomes Genet.* **2013**, *3*, 1481–1492. [CrossRef] [PubMed]
15. Kawamoto, M.; Jouraku, A.; Toyoda, A.; Yokoi, K.; Minakuchi, Y.; Katsuma, S.; Fujiyama, A.; Kiuchi, T.; Yamamoto, K.; Shimada, T. High-quality genome assembly of the silkworm, *Bombyx mori*. *Insect Biochem. Mol. Biol.* **2019**, *107*, 53–62. [CrossRef] [PubMed]
16. Bolger, A.M.; Lohse, M.; Usadel, B. Trimmomatic: A flexible trimmer for Illumina sequence data. *Bioinformatics* **2014**, *30*, 2114–2120. [CrossRef]
17. Kim, D.; Paggi, J.M.; Park, C.; Bennett, C.; Salzberg, S.L. Graph-based genome alignment and genotyping with HISAT2 and HISAT-genotype. *Nat. Biotechnol.* **2019**, *37*, 907–915. [CrossRef] [PubMed]
18. Pertea, M.; Kim, D.; Pertea, G.M.; Leek, J.T.; Salzberg, S.L. Transcript-level expression analysis of RNA-seq experiments with HISAT, StringTie and Ballgown. *Nat. Protoc.* **2016**, *11*, 1650–1667. [CrossRef]
19. Bray, N.L.; Pimentel, H.; Melsted, P.; Pachter, L. Near-optimal probabilistic RNA-seq quantification. *Nat. Biotechnol.* **2016**, *34*, 525–527. [CrossRef] [PubMed]
20. Ichino, F.; Bono, H.; Nakazato, T.; Toyoda, A.; Fujiyama, A.; Iwabuchi, K.; Sato, R.; Tabunoki, H. Construction of a simple evaluation system for the intestinal absorption of an orally administered medicine using *Bombyx mori* larvae. *Drug Discov. Ther.* **2018**, *12*, 7–15. [CrossRef] [PubMed]
21. Kobayashi, Y.; Nojima, Y.; Sakamoto, T.; Iwabuchi, K.; Nakazato, T.; Bono, H.; Toyoda, A.; Fujiyama, A.; Kanost, M.; Tabunoki, H. Comparative analysis of seven types of superoxide dismutases for their ability to respond to oxidative stress in *Bombyx mori*. *Sci. Rep.* **2019**, *9*, 2170. [CrossRef]
22. Kikuchi, A.; Nakazato, T.; Ito, K.; Nojima, Y.; Yokoyama, T.; Iwabuchi, K.; Bono, H.; Toyoda, A.; Fujiyama, A.; Sato, R. Identification of Functional Enolase Genes of the Silkworm *Bombyx mori* from Public Databases with a Combination of Dry and Wet Bench Processes. *BMC Genom.* **2017**, *18*, 1–12. [CrossRef]
23. Altschul, S.F.; Madden, T.L.; Schäffer, A.A.; Zhang, J.; Zhang, Z.; Miller, W.; Lipman, D.J. Gapped BLAST and PSI-BLAST: A new generation of protein database search programs. *Nucleic Acids Res.* **1997**, *25*, 3389–3402. [CrossRef]
24. Tabunoki, H.; Ono, H.; Ode, H.; Ishikawa, K.; Kawana, N.; Banno, Y.; Shimada, T.; Nakamura, Y.; Yamamoto, K.; Satoh, J.-I.; et al. Identification of Key Uric Acid Synthesis Pathway in a Unique Mutant Silkworm *Bombyx mori* Model of Parkinson's Disease. *PLoS ONE* **2013**, *8*, e69130. [CrossRef]
25. Kinsella, R.J.; Kahari, A.; Haider, S.; Zamora, J.; Proctor, G.; Spudich, G.; Almeida-King, J.; Staines, D.; Derwent, P.; Kerhornou, A.; et al. Ensembl BioMarts: A hub for data retrieval across taxonomic space. *Database* **2011**, *2011*, bar030. [CrossRef]
26. Zhou, Y.; Zhou, B.; Pache, L.; Chang, M.; Khodabakhshi, A.H.; Tanaseichuk, O.; Benner, C.; Chanda, S.K. Metascape provides a biologist-oriented resource for the analysis of systems-level datasets. *Nat. Commun.* **2019**, *10*, 1–10. [CrossRef]
27. Yang, C.-C.; Yokoi, K.; Yamamoto, K.; Jouraku, A. An update of KAIKObase, the silkworm genome database. *Database* **2021**, *2021*. [CrossRef] [PubMed]
28. Chang, H.; Cheng, T.; Wu, Y.; Hu, W.; Long, R.; Liu, C.; Zhao, P.; Xia, Q. Transcriptomic Analysis of the Anterior Silk Gland in the Domestic Silkworm (*Bombyx mori*)—Insight into the Mechanism of Silk Formation and Spinning. *PLoS ONE* **2015**, *10*, e0139424. [CrossRef]
29. Shi, R.; Ma, S.; He, T.; Peng, J.; Zhang, T.; Chen, X.; Wang, X.; Chang, J.; Xia, Q.; Zhao, P. Deep Insight into the Transcriptome of the Single Silk Gland of *Bombyx mori*. *Int. J. Mol. Sci.* **2019**, *20*, 2491. [CrossRef]
30. Ishikawa, E.; Suzuki, Y. Tissue- and Stage-Specific Expression of Sericin Genes in the Middle Silk Gland of *Bombyx mori*. (sericin mRNA/northern blotting/gene regulation). *Dev. Growth Differ.* **1985**, *27*, 73–82. [CrossRef]
31. Garel, A.; Deleage, G.; Prudhomme, J.C. Structure and organization of the *Bombyx mori sericin 1* gene and of the *sericins 1* deduced from the sequence of the *Ser 1B* cDNA. *Insect Biochem. Mol. Biol.* **1997**, *27*, 469–477. [CrossRef]
32. Takasu, Y.; Yamada, H.; Tsubouchi, K. Isolation of Three Main Sericin Components from the Cocoon of the Silkworm, *Bombyx mori*. *Biosci. Biotechnol. Biochem.* **2002**, *66*, 2715–2718. [CrossRef]
33. Zurovec, M.; Yonemura, N.; Kludkiewicz, B.; Sehnal, F.; Kodrík, D.; Vieira, L.C.; Kucerova, L.; Strnad, H.; Konik, P.; Sehadová, H. Sericin Composition in the Silk of *Antheraea yamamai*. *Biomacromolecules* **2016**, *17*, 1776–1787. [CrossRef]
34. Takasu, Y.; Yamada, H.; Tsubouchi, K. The Silk Sericin Component with Low Crystallinity. *Sanshi-Konchu Biotec* **2006**, *75*, 133–139.

35. Takasu, Y.; Yamada, H.; Tamura, T.; Sezutsu, H.; Mita, K.; Tsubouchi, K. Identification and characterization of a novel sericin gene expressed in the anterior middle silk gland of the silkworm *Bombyx mori*. *Insect Biochem. Mol. Biol.* **2007**, *37*, 1234–1240. [CrossRef]
36. Dong, Z.; Guo, K.; Zhang, X.; Zhang, T.; Zhang, Y.; Ma, S.; Chang, H.; Tang, M.; An, L.; Xia, Q.; et al. Identification of *Bombyx mori sericin 4* protein as a new biological adhesive. *Int. J. Biol. Macromol.* **2019**, *132*, 1121–1130. [CrossRef] [PubMed]
37. Zhu, Z.; Guan, Z.; Liu, G.; Wang, Y.; Zhang, Z. SGID: A comprehensive and interactive database of the silkworm. *Database* **2019**, *2019*. [CrossRef]
38. Zhao, D.; Woolner, S.; Bownes, M. The Mirror transcription factor links signalling pathways in Drosophila oogenesis. *Dev. Genes Evol.* **2000**, *210*, 449–457. [CrossRef]
39. Li, M.A.; Alls, J.D.; Avancini, R.M.; Koo, K.; Godt, D. The large Maf factor Traffic Jam controls gonad morphogenesis in Drosophila. *Nat. Cell Biol.* **2003**, *5*, 994–1000. [CrossRef] [PubMed]
40. Wang, Z.; Mann, R.S. Requirement for two nearly identical TGIF-related homeobox genes in Drosophila spermatogenesis. *Development* **2003**, *130*, 2853–2865. [CrossRef]
41. Nanda, S.; DeFalco, T.; Loh, S.H.Y.; Phochanukul, N.; Camara, N.; Van Doren, M.; Russell, S. Sox100B, a Drosophila Group E Sox-domain Gene, Is Required for Somatic Testis Differentiation. *Sex. Dev.* **2009**, *3*, 26–37. [CrossRef]
42. Vaskivuo, T.E.; Anttonen, M.; Herva, R.; Billig, H.; Dorland, M.; Velde, E.R.T.; Stenbäck, F.; Heikinheimo, M.; Tapanainen, J.S. Survival of Human Ovarian Follicles from Fetal to Adult Life: Apoptosis, Apoptosis-Related Proteins, and Transcription Factor GATA-41. *J. Clin. Endocrinol. Metab.* **2001**, *86*, 3421–3429. [CrossRef] [PubMed]
43. Mach, V.; Takiya, S.; Ohno, K.; Handa, H.; Imai, T.; Suzuki, Y. Silk gland factor-1 involved in the regulation of bombyx *sericin-1* gene contains fork head motif. *J. Biol. Chem.* **1995**, *270*, 9340–9346. [CrossRef]
44. Ohno, K.; Sawada, J.-I.; Takiya, S.; Kimoto, M.; Matsumoto, A.; Tsubota, T.; Uchino, K.; Hui, C.-C.; Sezutsu, H.; Handa, H.; et al. Silk Gland Factor-2, Involved in Fibroin Gene Transcription, Consists of LIM Homeodomain, LIM-interacting, and Single-stranded DNA-binding Proteins. *J. Biol. Chem.* **2013**, *288*, 31581–31591. [CrossRef]
45. Kimoto, M.; Tsubota, T.; Uchino, K.; Sezutsu, H.; Takiya, S. Hox transcription factor Antp regulates *sericin-1* gene expression in the terminal differentiated silk gland of *Bombyx mori*. *Dev. Biol.* **2014**, *386*, 64–71. [CrossRef]
46. Kimoto, M.; Tsubota, T.; Uchino, K.; Sezutsu, H.; Takiya, S. LIM-homeodomain transcription factor Awh is a key component activating all three fibroin genes, fibH, fibL and fhx, in the silk gland of the silkworm, *Bombyx mori*. *Insect Biochem. Mol. Biol.* **2015**, *56*, 29–35. [CrossRef]
47. Tsubota, T.; Tomita, S.; Uchino, K.; Kimoto, M.; Takiya, S.; Kajiwara, H.; Yamazaki, T.; Sezutsu, H. A Hox Gene, Antennapedia, Regulates Expression of Multiple Major Silk Protein Genes in the Silkworm *Bombyx mori*. *J. Biol. Chem.* **2016**, *291*, 7087–7096. [CrossRef] [PubMed]
48. Baumgartner, S.; Bopp, D.; Burri, M.; Noll, M. Structure of two genes at the gooseberry locus related to the paired gene and their spatial expression during Drosophila embryogenesis. *Genes Dev.* **1987**, *1*, 1247–1267. [CrossRef]
49. Grossniklaus, U.; Pearson, R.K.; Gehring, W.J. The Drosophila sloppy paired locus encodes two proteins involved in segmentation that show homology to mammalian transcription factors. *Genes Dev.* **1992**, *6*, 1030–1051. [CrossRef]
50. Russell, S.; Sanchez-Soriano, N.; Wright, C.; Ashburner, M. The Dichaete gene of Drosophila melanogaster encodes a SOX-domain protein required for embryonic segmentation. *Development* **1996**, *122*, 3669–3676. [CrossRef]
51. Couturier, L.; Mazouni, K.; Corson, F.; Schweisguth, F. Regulation of Notch output dynamics via specific E(spl)-HLH factors during bristle patterning in Drosophila. *Nat. Commun.* **2019**, *10*, 1–13. [CrossRef] [PubMed]
52. Yao, J.-G.; Weasner, B.M.; Wang, L.-H.; Jang, C.-C.; Weasner, B.; Tang, C.-Y.; Salzer, C.L.; Chen, C.-H.; Hay, B.; Sun, Y.H.; et al. Differential requirements for the Pax6(5a) genes eyegone and twin of eyegone during eye development in Drosophila. *Dev. Biol.* **2008**, *315*, 535–551. [CrossRef]
53. Ferree, P.M.; Fang, C.; Mastrodimos, M.; Hay, B.A.; Amrhein, H.; Akbari, O.S. Identification of Genes Uniquely Expressed in the Germ-Line Tissues of the Jewel Wasp Nasonia vitripennis. *G3 Genes Genomes Genet.* **2015**, *5*, 2647–2653. [CrossRef] [PubMed]
54. Kondo, S.; Vedanayagam, J.; Mohammed, J.; Eizadshenass, S.; Kan, L.; Pang, N.; Aradhya, R.; Siepel, A.; Steinhauer, J.; Lai, E.C. New genes often acquire male-specific functions but rarely become essential in Drosophila. *Genes Dev.* **2017**, *31*, 1841–1846. [CrossRef] [PubMed]
55. Kaessmann, H. Origins, evolution, and phenotypic impact of new genes. *Genome Res.* **2010**, *20*, 1313–1326. [CrossRef] [PubMed]
56. Li, W.; Cheng, T.; Hu, W.; Peng, Z.; Liu, C.; Xia, Q. Genome-wide identification and analysis of JHBP-domain family members in the silkworm *Bombyx mori*. *Mol. Genet. Genom.* **2016**, *291*, 2159–2171. [CrossRef]
57. Hu, X.; Zhu, M.; Wang, S.; Zhu, L.; Xue, R.; Cao, G.; Gong, C. Proteomics analysis of digestive juice from silkworm during *Bombyx mori* nucleopolyhedrovirus infection. *Proteomics* **2015**, *15*, 2691–2700. [CrossRef] [PubMed]
58. Horne, I.; Haritos, V.S. Multiple tandem gene duplications in a neutral lipase gene cluster in Drosophila. *Gene* **2008**, *411*, 27–37. [CrossRef] [PubMed]
59. Julien, E.; Coulon-Bublex, M.; Garel, A.; Royer, C.; Chavancy, G.; Prudhomme, J.-C.; Couble, P. Silk Gland Development and Regulation of Silk Protein Genes. *Compr. Mol. Insect Sci.* **2005**, 369–384. [CrossRef]

60. Akai, H. *Ultrastructural Morphology of Insects*; University of Tokyo Press: Tokyo, Japan, 1976.
61. Takasu, Y.; Hata, T.; Uchino, K.; Zhang, Q. Identification of *Ser2* proteins as major sericin components in the non-cocoon silk of *Bombyx mori*. *Insect Biochem. Mol. Biol.* **2010**, *40*, 339–344. [CrossRef]
62. Bello, B.; Horard, B.; Couble, P. The Selective Expression of Silk-Protein-Encoding Genes in *Bombyx mori* Silk Gland. *Bull. de l'Institut Pasteur* **1994**, *92*, 81–100.
63. Akai, H. Porous Cocoon Filaments-Their Characteristics and Formation. *Int. J. Wild Silkmoth Silk* **2005**, *10*, 57–74.

insects

Article

Evaluation of Antibacterial Drugs Using Silkworms Infected by *Cutibacterium acnes*

Yasuhiko Matsumoto *, Yuki Tateyama and Takashi Sugita

Department of Microbiology, Meiji Pharmaceutical University, 2-522-1 Noshio, Kiyose 204-8588, Tokyo, Japan; m196219@alm.my-pharm.ac.jp (Y.T.); sugita@my-pharm.ac.jp (T.S.)
* Correspondence: ymatsumoto@my-pharm.ac.jp

Simple Summary: *Cutibacterium acnes*, a common bacterium on human skin, is a causative agent of inflammatory skin diseases and systemic infections. Systemic infections caused by *C. acnes* are difficult to treat because of the drug resistance to typical macrolides. The development of a systemic infection model for *C. acnes* contributes to searching candidates for drug discovery. Silkworm is a useful model animal for evaluating the efficacy of compounds. Several silkworm infection models have been established for the identification of virulence-related genes and novel antibacterial drugs. However, the systemic infection model of *C. acnes* using silkworms was not yet established. We established a new silkworm infection model with *C. acnes* and evaluated the efficacy of antibacterial drugs using the silkworm infection model. The silkworm infection model might be used to identify drug candidates for the treatment against *C. acnes* infection.

Abstract: *Cutibacterium acnes* is a causative agent of inflammatory skin diseases and systemic infections. Systemic infections caused by *C. acnes* are difficult to treat, and the development of a systemic infection model for *C. acnes* would be useful for elucidating the mechanisms of infection and searching for therapeutic agents. In this study, we established a silkworm infection model as a new experimental system to evaluate the interaction between *C. acnes* and the host, and the efficacy of antibacterial drugs. Silkworms infected with *C. acnes* died when reared at 37 °C. The dose of injected bacterial cells required to kill half of the silkworms (LD_{50}) was determined under rearing conditions at 37 °C. The viable cell number of *C. acnes* was increased in the hemolymph and fat body of the infected silkworms. Silkworms injected with autoclaved *C. acnes* cells did not die during the study period. The survival time of silkworms injected with *C. acnes* was prolonged by the injection of antibacterial drugs such as tetracycline and clindamycin. These findings suggest that the silkworm *C. acnes* infection model can be used to evaluate host toxicity caused by *C. acnes* and the *in vivo* efficacy of antimicrobial drugs.

Keywords: silkworm; *Cutibacterium acnes*; infection; antibacterial drugs

Citation: Matsumoto, Y.; Tateyama, Y.; Sugita, T. Evaluation of Antibacterial Drugs Using Silkworms Infected by *Cutibacterium acnes*. *Insects* **2021**, *12*, 619. https://doi.org/10.3390/insects12070619

Academic Editors: Silvia Cappellozza, Morena Casartelli, Federica Sandrelli, Alessio Saviane and Gianluca Tettamanti

Received: 31 May 2021
Accepted: 5 July 2021
Published: 8 July 2021

1. Introduction

Cutibacterium acnes (formerly *Propionibacterium acnes*), a common bacterium on human skin, causes inflammatory skin diseases and systemic infections [1,2]. *C. acnes* is isolated as the predominant species in 34% [3] or 36.2% [4] of intervertebral discs removed from patients with chronic low back pain, such as disc herniation. Biofilm formation by *C. acnes* on implants and intervertebral discs causes bloodstream infections [5–7]. Because *C. acnes* forms a biofilm and more than 50% of clinically isolated *C. acnes* are resistant to typical macrolides, systemic infections caused by *C. acnes* are difficult to treat [8,9]. Therefore, the development of treatments for systemic infections caused by *C. acnes* is clinically important. Although mammalian animal models of *C. acnes* infection have been established, their use for the evaluation of antibacterial drugs is difficult due to the long duration of the infection [10,11]. Infection experiments using a large number of mammalian animals are also difficult to perform due to animal welfare issues [12].

Silkworms are useful animals for assessing host–pathogen interactions in systemic infections and for evaluating the therapeutic effects of antimicrobial drugs [12–14]. Because silkworms also have advantageous features, such as easy rearing in large numbers in a small space with few ethical issues, experiments with large numbers of silkworms can be performed [15]. Moreover, quantitative drug administration and monitoring of parameters in silkworm blood can be performed due to the ease of sample injection and blood collection [12,16,17]. The use of a silkworm infection model based on these advantageous features led to the discovery of virulence-related genes for pathogenic microorganisms such as *Staphylococcus aureus*, *Candida albicans*, and *Candida glabrata* [18–21]. Further, exploratory studies of antimicrobial drugs using a silkworm infection model led to the identification of compounds exhibiting therapeutic efficacy in mouse infection experiments [22–24]. Therefore, the silkworm infection model is useful for studies aimed at elucidating the infection mechanisms of pathogenic microorganisms and evaluating the efficacy of antimicrobial drugs.

In the present study, we attempted to establish an animal model of systemic infection by *C. acnes* using silkworms. We found that injection of *C. acnes* cells killed silkworms. Survival times of the infected silkworms were prolonged by injection of tetracycline and clindamycin. Our findings suggest that the silkworm infection model with *C. acnes* is useful for evaluating the efficacy of antimicrobial drugs.

2. Materials and Methods

2.1. Reagents

Tetracycline was purchased from FUJIFILM Wako Pure Chemical Corporation (Osaka, Japan). Clindamycin was purchased from Tokyo Chemical Industry Co., Ltd. (Tokyo, Japan). These reagents were dissolved in physiologic saline solution (0.9% NaCl). GAM agar was purchased from Nissui Pharmaceutical Co., Ltd. (Tokyo, Japan).

2.2. Culture of C. acnes

C. acnes ATCC6919 strain was used in this study. The *C. acnes* ATCC6919 strain was grown on GAM agar plates at 37 °C for 3 days under anaerobic conditions. Physiologic saline (2 mL) was added to the agar plate and suspended using a cell spreader. The *C. acnes* cell suspension was collected from the agar plate and measured absorbance at 600 nm using a spectrophotometer (U-1100; Hitachi Ltd., Tokyo, Japan). The *C. acnes* cells were suspended in physiologic saline to 5–10 of absorbance at 600 nm and the cell suspension was used in the experiments. For calculating the cell number in the cell suspension, the suspension of *C. acnes* cells was 10^7-fold diluted with saline, and a 100 μL aliquot was spread on a GAM agar plate. After incubation at 37 °C for 3 days under anaerobic conditions, the number of colonies was counted for determining the colony-forming unit (CFU).

2.3. Silkworm Rearing

The silkworm rearing was performed as previously described [25]. Eggs of silkworms (Hu Yo × Tukuba Ne) were purchased from Ehime-Sanshu Co., Ltd. (Ehime, Japan), disinfected, and hatched at 25–27 °C. The silkworms were fed an artificial diet, Silkmate 2S, containing antibiotics purchased from Ehime-Sanshu Co., Ltd. (Ehime, Japan). Fifth instar larvae were used in the infection experiments.

2.4. Silkworm Infection Experiments

The silkworm infection experiments were performed as previously described [25]. Silkworm fifth instar larvae were fed an artificial diet (1.5 g; Silkmate 2S; Ehime-Sanshu Co., Ltd., Ehime, Japan) overnight. A suspension (50 μL) of the *C. acnes* cells was injected into the silkworm hemolymph with a 1 mL tuberculin syringe (Terumo Medical Corporation, Tokyo, Japan). Silkworms injected with the *C. acnes* cells were placed in an incubator and survival was monitored.

2.5. Measurement of Viable Cell Number of C. acnes in Silkworms

Silkworms were injected with *C. acnes* cell suspension (3.3×10^8 cells/50 µL) and incubated at 37 °C. Hemolymph and fat body were harvested from the silkworms at 1 or 24 h after injection. Hemolymph was collected from the larva through a cut on the first proleg according to the previous report [16]. For calculating the *C. acnes* cell number in the hemolymph, the hemolymph was 10^3- or 10^4-fold diluted with saline, and a 100 µL aliquot was spread on GAM agar plates. The number of colonies was counted after incubation at 37 °C for 3 days under anaerobic conditions. The viable cell number of *C. acnes* in the hemolymph was calculated as the colony-forming unit (CFU) per 1 mL.

Fat body isolation was performed as previously described [26]. Silkworms were placed on ice for 15 min. The fat body was isolated from the dorsolateral region of each silkworm and rinsed in saline. The wet weight of the fat body was measured. The fat body was homogenized in 1 mL of saline, and the lysate was obtained. The lysate was 10^2 or 10^3-diluted with saline, and a 100 µL aliquot was spread on GAM agar plates. The number of colonies was counted after incubation at 37 °C for 3 days under anaerobic conditions. The viable cell number of *C. acnes* per wet weight of the fat body was determined.

2.6. LD_{50} Measurement

LD_{50} values, which is the dose of *C. acnes* required to kill half of the silkworms, were determined according to a previous report with slight modification [27]. The *C. acnes* ATCC6919 strain was grown on GAM agar plates at 37 °C under anaerobic conditions for 3 days. *C. acnes* cells grown on GAM agar plate were suspended with 2 mL of physiologic saline. A 2- or 4-fold dilution series of the bacterial suspension was prepared. The *C. acnes* cell suspension (Experiment 1: 2.5×10^7–1.6×10^9 cells/50 µL, Experiment 2: 2.1×10^7–1.3×10^9 cells/50 µL, or Experiment 3: 7.8×10^5–4.1×10^8 cells/50 µL) was injected into the silkworm hemolymph and incubated at 37 °C. Survival of the silkworms at 48 h was monitored. The LD_{50} was determined by a nonlinear regression model using Prism 9 (GraphPad Software, LLC, San Diego, CA, USA, https://www.graphpad.com/scientific-software/prism/, accessed on 25 April 2021).

2.7. Evaluation of Therapeutic Activities of Antibacterial Drugs Using Silkworms

Therapeutic activity tests using silkworms were performed according to a previous report with slight modification [25]. Saline (50 µL) or *C. acnes* suspension (3.6×10^8 cells/50 µL) was injected into the hemolymph of silkworms. Immediately after inoculation with the *C. acnes* suspension, 50 µL of an antibacterial drug (50 µg/g larva or 10 µg/g larva) was injected into the hemolymph of the silkworms.

2.8. Statistical Analysis

The significance of differences between groups on survival time of silkworms was calculated using a log-rank test based on the Kaplan–Meier method using Prism 9 (GraphPad Software, LLC, San Diego, CA, USA, https://www.graphpad.com/scientific-software/prism/, accessed on 25 April 2021). The statistically significant differences between groups on viable cell number of *C. acnes* in silkworm body were evaluated using the Student *t*-test. $p < 0.05$ was considered a statistically significant difference. Raw data are shown in Table S1.

3. Results

3.1. Pathogenicity of C. acnes against Silkworms

Silkworm models of infection by various microorganisms have been established [15]. The body temperature of silkworms, which can be regulated by changing the rearing temperature, is important for bacterial pathogenicity against silkworms [15]. In the case of *Aspergillus fumigatus* and dermatophytes, silkworms do not die at 37 °C by depending on their infection [24,28]. Therefore, it is important to consider the temperature at which the experiment is performed for each pathogen. *C. acnes* grows at approximately 32 °C on the skin and 37 °C in the human body. We examined the rearing temperatures that caused death in C. *acnes*-infected silkworms. Silkworms injected with C. *acnes* (1.6×10^9 cells) reared at 37 °C after injection died within 40 h, whereas infected silkworms reared at 32 °C survived longer (Figure 1). The time required for half of the infected silkworms to die (LT_{50}) was 48 h for silkworms reared at 32 °C and 27 h for those reared at 37 °C (Figure 1). The LD_{50} value, which is the bacterial number required to kill half of the silkworms, was 1–3 × 10^8 cells when infected silkworms were reared at 37 °C (Figure 2). These results suggest that rearing C. *acnes*-infected silkworms at 37 °C decreased survival and that the pathogenicity of C. *acnes* can be quantitatively assessed based on the LD_{50} value.

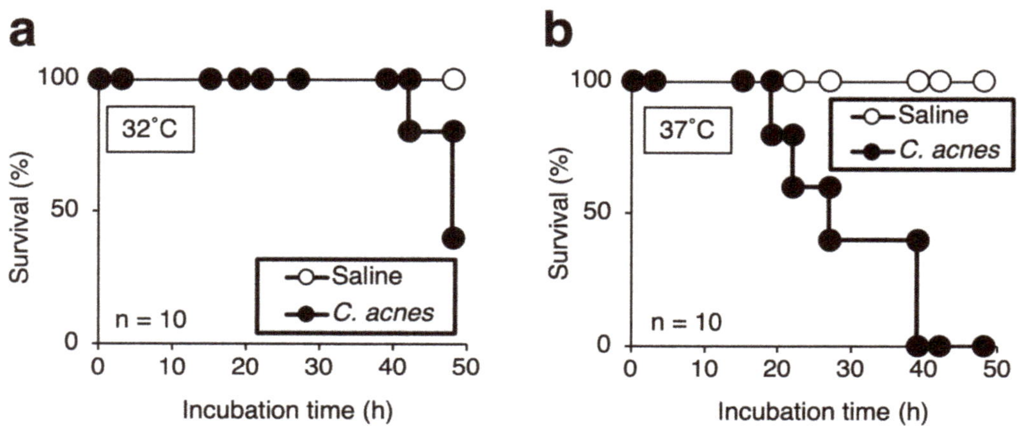

Figure 1. Toxic effects of C. *acnes* injected into silkworms. Silkworms were injected with saline (50 μL) or C. *acnes* cell suspension (1.6×10^9 cells/50 μL) and incubated at 32 °C (**a**) and 37 °C (**b**). The number of surviving silkworms was measured for 48 h. n = 10/group.

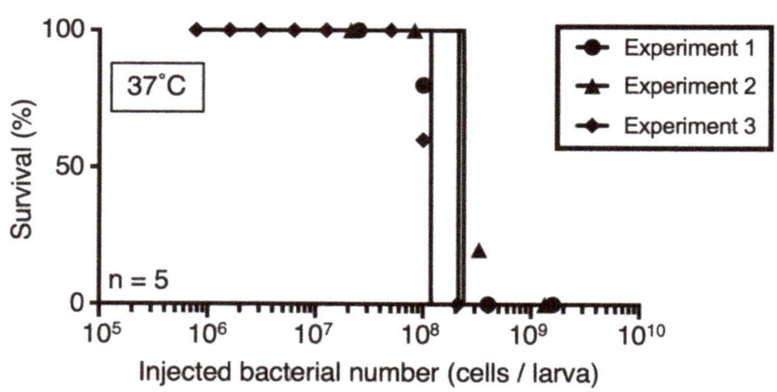

Figure 2. Dose response of C. *acnes* in the silkworm infection model. Silkworms were injected with saline or C. *acnes* cell suspension (Experiment 1: 2.5×10^7–1.6×10^9 cells/50 μL, Experiment 2: 2.1×10^7–1.3×10^9 cells/50 μL, or Experiment 3: 7.8×10^5–4.1×10^8 cells/50 μL) and incubated at 37 °C. The number of surviving silkworms was measured at 48 h after injection. n = 5/group.

3.2. Increase in Viable Cell Number of C. acnes in Silkworms

We next investigated whether C. *acnes* cells can grow in the silkworm body. The scheme of the experiment is shown in Figure 3a. The viable cell number of C. *acnes* was increased in hemolymph, the blood of silkworm (Figure 3b). Silkworm has a fat body, which is a soft organ that has a similar function to the liver and adipose tissue in mammals. The viable cell number of *C. acnes* was also increased in the fat body of silkworms (Figure 3c). The result suggests that *C. acnes* can proliferate in the silkworm body.

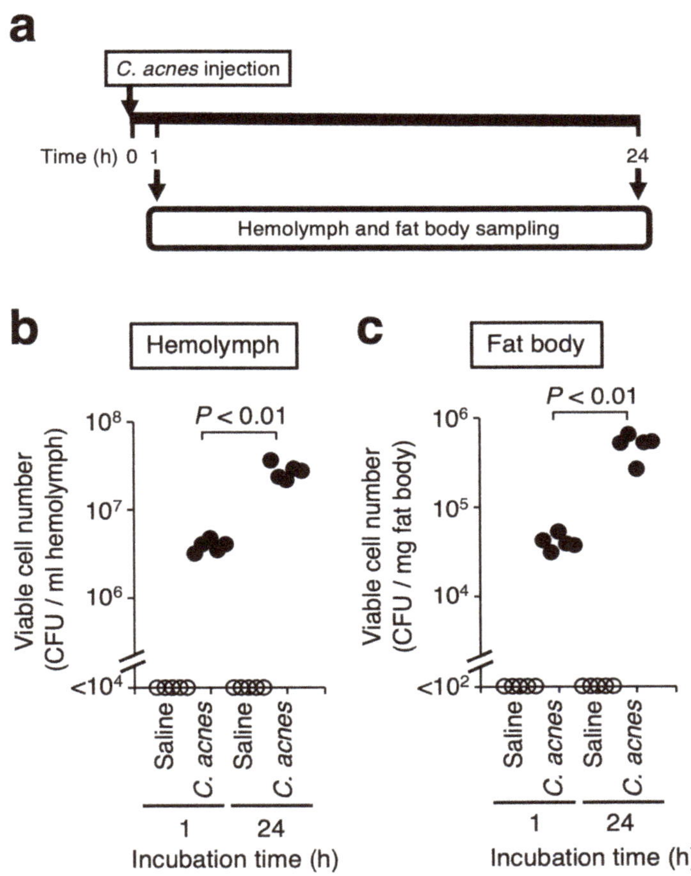

Figure 3. Increase in viable cell number of *C. acnes* in silkworms. (**a**) Scheme of the experiment. Silkworms were injected with saline or *C. acnes* cell suspension (3.3 × 10^8 cells/50 μL) and incubated at 37 °C. Hemolymph (**b**) and fat body (**c**) were harvested from the silkworms at 1 or 24 h after injection. Viable cell number of *C. acnes* in the samples was measured by counting the colony-forming unit (CFU). Statistically significant differences between groups were evaluated using Student *t*-test. n = 5 / group.

3.3. Effect of Heat-Killed C. acnes Cells on Silkworms

Bacterial components such as peptidoglycans of *Porphyromonas gingivalis* lead to shock in silkworms, resulting in their death [29]. Injection of viable or heat-killed *P. gingivalis* cells causes silkworm death [29]. Under conditions in which shock occurs, silkworms cannot be treated with antibiotics [29]. Therefore, it is important to determine whether the death of silkworms by injection of a pathogen is shock by the bacterial components. We next examined the heat-killed cells to evaluate whether *C. acnes* cells must be alive to exert pathogenicity against silkworms. Injection of *C. acnes* cells killed silkworms, but injection of autoclaved C. acnes cells did not (Figure 4). These results suggest that the heat-stable components of C. acnes do not cause acute silkworm death in the experimental condition.

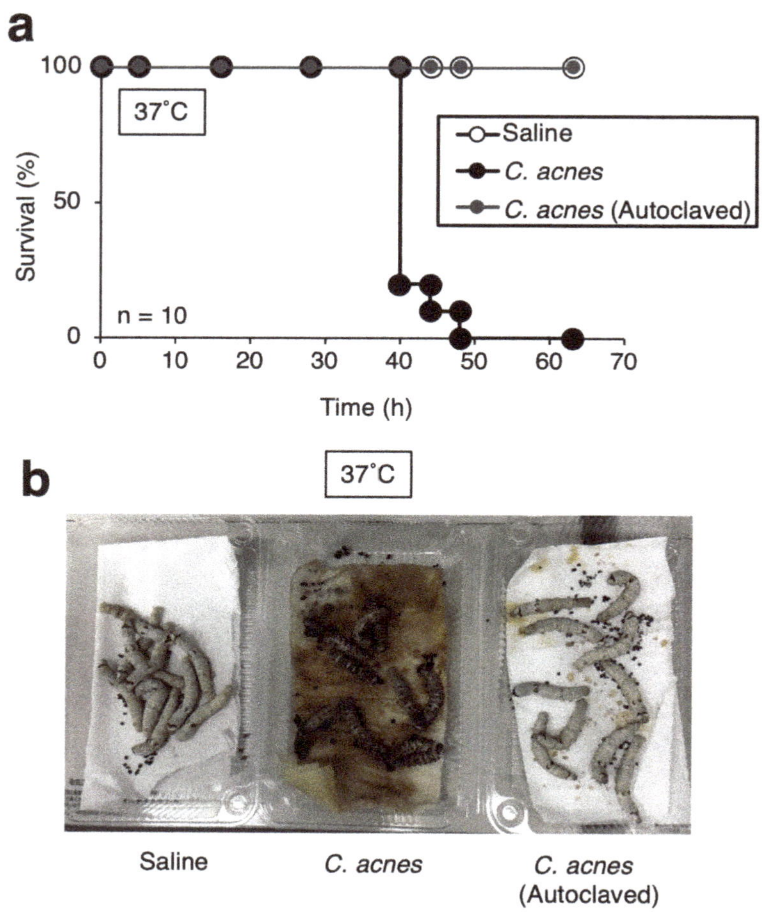

Figure 4. Effects of injecting autoclaved C. acnes cells in silkworms. (**a**) Silkworms were injected with saline (50 μL), C. acnes cell suspension (7.8×10^8 cells/50 μL), or autoclaved C. acnes cell suspension and incubated at 37 °C. The number of surviving silkworms was measured for 63 h. n = 10/group. (**b**) Picture at 48 h after injection is shown.

3.4. Therapeutic Effects of Antibacterial Drugs against Silkworms Infected with C. acnes

Tetracycline and clindamycin are used for the treatment of dermatologic disorders including acne vulgaris and are often administrated orally [30–32]. We next examined the efficacy of these antibacterial drugs in the silkworm C. acnes infection model. In the clinic, tetracycline and clindamycin are administered 250–500 mg/person/day and 600 mg/person/day, respectively [30–32]. Based on the information, we determined injection amounts of the antibacterial drugs in the experiments. Administration of tetracycline and clindamycin to silkworms infected with C. acnes prolonged the survival time (Figure 5). These results suggest that the efficacy of antibacterial drugs can be evaluated using the silkworm infection model with C. acnes.

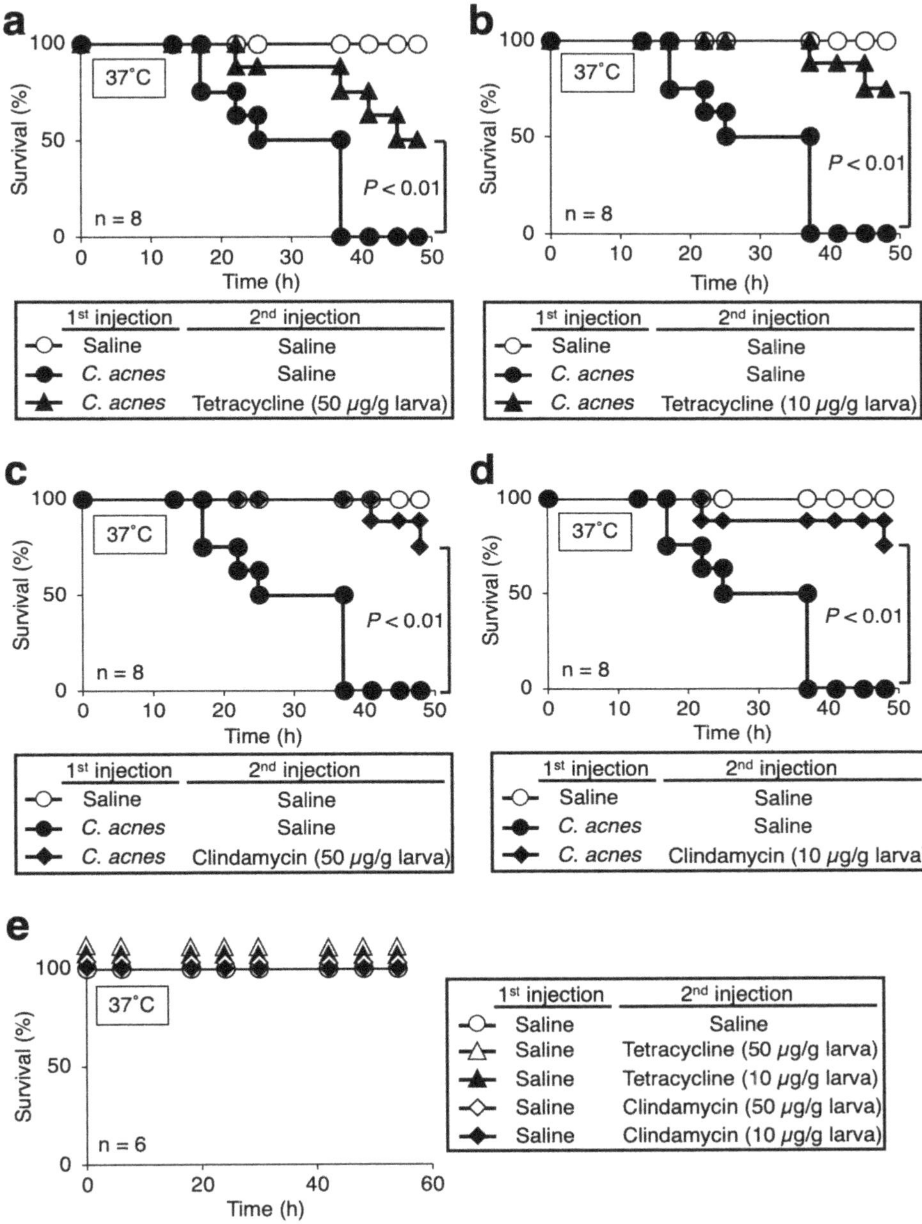

Figure 5. Effect of antibacterial drugs in silkworms infected by *C. acnes*. Silkworms were injected with saline (50 μL) or *C. acnes* cell suspension (3.6×10^8 cells/50 μL), and then with tetracycline (50 μg/g larva) (**a**), tetracycline (10 μg/g larva) (**b**), clindamycin (50 μg/g larva) (**c**), or clindamycin (10 μg/g larva) (**d**). The number of surviving silkworms under incubation at 37 °C was measured for 48 h. Statistically significant differences between groups were evaluated using a log-rank test. n = 8/group. (**e**) Silkworms were injected with saline (50 μL), and then with tetracycline (50 μg/g larva), tetracycline (10 μg/g larva), clindamycin (50 μg/g larva), or clindamycin (10 μg/g larva). The number of surviving silkworms under incubation at 37 °C was measured for 54 h. n = 6/group.

4. Discussion

In this study, we demonstrated that *C. acnes* kills silkworms reared at 37 °C and that the silkworm infection model can be used to evaluate the efficacy of antibacterial drugs. Our findings suggest that the silkworm infection model is useful for assessing pathogenicity and the efficacy of antimicrobial drugs against systemic infection by *C. acnes*

Because *C. acnes* causes bloodstream infections, we selected injection the *C. acnes* into the hemolymph of the silkworms. *C. acnes*-infected silkworms under the rearing condition at 37 °C were more sensitive than that at 32 °C. We assumed that this is due to two factors: the effect of high-temperature stress on the silkworm and the optimal temperature at which the pathogens have high virulence. The silkworms injected with the autoclaved *C. acnes* cells seem to be unhealthy compared with those in the saline group (Figure 4b). We assumed the reason is that the hemolymph of silkworms was melanized and the melanization might affect the silkworm condition. The *C. acnes* cells grow in the hemolymph and on the tissues of the silkworms. *C. acnes* use the nutrition of the silkworms and might cause toxicity to the tissues. The melanization of the hemolymph of silkworms caused by *C. acnes* might affect tissue toxicity. We assumed that the silkworms infected with *C. acnes* cause sepsis by the proliferation of *C. acnes* cells in the hemolymph and on tissues of the silkworms.

C. acnes ATCC6919 used in this study is a type strain whose genome is publicly available and can invade osteoclasts and osteoblasts [33]. Therefore, we used the *C. acnes* ATCC6919 for systemic infection experiments using silkworms. The viable cell number of *C. acnes* was increased in the hemolymph and fat body of silkworms. The results suggest that *C. acnes* causes bacteremia by growth in the silkworm body. *C. acnes* forms a biofilm on the intervertebral discs [5]. We assumed that *C. acnes* might form a biofilm on the fat body of silkworms. Future work will be to evaluate the difference in pathogenicity and drug-resistant in clinical isolates from patients with disc herniation. Moreover, identification of high pathogenic *C. acnes* clinical strain using the silkworm infection model and genetic study of the high pathogenic strain and drug-resistant strain will be an important subject.

C. acnes causes superficial inflammatory diseases and systemic infections. In this study, we focused on the systemic infections caused by *C. acnes* and established the systemic infection model using silkworms. We assumed that the silkworm *C. acnes* infection model established in this study is not a superficial inflammatory disease model. The establishment of a superficial inflammatory disease model using silkworms is a future subject for the evaluation of topical drugs.

In a previous study using nucleus pulposus-derived disc cells in vitro, *C. acnes* cells (10^7–10^8 cells/mL) were co-cultured with the disc cells and proinflammatory responses of the disc cells were induced by the addition of the *C. acnes* cells [34]. We injected *C. acnes* cells (3.3×10^8 cells/larva), and the range of viable cell number of *C. acnes* in silkworm hemolymph at 1 h after the injection was 3.2–4.8×10^7 cells/mL (Figure 3). Moreover, the LD_{50} was 1–3×10^8 cells when infected silkworms were reared at 37 °C (Figure 2). Therefore, we used *C. acnes* cells (10^8–10^9 cells/larva) in this study, because the amounts are needed for silkworm death and the similar range of *C. acnes* cells in previous *in vitro* study.

In the silkworm *C. acnes* infection model established in this study, administration of heat-killed bacteria did not kill the silkworm. Further, *C. acnes*-infected silkworms could be effectively treated with antibacterial drugs, suggesting that the growth of *C. acnes* in the body of silkworms is important for its pathogenicity. In the clinic, tetracycline and clindamycin are administered 250–500 mg/person/day and 600 mg/person/day, respectively [30–32]. Given that the average weight of humans is 60 kg, the calculated values of tetracycline and clindamycin are 4.2–8.3 and 10 mg/kg, respectively, for humans. The bodyweight of the silkworm used in this study was approximately 2 g. One-hundred micrograms (50 µg/g larva) and 20 µg (10 µg/g larva) of tetracycline and clindamycin were administered to each silkworm (Figure 5). Therefore, doses of tetracycline and clindamycin were 50 mg/kg or 10 mg/kg. The 10 mg/kg dose is a similar dose to human clinical use. The results support that the silkworm infection model is useful for the evaluation of antibacterial drugs against systemic infection by *C. acnes*. Silkworm infection models are useful for identifying virulence factors of pathogenic microorganisms [13,15]. Further studies are needed to determine which factors in *C. acnes* are responsible for the pathogenicity against silkworms.

The pharmacokinetics of antimicrobial agents are similar between silkworms and mammals, and antimicrobial drug efficacy in the silkworm infection model can be evaluated based on the pharmacokinetics [23,35–37]. A previous study demonstrated that total body clearance (CL_{tot}), distribution volume in steady state (Vd_{ss}), and the elimination half-life ($t_{1/2}$) of antibiotics in silkworm were calculated [37]. For example, CL_{tot}, Vd_{ss}, and $t_{1/2}$ of tetracycline in silkworm were 0.07 mL/g/h, 0.6 mL/g, and 5.5h, respectively. In rabbit, these values of tetracycline were 0.35 mL/g/h, 1.0 mL/g, and 2.0 h, respectively. The pharmacokinetic parameters of tetracycline in silkworms and rabbits are not different at least 10-fold [37]. Moreover, silkworm infection models are useful to develop for in vivo screening of new antimicrobial agents [22,24]. The silkworm *C. acnes* infection model may be useful for identifying effective antibacterial compounds against systemic infection by *C. acnes*. Azole antifungals exhibit antimicrobial activity against *C. acnes* in vitro, and ketoconazole inhibits the lipase activity of *C. acnes* [38,39]. Further studies are needed to identify effective compounds against systemic *C. acnes* infections from among clinically applied drugs using the silkworm infection model.

Recently, an infection model with *C. acnes* using a nematode, *Caenorhabditis elegans*, was reported [40]. *C. elegans* is useful for identifying host factors against *C. acnes* infection based on genetic approaches [40]. The differences between the silkworm system and the *C. elegans* system are that the silkworm blood can be directly injected with *C. acnes* and its pathogenicity at 37 °C, the same temperature as the human body, can be verified. *C. elegans* is difficult to inject quantitatively into body fluids and cannot grow at 37 °C [41]. The silkworm infection model might allow us to identify virulence factors of *C. acnes* at the body temperature of humans.

5. Conclusions

We established a silkworm infection model with *C. acnes* and found the system to be useful for evaluating antibacterial drug efficacy. Further studies are needed to determine the clinical applicability of research using the silkworm *C. acnes* infection model.

Supplementary Materials: The following are available online at https://www.mdpi.com/article/10.3390/insects12070619/s1, Table S1: Raw data of each experiment in this study.

Author Contributions: Conceptualization, Y.M.; methodology, Y.M.; validation, Y.M.; formal analysis, Y.M.; investigation, Y.M. and Y.T.; resources, Y.M. and T.S.; data curation, Y.M. and Y.T.; writing—original draft preparation, Y.M.; writing—review and editing, Y.M. and T.S.; visualization, Y.M.; supervision, Y.M.; project administration, Y.M.; funding acquisition, Y.M. All authors have read and agreed to the published version of the manuscript.

Funding: This research was funded by Kose Cosmetology Research Foundation (No. 711 to Y.M.) and by JSPS KAKENHI grant number JP20K07022 (Scientific Research (C) to Y.M.).

Institutional Review Board Statement: Not applicable.

Informed Consent Statement: Not applicable.

Data Availability Statement: Data are contained within the article and Supplementary Materials.

Acknowledgments: We thank Tae Nagamachi, Asami Yoshikawa, Yu Sugiyama, Eri Sato, and Asuka Toshima (Meiji Pharmaceutical University) for their technical assistance rearing the silkworms.

Conflicts of Interest: The authors declare no conflict of interest.

References

1. Byrd, A.L.; Belkaid, Y.; Segre, J.A. The human skin microbiome. *Nat. Rev. Microbiol.* **2018**, *16*, 143–155. [CrossRef] [PubMed]
2. Stirling, A.; Worthington, T.; Rafiq, M.; Lambert, P.A.; Elliott, T.S. Association between sciatica and *Propionibacterium acnes*. *Lancet* **2001**, *357*, 2024–2025. [CrossRef]
3. Urquhart, D.M.; Zheng, Y.; Cheng, A.C.; Rosenfeld, J.V.; Chan, P.; Liew, S.; Hussain, S.M.; Cicuttini, F.M. Could low grade bacterial infection contribute to low back pain? A systematic review. *BMC Med.* **2015**, *13*, 13. [CrossRef] [PubMed]
4. Ganko, R.; Rao, P.J.; Phan, K.; Mobbs, R.J. Can bacterial infection by low virulent organisms be a plausible cause for symptomatic disc degeneration? A systematic review. *Spine* **2015**, *40*, E587–E592. [CrossRef] [PubMed]
5. Capoor, M.N.; Ruzicka, F.; Schmitz, J.E.; James, G.A.; Machackova, T.; Jancalek, R.; Smrcka, M.; Lipina, R.; Ahmed, F.S.; Alamin, T.F.; et al. *Propionibacterium acnes* biofilm is present in intervertebral discs of patients undergoing microdiscectomy. *PLoS ONE* **2017**, *12*, e0174518. [CrossRef] [PubMed]
6. Achermann, Y.; Goldstein, E.J.C.; Coenye, T.; Shirtliff, M.E. *Propionibacterium acnes*: From commensal to opportunistic biofilm-associated implant pathogen. *Clin. Microbiol. Rev.* **2014**, *27*, 419–440. [CrossRef] [PubMed]
7. Portillo, M.E.; Corvec, S.; Borens, O.; Trampuz, A. *Propionibacterium acnes*: An underestimated pathogen in implant-associated infections. *Biomed Res. Int.* **2013**, *2013*, 804391. [CrossRef] [PubMed]
8. Walsh, T.R.; Efthimiou, J.; Dréno, B. Systematic review of antibiotic resistance in acne: An increasing topical and oral threat. *Lancet Infect. Dis.* **2016**, *16*, e23–e33. [CrossRef]
9. Tiltnes, T.S.; Kehrer, M.; Hughes, H.; Morris, T.E.; Justesen, U.S. Ceftriaxone treatment of spondylodiscitis and other serious infections with *Cutibacterium acnes*. *J. Antimicrob. Chemother.* **2020**, *75*, 3046–3048. [CrossRef]
10. Shinohara, D.B.; Vaghasia, A.M.; Yu, S.-H.; Mak, T.N.; Brüggemann, H.; Nelson, W.G.; De Marzo, A.M.; Yegnasubramanian, S.; Sfanos, K.S. A mouse model of chronic prostatic inflammation using a human prostate cancer-derived isolate of *Propionibacterium acnes*. *Prostate* **2013**, *73*, 1007–1015. [CrossRef]
11. Olsson, J.; Drott, J.B.; Laurantzon, L.; Laurantzon, O.; Bergh, A.; Elgh, F. Chronic prostatic infection and inflammation by *Propionibacterium acnes* in a rat prostate infection model. *PLoS ONE* **2012**, *7*, e51434.
12. Matsumoto, Y. Facilitating Drug Discovery in Human Disease Models Using Insects. *Biol. Pharm. Bull.* **2020**, *43*, 216–220. [CrossRef]
13. Kaito, C.; Murakami, K.; Imai, L.; Furuta, K. Animal infection models using non-mammals. *Microbiol. Immunol.* **2020**, *64*, 585–592. [CrossRef]
14. Montali, A.; Berini, F.; Brivio, M.F.; Mastore, M.; Saviane, A.; Cappellozza, S.; Marinelli, F.; Tettamanti, G. A Silkworm Infection Model for In Vivo Study of Glycopeptide Antibiotics. *Antibiotics* **2020**, *9*, 300. [CrossRef]
15. Matsumoto, Y.; Sekimizu, K. Silkworm as an experimental animal to research for fungal infections. *Microbiol. Immunol.* **2019**, *63*, 41–50. [CrossRef]
16. Matsumoto, Y.; Sumiya, E.; Sugita, T.; Sekimizu, K. An invertebrate hyperglycemic model for the identification of anti-diabetic drugs. *PLoS ONE* **2011**, *6*, e18292. [CrossRef]
17. Matsumoto, Y.; Ishii, M.; Hasegawa, S.; Sekimizu, K. *Enterococcus faecalis* YM0831 suppresses sucrose-induced hyperglycemia in a silkworm model and in humans. *Commun. Biol.* **2019**, *2*, 157. [CrossRef]
18. Kaito, C.; Kurokawa, K.; Matsumoto, Y.; Terao, Y.; Kawabata, S.; Hamada, S.; Sekimizu, K. Silkworm pathogenic bacteria infection model for identification of novel virulence genes. *Mol. Microbiol.* **2005**, *56*, 934–944. [CrossRef]
19. Hanaoka, N.; Takano, Y.; Shibuya, K.; Fugo, H.; Uehara, Y.; Niimi, M. Identification of the putative protein phosphatase gene *PTC1* as a virulence-related gene using a silkworm model of *Candida albicans* infection. *Eukaryotic Cell* **2008**, *7*, 1640–1648. [CrossRef]
20. Ueno, K.; Matsumoto, Y.; Uno, J.; Sasamoto, K.; Sekimizu, K.; Kinjo, Y.; Chibana, H. Intestinal resident yeast *Candida glabrata* requires Cyb2p-mediated lactate assimilation to adapt in mouse intestine. *PLoS ONE* **2011**, *6*, e24759.
21. Paudel, A.; Hamamoto, H.; Panthee, S.; Matsumoto, Y.; Sekimizu, K. Large-Scale Screening and Identification of Novel Pathogenic *Staphylococcus aureus* Genes Using a Silkworm Infection Model. *J. Infect. Dis.* **2020**, *221*, 1795–1804. [CrossRef] [PubMed]
22. Hamamoto, H.; Urai, M.; Ishii, K.; Yasukawa, J.; Paudel, A.; Murai, M.; Kaji, T.; Kuranaga, T.; Hamase, K.; Katsu, T.; et al. Lysocin E is a new antibiotic that targets menaquinone in the bacterial membrane. *Nat. Chem. Biol.* **2015**, *11*, 127–133. [CrossRef] [PubMed]
23. Paudel, A.; Panthee, S.; Urai, M.; Hamamoto, H.; Ohwada, T.; Sekimizu, K. Pharmacokinetic parameters explain the therapeutic activity of antimicrobial agents in a silkworm infection model. *Sci. Rep.* **2018**, *8*, 1578–1588. [CrossRef] [PubMed]
24. Nakamura, I.; Kanasaki, R.; Yoshikawa, K.; Furukawa, S.; Fujie, A.; Hamamoto, H.; Sekimizu, K. Discovery of a new antifungal agent ASP2397 using a silkworm model of *Aspergillus fumigatus* infection. *J. Antibiot.* **2017**, *70*, 41–44. [CrossRef]
25. Kaito, C.; Akimitsu, N.; Watanabe, H.; Sekimizu, K. Silkworm larvae as an animal model of bacterial infection pathogenic to humans. *Microb. Pathog.* **2002**, *32*, 183–190. [CrossRef]
26. Matsumoto, Y.; Ishii, M.; Hayashi, Y.; Miyazaki, S.; Sugita, T.; Sumiya, E.; Sekimizu, K. Diabetic silkworms for evaluation of therapeutically effective drugs against type II diabetes. *Sci. Rep.* **2015**, *5*, 10722. [CrossRef]
27. Miyazaki, S.; Matsumoto, Y.; Sekimizu, K.; Kaito, C. Evaluation of *Staphylococcus aureus* virulence factors using a silkworm model. *FEMS Microbiol. Lett.* **2012**, *326*, 116–124. [CrossRef]
28. Ishii, M.; Matsumoto, Y.; Yamada, T.; Abe, S.; Sekimizu, K. An invertebrate infection model for evaluating anti-fungal agents against dermatophytosis. *Sci. Rep.* **2017**, *7*, 12289. [CrossRef]

29. Ishii, K.; Hamamoto, H.; Imamura, K.; Adachi, T.; Shoji, M.; Nakayama, K.; Sekimizu, K. *Porphyromonas gingivalis* peptidoglycans induce excessive activation of the innate immune system in silkworm larvae. *J. Biol. Chem.* **2010**, *285*, 33338–33347. [CrossRef]
30. Van der Zee, H.H.; Boer, J.; Prens, E.P.; Jemec, G.B.E. The effect of combined treatment with oral clindamycin and oral rifampicin in patients with hidradenitis suppurativa. *Dermatology* **2009**, *219*, 143–147. [CrossRef]
31. Armstrong, A.W.; Hekmatjah, J.; Kircik, L.H. Oral Tetracyclines and Acne: A Systematic Review for Dermatologists. *J. Drugs Dermatol.* **2020**, *19*, s6–s13.
32. Del Rosso, J.Q. Sarecycline and the Narrow-spectrum tetracycline concept: Currently Available Data and Potential Clinical Relevance in Dermatology. *J. Clin. Aesthet. Dermatol.* **2020**, *13*, 45–48.
33. Aubin, G.G.; Baud'huin, M.; Lavigne, J.-P.; Brion, R.; Gouin, F.; Lepelletier, D.; Jacqueline, C.; Heymann, D.; Asehnoune, K.; Corvec, S. Interaction of *Cutibacterium* (formerly *Propionibacterium*) *acnes* with bone cells: A step toward understanding bone and joint infection development. *Sci. Rep.* **2017**, *7*, 42918–42919. [CrossRef]
34. Capoor, M.N.; Konieczna, A.; McDowell, A.; Ruzicka, F.; Smrcka, M.; Jancalek, R.; Maca, K.; Lujc, M.; Ahmed, F.S.; Birkenmaier, C.; et al. Pro-Inflammatory and Neurotrophic Factor Responses of Cells Derived from Degenerative Human Intervertebral Discs to the Opportunistic Pathogen *Cutibacterium acnes*. *Int. J. Mol. Sci.* **2021**, *22*, 2347. [CrossRef]
35. Hamamoto, H.; Kurokawa, K.; Kaito, C.; Kamura, K.; Manitra Razanajatovo, I.; Kusuhara, H.; Santa, T.; Sekimizu, K. Quantitative evaluation of the therapeutic effects of antibiotics using silkworms infected with human pathogenic microorganisms. *Antimicrob. Agents Chemother.* **2004**, *48*, 774–779. [CrossRef]
36. Hamamoto, H.; Tonoike, A.; Narushima, K.; Horie, R.; Sekimizu, K. Silkworm as a model animal to evaluate drug candidate toxicity and metabolism. *Comp. Biochem. Physiol. C Toxicol. Pharmacol.* **2009**, *149*, 334–339. [CrossRef]
37. Hamamoto, H.; Horie, R.; Sekimizu, K. Pharmacokinetics of anti-infectious reagents in silkworms. *Sci. Rep.* **2019**, *9*, 9451–9458. [CrossRef]
38. Sugita, T.; Miyamoto, M.; Tsuboi, R.; Takatori, K.; Ikeda, R.; Nishikawa, A. *In vitro* activities of azole antifungal agents against *Propionibacterium acnes* isolated from patients with acne vulgaris. *Biol. Pharm. Bull.* **2010**, *33*, 125–127. [CrossRef]
39. Unno, M.; Cho, O.; Sugita, T. Inhibition of *Propionibacterium acnes* lipase activity by the antifungal agent ketoconazole. *Microbiol. Immunol.* **2017**, *61*, 42–44. [CrossRef]
40. Huang, X.; Pan, W.; Kim, W.; White, A.; Li, S.; Li, H.; Lee, K.; Fuchs, B.B.; Zeng, K.; Mylonakis, E. *Caenorhabditis elegans* mounts a p38 MAPK pathway-mediated defence to *Cutibacterium acnes* infection. *Cell. Microbiol.* **2020**, *22*, e13234. [CrossRef]
41. Ishii, M.; Matsumoto, Y.; Sekimizu, K. Usefulness of silkworm as a host animal for understanding pathogenicity of *Cryptococcus neoformans*. *Drug Discov. Ther.* **2016**, *10*, 9–13. [CrossRef] [PubMed]

insects

MDPI

Article

Co-Expression Network and Time-Course Expression Analyses to Identify Silk Protein Regulatory Factors in *Bombyx mori*

Yudai Masuoka [1,2,*], Wei Cao [2], Akiya Jouraku [1], Hiroki Sakai [3], Hideki Sezutsu [3] and Kakeru Yokoi [1,2,*]

1 Insect Design Technology Module, Division of Insect Advanced Technology, Institute of Agrobiological Sciences, National Agriculture and Food Research Organization (NARO), 1-2 Owashi, Tsukuba 305-8634, Ibaraki, Japan; joraku@affrc.go.jp

2 Research Center for Agricultural Information Technology (RCAIT), National Agriculture and Food Research Organization (NARO), 1-31-1 Kannondai, Tsukuba 305-0856, Ibaraki, Japan; sou197@affrc.go.jp

3 Silkworm Research Module, Division of Silk-Producing Insect Biotechnology, Institute of Agrobiological Sciences, National Agriculture and Food Research Organization (NARO), 1-2 Owashi, Tsukuba 305-8634, Ibaraki, Japan; sakaih786@affrc.go.jp (H.S.); hsezutsu@affrc.go.jp (H.S.)

* Correspondence: masuokay781@affrc.go.jp (Y.M.); yokoi123@affrc.go.jp (K.Y.); Tel.: +81-29-838-6129 (Y.M. & K.Y.)

Simple Summary: Previous studies have reported how the silk production ability of *Bombyx mori* can be enhanced, but the mechanism that regulates silk protein genes remains unclear. We performed co-expression network analysis using *networkz*, an in-house program, which led to the identification of 91 transcription factors were co-expressed with silk protein genes. Of them, 13 transcripts were identified to be novel regulatory factors by time-course expression analysis during the fifth instar larvae stage. Their expression patterns were highly relevant to those of silk protein genes. Our results suggest that the two-step expression screening was robust and highly sensitive to screen relative genes, and a complex mechanism regulates silk protein production in *B. mori*. The novel candidates that were identified herein can serve as key genes to develop methods to enhance the silk protein production ability of *B. mori*.

Abstract: *Bombyx mori* is an important economic insect and an animal model in pharmacomedical research. Although its physiology has been studied for many years, the mechanism via which silk protein genes are regulated remains unclear. In this study, we performed two-step expression screening, namely co-expression network and time-course expression analyses to screen silk protein regulation factors. A co-expression network analysis using RNA-seq data that were obtained from various tissues, including the silk glands of *B. mori*, was performed to identify novel silk protein regulatory factors. Overall, 91 transcription factors, including some known ones, were found to be co-expressed with silk protein genes. Furthermore, time-course expression analysis during the fifth instar larvae stage revealed that the expression pattern of 13 novel transcription factors was highly relevant to that of silk protein genes and their known regulatory factor genes. In particular, the expression peak of several transcription factors (TFs) was detected before the expression of silk protein genes peak. These results indicated that a larger number of genes than expected may be involved in silk protein regulation in *B. mori*. Functional analyses of function-unknown transcription factors should enhance our understanding of this system.

Keywords: co-expression network analysis; *Bombyx mori*; silk protein; sericin; fibroin; transcription factor

Citation: Masuoka, Y.; Cao, W.; Jouraku, A.; Sakai, H.; Sezutsu, H.; Yokoi, K. Co-Expression Network and Time-Course Expression Analyses to Identify Silk Protein Regulatory Factors in *Bombyx mori*. *Insects* **2022**, *13*, 131. https://doi.org/10.3390/insects13020131

Academic Editors: Silvia Cappellozza, Morena Casartelli, Federica Sandrelli, Alessio Saviane and Gianluca Tettamanti

Received: 24 December 2021
Accepted: 25 January 2022
Published: 26 January 2022

1. Introduction

Silkworms (*Bombyx mori*) generate silk proteins; they are an economically important insect in sericulture and have proved their value in biotechnology as a bioreactor for the production of recombinant proteins and silk-based biomaterials. Silk proteins can be broadly classified into sericin and fibroin, which are secreted from the middle and posterior silk glands (SGs), respectively. The SG consists of endomitotic cells [1], and the expression of genes encoding these proteins shows a considerably increase in the fifth (last) instar larvae stage. The elucidation of mechanisms that regulate the expression of such genes is necessary to further enhance the ability of this insect to produce silk.

In previous studies, it has been reported that some transcription factors (TFs), including homeobox genes, regulate the expression of silk protein genes [2,3]. For instance, *Antennapedia* (*Antp*), a Hox gene that controls leg formation, directly regulates the expression of *sericin1* in the middle SG [4,5]. Further, silk gland factor-2 (SGF2), a protein complex containing the homeodomain protein Arrowhead (Awh), LIM domain-binding protein, and sequence-specific single-stranded DNA-binding protein, evidently regulates the expression of genes encoding fibroin in the posterior SG [6,7]. The *silk gland factor-1* (*SGF1*), containing a forkhead domain, and *silk gland factor-3* (*SGF3*) genes are involved in regulating sericin1 expression [8–10]. Besides, sage, encoding a basic helix-loop-helix TF, is involved in regulating the expression of fibroin heavy-chain along with *SGF1* [11].

Although some genes have been identified to function as expression regulators of silk protein genes, the pertinent regulatory mechanism and pathways still remain unclear. Furthermore, these regulatory factors, such as hox genes, have been known to possess other functions [3] which can lead to lethal effects when they are genetically modified. To avoid the risk as much as possible, the factors that are specific to the silk gene regulation in the silk gland are desirable as targets for genetic modification to increase silk yield. Thus, a co-expression relationship among silk proteins and their regulatory genes (known and unknown) needs to be elucidated. For this purpose, gene expression network analysis using large-scale transcriptome data is essential. Co-expression network analysis is an effective approach to elucidate groups of genes that are showing distinct co-expression patterns among phenotypes. This approach has been widely adopted for various purposes, including to predict diseases in humans [12], detect metabolic pathways involving organic compounds and stress-responsive genes in plants [13–15], and determine gene sets that are related to biological processes in bacteria [16]. In insects, co-expression network analysis has been mainly used in model species considering the availability of abundant transcriptome data. Co-expressed genes at different stages, including young lncRNA genes, have been detected in *Drosophila melanogaster* [17]. In mosquitoes (*Aedes aegypti*), infection-responsive genes were identified using genome-wide transcriptome profiling, including co-expression network analysis [18]. In *B. mori*, lncRNA and domestication-related genes including silk gland-related genes were identified by co-expression network analysis [19,20]. Although co-expression network analysis is actually useful for identifying relevant gene groups, further detailed analysis, such as time-course expression analysis, is necessary to detect more important genes. Functional analysis of screened candidates is thus required to understand the mechanisms regulating silk protein genes.

Herein we attempted to identify genes regulating the expression of silk protein genes using co-expression network as well as time-course expression analyses. Screening precision is dependent on the input data volume and variation, and standard Java-based tools that are used in co-expression network analysis (e.g., Gephi and Cytoscape) take a long time to process large quantities of expression data. Accordingly, we developed a fast C++-based tool to quickly process large expression datasets. Co-expression network analysis was performed using published transcriptome data [21–24] comprising five SG regions [anterior SG (ASG), anterior-middle SG (A-MSG), middle-middle SG (M-MSG), posterior-middle SG (P-MSG), and posterior SG (PSG)], Malpighian tubule (MT), testis (TT), and ovary (OV). A total of six silk protein genes [*sericin1*, *sericin2*, *sericin3*, *fibroin heavy-chain* (*fibroin-H*), *fibroin light-chain* (*fibroin-L*), and *fibrohexamerin* (*P25*)] were selected as target genes to search for regulatory factors. There were also five existing regulatory genes [*SGF1*, *SGF3*, *sage*, *Antp*, and *Awh* (main isoform PA)] that were also chosen as target genes. TFs that showed expression patterns that were similar to those of the target genes were subjected to time-course expression analysis, which was performed at A-MSG, M-MSG, P-MSG, and PSG on every day during last instar larva (day zero to seven). Further, TFs with expression patterns that were related to those of target genes were shortlisted as candidates of silk protein regulatory genes. Our results provide insights into how silk protein genes are regulated; moreover, the genes that are discussed herein can be used as targets to improve silk protein production ability.

2. Materials and Methods

2.1. Constructing a Gene Co-Expression Network and Detecting Modules

We developed a command line tool named networkz to handle large gene co-expression datasets (or gene expression profiles) and to perform co-expression network analysis. networkz was written in C++ and the source code is available at https://github.com/davecao/networkz.git (accessed on 23 December 2021); it is based on Boost Graph Library v1.70 [https://www.boost.org (accessed on 23 December 2021)] for graph data structure operations and Eigen Library v3.3.90 [(https://gitlab.com/libeigen/eigen/-/releases (accessed on 23 December 2021)] for matrix operations.

The relationships among genes in the co-expression dataset can be represented by a network, which is an undirected and weighted graph consisting of vertices and edges; herein genes are referred to as vertices while their edges represent the pairwise co-expression measure. To construct an initial co-expression network, we selected a significance measure threshold to determine the connected gene pairs with a significant co-expression relationship, and then modules (or hub genes) that were highly connected with others were detected in the subsequent analysis.

In this study, a gene profile is denoted as a vector with m components; $\mathrm{x}_i = (x_{i,1}, x_{i,2}, \ldots, x_{i,m})$. Then n gene expression profiles were represented by an $n \times m$ matrix; $\mathrm{X} = (\mathrm{x}_1, \mathrm{x}_2 \ldots, \mathrm{x}_n)^T$. The expression measure between the genes p and q ($d_{p,q}$,) was defined as follows:

$$d_{p,q} = 1 - |corr(p,q)|$$

$$corr(p,q) = corr(\mathrm{x}_i, \mathrm{x}_j) = \frac{\sum_{k=1}^{m}(x_{i,k} - \overline{x_i})(x_{j,k} - \overline{x_j})}{\sqrt{\sum_{k=1}^{m}(x_{i,k} - \overline{x_i})^2 \sum_{k=1}^{m}(x_{j,k} - \overline{x_j})^2}},\ i,\ j = 1,\ \ldots,\ n,\ i \neq j$$

wherein $|corr(p,q)|$ represents the absolute value of Pearson's correlation coefficient between the expression profiles of p and q; $\overline{\mathrm{x_i}}$ and $\overline{\mathrm{x_j}}$ present mean of x_i and $\mathrm{x_j}$, respectively. The smaller the value of $d_{p,q}$ is, the higher the likelihood of the two genes (p and q) in the network being interconnected (i.e., showing high correlation in terms of pairwise gene similarity). The threshold of 0.1 was selected via trial and error.

To detect modules in the initially constructed network, we further employed the Kruskal's algorithm [25], as vertices were much more than edges, to find a minimum spanning tree (MST) with minimum sum of edge weights; then, the Louvain method [26] was performed on the MST to assign each gene with a community ID. Finally, modules of interest were found.

2.2. Co-Expression Network Analysis

For co-expression network analysis with *networkz*, we used transcript-level transcripts per million (TPM) values as expression data of two RNA-seq data series, which were used for the assembly and verification of the current reference transcriptome dataset of *B. mori* in our previous study [24] The first RNA-seq data series (SRA Run ID: DRR068893-068895 and DRR095105-095116) was obtained from the fat body (FB), midgut (MG), MT, whole SG (SG), and TT of the o751 strain last instar larvae on third day (Table 1) [21–23]. The second RNA-seq data series (SRA Run ID: DRR186474-186503) was obtained from the aforementioned five SG regions (ASG, A-MSG, M-MSG, P-MSG, and PSG), FB, MG, MT, TT, and OV of p50T strain last instar larvae on third day (Table 1) [24]. The transcript-level TPM expression data that were used in this study are available at "expression data of each transcript in multiple tissues" in the study by Yokoi et al. 2021 (doi: 10.18908/lsdba.nbdc02443-002.V001), in which 51,926 transcripts were used as reference sequences for TPM calculation [24].

Herein we used the same transcript ID as that of reference transcript sequences. The silk protein genes (*sericin1*, *sericin2*, *sericin3*, *fibroin-H*, *fibroin-L*, and *P25*) and five existing regulatory genes (*SGF1*, *SGF3*, *sage*, *Antp*, and *Awh*) served as target genes. Target network modules containing transcripts (isoforms) of the target genes were identified from network modules that were constructed by *networkz*. As the target genes showed multiple isoforms multiple target network modules were identified for each target gene. The transcripts that were annotated with major TF-specific motif in target network modules were screened as candidate TFs.

Table 1. RNA-seq datasets using co-expression network analysis.

Series	Tissue	Strain	SRA Run ID	Replicate	Reference
RNA-seq 1	testis (TT)	o751	DRR068893-068895	3	Kikuchi et al., 2017 [21]
	fat body (FB)	o751	DRR095105-095107	3	Kobayashi et al., 2019 [23]
	midgut (MG)	o751	DRR095108-095110	3	Ichino et al., 2018 [22]
	Malpighian tubule (MT)	o751	DRR095111-095113	3	Kobayashi et al., 2019 [23]
	whole silk gland (SG)	o751	DRR095114-095116	3	Kobayashi et al., 2019 [23]
RNA-seq 2	anterior SG (ASG)	p50T	DRR186474-186476	3	Yokoi et al., 2021 [24]
	anterior middle SG (A-MSG)	p50T	DRR186477-186479	3	Yokoi et al., 2021 [24]
	middle middle SG (M-MSG)	p50T	DRR186480-186482	3	Yokoi et al., 2021 [24]
	posterior middle SG (P-MSG)	p50T	DRR186483-186485	3	Yokoi et al., 2021 [24]
	posterior SG (PSG)	p50T	DRR186486-186488	3	Yokoi et al., 2021 [24]
	fat body (FB)	p50T	DRR186489-186491	3	Yokoi et al., 2021 [24]
	midgut (MG)	p50T	DRR186492-186494	3	Yokoi et al., 2021 [24]
	Malpighian tubule (MT)	p50T	DRR186495-186497	3	Yokoi et al., 2021 [24]
	testis (TT)	p50T	DRR186498-186500	3	Yokoi et al., 2021 [24]
	ovary (OV)	p50T	DRR186501-186503	3	Yokoi et al., 2021 [24]

2.3. RNA Extraction

To extract total RNA, fifth instar larva of the w-1 pnd strain of *B. mori* were kept on an artificial diet (Nihon Nosan Kogyo, Yokohama, Japan) at 25 °C under LD 12:12 h. The SGs of three male and female insects were then extracted every day during the last instar period (day 0–7). Total RNA was isolated from one pair of SGs for each individual using TRIzol (Invitrogen, Carlsbad, CA, USA) and RNeasy Plus Mini Kit (Qiagen, Hilden, Germany), and the wet weight of the whole SG was measured using the other pair of SG.

2.4. Gene Expression Analysis

For quantitative RT-PCR (qRT-PCR), cDNAs were synthesized from 500 ng RNA using the Prime Script® RT reagent kit (Takara, Tokyo, Japan). *Elongation factor-2* (*EF-2*) was used as a reference gene to calculate the relative expression levels [27,28]. Except EF-2, the specific primers were newly designed for each gene using Primer3Plus (Table S1) [29]. The expression levels of each gene were quantified using TB Green™ Premix Ex Taq™ II (Takara, Tokyo, Japan) on a Light Cycler 480 (Roche Diagnostics, Mannheim, Germany). Biological triplicates were subjected to qRT-PCR, and each sample contained cDNA from each tissue of a male and female pair. The relative expression levels of each gene were calculated by adopting the standard curve method. Statistical analysis was performed using ANOVA and the Tukey–Kramer test for comparisons among the last instar period. These statistical analyses were performed using the statistical software Mac Statistical Analysis ver. 2.0.

3. Results

3.1. Co-Expression Network Analysis with Tissue Expression Data

Co-expression network analysis was performed with *networkz* to detect the candidate genes that regulate silk protein genes or the known regulatory factors of silk proteins. The program (*networkz*) allocates each transcript to the single most plausible network module. In total, 1022 network modules were generated, and the transcripts of the target genes were identified in 20 network modules (Table 2, Data S1). Of these, two target genes, *P25* and *Awh*, belonged to the *fibroin-L* and *fibroin-H* modules, respectively, whereas four known TFs (*SGF1*, *SGF3*, *sage*, and *Antp*) were sorted into different modules. Overall, 91 TFs were detected in the above 20 modules. The sericin1 modules, which showed a specific expression pattern in the M-MSG and P-MSG, contained 39 TFs among 565 transcripts. In addition, the sericin2 modules, which showed a specific expression pattern in the A-MSG, contained 11 TFs among 289 transcripts, and the sericin3 module, which showed a specific expression pattern in the whole SG of RNA-seq data-1 and M-MSG, contained two TFs among 36 transcripts (Figure 1A, Table 2). Although *fibroin-H* and *fibroin-L* showed PSG-specific expression patterns, they were separated into different network modules because of differences in TPM values. These modules contained nine TFs among 122 transcripts (Figure 1A, Table 2). The modules of four known TFs (*SGF1*, *SGF3*, *sage*, and *Antp*) contained >100 transcripts, including 5–11 TFs (Figure 1B, Table 2). All the obtained TFs were similarly expressed at one or more tissues with each target gene. Collectively, 91 transcripts were screened as candidate TFs that seem to regulate target gene expression.

Table 2. Total transcripts and TFs in each gene module.

Target Gene	Modules	Total Transcripts	Transcription Factor
total	1022		
sericin1	7	565	39
sericin2	6	289	11
sericin3	1	36	2
fibroin-H	1	42	5 (including *Arrowhead*)
fibroin-L	1	80 (including *P25*)	4
SGF1	1	119	11
SGF3	1	120	6
sage	1	114	8
Antennapedia	1	126	5

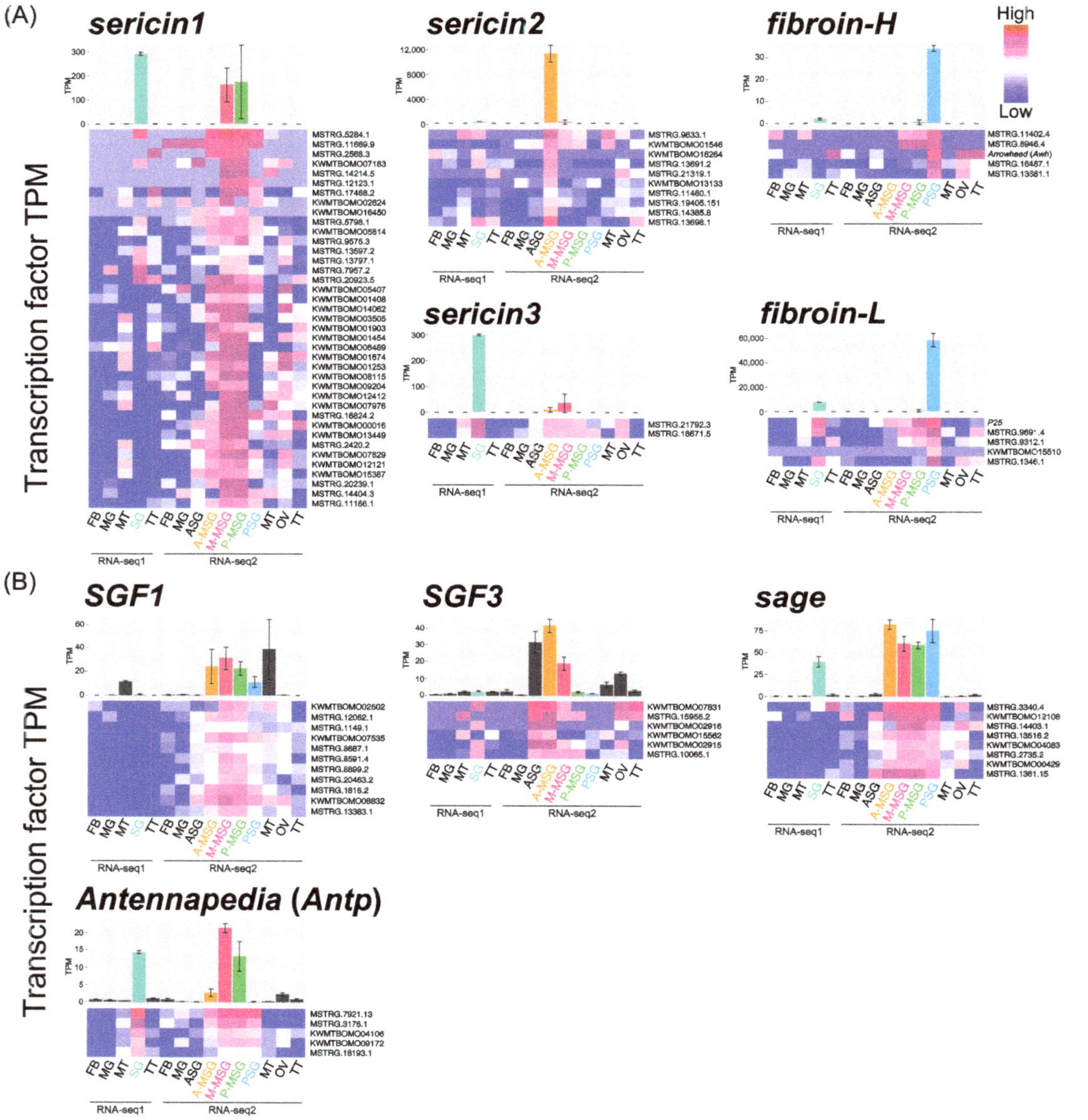

Figure 1. TPM (mean ± SE, biological triplicates) of silk protein genes (**A**) and TFs (**B**) from RNA-seq analysis and heatmap that was based on TPM of each module gene. The *sericin1* and *sericin2* graphs were drawn based on TPM values of main transcripts (*sericin1*: *KWMTBOMO06216*, *sericin2*: *KWMTBOMO06334*). Transcript ID is indicated on the right. Tissues that were used for RNA-seq are indicated under the heatmap.

3.2. Time-Course Expression Analysis of the Four SG Regions during the Last Instar Period

It is notable that the SG developed for seven days, with the wet weight reaching a peak on the fifth day of last instar (Figure 2A). To narrow down the candidate regulatory genes, we evaluated the time-course expression pattern of TFs that were screened by co-expression network analysis in the four SG regions (A-MSG, M-MSG, P-MSG, and PSG) during the last instar period using qRT-PCR (Figure 2B–D, Figure S1A, Data S2). The expression levels of 45 TFs, showing high transcript-level TPM values among the *sericin1–3*, *SGF1*, *SGF3*, *sage*, and *Antp* modules were quantified in three regions of the MSG. *sericin1* was mainly expressed in the M-MSG and its expression level reached a peak on fourth day (Figures 2E and 3A). *Antp* was also mainly expressed in the M-MSG, but its expression level reached a peak before that of *sericin1* (Figures 3H and S1B). Similar to *Antp*, the expression level of five TFs belonging to the *sericin1* module (*KWMTBOMO00016*, *KWMTBOMO14062*, *MSTRG.11166.1*, *MSTRG.14404.3*, and *MSTRG.16824.2*) including homeobox domain-containing genes (Table 3) and that of a TF belonging to the *Antp* module (*MSTRG.3176.1*) reached a peak before that of *sericin1* (Figure 3A,H). *sericin2* was mainly expressed in the A-MSG, and its expression level markedly decreased on the fifth day (Figures 2E and 3B). *sericin2* and four TFs (*MSTRG.11460.1*, *MSTRG.13691.2*, *MSTRG.14385.8*, and *MSTRG.19405.151*) showed similar expression patterns (Figure 3B). *sericin3* was also mainly expressed in the A-MSG, and its expression level increased over the later period (Figures 2E and 3C). *MSTRG.18671.5* and *sericin3* showed similar expression patterns (Figure 3C). The expression of *SGF1* showed a similar pattern among all regions of the MSG, with the expression level decreasing on the fifth day (Figure S1B). *KWMTBOMO08832* belonging to the *SGF1* module, showed similar expression pattern to SGF1 (Figure 3E). *SGF3* was primarily expressed in the A-MSG, with its expression peaking on the fifth day (Figures 3F and S1B). Although the forkhead domain-containing gene *KWMTBOMO02915* belonged to the *SGF3* module, its expression pattern was similar to that of *sericin3* (Figure 3F, Table 3). *sage* was also mainly expressed in the A-MSG, and its expression pattern was similar to that of *sericin3* (Figures 3G and S1B). In contrast, *KWMTBOMO12108*, belonging to the *sage* module, showed a high expression level in the earlier period, with its expression level markedly decreasing on the fifth day. This was similar to the pattern that was exhibited by *sericin2* (Figure 3G). Furthermore, the expression levels of nine TFs belonging to the *fibroin-H* and *fibroin-L* modules were quantified in the PSG. *fibroin-H*, *fibroin-L*, and *P25* expression levels were found to be elevated through the last instar period, along with SG development (Figures 2A and 3D). Although both the fibroin modules contained no TFs with expression patterns that were similar to those of *fibroin-H* and fibroin-L (Figure 2D), three TFs (*MSTRG.11402.4*, MSTRG.9312.1, and *MSTRG.1346.1*) were expressed during the earlier period, in contrast to the pattern that was exhibited by *fibroin-H* and *fibroin-L* (Figure 3D, Table 3). *Awh* was expressed through the mid-phase of the last instar period (Figure 2D, Table 3). In total, 17 TFs were eventually detected and found to be related to silk protein genes; they contained not only known regulatory factors such as the *Awh* isoform PB (*MSTRG.1346.1*) but also uncharacterized or function-unknown genes (Table 3).

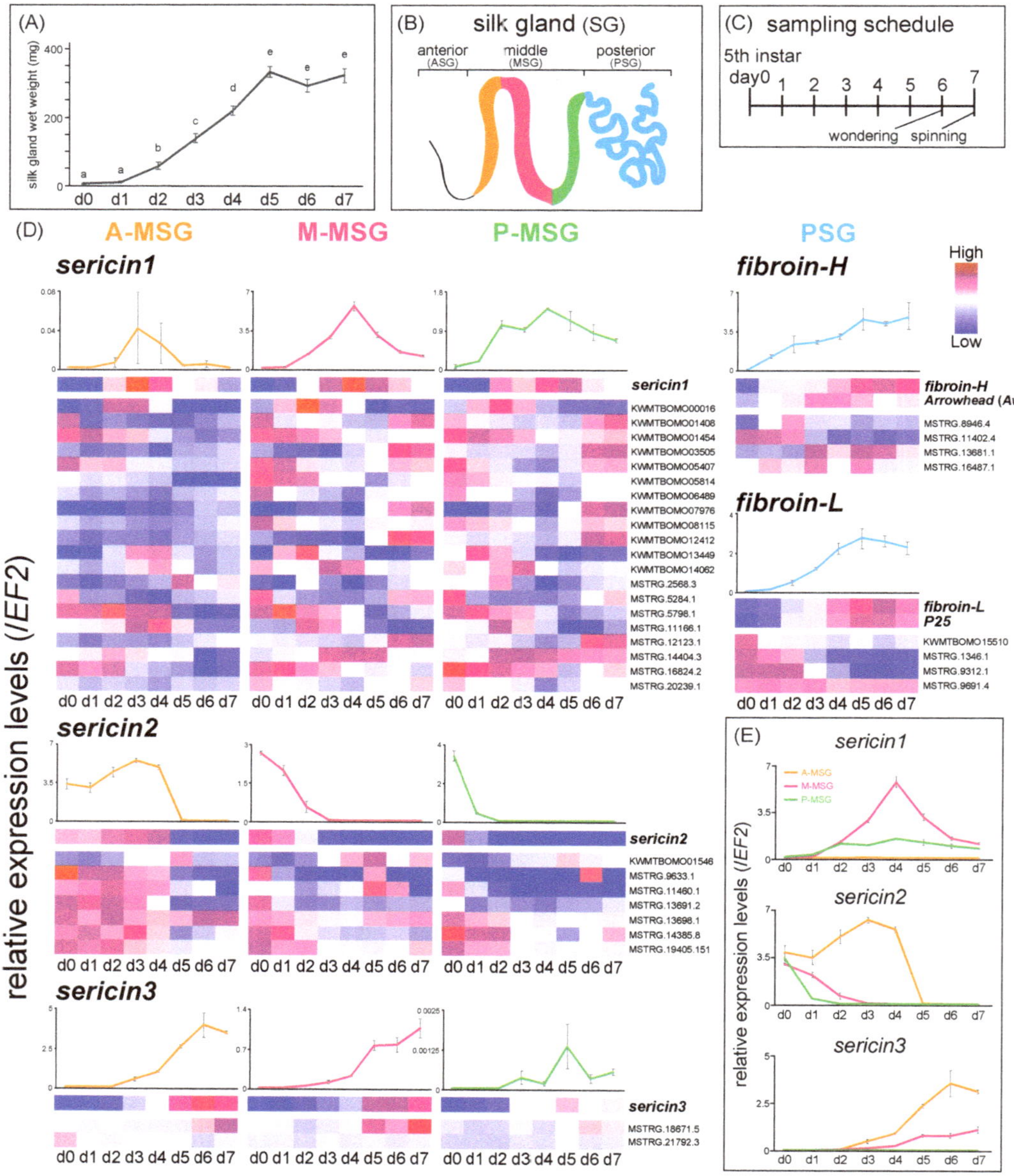

Figure 2. Wet weight transition of the whole SG during last instar larva (**A**). Different letters indicate significant differences in each gene (Tukey–Kramer test, $p < 0.05$). Schematic of the whole SG (**B**). Sampling schedule for qRT-PCR during last instar larva (**C**). The relative expression levels (mean ± SE, biological triplicates) of silk protein genes at each SG region during the last instar period, and a heatmap that was based on the expression of silk protein genes and their module TFs (**D**). Integrated graphs (mean ± SE, biological triplicates) showing sericin expression at each MSG region (**E**). Transcript ID is indicated on the right.

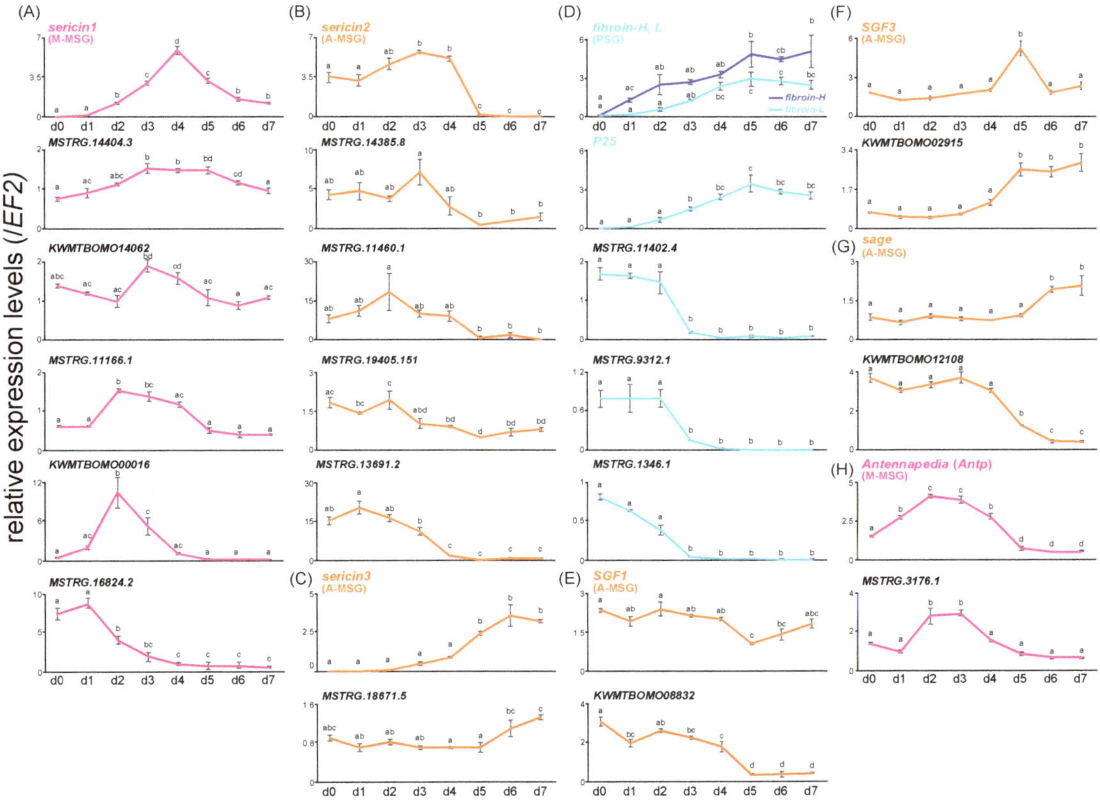

Figure 3. Relative expression levels (mean ± SE, biological triplicates) of each target gene and TFs at each SG region during last instar period [sericin1 (**A**), *sericin2* (**B**), *sericin3* (**C**), *fibroins* (**D**), *SGF1* (**E**), *SGF3* (**F**), *sage* (**G**), and *Antp* (**H**)]. Different letters indicate significant differences in each gene (Tukey–Kramer test, $p < 0.05$).

Table 3. Domain and description of focused TFs.

Transcript ID	Module	Domain (PfamID)	Description (NCBI-nr)
KWMTBOMO00016	*sericin1*	zf-CCHC (PF00098), RT_RNaseH (PF17917), RVT_1 (PF00078), rve (PF00665)	unnamed protein product [*Plutella xylostella*]
KWMTBOMO14062	*sericin1*	zf-C2H2_4 (PF13894), PI-PLC-Y,X (PF00387, 00388), SH2 (PF00017), SH3_1 (PF00018), C2 (PF00168)	endonuclease-reverse transcriptase [*Bombyx mori*]
MSTRG.11166.1	*sericin1*	bZIP_1 (PF000170)	uncharacterized protein LOC101735428 isoform X2 [*Bombyx mori*]
MSTRG.14404.3	*sericin1*	Homeobox_KN (PF05920)	homeobox protein homothorax-like [*Bombyx mori*]

Table 3. *Cont.*

Transcript ID	Module	Domain (PfamID)	Description (NCBI-nr)
MSTRG.16824.2	*sericin1*	zf-C2HC_2 (PF13913)	homeobox protein 5 isoform X8 [*Bombyx mori*]
MSTRG.11460.1	*sericin2*	NCU-G1 (PF15065)	glycosylated lysosomal membrane protein B [*Bombyx mori*]
MSTRG.13691.2	*sericin2*	CENP-F_leu_zip (PF10473)	uncharacterized protein LOC114240082 [*Bombyx mandarina*]
MSTRG.14385.8	*sericin2*	Bromodomain (PF00439)	bromodomain adjacent to zinc finger domain protein 1A isoform X3 [*Bombyx mori*]
MSTRG.19405.151	*sericin2*	FLYWCH_zf (PF04500), BTB/POZ (PF00651)	Mod(mdg4)-heS00531 [*Bombyx mori*]
MSTRG.18671.5	*sericin3*	HSF_DNA-bind (PF00447)	heat shock factor-d isoform X4 [*Bombyx mori*]
MSTRG.11402.4	*fibroin-H*	MBF2 (PF15868)	MBF2, partial [*Bombyx mori*]
MSTRG.1346.1	*fibroin-L*	LIM (PF00412)	arrowhead PB [*Bombyx mori*]
MSTRG.9312.1	*fibroin-L*	Myb_DNA-bind_7 (PF15963)	transcription factor TFIIIB component B″ [*Bombyx mori*]
KWMTBOMO08832	*SGF1*	zf-CCHC (PF00098), rev (PF00665), Integrase_H2C2 (PF17921), Asp_protease_2 (PF13650)	uncharacterized protein LOC114250529 isoform X1 [*Bombyx mandarina*]
KWMTBOMO02915	*SGF3*	Forkhead (PF00250)	fork head domain-containing protein FD4 [*Bombyx mori*]
KWMTBOMO12108	*sage*	Histone (PF00125), CBFD_NFYB_HMF (PF00808)	nuclear Y/CCAAT-box binding factor C subunit NF/YC isoform X1 [*Bombyx mori*]
MSTRG.3176.1	*Antennapedia*	MTABC_N (PF16185)	transcriptional regulator ATRX homolog [*Bombyx mandarina*]

4. Discussion

In previous studies, some genes or gene groups that are specifically expressed in the SG were identified using RNA-seq and microarray [30–32]. Despite this, a comprehensive screening strategy is much needed to identify the key factors that regulate silk proteins. Although *B. mori* has been previously used for co-expression network analysis [19,20], the mechanisms underlying the regulation of silk protein genes remain unclear. Therefore, in this study, we performed co-expression network as well as time-course expression analyses to identify the genes that regulate silk protein genes in *B. mori*. The co-expression network analysis was performed using an in-house program called *networkz*; consequently, 20 network modules that were related to 11 target genes were identified. The obtained TFs exhibited tissue expression patterns that were similar to those of each target gene (Figure 1), whereas, the majority of known TFs (*SGF1*, *SGF3*, *sage*, and *Antp*) formed a module that was distinct from the silk genes, respectively. Although the known TFs are co-expressed with the silk genes in the silk glands, they showed different expression patterns in other tissues, which led to the different modules. The different tissue expression patterns may be due to additional functions of these TFs which are not related with the silk gene regulation in the silk glands. These results indicated that *networkz* could successfully identify the related transcripts of each target from transcriptome data. *sericin1* and *sericin2* showed multiple modules as their mRNAs encode multiple isoforms with slightly different expression patterns at the tissue level (Table 2, Figure 1A) [4,33–35]. It, therefore, seems possible that diverse genes regulate *sericin1* and *sericin2* expression.

Time-course expression analysis led to the identification of 17 TFs that showed specific expression patterns and were related to target genes in the MSG and PSG during the last instar period (Figures 2 and 3). The *sericin1* module contained two homeobox domain-containing genes (*MSTRG.14404.3* and *MSTRG.16824.2*), the expression of which appeared before the expression of sericin1 peaked (Figure 3A). *MSTRG.14404.3* possessed a *homothorax* (*Hth*)-like motif (Table 3). *Hth* is a known cofactor of *Antp* and is thus related to regulating sericin1 expression [4]; therefore, it appears that *MSTRG.14404.3* is also involved in *sericin1* expression regulation. Although *SGF3*, as with *SGF1*, is also involved in regulating *sericin1* expression [8–10], it is possible that *KWMTBOMO02915* (Figure 2D) regulates *sericin3* expression as its expression pattern was similar to that of *sericin3* in the A-MSG (Figure 3C,F). Furthermore, *KWMTBOMO02915* was already recognized as MSG-specific expression TF in a previous study [24]. The expression level of the histone superfamily gene *KWMTBOMO12108* decreased in the later period of last instar, and it was similar to that of *sericin2*. It has been reported that 20-hydroxyecdysone (20E) titer increases in the later period of last instar [36], and that 20E treatment has a repressive effect on histone gene expression [37]. Hence, it is possible that *KWMTBOMO12108* regulates *sericin2* expression via 20E titer transition. The *fibroin*s modules contained *MSTRG.11402.4*, *MSTRG.9312.1*, and *MSTRG.1346.1*, which showed high expression levels during the earlier period, in contrast to the *fibroin*s expression pattern. *MSTRG.11402.4*, MBF2 partial transcript, is reportedly involved in *fibroin-H* expression regulation [38] and is also involved in nuclear transport in the SG along with *FTZ-F1* [39]. Although *Awh* isoform PA (*KWMTBOMO00651*) and *Awh* isoform PB (*MSTRG.1346.1*) belonged to the *fibroin*s modules, their expression patterns were different during the last instar period (Figures 2D and 3D). Besides, although the TFs *KWMTBOMO02915* and *KWMTBOMO12108* showed similar expression patterns with their target genes at the tissue level (Figure 1A), different expression patterns from their target genes were observed in the time-course expression (Figure 3F,G). These results suggested that when designing a screening strategy, including both co-expression network and time-course expression analyses is pivotal. As stated earlier, the TFs *MSTRG.11402.4*, *MSTRG.14404.3*, *MSTRG.1346.1*, and *KWMTBOMO02915* are known to be related to silk protein genes, while 13 novel function-unknown TFs were recognized as candidates of silk proteins regulation factor. Herein we performed time-course expression analysis to screen related TFs by qRT-PCR focusing on the candidates. Extending this approach to co-expression network analysis using RNA-seq data will help to provide insights into full extent of silk protein genes regulation.

5. Conclusions

In this study, silk protein regulatory genes in *B. mori* were identified using a two-step screening strategy. In the first step, 20 network modules including 91 TFs were screened by co-expression network analysis using the in-house program *networkz*, and in the second step, 17 transcripts were screened as silk protein-related genes by time-course expression analysis of the MSG and PSG during the last instar period. Since four of these TFs were already known to be related with the silk gene, we found 13 TFs as candidates for novel silk regulatory factors. As we identified both known as well as function-unknown TFs, we believe that our strategy is robust and highly sensitive to screen relative genes. Furthermore, screening results indicated that a larger number of genes than expected may be involved in silk protein regulation in *B. mori*. Functional analyses of function-unknown TFs should further our understanding of the mechanisms underlying silk protein regulation.

Supplementary Materials: The following materials are available online at figshare. Figure S1: The relative expression levels of known regulatory factors at each SG region during the last instar period, and a heatmap that was based on the expression of known regulatory factors and their module TFs. (doi:10.6084/m9.figshare.17141888). Table S1: Primer sets that were used in this paper (doi:10.6084/m9.figshare.17141969). Data S1: Output raw data of co-expression network analysis using *networkz* (doi:10.6084/m9.figshare.17427359). Data S2: Relative expression levels of all tested transcripts (doi:10.6084/m9.figshare.17427362).

Author Contributions: Y.M. and K.Y. designed the experiments; Y.M. collected samples and performed all experiments; W.C. produced tool program for network analysis; H.S. (Hiroki Sakai) and H.S. (Hideki Sezutsu) supplied the silkworms; Y.M., A.J. and K.Y. wrote the manuscript; Y.M. and K.Y. conceived, designed, and coordinated the study. All authors discussed the data and cooperated in preparing the manuscript. All authors have read and agreed to the published version of the manuscript.

Funding: This work was supported by the Cabinet Office, Government of Japan, Cross-ministerial Strategic Innovation Promotion Program (SIP), "Technologies for Smart Bio-industry and Agriculture" (funding agency: Bio-oriented Technology Research Advancement Institution, NARO) to H.S. and K.Y. This work was also supported by a grant from the Ministry of Agriculture, Forestry and Fisheries of Japan (Research Project for Sericultural Revolution) to Y.M., H.S. and K.Y.

Institutional Review Board Statement: Not applicable.

Data Availability Statement: Not applicable.

Acknowledgments: We are grateful to Ken-ichiro Tatematsu, Takuya Tsubota, and Kimiko Yamamoto for productive discussions. We also thank Nobuto Yamada, Hiroko Hoshida, and Seigo Kuwazaki for help during laboratory work. Computations were partially performed on the SHIHO supercomputer at the National Agriculture and Food Research Organization (NARO), Japan.

Conflicts of Interest: The authors declare no conflict of interest. The funders had no role in the design of the study; in the collection, analyses, or interpretation of data; in the writing of the manuscript, or in the decision to publish the results.

References

1. Dhawan, S.; Gopinathan, K.P. Cell cycle events during the development of the silk glands in the mulberry silkworm *Bombyx mori*. *Dev. Genes Evol.* **2003**, *213*, 435–444. [CrossRef] [PubMed]
2. Kimoto, M.; Yamaguchi, M.; Fujimoto, Y.; Takiya, S. Expression profiles of the genes for nine transcription factors and their isoforms in the posterior silk gland of the silkworm Bombyx mori during the last and penultimate instars. *J. Insect Biotechnol Sericology* **2011**, *79*, 31–43.
3. Takiya, S.; Tsubota, T.; Kimoto, M. Regulation of silk genes by hox and homeodomain proteins in the terminal differentiated silk gland of the silkworm *Bombyx mori*. *J. Dev. Biol.* **2016**, *4*, 19. [CrossRef] [PubMed]
4. Kimoto, M.; Tsubota, T.; Uchino, K.; Sezutsu, H.; Takiya, S. Hox transcription factor Antp regulates *sericin-1* gene expression in the terminal differentiated silk gland of *Bombyx mori*. *Dev. Biol.* **2014**, *386*, 64–71. [CrossRef] [PubMed]
5. Tsubota, T.; Tomita, S.; Uchino, K.; Kimoto, M.; Takiya, S.; Kajiwara, H.; Yamazaki, T.; Sezutsu, H. A hox gene, Antennapedia, regulates expression of multiple major silk protein genes in the silkworm *Bombyx mori*. *J. Biol. Chem.* **2016**, *291*, 7087–7096. [CrossRef]
6. Ohno, K.; Sawada, J.; Takiya, S.; Kimoto, M.; Matsumoto, A.; Tsubota, T.; Uchino, K.; Hui, C.C.; Sezutsu, H.; Handa, H.; et al. Silk gland factor-2, involved in fibroin gene transcription, consists of LIM homeodomain, LIM-interacting, and single-stranded DNA-binding proteins. *J. Biol. Chem.* **2013**, *288*, 31581–31591. [CrossRef]
7. Kimoto, M.; Tsubota, T.; Uchino, K.; Sezutsu, H.; Takiya, S. LIM-homeodomain transcription factor Awh is a key component activating all three fibroin genes, *fibH*, *fibL* and *fhx*, in the silk gland of the silkworm, *Bombyx mori*. *Insect Biochem. Mol. Biol.* **2015**, *56*, 29–35. [CrossRef]
8. Mach, V.; Takiya, S.; Ohno, K.; Handa, H.; Imai, T.; Suzuki, Y. Silk gland factor-1 involved in the regulation of *Bombyx* sericin-1 gene contains fork head motif. *J. Biol. Chem.* **1995**, *270*, 9340–9346. [CrossRef]
9. Matsuno, K.; Takiya, S.; Hui, C.C.; Suzuki, T.; Fakuta, M.; Ueno, K.; Suzuki, Y. Transcriptional stimulation via SG site of *Bombyx* sericin-1 gene through an interaction with a DNA binding protein SGF-3. *Nucleic Acids Res.* **1990**, *18*, 1853–1858. [CrossRef]
10. Matsunami, K.; Kokubo, H.; Ohno, K.; Suzuki, Y. Expression pattern analysis of SGF-3/POU-M1 in relation to sericin-1 gene expression in the silk gland. *Dev. Growth Differ.* **1998**, *40*, 591–597. [CrossRef]
11. Zhao, X.M.; Liu, C.; Li, Q.Y.; Hu, W.B.; Zhou, M.T.; Nie, H.Y.; Zhang, Y.X.; Peng, Z.C.; Zhao, P.; Xia, Q.Y. Basic helix-loop-helix transcription factor bmsage is involved in regulation of *fibroin H-chain* gene via interaction with SGF1 in *Bombyx mori*. *PLoS ONE* **2014**, *9*, e94091. [CrossRef] [PubMed]
12. Dam, S.; Vosa, U.; Graaf, A.; Franke, L.; Magalhaes, J.P. Gene co-expression analysis for functional classification and gene-disease predictions. *Brief. Bioinform.* **2018**, *19*, 575–592. [PubMed]
13. Tokimatsu, T.; Sakurai, N.; Suzuki, H.; Ohta, H.; Nishitani, K.; Koyama, T.; Umezawa, T.; Misawa, N.; Saito, K.; Shibata, D. KPPA-view. A web-based analysis tool for integration of transcript and metabolite data on plant metabolic pathway maps. *Plant Physiol.* **2005**, *138*, 1289–1300. [CrossRef] [PubMed]
14. Aoki, K.; Ogata, Y.; Shibata, D. Approaches for extracting practical information from gene co-expression networks in plant biology. *Plant Cell Physiol.* **2007**, *48*, 381–390. [CrossRef]

15. Amrine, K.C.H.; Blanco-Ulate, B.; Cantu, D. Discovery of core biotic stress responsive genes in Arabidopsis by weighted gene co-expression network analysis. *PLoS ONE* **2015**, *10*, e0118731. [CrossRef]
16. Liu, W.; Li, L.; Long, X.; You, W.; Zhong, Y.; Wang, M.; Tao, H.; Lin, S.; He, H. Construction and analysis of gene co-expression networks in *Escherichia coli*. *Cells* **2018**, *7*, 19. [CrossRef]
17. Liu, H.Q.; Li, Y.; Irwin, D.M.; Zhang, Y.P.; Wu, D.D. Integrative analysis of young genes, positively selected genes and lncRNAs in the development of *Drosophila melanogaster*. *BMC Evol. Biol.* **2014**, *14*, 241. [CrossRef]
18. Behura, S.K.; Gomez-Machorro, C.; Harker, B.W.; deBruyn, B.; Lovin, D.D.; Hemme, R.R.; Mori, A.; Romero-Severson, J.; Severso, D.W. Global cross-talk of genes of the mosquito *Aedes aegypti* in response to dengue virus infection. *PLoS Negl. Trop. Dis.* **2011**, *5*, e1385. [CrossRef]
19. Wu, Y.; Cheng, T.; Liu, C.; Liu, D.; Zhang, Q.; Long, R.; Zhao, P.; Xia, Q. Systematic identification and characterization of long non-cording RNAs in the silkworm, *Bombyx mori*. *PLoS ONE* **2016**, *11*, e0147147.
20. Zhou, Q.Z.; Fu, P.; Li, S.S.; Zhang, C.J.; Yu, Q.Y.; Qiu, C.Z.; Zhang, H.B.; Zhang, Z. A comparison of co-expression networks in silk gland reveals the causes of silk yield increase during silkworm domestication. *Front. Genet.* **2020**, *11*, 225. [CrossRef]
21. Kikuchi, A.; Nakazato, T.; Ito, K.; Nojima, Y.; Yokoyama, T.; Iwabuchi, K.; Bono, H.; Toyoda, A.; Fujiyama, A.; Sato, R. Identification of functional enolase genes of the silkworm bombyx mori from public databases with a combination of dry and wet bench processes. *BMC Genom.* **2017**, *18*, 83. [CrossRef] [PubMed]
22. Ichino, F.; Bono, H.; Nakazato, T.; Toyoda, A.; Fujiyama, A.; Iwabuchi, K.; Sato, R.; Tabunoki, H. Construction of a simple evaluation system for the intestinal absorption of an orally administered medicine using *Bombyx mori* larvae. *Drug Discov. Ther.* **2018**, *12*, 7–15. [CrossRef] [PubMed]
23. Kobayashi, Y.; Nojima, Y.; Sakamoto, T.; Iwabuchi, K.; Nakazato, T.; Bono, H.; Toyoda, A.; Fujiyama, A.; Kanost, M.; Tabunoki, H. Comparative analysis of seven types of superoxide dismutases for their ability to respond to oxidative stress in *Bombyx mori*. *Sci. Rep.* **2019**, *9*, 2170. [CrossRef] [PubMed]
24. Yokoi, K.; Tsubota, T.; Jouraku, A.; Sezutsu, H.; Bono, H. Reference transcriptome data in silkworm *Bombyx mori*. *Insects* **2021**, *12*, 519. [CrossRef] [PubMed]
25. Kruskal, J.B. On the shortest spanning subtree of a graph and the traveling salesman problem. *Proc. Am. Math. Soc.* **1956**, *7*, 48–50. [CrossRef]
26. Campigotto, R.; Cespedes, P.C.; Guillaume, J.L. A generalized and adaptive method for community detection. *arXiv* **2014**, arXiv:1406.2518.
27. Koike, Y.; Mita, K.; Suzuki, M.G.; Maeda, S.; Abe, H.; Osoegawa, K.; deJong, P.J.; Shimada, T. Genomic sequence of a 320-kb segment of the Z chromosome of *Bombyx mori* containing a *kettin* ortholog. *Mol. Genet. Genom.* **2003**, *269*, 137–149. [CrossRef]
28. Sakai, H.; Sumitani, M.; Chikami, Y.; Yahata, K.; Uchino, K.; Kiuchi, T.; Katsuma, S.; Aoki, F.; Sezutsu, H.; Suzuki, M.G. Transgenic expression of the piRNA-resistant *Masculinizer* gene induces female-specific lethality and partial female-to-male sex reversal in the silkworm, *Bombyx mori*. *PLoS Genet.* **2016**, *12*, e1006203. [CrossRef]
29. Untergasser, A.; Nijveen, H.; Rao, X.; Bisseling, T.; Geurts, R.; Leunissen, J.A. Primer3Plus, an enhanced web interface to Primer3. *Nucleic Acids Res.* **2007**, *35*, W71–W74. [CrossRef]
30. Fang, S.M.; Hu, B.L.; Zhou, Q.Z.; YU, Q.Y.; Zhang, Z. Comparative analysis of the silk gland transcriptomes between the domestic and wild silkworms. *BMC Genom.* **2015**, *16*, 60. [CrossRef]
31. Chang, H.; Cheng, T.; Wu, Y.; Hu, W.; Long, R.; Liu, C.; Zhao, P.; Xia, Q. Transcriptomic analysis of the anterior silk gland in the domestic silkworm (*Bombyx mori*)–insight into the mechanism of silk formation and spinning. *PLoS ONE* **2015**, *10*, e0139424. [CrossRef] [PubMed]
32. Ma, Y.; Sun, Q.; Huang, L.; Luo, Q.; Zeng, W.; Ou, Y.; Ma, J.; Xu, H. Genome-wide survey and characterization of transcription factors in the silk gland of the silkworm, *Bombyx mori*. *PLoS ONE* **2021**, *16*, e0259870. [CrossRef] [PubMed]
33. Couble, P.; Michaille, J.J.; Garel, A.; Couble, M.L.; Prudhomme, J.C. Developmental switches of sericin mRNA splicing in individual cells of *Bombyx mori* silkgland. *Dev. Biol.* **1987**, *124*, 431–440. [CrossRef]
34. Michaille, J.J.; Garel, A.; Prudhomme, J.C. Cloning and characterisation of the highly polymorphic Ser2 gene of *Bombyx mori*. *Gene* **1990**, *86*, 177–184. [CrossRef]
35. Garel, A.; Deleage, G.; Prudhomme, J.C. Structure and organization of the *Bombyx mori* sericin1 gene and of the sericin1 deduced from the sequence of the ser 1B cDNA. *Insect Biochem. Mol. Biol.* **1997**, *27*, 469–477. [CrossRef]
36. Kaneko, Y.; Takaki, K.; Iwami, M.; Sakurai, S. Developmental profile of annexin IX and its possible role in programmed cell death of the *Bombyx mori* anterior silk gland. *Zool. Sci.* **2006**, *23*, 533–542. [CrossRef]
37. Furukawa, S.; Sagisaka, A.; Tanaka, H.; Ishibashi, J.; Kaneko, Y.; Yamaji, K.; Yamanaka, M. Molecular cloning and characterization of histone *H2A.Z* gene of the silkworm, *Bombyx mori*. *J. Insect Biotechnol. Sericology* **2007**, *76*, 121–127.
38. Zhou, C.; Zha, X.; Shi, P.; Wei, S.; Wang, H.; Zheng, R.; Xia, Q. Multiprotein bridging factor 2 regulates the expression of the fibroin heavy chain gene by interacting with Bmdimmed in the silkworm *Bombyx mori*. *Insect Mol. Biol.* **2016**, *25*, 509–518. [CrossRef]
39. Liu, Q.X.; Ueda, H.; Hirose, S. MBF2 is a tissue-and stage-specific coactivator that is regulated at the step of nuclear transport in the silkworm *Bombyx mori*. *Dev. Biol.* **2000**, *225*, 437–446. [CrossRef]

insects

MDPI

Technical Note

Products of Sericulture and Their Hypoglycemic Action Evaluated by Using the Silkworm, *Bombyx mori* (Lepidoptera: Bombycidae), as a Model

Salvador D. Aznar-Cervantes *, Beatriz Monteagudo Santesteban and José L. Cenis

Departamento de Biotecnología, Genómica y Mejora Vegetal, Instituto Murciano de Investigación y Desarrollo Agrario y Medioambiental (IMIDA), La Alberca, 30150 Murcia, Spain; bea.monteagudo1@gmail.com (B.M.S.); josel.cenis@carm.es (J.L.C.)
* Correspondence: sdac1@um.es; Tel.: +34-968368568

Simple Summary: The use of invertebrates as animal models is gaining attention within the scientific community due to numerous advantages during the development of the experiments, low cost of rearing, and fewer ethical problems. The well-documented biology of the silkworm (*Bombyx mori*) makes this insect an ideal candidate to be used in different fields of research. In this study, we demonstrated the feasibility of using the silkworm to evaluate the hypoglycemic action of various products of sericulture included in the diet after promoting glucose or sucrose-induced hyperglycemia in silkworms. The postprandial antihyperglycemic activity of fibroin, sericin, and powder made from pupae of silkworms is confirmed. These natural products are therefore ideal candidates for the prevention and treatment of diabetes, obesity, and other lifestyle-related diseases.

Abstract: Sericulture generates different natural products with potential medical applications. Silk peptides, worms, or even pupae are commonly employed in traditional Asian medicine with a wide variety of purposes, and some scientific work has been focused on their antidiabetic properties. This work evaluates the postprandial antihyperglycemic activity of fibroin, sericin, and powder made from either larvae or pupae of silkworms, and *Bombyx mori* L. (Lepidoptera: Bombycidae), employing the silkworm itself as an animal model. The results indicate a reduction in the glucose levels in hemolymph after sucrose or glucose-induced hyperglycemia when these products are included in the diet of the worms.

Keywords: fibroin; sericin; silkworm; chrysalis; diabetes

Citation: Aznar-Cervantes, S.D.; Monteagudo Santesteban, B.; Cenis, J.L. Products of Sericulture and Their Hypoglycemic Action Evaluated by Using the Silkworm, *Bombyx mori* (Lepidoptera: Bombycidae), as a Model. *Insects* **2021**, *12*, 1059. https://doi.org/10.3390/insects12121059

Academic Editors: Silvia Cappellozza, Morena Casartelli, Federica Sandrelli, Alessio Saviane and Gianluca Tettamanti

Received: 3 November 2021
Accepted: 24 November 2021
Published: 25 November 2021

1. Introduction

The growing number of studies targeting natural compounds to treat various pathologies suggests the importance of this field of research within the scientific community. Traditional Asian medicine has inspired many of these studies, being a rich source of ideas in exploring new bioactive compounds. Sericulture is a production cycle that involves the use of mulberry to feed the worms until the cocoon is spun by the larvae. Subsequently, the silk is degummed, and it is used for the textile industry or some incipient biomedical applications. Throughout this process, several potential products of medical interest are generated. This is not surprising, given the multiple applications attributed to the silkworms or the silk proteins (fibroin and sericin) in Asia. On the one hand, the powder produced from dehydrated silkworm chrysalides (pupae) has demonstrated potential medical uses presenting notable activities, such as increasing fat metabolism in rats [1], increasing levels of nitrite and nitric oxide synthase expression in a model of erectile dysfunction in rats [2], inducing apoptosis in human gastric cancer cells [3], and constituting a healthy nutritional source of protein and fat [4,5]. On the other hand, the powder made from silkworm larvae induces a reduction of plasma glucose level [6,7] due to the presence of 1-deoxynojirimycin (DNJ) ingested and accumulated during the feeding with mulberry leaves. DNJ presents α-glucosidase inhibitory effect [8], and the silkworm powder also inhibits the expression of glucose transporter (SGLT1) of human intestinal epithelial cell line Caco-2 [9]. Moreover, silk proteins present interesting biomedical properties when administered as nutritional supplements. An increase in fat oxidation and exercise performance has been demonstrated in mice after ingesting silk peptides in the diet [10,11] Furthermore, fibroin, sericin or peptides, and hydrolysates derived from cocoons also present α-glucosidase inhibitory activity as previously stated in several studies [6,12–14].

The use of invertebrates as animal models is gaining attention within the scientific community due to several advantages in the design and execution of the experiments, low cost of rearing, and fewer ethical problems. The well-documented biology of the silkworm (*B. mori*), as well as its ancestral domestication, make this insect an ideal candidate to be used as an animal model in different fields of research. In fact, its utility as a model for evaluating antidiabetic drugs has already been investigated [15,16], as well as its use in the study of the pathogenicity of bacteria [17,18], fungi [19], and drugs for the treatment of different pathologies [20,21]. In this context, this work aims to demonstrate the feasibility of using the silkworm to evaluate the hypoglycemic action of various products of sericulture by including them in the diet after promoting glucose or sucrose-induced hyperglycemia in silkworms. As far as we know, this is the first time that the silkworm is used as a model to evaluate the products derived from its rearing for the potential use in the prevention or treatment of diabetes.

2. Materials and Methods

2.1. Silkworm Rearing

A Spanish hybrid of silkworm races (*Sierra Morena X Bagdad)* was used for this experiment. The larvae were fed until the first day of the fifth instar with an artificial diet provided by the Padua Sericulture Station [22] at 23–25 °C and 50–60% relative humidity, under the normal photoperiod in our location in Spain during August (14 h light:10 h darkness) These worms were subsequently used to induce the sucrose or glucose-derived hyperglycemia (except the negative control) and to administer the different products in the same hyperglycemic diets.

2.2. Fibroin and Sericin Extraction

The purification of silk fibroin involved degumming in order to remove the sericin White cocoons were boiled for 30 min in an aqueous solution of 0.02 M Na_2CO_3 (Panreac, Barcelona, Spain) and then rinsed thoroughly with distilled water to extract the glue-like sericin proteins. The extracted fibroin was then dried at room temperature for three days and dissolved in LiBr (Acros Organics, Fairlawn, NJ, USA) 9.3 M for 3 h at 60 °C to generate a 20% w/v solution. Then it was dialyzed against distilled water for three days (performing at least eight water changes). The resultant aqueous solution was lyophilized in order to have purified fibroin ready to be dissolved in the water employed to prepare the diets prior to feeding the worms [23].

The extraction of silk sericin was performed by autoclaving the cocoons in distilled water (25 g/L) at 120 °C for 1 h [24] to obtain pure sericin aqueous solutions after filtering them. Then, these solutions were freeze-dried, obtaining a sericin powder ready to be included in the diets. Both sericin and fibroin were stored dry at room temperature until use.

2.3. Preparation of Silkworm and Chrysalis Powder

Prior to hyperglycemic induction experiments, the first batch of worms (N = 50) was fed ad libitum with mulberry leaves to enrich silkworm larvae and pupae with the components derived from the foliage and DNJ, among others. Part of the fifth instar larvae (N = 25) was frozen and subsequently freeze-dried; the same procedure was performed using chrysalides (N = 25). The entire body of lyophilized larvae or pupae was ground using a waring® commercial laboratory blender and incorporated as a supplement to the artificial diet, as detailed below. Both powders were stored in a dry container at room temperature, protected from light until use.

2.4. Induction of Hyperglycemia and Treatment with Different Products

Eleven types of diet were prepared to feed worms from the eleven different batches (N = 10) during 24 h ad libitum (Figure 1). After this period, the glucose levels in hemolymph were determined, as explained below.

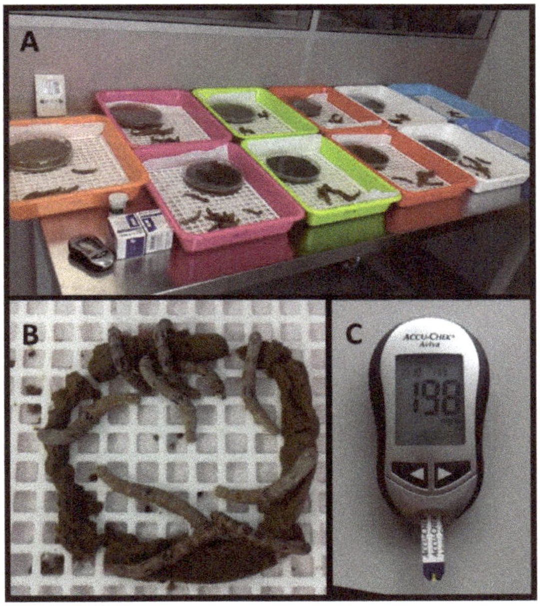

Figure 1. Illustrative images of the different batches of worms with their respective diets prepared immediately before feeding (**A**), silkworms actively feeding (**B**) and example of measurement of glucose levels in the hemolymph (**C**).

The negative control diet (considered normal) was prepared by mixing 25 g of powdered diet, supplied by the Padua Sericulture Station, with 75 mL of boiling water (ratio 1:3) until a consistent homogeneous paste was obtained that was allowed to cool down before use.

Following a similar procedure, two types of hyperglycemic diets (considered positive controls) were prepared separately, one including glucose and the other sucrose (Sigma-Aldrich, St. Louis, MO, USA), both at 10% (*w*/*w*) as previously stated by several authors [25]. Briefly, glucose or sucrose was dissolved in the water that would later be used to add and mix the powdered diet.

In order to evaluate the potential postprandial antihyperglycemic activity of fibroin, sericin, and powder made from either larvae or pupae of silkworms, these four products were added at 5% (*w*/*w*) to eight different diets, four containing glucose and four containing sucrose, at 10% (*w*/*w*) in both cases, comparable to the diets considered positive controls of hyperglycemic induction. Sericin and fibroin were dissolved in the water of their respective diets prior to mixing them with the commercial powdered diet. On the other hand, the silkworm or chrysalis powder was directly homogenized in the diet.

2.5. Evaluation of Glucose Level in Haemolymph

Glucose levels in the hemolymph were determined immediately 24 h after the start of the feeding as described by other authors [16,25], collecting it from the silkworms through a cut on the first proleg, making this determination in at least three specimens per treatment and using a glucometer (Accu-Check, Roche, Basel, Switzerland) for this purpose. Briefly, the glucometer test strips are filled by capillary action when they come into contact with the drop of hemolymph produced after making the incision, without the need to carry out any type of volumetric measurement or dilution of the sample. The choice of the feeding period is based on previous experiences in our laboratory, in which we confirmed that it is an adequate time to establish high glucose levels in the hemolymph.

2.6. Statistical Analysis

For the statistical analyses, SPSS 25 software was used. When the data were compiled with the normality and homogeneity of variance requirements, they were compared using parametric tests: ANOVA followed by Bonferroni's post hoc multiple *t*-test. When the assumption of homoscedasticity was not satisfied, the statistical significance was determined using Mann–Whitney test (nonparametric). In every situation, the significance level was set to $p < 0.05$.

3. Results and Discussion

3.1. Effective Induction of Hyperglycemia in Silkworms Using Glucose or Sucrose as Additives in the Artificial Diet

As expected, feeding silkworms with an artificial diet enriched with glucose or sucrose at 10% (w/w) significantly increased glucose levels in the hemolymph (Figure 2). These differences were statistically significant in both comparisons (Bonferroni, $p < 0.05$), with the negative control (standard diet), and between both hyperglycemic treatments. The mean value obtained in the case of worms fed using a diet enriched in glucose (501 mg/dL) was higher than that using diet enriched in sucrose (353 mg/dL). Moreover, both are far superior to the average value of the negative control diet (46 mg/dL). These hemolymph glucose levels are consistent with those reported in similar studies after inducing hyperglycemia by feeding sucrose or glucose-enriched diets, consistently higher in the latter case [15,25,26].

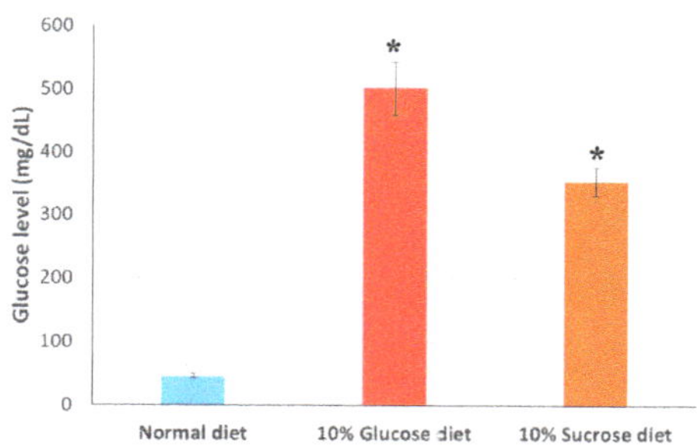

Figure 2. Glucose levels in hemolymph detected in silkworms fed a standard artificial diet (regular diet) and diets enriched in glucose or sucrose (10% w/w) 24 h after the start of the feeding. Data are expressed as mean ± SEM. * Indicates significantly higher values compared to the control diet ($p < 0.05$).

3.2. Hypoglycemic Action of Products Derived from Sericulture under Glucose-Induced Hyperglycemia

As it has been previously stated, there is a similar mechanism, in terms of uptake and storage of sugars, in silkworms and mammals, displaying common features that make feasible the use of silkworm models for the assessment of antidiabetic drugs for both types of diabetes (I and II) [26]. In this work, two types of hyperglycemia have been induced in silkworms; on the one hand, using a diet enriched in glucose, and on the other, employing sucrose. This helps us discern, at least partly, whether the mechanism of action of this potential hypoglycemic effect is related to inhibition of glucose absorption, or on the contrary, is more linked to inhibition of α-glucosidase activity.

In this sense, the data obtained from the experiment using glucose (monosaccharide) revealed a hypoglycemic effect in all the treatments studied, decreasing the average values of glucose. These values were significantly lower than those obtained in worms fed the positive control diet in the case of feeding with chrysalis powder, fibroin, or sericin ($p < 0.05$). In this last case recovering glucose values that were statistically equivalent to the negative control, which simulates a nonpathological condition (Bonferroni, $p > 0.05$). Figure 3 clearly shows a larger SEM in the glucose values of silkworms fed with worm powder (442 ± 102 mg/dL). This result could be due to the lower homogeneity in terms of its distribution in the artificial diet. It was observed that the powder resulting from the grinding of the worms, and its distribution in the diet, were more heterogeneous than in the case of the chrysalis. This fact could lead to a differential intake of this product by the different specimens, thus increasing the variability in the results. Similarly, previous work that compared the antihyperglycemic effect of worm powder, fibroin, and sericin using rats as an animal model, obtaining better results with fibroin or sericin than with worm powder [6]. This result is analogous to the findings in our study us using *B. mori* as an animal model. Furthermore, silkworm powder has been described as an inhibitor of the expression of glucose transporter (SGLT1) of human intestinal epithelial cell line Caco-2 [9], and this type of transporters are sometimes detected in insects [27]. Therefore, it can be suggested that part of the hypoglycemic effect, in this case, could be due to this mechanism, both for worm and chrysalis powder. However, this aspect should be further studied to confirm this analogy.

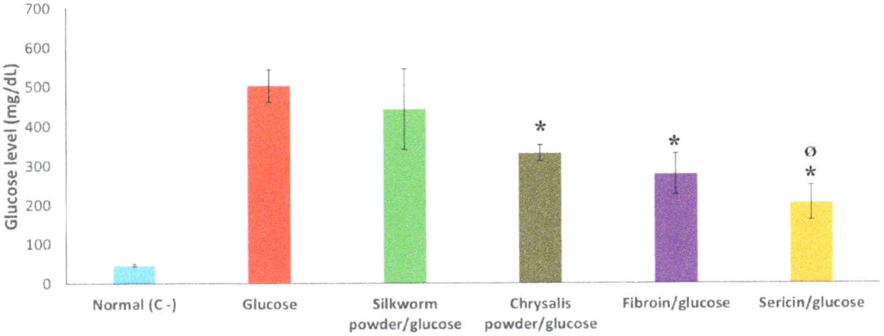

Figure 3. Glucose levels in hemolymph detected in silkworms fed a standard artificial diet (regular diet), a diet enriched in glucose (10% *w*/*w*), and different diets containing the same amount of glucose and the products studied at 5% *w*/*w* (silkworm powder, chrysalis powder, fibroin or sericin). Determinations were carried out 24 h after the start of feeding. Data are expressed as mean ± SEM. * Indicates values significantly lower than the glucose diet considered positive control for induction of hyperglycemia ($p < 0.05$). Ø Indicates values equivalent to the negative control of hyperglycemia (regular diet) considered as a nonpathological condition ($p > 0.05$).

Silk peptides have been studied previously due to their antidiabetic properties. Some studies have shown, for example, how the consumption of silk hydrolysates in rats promotes improvements in antidiabetic symptoms by potentiating insulin secretion [28]. Taking into account the presence of a peptide hormone in the silkworm, similar in structure to the insulin, called bombyxin, as well as the existence of an insulin signaling pathway it activates with the same mode of action as that of mammals [16], we can begin to understand at least part of the mechanism involved in the decrease in the hemolymph glucose values detected in our experiment, especially when fibroin or sericin were used in the artificial diet. It is important to mention that hydrolyzed silk fibroin also promotes the regeneration of pancreatic β-Cells, as stated by Park et al. [29].

3.3. Hypoglycemic Action of Products Derived from Sericulture under Sucrose-Induced Hyperglycemia

Sucrose is a sweetener widely used in foods and beverages whose abuse can lead to postprandial hyperglycemia, which in the long term implies the development of diseases such as obesity or diabetes [30]. Therefore, it is important to minimize these effects of the blood glucose caused by sucrose and thus avoid the emergence of these lifestyle-related diseases [15]. Sucrose is divided into glucose and fructose under the action of the enzyme α-glycosidase in the intestine to be absorbed, subsequently causing undesirable increases in blood glucose [31]. Thus, the inhibition of the catalytic activity of this enzyme is important for the prevention and treatment of diabetes, and several compounds used as antidiabetic drugs are based on precisely this inhibitory mechanism (acarbose and voglibose, among others).

Figure 4 shows the average values of glucose obtained from the hemolymph of worms fed different diets. In this case, the induction of hyperglycemia was carried out by adding sucrose at 10% (w/w) to both the positive control diet and the diets including the study products, the latter added at 5% w/w (explained in the Section 2). Furthermore, glucose measurements were also conducted in worms fed a standard artificial diet, named regular diet, and those considered negative control of hyperglycemia. Sucrose is a disaccharide divided in the intestines into glucose and fructose, its two basic units, to allow the absorption of this glucose later and finally raise its levels in hemolymph in a way analogous to what that occurs in human blood. The fact that this increase in hemolymph glucose levels varies after feeding with sucrose-enriched diets and different compounds would suggest varying degrees of inhibition of the α-glucosidase activity (respectively) since this enzyme is responsible for the breakdown of the disaccharide towards its component monosaccharides.

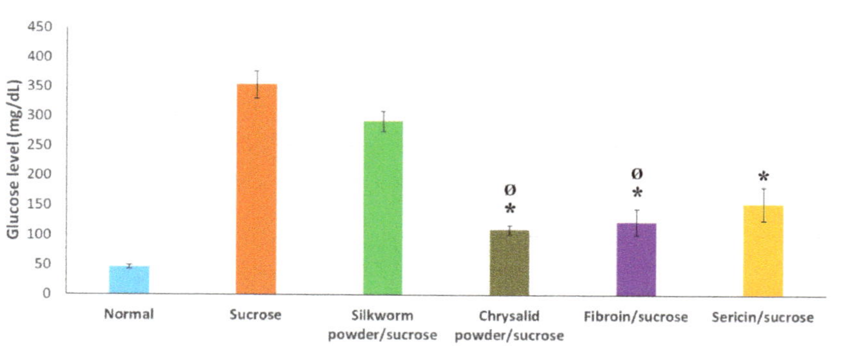

Figure 4. Glucose levels in hemolymph detected in silkworms fed a standard artificial diet (regular diet), a diet enriched in sucrose (10% w/w), and different diets containing the same amount of sucrose and the products studied at 5% w/w (silkworm powder, chrysalis powder, fibroin or sericin). Determinations were carried out 24 h after the start of feeding. Data are expressed as mean ± SEM. * Indicates values significantly lower than the sucrose diet considered positive control for induction of hyperglycemia ($p < 0.05$). Ø Indicates values equivalent to the negative control of hyperglycemia (regular diet) considered as a nonpathological condition ($p > 0.05$).

As shown in Figure 4, the trend in terms of the hypoglycemic effect of the different products studied was similar to that obtained in the case of glucose-induced hyperglycemia (Figure 3) with some variations. Once again, the silkworm powder showed a certain reduction in the average glucose value, but this was not detected as statistically significant with respect to the positive control of hyperglycemia in the sucrose diet (Bonferroni, $p > 0.05$). On the other hand, significantly lower glucose values were detected in the case of feeding with chrysalis powder, fibroin, and sericin compared with the positive control

(Bonferroni, $p < 0.05$). These values were equivalent to the negative control diet (normal diet) in the case of chrysalis powder and fibroin (Bonferroni, $p > 0.05$). Therefore, both are the most effective candidate products in the approach of hyperglycemia induced feeding with sucrose.

The inhibitory effect of α-glucosidase activity in silkworm powder obtained from larvae has been previously stated [7–9], and it is related to its high content in DNJ, even higher than that of mulberry leaves on which they feed [8]. Therefore, it would accumulate to a greater extent the longer the worm spends feeding throughout its life cycle. Hence, the hypoglycemic effect is more significant in the case of chrysalis powder. Moreover, this inhibition of α-glucosidase activity has also been indicated for sericin [14,32] and fibroin [13], with fibroin reporting a potential regenerative effect of pancreatic cells [29].

4. Conclusions

In this work, we demonstrated the feasibility of using the silkworm to evaluate the hypoglycemic action of various products resulting from sericulture by including them in the diet after promoting glucose or sucrose-induced hyperglycemia. As far as we know, this is the first time that the silkworm is used as a model to evaluate the products derived from its rearing in relation to the potential use in the prevention or treatment of diabetes.

The postprandial antihyperglycemic activity of fibroin, sericin, and powder made from pupae of silkworms (*B. mori*) is confirmed after its addition to hyperglycemic artificial diets. The inhibitory effect of α-glucosidase activity, confirmed in these products [7–9,12–14,32], as well as other previously described mechanisms, such as the inhibition of the expression of intestinal glucose transporters [9], or even the activation of insulin-like signaling pathway by different peptides derived from silk [28], seem to be responsible for this interesting hypoglycemic effect. Therefore, these products derived from sericulture could be ideal candidates for the prevention and treatment of diabetes, obesity, and other lifestyle-related diseases.

Author Contributions: S.D.A.-C. designed the experiment and oversaw silkworm rearing, fibroin and sericin extraction, preparation of silkworm and chrysalis powder, induction of hyperglycemia, and treatment with different products, evaluation of glucose level in hemolymph, and the statistical analysis and drafting of the manuscript. B.M.S., as a student in practices, collaborated in the development of the experiment, being an active part, especially in the silkworm rearing and the evaluation of glucose levels in the hemolymph. J.L.C.—the head of the research team—supervised all the work during its development and contributed to the writing and corrections. All authors have read and agreed to the published version of the manuscript.

Funding: This work has been supported by the European Commission ERDF/FEDER Operational Programme 'Murcia' CCI N° 2007ES161PO001 (Project No. 14–20/20).

Institutional Review Board Statement: Not applicable.

Informed Consent Statement: Not applicable.

Data Availability Statement: Not applicable.

Acknowledgments: S.D.A.-C. acknowledges the financial support of his former research contract, program INIA-CCAA (DOC INIA 2015), announced by the National Institute for Agricultural and Food Research and Technology (INIA) and supported by The Spanish State Research Agency (AEI) under the Spanish Ministry of Economy, Industry, and Competitiveness.

Conflicts of Interest: The authors declare no conflict of interest.

References

1. Ryu, S.P. Silkworm pupae powder ingestion increases fat metabolism in swim-trained rats. *J. Exerc. Nutr. Biochem.* **2014**, *18*, 141–149. [CrossRef] [PubMed]
2. Oh, H.-G.; Lee, H.-Y.; Kim, J.-H.; Kang, Y.-R.; Moon, D.-I.; Seo, M.-Y.; Back, H.-I.; Kim, S.-Y.; Oh, M.-R.; Park, S.-H.; et al. Effects of male silkworm pupa powder on the erectile dysfunction by chronic ethanol consumption in rats. *Lab. Anim. Res.* **2012**, *28*, 83–90. [CrossRef] [PubMed]

3. Li, X.; Xie, H.; Chen, Y.; Lang, M.; Chen, Y.; Shi, L. Silkworm pupa protein hydrolysate induces mitochondria-dependent apoptosis and s phase cell cycle arrest in human gastric cancer SGC-7901 cells. *Int. J. Mol. Sci.* **2018**, *19*, 1013. [CrossRef] [PubMed]
4. Mishra, N.; Hazarika, N.C.; Narain, K.; Mahanta, J. Nutritive value of non-mulberry and mulberry silkworm pupae and consumption pattern in Assam, India. *Nutr. Res.* **2003**, *23*, 1303–1311. [CrossRef]
5. Tomotake, H.; Katagiri, M.; Yamato, M. Silkworm pupae (*Bombyx mori*) are new sources of high quality protein and lipid. *J. Nutr. Sci. Vitaminol.* **2010**, *56*, 446–448. [CrossRef]
6. Rattana, S.; Katisart, T.; Butiman, C.; Sungthong, B.; Rattana, S.; Katisart, T.; Butiman, C. Antihyperglycemic Effect of Silkworm Powder, Fibroin and Sericin from Three Thai Silkworm (*Bombyx mori* Linn.) in Streptozotocin-Induced Diabetic Rats. *Pharmacogn. J.* **2017**, *9*, 559–564. [CrossRef]
7. Ryu, K.S.; Lee, H.S.; Kim, I. Effects of silkworm powder as a blood glucose lowering agent. *Int. J. Indust. Entomol.* **2002**, *4*, 93–100.
8. Yatsunami, K.; Murata, K.; Kamei, T. 1-Deoxynojirimycin Content and Alfa-Glucosidase Inhibitory Activity and Heat Stability of 1-Deoxynojirimycin in Silkworm Powder. *Food Nutr. Sci.* **2011**, *2*, 87–89. [CrossRef]
9. Han, J.; Inoue, S.; Isoda, H. Effects of silkworm powder on glucose absorption by human intestinal epithelial cell line Caco-2. *J. Nat. Med.* **2007**, *61*, 390. [CrossRef]
10. Kim, J.; Hwang, H.; Park, J.; Yun, H.-Y.; Suh, H.; Lim, K. Silk peptide treatment can improve the exercise performance of mice. *J. Int. Soc. Sports Nutr.* **2014**, *11*, 35. [CrossRef]
11. Kim, J.; Hwang, H.; Yun, H.-Y.; Kim, B.; Lee, C.-H.; Suh, H.; Lim, K. Silk Peptide intake increases fat oxidation at rest in exercised mice. *J. Nutr. Sci. Vitaminol.* **2013**, *59*, 250–255. [CrossRef] [PubMed]
12. Lee, H.J.; Lee, H.S.; Choi, J.W.; Ra, K.S.; Kim, J.M.; Suh, H.J. Novel Tripeptides with α-Glucosidase Inhibitory Activity Isolated from Silk Cocoon Hydrolysate. *J. Agric. Food Chem.* **2011**, *59*, 11522–11525. [CrossRef]
13. Hu, C.; Cui, J.; Ren, F.; Peng, C. Enzyme hydrolysis of silk fibroin and the anti-diabetic activity of the hydrolysates. *Int. J. Food Eng.* **2008**, *4*, 1–16. [CrossRef]
14. Fang, Y.; Wang, S.; Wu, J.; Zhang, L.; Wang, Z.; Gan, L.; He, J.; Shi, H.; Hou, J. The kinetics and mechanism of α-glucosidase inhibition by F5-SP, a novel compound derived from sericin peptides. *Food Funct.* **2017**, *8*, 323–332. [CrossRef] [PubMed]
15. Matsumoto, Y.; Ishii, M.; Sekimizu, K. An in vivo invertebrate evaluation system for identifying substances that suppress sucrose-induced postprandial hyperglycemia. *Sci. Rep.* **2016**, *6*, 26354. [CrossRef]
16. Kumar, S.K.P.; Mallappa, S.; Polekar, P.K.; Deepika, M.; Naik, M.M. Silk worm as a model to evaluate hypoglycemic action. *Indian J. Sci. Res.* **2017**, *16*, 1–9.
17. Kaito, C.; Akimitsu, N.; Watanabe, H.; Sekimizu, K. Silkworm larvae as an animal model of bacterial infection pathogenic to humans. *Microb. Pathog.* **2002**, *32*, 183–190. [CrossRef]
18. Matsumoto, Y.; Sekimizu, K. Silkworm Infection Model for Evaluating Pathogen Virulence. In *Immunity in Insects*; Springer: New York, NY, USA; Humana: Louisville, KY, USA, 2020; pp. 233–240. [CrossRef]
19. Matsumoto, Y.; Sekimizu, K. Silkworm as an experimental animal for research on fungal infections. *Microbiol. Immunol.* **2019**, *63*, 41–50. [CrossRef]
20. Nwibo, D.D.A.; Hamamoto, H.; Matsumoto, Y.; Kaito, C.; Sekimizu, K. Current use of silkworm larvae (*Bombyx mori*) as an animal model in pharmaco-medical research. *Drug Discov. Ther.* **2015**, *9*, 133–135. [CrossRef] [PubMed]
21. Panthee, S.; Paudel, A.; Hamamoto, H.; Sekimizu, K. Advantages of the silkworm as an animal model for developing novel antimicrobial agents. *Front. Microbiol.* **2017**, *8*, 373. [CrossRef]
22. Cappellozza, L.; Cappellozza, S.; Saviane, A.; Sbrenna, G. Artificial diet rearing system for the silkworm *Bombyx mori* (Lepidoptera: Bombycidae): Effect of vitamin C deprivation on larval growth and cocoon production. *Appl. Entomol. Zool.* **2005**, *40*, 405–412. [CrossRef]
23. Aznar-Cervantes, S.D.; Vicente-Cervantes, D.; Meseguer-Olmo, L.; Cenis, J.L.; Lozano-Pérez, A. Influence of the protocol used for fibroin extraction on the mechanical properties and fiber sizes of electrospun silk mats. *Mater. Sci. Eng. C* **2013**, *33*, 1945–1950. [CrossRef]
24. Chlapanidas, T.; Faragò, S.; Lucconi, G.; Perteghella, S.; Galuzzi, M.; Mantelli, M.; Avanzini, M.A.; Tosca, M.C.; Marazzi, M.; Vigo, D.; et al. Sericins exhibit ROS-scavenging, anti-tyrosinase, anti-elastase, and in vitro immunomodulatory activities. *Int. J. Biol. Macromol.* **2013**, *58*, 47–56. [CrossRef]
25. Ishii, M.; Matsumoto, Y.; Sekimizu, K. Inhibitory effects of alpha-cyclodextrin and its derivative against sucrose-induced hyperglycemia in an in vivo evaluation system. *Drug Discov. Ther.* **2018**, *12*, 122–125. [CrossRef] [PubMed]
26. Matsumoto, Y.; Sekimizu, K. Evaluation of anti-diabetic drugs by using silkworm, *Bombyx mori*. *Drug Discov. Ther.* **2016**, *10*, 19–23. [CrossRef] [PubMed]
27. Caccia, S.; Casartelli, M.; Grimaldi, A.; Losa, E.; De Eguileor, M.; Pennacchio, F.; Giordana, B. Unexpected similarity of intestinal sugar absorption by SGLT1 and apical GLUT2 in an insect (*Aphidius ervi*, Hymenoptera) and mammals. *Am. J. Physiol.-Regul. Integr. Comp. Physiol.* **2007**, *292*, 2284–2291. [CrossRef]
28. Park, S.; Zhang, T.; Qiu, J.Y.; Wu, X.; Lee, J.Y.; Lee, B.Y. Acid hydrolyzed silk peptide consumption improves anti-diabetic symptoms by potentiating insulin secretion and preventing gut microbiome dysbiosis in non-obese type 2 diabetic animals. *Nutrients* **2020**, *12*, 311. [CrossRef]
29. Park, S.Y.; Kim, B.; Lee, Y.K.; Lee, S.; Chun, J.M.; Suh, J.G.; Park, J.H. Silk fibroin promotes the regeneration of pancreatic β-cells in the C57BL/KsJ-Lepr$^{db/db}$ Mouse. *Molecules* **2020**, *25*, 3259. [CrossRef]

30. Johnson, R.J.; Nakagawa, T.; Sanchez-Lozada, L.G.; Shafiu, M.; Sundaram, S.; Le, M.; Ishimoto, T.; Sautin, Y.Y.; Lanaspa, M.A. Sugar, uric acid, and the etiology of diabetes and obesity. *Diabetes* **2013**, *62*, 3307–3315. [CrossRef]
31. Bischoff, H. Pharmacology of alpha-glucosidase inhibition. *Eur. J. Clin. Investig.* **1994**, *24* (Suppl. S3), 3–10. [CrossRef]
32. Xie, F.; Wang, S.; Zhang, L.; Wu, J.; Wang, Z. Investigating inhibitory activity of novel synthetic sericin peptide on α-D-glucosidase: Kinetics and interaction mechanism study using a docking simulation. *J. Organ. Behav.* **2007**, *28*, 303–325. [CrossRef] [PubMed]

MDPI
St. Alban-Anlage 66
4052 Basel
Switzerland
Tel. +41 61 683 77 34
Fax +41 61 302 89 18
www.mdpi.com

Insects Editorial Office
E-mail: insects@mdpi.com
www.mdpi.com/journal/insects

www.ingramcontent.com/pod-product-compliance
Lightning Source LLC
Chambersburg PA
CBHW061555220926
51850CB00005B/340
* 9 7 8 3 0 3 6 5 6 8 5 8 4 *